Biopolymers

Edited by A. Steinbüchel

Volume 4
Polyesters III
Applications and Commercial Products

Edited by Y. Doi and A. Steinbüchel

Biopolymers

Edited by A. Steinbüchel

Volume 4
Polyesters III
Applications and Commercial Products

Edited by Y. Doi and A. Steinbüchel

WILEY-VCH

Editors:

Prof. Dr. Alexander Steinbüchel
Institut für Mikrobiologie
Westfälische Wilhelms-Universität
Corrensstrasse 3
D-48149 Münster
Germany

Prof. Dr. Yoshiharu Doi
Polymer Chemistry Laboratory
RIKEN Insitute
Hirosawa, Wako-shi
Saitama 351-0198
Japan

Library of Congress Card No: applied for
British Library Cataloguing-in-Publication Data: A catalogue record for this book is available from the British Library

Die Deutsche Bibliothek –
CIP-Cataloguing-in-Publication Data
A catalogue record for this publication is available from Die Deutsche Bibliothek

Printed in the Federal Republic of Germany
Printed on acid-free paper.

Composition, Printing, and Bookbinding:
Konrad Triltsch
Print und digitale Medien GmbH,
Ochsenfurt-Hohestadt

ISBN   3-527-30225-5

Preface

Biopolymers and their derivatives are diverse, abundant, important for life, they exhibit fascinating properties and are of increasing importance for various applications. Living matter is able to synthesize an overwhelming variety of polymers, which can be divided into eight major classes according to their chemical structure: (1) nucleic acids such as ribonucleic acids and deoxyribonucleic acids, (2) polyamides such as proteins and poly(amino acids), (3) polysaccharides such as cellulose, starch and xanthan, (4) organic polyoxoesters such as poly(hydroxyalkanoic acids), poly(malic acid) and cutin, (5) polythioesters, which were reported only recently, (6) inorganic polyesters with polyphosphate as the only example, (7) polyisoprenoids such as natural rubber or Gutta Percha and (8) polyphenols such as lignin or humic acids.

Biopolymers occur in any organism, and in most organisms they contribute to the by far major fraction of the cellular dry matter. Biopolymers possess a wide range of different essential or beneficial functions for the organisms: conservation and expression of genetic information, catalysis of reactions, storage of carbon, nitrogen, phosphorus and other nutrients and of energy, defense and protection against the attack of other cells or hazardous environmental or intrinsic factors, sensors of biotic and abiotic factors, communication with the environment and other organisms, mediators of adhesion to surfaces of other organisms or of non-living matter and many more. In addition, many biopolymers are structural components of cells, tissues, and whole organisms.

To fulfil all these different functions, biopolymers must exhibit rather diverse properties. They must very specifically interact with a large variety of different substances, components and materials, and often they must have extraordinarily high affinities to them. Finally, many of them must have a high strength. Some of these properties are utilized directly or indirectly for various applications. This and the possibility to produce them from renewable resources, as living matter mostly does, make biopolymers interesting candidates to industry.

Basic and applied research have already revealed much knowledge on the enzyme systems catalyzing biosynthesis, degradation and modification of biopolymers as well as on the properties of biopolymers. This has also resulted in an increased interest in biopolymers for various applications in industry, medicine, pharmacy, agriculture, electronics and various other areas. However, considering the developments during the last two decades and reviewing the literature shows that our knowledge is still scarce. The genes for the biosynthesis pathways of many biopolymers are still not available or were identified only recently, many new biopolymers have just been described, and from only a minor fraction of biopolymers the biological, chemical, physical and material properties have been investigated. Often promising biopolymers are not available in sufficient amounts. Nevertheless, polymer chemists, engineers and material scientists in academia and industry have discovered biopolymers as chemicals and materials for many new applications, or they consider biopolymers as models to design novel synthetic polymers.

The first edition of this multivolume handbook comprehensively reviews and compiles information on biopolymers in 10 volumes covering (a) occurrence, synthesis, isolation and production, (b) properties and applications, (c) biodegradation and modification not only of natural but also of synthetic polymers, and (e) the relevant analysis methods to reveal the structures and properties. Volumes 1-8 are structured according to the chemical classes of biopolymers, whereas Volume 9 focusses on aspects of the biodegradation of synthetic polymers and Volume 10 deals with general aspects related to biopolymers.

This book series will hopefully be helpful to many scientists, physicians, pharmaceutics, engineers and other experts in a wide variety of different disciplines, in academia and in industry. It may not only support research and development but may be also suitable for teaching.

Publishing of this book series was achieved by chosing volume editors and authors of the individual volumes and chapters for their recognized expertise and for their excellent contributions to the various fields of research. I am very grateful to these scientists for their willingness to contribute to this reference work and for their engagement. Without them and without their comitment and enthusiasm it would have not been possible to compile such a book series.

I am also very grateful to the publisher WILEY-VCH for recognizing the demand for such a book series, for taking the risk to start such a big new project and for realizing the publication of *Biopolymers* in excellent quality. Special thanks are due to Karin Dembowsky and many of her WILEY-VCH colleagues, especially from production and marketing, for their constant effort, their helpful suggestions, constructive criticism, and wonderful ideas.

Last but not least I would like to thank my family for their patience, and I have to excuse for the many hours the preparation of this book series kept me away from them.

Münster, February 2001   Alexander Steinbüchel

Introduction

Living systems are capable of the synthesis of a wide range of different polyesters. Most of them are synthesized by plants as structural components of the cuticle covering the aerial parts of plants, such as cutin and suberin, or by prokaryotic microorganisms as intracellular storage compounds. These PHAs are water-insoluble. Furthermore, water-soluble polyesters are synthesized by a few eukaryotic organisms. In addition, polymers of, e.g., 3-hydroxybutyrate exhibiting a rather low degree of polymerization were detected, which are complexed with other biopolymers such as calcium polyphosphate or proteins. The latter were found in almost any living system investigated. However, the physiological function(s) were not revealed yet.

The chemical composition of insoluble cytoplasmic inclusions in the Gram-positive bacterium *Bacillus megaterium* was identified by Lemoigne in 1926 as poly(3-hydroxybutyric acid). By the end of the 1950s, sufficient evidence was accumulated from physiological studies to suggest that this biopolymer functions as an intracellular reservoir of carbon and energy. Meanwhile, it is known that this or structurally related (see below) storage polyesters are synthesized by members of almost any phylogenetic taxon of prokaryotes. In 1974, the identification of 3-hydroxyalkanoates other than 3-hydroybutyrate, such as 3-hydroxyvalerate and 3-hydroxyhexanoate, was reported in chloroform extracts of activated sludge. Since the 1980s, many bacteria were demonstrated to synthesize various types of polyesters containing 3-, 4-, and 5-hydroxyalkanoate units. To date, approximately 150 different hydroxyalkanoates are known as constituents of these polyesters which are, therefore, generally referred to as polyhydroxyalkanoates (PHA).

The onset of the molecular biology revolution during the late 1970s provided new tools for biological research, which were successfully used to decipher genetic information and to better understand the principles behind PHA biosynthesis at the molecular level. By the end of the 1980s, the genes coding for enzymes involved in PHA biosynthesis were cloned from *Ralstonia eutropha* (formerly known as *Alcaligenes eutrophus*), and the genes were also shown to be functionally active in *Escherichia coli*. So far, about 60 PHA synthase structural genes have been cloned from different bacteria. In addition, many genes encoding enzymes and proteins relating to PHA biosynthesis were cloned and characterized at the molecular level. This strongly stimulated research and provided new perspectives for biotechnological production of PHAs. This knowledge has been utilized to establish PHA biosynthesis in many prokaryotic organisms and plants. The methodology of metabolic engineering was successfully applied for effective production of various PHAs by fermentation biotechnology or agriculture in economically feasible processes. In particular transgenic plants expressing PHA biosynthesis pathways may provide potential producers of PHAs in the future. One important aspect is the large-scale biotechnological production of PHAs by fermentative processes and by agriculture from renewable carbon sources and CO_2, respectively.

The PHA family of polyesters is thermoplastic with biodegradable and biocompatible properties. Many of these water-insoluble polyesters can be thermoformed to various types of products such as bottles, films and fibers like established petrochemical-based thermoplastics by using conventional extrusion and molding equipment. Some PHAs have become commercially attractive for applications in various areas. Applications are also known for the water-soluble polyester poly(malic acid). This explains the interest of industry in PHAs and other polyesters as large-scale biotechnological products. The physical and mechanical properties can be regulated by varying the composition of the polyesters. As a result, PHA can be made in a wide variety of polymeric materials, from hard crystalline plastics to very elastic rubber. Besides thermoplasticity, one of the most important characteristics of PHA products is their biodegradability. PHA products such as films and fibers are degraded in soil, sludge and seawater. Under optimum conditions the degradation rate is extremely fast. Many prokaryotic and eukaryotic microorganisms excrete extracellular PHA depolymerases to hydrolyze PHA products, and they utilize the decomposed compounds as nutrients. These genes have also been cloned and characterized at a molecular level. Today, interdisciplinary research and development of biological polyesters are rapidly expanding in both the biological and polymer sciences. Concerted multidisciplinary scientific approaches have been directed to elucidate various new aspects of PHA. One important impact of studying and introducing natural polyesters was that efforts to establish new synthetic biodegradable materials were strongly stimulated. As a consequence, many new biodegradable packaging materials were developed by the chemical industry, and production processes for existing synthetic polyesters were highly optimized. For example, polylactides, which were formerly only affordable for medical applications, will now become also available for bulk applications.

This third volume on polyesters will focus on the various applications and commercial products of biosynthetic, semisynthetic and fully synthetic polyesters. Properties and general applications are summarized at the beginning (Chapters 1 and 2). Bulk applications of microbial PHAs (Chapter 3) and applications of PHAs in medicine and pharmacy (Chapter 4) are presented. The next eight chapters focus on synthetic polyesters and structurally related polymers such as polylactides (Chapter 5), polyglycolide and copolyesters with lactide (Chapter 6), polyanhydrides (Chapter 7), commercial products of polylactides such as Natureworks™ PLA (Chapter 8) and Lacea (Chapter 9), the aliphatic polyester Bionolle (Chapter 10), the aliphatic-aromatic polyester Ecoflex (Chapter 11) and the polyesteramides BAK (Chapter 12). The last two chapters of this volume describe how natural and synthetic polyesters are chemically modified (Chapter 13) and how chiral and other valuable compounds can be produced from microbial polyesters (Chapter 14).

These topics reflect the recent progress on polyester research, and we hope that these three volumes will provide useful new information and knowledge for scientists of many disciplines and to engineers from industry and for all others who want to gain deeper knowledge on the biology of naturally occurring and synthetic polyesters and their properties. Another aim of this book is to support ongoing interdisciplinary efforts to stimulate and improve the commercial production of polyesters and to broaden the uses and applications of these polymers.

We are very grateful to the many authors and experts in this field who contributed these excellent chapters to the three volumes on polyesters. The expertise, enthusiasm and the costly time, which they devoted to their chapters, is highly appreciated. We are well aware that all of them have many other obligations and duties. Without such committed individuals and scientists such a book could never have been prepared.

Last but not least, we would like to thank the publisher WILEY-VCH for publishing *Biopolymers* with the customary professionality and excellence of the many employees involved in this project. Special thanks are due to Karin Dembowsky; without her constant effort, this book could not have been published.

Saitama and Münster   Y. Doi
October 2001   A. Steinbüchel

Contents

1
Properties and Application of Aliphatic Polyester Products

Dr. Masatsugu Mochizuki
Technology Development Division, Unitika Ltd., 4-1-3 Kyutaro-machi, Chuo-ku, Osaka, 541-8566, Japan; Tel.: + 81-6-6281-5245; Fax: + 81-6-6281-5256; E-mail: masa-mochizuki@unitika.co.jp

BS	butylene succinate
d	denier
DR	draw ratio
DSC	differential scanning calorimetry
E'	dynamic storage modulus
ES	ethylene succinate
f	filament
GRAS	"Generally Recognized As Safe"
3HV	3-hydroxyvalerate
Mn	number-average molecular weight
OPS	oriented polystyrene
P[(R)-3HB], or simply P(3HB)	poly[(R)-3-hydroxybutyrate]
PBES	poly(butylene succinate-*co*-ethylene succinate)
PBS	poly(butylene succinate)
PBSA	poly(butylene succinate-*co*-butylene adipate)
PCL	poly(ε-caprolactone)
PET	polyethylene terephthalate
PGA	poly(glycolic acid)
PHA	polyhydroxyalkanoate
P(3HB-*co*-3HV) or simply PHBV	poly[(R)-3-hydroxybutyrate-*co*-(R)-hydroxyvalerate]
PLA	poly(lactic acid)
PLLA	poly(L-lactic acid)
P[(S)-3HB]	poly[(S)-3-hydroxybutyrate]
SAXS	small angle X-ray scattering
SEM	scanning electron microscopy
T_c	crystallization temperature
T_g	glass transition temperature
T_m	melting point
TOC	total organic carbon
WAXD	wide-angle X-ray diffraction

1
Introduction

For the past 50 years, films and textile materials made from synthetic polymers such as nylon, polyester and polyolefin have been used in industrial applications as well as for clothing. In some instances, the excellent durability of synthetic polymers has proved to be a major advantage, but in others the great durability of these materials has led to adverse effects. The current crisis in solid waste management has, in particular, focused attention on the development of readily compostable materials, especially those used for packaging and for single-use disposable products. In geotextile and agricultural applications such as the erosion protection of soil with slope stabilization, mulch films for ground cover, and root covers, it is desirable that such materials are degraded in soil once their service life has been completed. Biodegradable fishing lines and nets would also be valuable in order to prevent their becoming a hazard to sea-living animals, for example when seals become entangled with fishing nets that have been lost at sea.

Underlying the rapid decline in landfill availability and growing concerns relating to incinerator emissions (e.g., dioxin), there is a growing demand for biodegradable plastics to be used to solve the problems of solid waste management. In this respect, aliphatic polyesters represent a structural material of great promise for the production of biodegradable fibers and films that meet requirements with regard to easy processability, excellent physical properties, and biodegradability.

In this chapter, the classification of aliphatic polyesters with regard to their chemical structure, processability (e.g., spinnability for fiber-making), and the yarn properties of the fibers obtained is outlined. Subsequently, the structural effects of aliphatic polyester fibers or films on biodegradation, including both environmental and enzymatic degradation, is discussed. The biodegradation behavior of these materials is influenced greatly by not only their primary chemical structure but also their solid-state morphology and highly ordered structures such as crystallinity and orientation. The biodegradation mechanism of these aliphatic polyesters is also discussed with regard to whether the major route of their hydrolysis is enzyme-catalyzed, or not.

Finally, the profile of poly(lactic acid) (PLA) – which perhaps offers the greatest potential among biodegradable aliphatic polyesters – is outlined, notably with regard to its manufacture, properties, applications, and degradation behavior, both in the natural environment and under conditions of composting. PLA has been highlighted as a "green plastic" whose monomer is obtained from renewable resources in agricultural materials that are grown annually; this contrasts with conventional synthetic polymers, which rely on natural reserves of oil for their monomer resource. Details of a wide variety of PLA products such as films, sheets, fibers, and nonwovens, are also introduced, and the key performance features and potential applications of the PLA products described.

2
Classification and Properties of Aliphatic Polyesters

Although most aliphatic polyesters are biodegradable, most aromatic polyesters are not (Potts et al., 1973; Diamond et al., 1975; Fields and Rodriguez, 1974; Tokiwa and Suzuki, 1981). One explanation proposed for this is that aliphatic polyesters are flexible enough to fit into the active site of the enzymes, whereas aromatic polyesters are too rigid to permit such a fit. Synthetic polymers such as aliphatic polyesters, which have structure and functional groups similar to those of natural organic materials, are expected to be degraded nonspecifically by some enzymes, such as lipase (Tokiwa and Suzuki, 1986).

Aliphatic polyesters may form two groups with regard to the mode of bonding of constituent monomers:

- poly(hydroxy acid)s, i.e., polyhydroxyalka-noates, which are polymers of hydroxy acids, with HO–R–COOH as repeating units. Hydroxy acids may be further classified into, α-, β-, and ω-hydroxy acids with respect to the bonding position of the OH group from the COOH end group.
- poly(alkylene dicarboxylate)s, which are synthesized by the polycondensation reaction of diols, HO–R$_1$–OH, and diacids, HOOC–R$_2$–COOH.

With regard to the mechanism of biodegradation, aliphatic polyesters are classified in two groups on the basis of their hydrolysis reaction – whether enzyme-catalyzed or not – although they are generally susceptible to both enzymatic and non-enzymatic hydrolysis (Mochizuki and Hirami, 1997a).

Four types of aliphatic polyesters, together with their currently available commercial products, are listed in Table 1, while the scheme of synthesis, mode of hydrolysis, and physical and thermal properties of the polymers are reviewed in the following sections.

2.1
Poly(α-hydroxy acid)

Poly(α-hydroxy acid) such as poly(glycolic acid), PGA, or poly(lactic acid), PLA, are crystalline polymers with relatively high melting points. Among the number of carbon atoms between ester bonds in the main chain, only one may be responsible for the major nonenzymatic hydrolytic degradation of poly(α-hydroxy acid), although microorganisms or enzymes capable of degrading PLA have recently been reported (Williams, 1981; Reeve et al., 1994; Torres et al., 1996; Tomita et al., 1999).

PLA has also been highlighted because of its availability from renewable resources such as corn. An important feature of lactic acid is its ability to exist in two optically active forms. Fermentation-derived lactic acid usually comprises 99.5% of the L-isomer (which rotates the plane of polarized light in a

Tab. 1 Classification of aliphatic polyesters

Chemical structure	Biodegrad-ability	Examples (trade mark/producer)	
R O | || –(O–CH–C)$_x$– Poly(α-hydroxy acid)	Chemical hydrolysis	R = H R = CH$_3$	Poly(glycolic acid) (PGA) Poly(L-lactic acid) (PLLA) Nature Works/Cargill Daw Lacty/Shimadzu
R O | || –(O–CH–CH$_2$ C)$_x$– Poly(β-hydroxyalkanoate)	Enzymatic hydrolysis	R = CH$_3$ R = CH$_3$, C$_2$H$_5$	Poly(β-hydroxybutyrate) (PHB) Poly(β-hydroxybutyrate-*co*- β-hydroxyvalerate) (PHBV) Biopol/Monsanto
O || –[O–(CH$_2$)$_n$ C]$_x$– Poly(ω-hydroxybutyrate)	Enzymatic hydrolysis	n = 3 n = 5	Poly(β-propiolactone) (PPL) Poly(ε-caprolactone) (PCL) Tone/Union Carbide Placcel/Daicell Chemical Industries
O O || || –[O–(CH$_2$)$_m$O —C–(CH$_2$)$_n$C]$_x$– Poly(alkylene dicarboxylate)	Enzymatic hydrolysis	m = 2, n = 2 m = 4, n = 2 m = 4, n = 2,4	Poly(ethylene succinate) (PES) Poly(butylene succinate) (PBS) Poly(butylene succinate-*co*- butylene adipate) (PBSA) Bionolle/Showa Highpolymer

clockwise direction), in contrast to the D form. The control of monomer stereochemistry provides a versatile family of polymers, ranging from amorphous (60 °C softening point) to a crystalline melting point of 175 °C. Polymers with high levels of the L-isomer may be used to produce crystalline products, whereas materials with higher levels of the D-isomer ($>15\%$) are amorphous (Lunt, 1998).

PLA is a hydrophobic polymer because of incorporation of the CH_3 side groups when compared with PGA, and the rate of chemical hydrolysis of PLA in buffer solution at 37 °C is much slower than that of PGA.

There are two routes by which PLA may be manufactured from the lactic acid monomer. The first is by ring-opening polymerization of a cyclic dimer intermediate; this is referred to as lactide, without solvent. The second route is direct polycondensation involving the use of solvent. The Mitsui Toatsu Chemical Inc. (presently, Mitsui Chemical Inc.) recently developed a new direct polymerization process (Ajioka et al., 1995). Advanced industrial technologies of polymerization have been developed which produce high-molecular weight, pure PLA without a need for either monomers, oligomers, or a catalyst. This leads to a potential for structural materials that have a sufficiently long service life in terms of maintaining their mechanical properties without rapid hydrolysis (even under a humid atmosphere), while retaining good compostability.

Both the physical properties and biodegradability of PLA can be regulated by employing a comonomer component of hydroxy acids or racemization of the D- or L-isomers, whereas a PLA homopolymer such as poly(L-lactic acid), PLLA, is a hard, transparent and crystalline polymer having a melting point (T_m) of 178 °C and a glass transition temperature (T_g) of ~ 57 °C.

2.2
Poly(β-hydroxyalkanoate)

It has been found that an optically active polymer of (R)-3-hydroxybutyrate, poly[(R)-3HB] or simply poly(3HB) (or more simply, PHB), and other hydroxyalkanoates, are synthesized by a wide variety of microorganisms as intracellular storage materials of carbon and energy – a finding originally reported by Lemoigne in 1925 (Doi, 1990). Poly(3HB) is a crystalline thermoplastic with a T_m of 177 °C. A copolymer of poly(3HB) and poly(3HV), poly(3HB-co-3HV) (or simply PHBV), is produced commercially using a fermentation process that employs glucose and propionic acid as carbon sources for *Ralstonia eutropha*, and was first marketed under the trade name Biopol™ by ICI in UK. Poly(3HB-co-3HV) is a flexible and tough thermoplastic, while poly(3HB) is brittle.

These materials, polyhydroxyalkanoate (PHA), are degraded by the extracellular PHA depolymerases secreted by various microorganisms present in the environment. PHA depolymerases have such substrate specificities that they can hydrolyze poly[(R)-3HB], but not poly[(S)-3HB]. No lipases are able to hydrolyze poly[(R)-3HB].

2.3
Poly(ω-hydroxyalkanoate)

A poly(ω-hydroxyalkanoate) such as poly(ε-caprolactone), PCL, is prepared from the cyclic ester monomer, lactone, by a ring-opening reaction with a catalyst such as stannous octanoate in the presence of an initiator that contains an active hydrogen atom.

High-molecular weight PCL is an almost crystalline polymer with a moderately low T_m (60 °C), and is a tough and semi-rigid material at room temperature, having a

Young's modulus between those of low- and high-density polyethylenes. PCL has also been shown to be degraded by lipases secreted by microorganisms.

2.4
Poly(alkylene dicarboxylate)

The polycondensation process is a classical and popular method of synthesizing current commercially available polyesters such as polyethylene terephthalate (PET); however, the process was not of any help in the successful development of high-molecular weight aliphatic polyesters that can be used as biodegradable plastics.

From data showing the relationship between composition and T_m of aliphatic polyesters from dicarboxylic acids and diols, only five compositions (in which oxalic acid or succinic acid is used as the dicarboxylic acid) possess a $T_m > 100\,°C$.

Recently, new biodegradable polyester-urethanes have been developed by producing hydroxy-terminated polyester oligomers from aliphatic diols and dicarboxylic acids in the first step, and then coupling these oligomers using a small amount of diisocyanate as a chain extender. Showa High-polymer developed a family of such polyesters (e.g., Bionolle™) that was composed of

poly(butylene succinate), PBS, or poly(butylene succinate-*co*-butylene adipate), PBSA (Fujimaki, 1998).

The biodegradation of poly(alkylene dicarboxylate) could be catalyzed by an enzyme such as lipase.

2.5
Processability of Aliphatic Polyesters

As in the case of biodegradable structural materials prepared for fiber production by melt spinning, aliphatic polyesters may be the only group that meets the requirements regarding processability, physical properties, and biodegradability of the fibers obtained. Biodegradable fibers made from aliphatic polyesters were manufactured using a conventional melt spinning process under optimized spinning conditions (Mochizuki and Hirami, 1997b). The thermomechanical properties of typical aliphatic polyesters, and yarn properties of monofilament and multifilament are shown in Table 2, in comparison with PET.

PLLA is a crystalline thermoplastic having a higher T_m of 178 °C, a higher crystallization temperature (T_c) of 103 °C, and a higher T_g of 57 °C than other aliphatic polyesters, leading to excellent spinnability. PLA can be processed into fibers on conventional PET melt-

Tab. 2 Thermal properties, spinnability, and yarn properties of aliphatic polyesters

Materials	Thermal Properties				Yarn Properties		
	T_m (°C)	T_c (°C)	T_g (°C)	Spinn-ability	Tensile Strength (g/d)		Young's Modulus (g/d)
					Mono-filament	Multi-filament	
Poly(L-lactic acid) (PLLA)	178	103	57	Excellent	4.0–6.0	4.0–6.0	55–65
Poly(β–hydroxybutyrate) (PHB)	175	60	4	Poor	2.5–3.5	–	10–20
Poly(ε–caprolactone) (PCL)	60	22	– 60	Fair	7.5–8.5	4.0–5.5	10–20
Poly(butylene succinate) (PBS)	116	77	– 32	Good	5.5–6.5	4.5–5.5	15–25
Poly(ethylene terephthalate) (PET)	256	170	69	Excellent	5.5–6.0	4.5–9.5	100–100

spinning equipment and the PLA fibers obtained provide performance comparable with that of PET, because the T_g of both materials is similar.

PHB has poor spinnability of conventional multifilament by air quenching, most likely due to lower T_g and T_c, and a narrow window of crystallization. It is possible to produce only relatively weak monofilament using an in-line, spin-draw process with water bath quenching.

Using high-molecular weight PCL, tough and extremely high strength monofilaments have been obtained by melt-spinning followed by cold-drawing (seven or eight times drawing), although the spinnability of the multifilament is fair because of the lower T_c and T_g values.

PBS has a relatively higher T_c, with good spinnability. However, the strength of the multifilament yarns is slightly less because of the limited molecular weight obtained during polymerization; the lower Young's modulus of the yarns also leads to poorer crimping properties of the staple yarns.

3
Structural Effects on Biodegradation of Aliphatic Polyester Fibers

3.1
PCL fiber

3.1.1
Structure and Properties

PCL polymer is available from Union Carbide Corporation, under the tradename TONE™. TONE™ P-787 is a tough, extensible polymer with the number-average molecular weight (Mn) of ~80,000, and consequently it was used to prepare a high-tenacity fiber. Tough and high-tenacity PCL monofilament (740 denier) and multifilament (72 denier/24 filament) of TONE were prepared by melt-spinning at 210 °C. The

monofilament was manufactured by the so-called spin-draw process, in which as-spun yarns were immediately introduced into a water bath at 25 °C for quenching, then passed to drawing zones at room temperature to 45 °C, and subsequently passed to a relaxing zone. Monofilaments with almost the same diameters (280 ± 5 μm) and different draw ratios (undrawn to nine times drawn) were prepared (Mochizuki et al., 1997b).

Thermal, mechanical, and dynamic mechanical properties of the PCL fibers were affected significantly by the draw ratio (DR). Differential scanning calorimetry (DSC) thermograms showed that the T_m increased linearly from 59 °C to 64 °C, and the enthalpy of fusion increased rapidly with DR. These results imply that the size of crystallites and crystallinity increased with drawing.

With regard to the tensile properties of the PCL fibers, tensile tenacity increased with an increase in DR, reaching more than 8 g/denier at DR = 9, whereas the tendency of ultimate elongation was to decrease as the draw ratio increased. This is a well-known feature that is characteristic of crystalline polymers, and suggests an increase in both molecular orientation and crystallinity along the fiber axis, with drawing.

Curves of the temperature dependence of dynamic mechanical properties of PCL fibers, measured by using a non-resonant, forced vibration method, are shown in Figure 1. The tanδ curve for the undrawn PCL fiber demonstrates two loss peaks which are attributed to the α and β dispersions of PCL at –60 °C and –125 °C, respectively. The α dispersion corresponds to the T_g of the fibers, and the T_g (temperature at tan δ peak of α dispersion) is shifted towards higher temperatures, with an increase in DR. As seen in Figure 1, stepwise decreases of the dynamic storage modulus

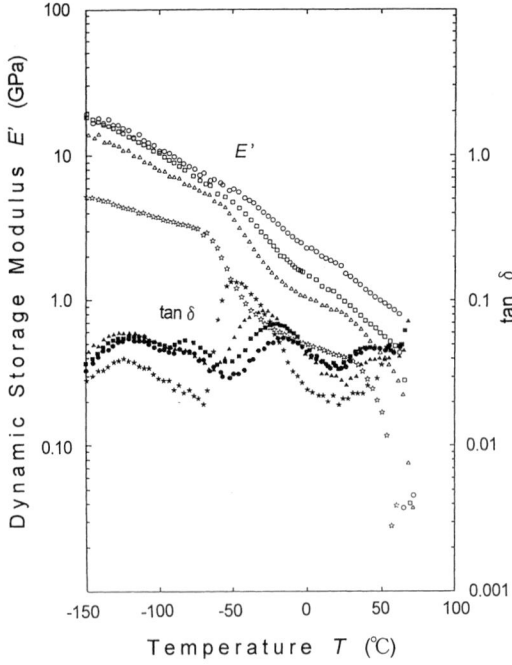

Fig. 1 Temperature dependence of dynamic storage modulus (E'; open symbols) and loss tangent (tan δ; filled symbols) for undrawn and drawn PCL fibers at 3.5 Hz: stars, undrawn PCL; triangles, DR=5; squares, DR=7; circles, DR=9.

(E') with temperature for undrawn and drawn fibers are observed in the T_g regions; these are marked for the undrawn fiber as compared with the highly drawn fibers.

In order to investigate the development of fine structure during the spin-draw process, a wide-angle X-ray diffraction (WAXD) study was employed (Table 3). The crystallinity of the PCL fibers was evaluated using the Ruland method, and the degree of orientation was evaluated using the Hermans orientation function. WAXD data show that crystallite size along the fiber axis would appear to increase, crystallite sizes perpendicular to the fiber axis to decrease, and crystal volume to decrease with increasing DR, while crystallinity showed a tendency to increase. The crystallinity index of the undrawn fiber was 40.0%. Crystallinity increased with the increase in DR, reaching 69.3% at DR = 9. The increase in crystallinity with drawing is supported by increases in density and heat of fusion. An effect of DR on molecular orientation was similar to that of crystallinity.

The crystalline structures (mainly consisting of extended chain crystals) developed during the drawing process since PCL is a crystalline polymer, and this accounts for the fact that crystallinity and crystalline orientation increased with increasing DR. It is also considered that tenacity increases due to an increase of extended chain crystals and tie molecules in the amorphous regions between crystals.

3.1.2
Environmental Degradation
The PCL monofilaments (diameter 280 ± 5 μm) with various DRs were degraded under environmental conditions, including soil burial, seawater exposure, and activated sludge exposure. The extent of degradation was examined by weight loss, loss of me-

Tab. 3 Fine structures of PCL fibers before and after enzymatic degradation, measured by wide-angle X-ray diffractometry

Sample		Crystallite Sizes (Å)			Crystal Volume (Å³)	Crystallinity (%)	Orientation (%)
Draw Ratio (DR)	Degradation	L(110)	L(200)	L(0014)			
DR = 1 (undrawn)	Before	79	74	61	357,000	40.0	65.3
	After	73	62	57	258,000	38.7	63.3
DR = 5 (drawn)	Before	74	64	62	294,000	63.3	94.4
	After	81	59	60	287,000	57.3	95.3
DR = 7 (drawn)	Before	73	53	66	255,000	65.0	95.3
	After	76	54	50	207,000	61.8	95.1
DR = 9 (drawn)	Before	68	49	68	227,000	69,3	95.7
	After	73	48	59	205,000	63,9	95.3

chanical properties (e.g., tensile strength and ultimate elongation decrease), and visual observation by scanning electron microscopy (SEM). In order to compare the rate of degradation of PCL fibers in soil with other materials, drawn multifilaments of almost the same diameter (17.0 ± 2.5 μm) of nylon 6, PET, poly(vinyl alcohol), and rayon were also evaluated (Mochizuki et al., 1999).

Soil burial was carried out in the garden (soil: coarse, sandy loam). A 10-m length of each test specimen of the fibers was wound onto a skein and buried at a depth of 10 cm from the ground surface. The test specimens were periodically removed from the soil and gently washed to remove attached soil and dust. After drying in a vacuum oven, the extent of degradation was examined by weight loss, relative viscosity reduction, tensile strength decrease, and microscopic observation.

The rate of degradation of multifilament (3 denier per filament) was much faster than that of monofilament (740 denier). This might be attributed to the greater surface-to-volume ratio of the multifilament. The half-degradation time for the fibers to lose 50% of their initial tensile strength was about one month for the monofilament, and about one week for the multifilament. In comparison with other existing fibers, both PCL fiber and rayon lost almost 100% of the tensile strength within one week, while fibers of nylon 6, PET, and poly(vinyl alcohol) retained 60–90% of their initial tensile strength even after 50 days, suggesting that these fibers were resistant to microbial attack.

SEM photographs of the PCL monofilaments before and after soil burial are shown in Figure 2a–c. Figure 2b shows clear degradation with significant discrete depression at the area around the traces of microbial colonies after one month of burial. The monofilament developed severely hollow surfaces with mycelial growth after two months of burial (Figure 2c). Filamentous fungi are widely recognized as being a major cause of biodegradation.

The initial step of biodegradation is assumed to be a random scission of the in chain ester linkage of the polymer by extracellular enzymes secreted by microorganisms; this results ultimately in the formation of water-soluble products that are then removed. This surface erosion mechanism was supported by the SEM observations and almost constant molecular weight of the residual sample during biodegradation.

In order to understand the effect of the solid-state morphology of the fibers on

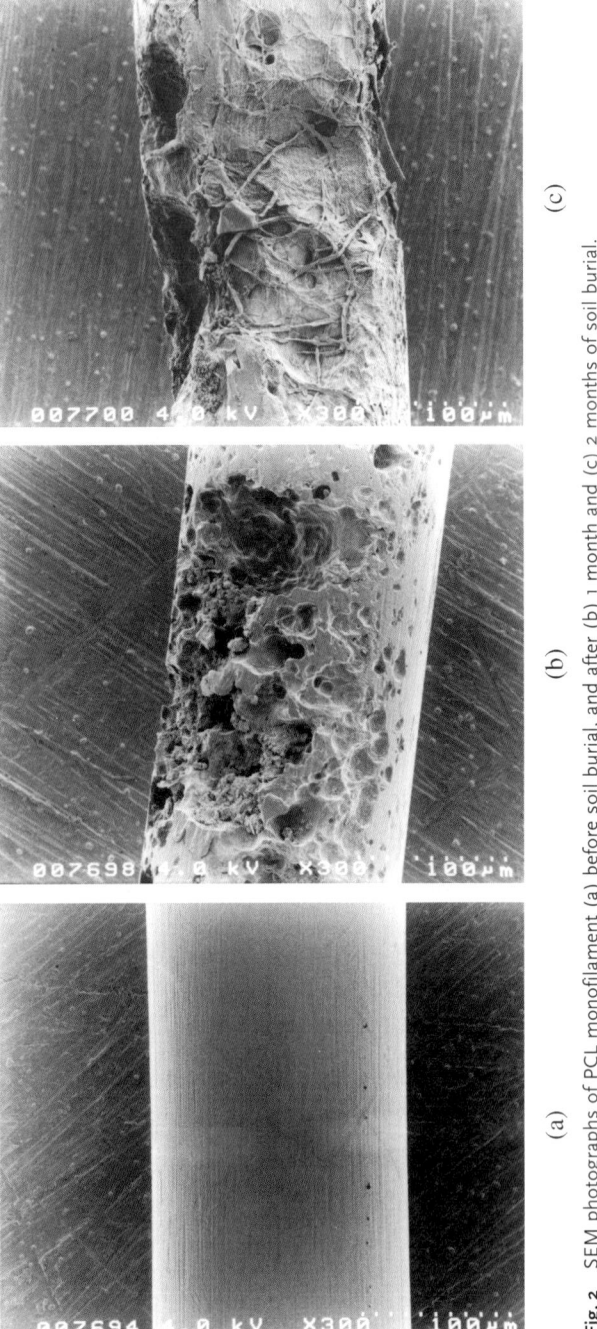

Fig. 2 SEM photographs of PCL monofilament (a) before soil burial, and after (b) 1 month and (c) 2 months of soil burial.

biodegradation, the rate of biodegradation was plotted against crystallinity, and an inverse linear relationship obtained (Figure 3). The initial rate of biodegradation of the undrawn monofilament is 10-fold faster than that of a nine times drawn monofilament, probably due to lower crystallinity. This suggests that enzymes secreted by microorganisms attack preferentially the amorphous or less-ordered region rather than the crystalline or more-ordered region, probably because such enzymes can migrate more readily into the less-ordered region. This is the reason that the degradability of the PCL fibers was inversely proportional to DR.

The unique pattern of surface erosion of the PCL multifilament subjected to two weeks of soil burial reveals a number of fine cracks perpendicular to the fiber axis (Figure 4). With regard to the morphology of the fiber structure, the microfibrillar model with chain folds and the interlocking shish-kebab structure, in which alternate crystalline and amorphous regions are arranged in the

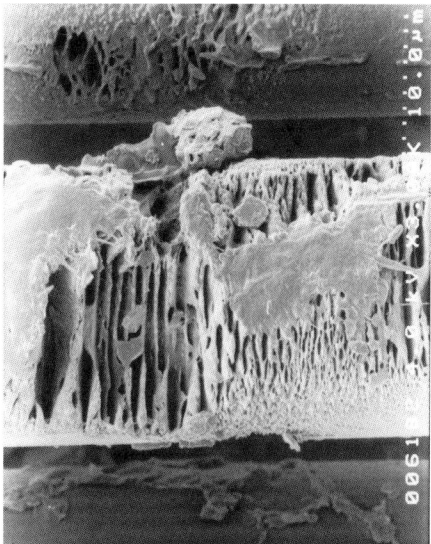

Fig. 4 SEM photograph of PCL multifilament after 2 weeks of soil burial.

direction of the fibers, are well known (Bashir et al., 1986). Therefore, it is believed that hydrolysis occurs initially in the amorphous regions sandwiched between two crystalline zones, as tie-chain segments, free chain ends, and chain folds in these regions degrade into water-soluble fragments.

The rate of degradation in activated sludge is much slower than that in soil burial, although the patterns of degradation were similar. The half-degradation time of the monofilament is expected to be 120–150 days.

In seawater exposure tests, the rate of degradation decreased with an increase in DR in the same manner as in the soil burial tests, although the rate of degradation in the marine environment was faster than that in soil.

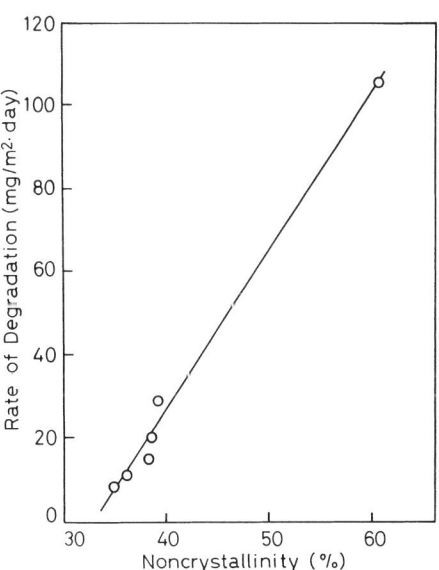

Fig. 3 Relationship between the rate of degradation in soil and crystallinity of PCL monofilaments.

3.1.3
Enzymatic Degradation

Tokiwa and Suzuki have studied analogous polymer series containing aliphatic, heterocyclic, and aromatic polyesters in a systematic manner with regard to the hydrolytic ability of lipases. These investigators found that aliphatic polyesters were degraded by lipases, whereas heterocyclic polyesters showed only limited lipase degradation; aromatic polyesters were hardly degraded by lipases (Tokiwa and Suzuki, 1981). Enzymes can be either totally specific for a given substrate, or they can be broadly specific for a given type of functional group. An example of the latter is an esterase-like lipase which catalyzes both the hydrolysis of lipids to fatty acids and glycerol and the hydrolysis of aliphatic polyesters to their repeating units.

Enzymatic degradability of the PCL fibers with different DR was followed by monitoring water-soluble total organic carbon (TOC) formation and weight loss after 16 h in an aqueous solution containing lipase from *Rhizopus arrhizus* at 30 °C (Mochizuki et al., 1995). Figure 5 shows the amount of TOC products in the reaction mixture which originated from water-soluble organic compound (monomer and oligomer). It was found that TOC values decreased with increase in DR; that is, the degradability of

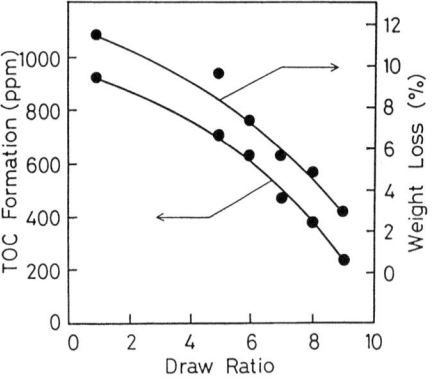

Fig. 5 TOC formation and weight loss of PCL fibers with different draw ratio after 16 h in aqueous solution containing lipase of *Rhizopus arrhizus* at 30 °C.

the PCL fibers was inversely proportional to DR. Figure 5 also shows plots of weight loss after degradation as a function of DR, the profile being similar to that for TOC values. It is supposed that enzymatic degradation of the PCL fibers proceeds preferentially in the amorphous regions, where there is a sufficient spatial degree of freedom that allows the enzyme to be physically adjacent to, and thus penetrate, the polymer molecule. The crystalline regions can be also degraded moderately, however (see Table 3).

Using SEM photographs (Figure 6a–c) which show the enzyme to attack preferentially the amorphous or less-ordered regions of the fiber (rather than the crystalline or more-ordered regions), spherulites were observed in the undrawn fiber (Figure 6a), while in the drawn fibers fibrillar stripes were seen along the fiber axis (Figure 6b,c). This suggests that the spherulites in the undrawn fiber were extended to be broken, i.e., unfolding of lamellar crystals, and fibril structure were formed during the drawing process. These stripes became finer as the DR became higher due to an increase in orientation of the fibrils and inter-fibrillar extended amorphous chain segments. The diameters of all fibers became smaller uniformly as degradation proceeded, although the higher the DR, the slower the rate of thinning.

3.2
Poly(butylene succinate) Fiber

3.2.1
Properties and Structure
High-molecular weight poly(butylene succinate) (PBS) or poly(butylene succinate-*co*-ethylene succinate)s (PBES) were obtained from succinic acid, 1,4-butanediol, and/or ethylene glycol through a polycondensation process (Mochizuki et al., 1997a) Fibers made from PBS and PBES with ethylene

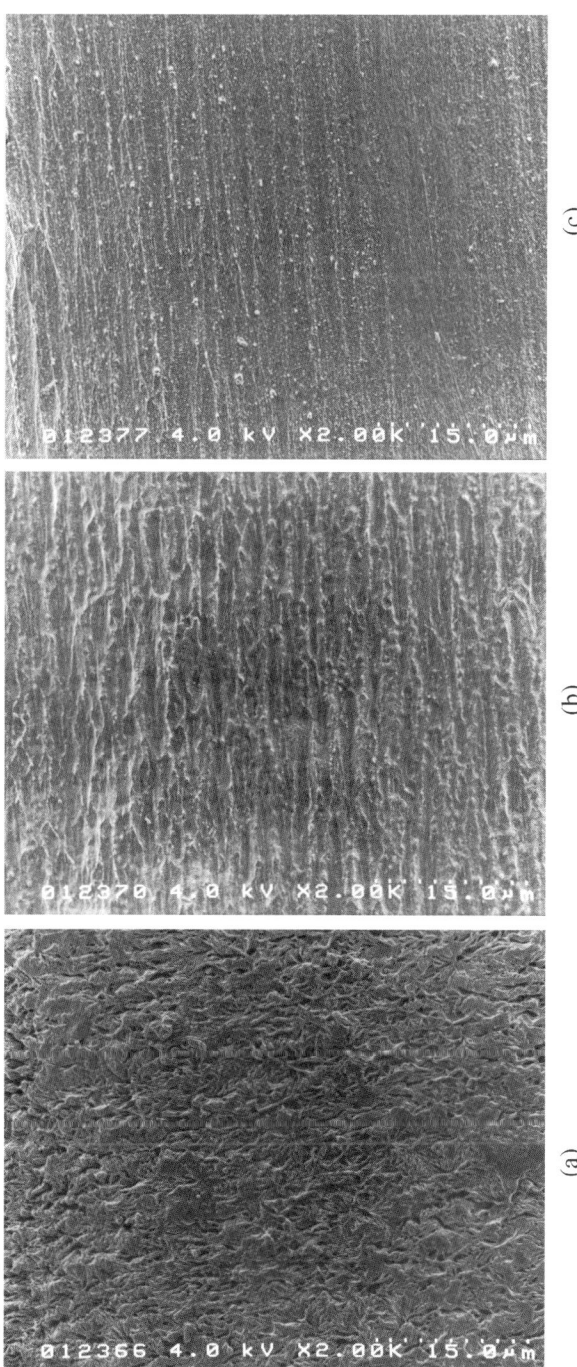

Fig. 6 SEM photographs of PCL fibers undrawn, and drawn in horizontal directions after enzymatic degradation. (a) DR = 1 (undrawn); (b) DR = 5 (drawn); (c) DR = 9 (drawn).

Tab. 4 Monofilament yarn properties of PBS and PBES

Sample (BS*/ES molar ratio)	PBS (100/0)		PBES-1 (91/9)		PBES-2 (78/22)	
Drawing	As-spun	Drawn	As-spun	Drawn	As-spun	Drawn
Draw ratio (times)	1	7.1	1	7.3	1	7.3
Denier	753	731	777	732	802	739
Yarn diameter (μm)	282	282	284	279	285	278
Tensile tenacity (10^{-3}N/tex)	0.90	6.01	0.73	4.99	0.37	2.74
Elongation at break (%)	737	27.4	779	27.2	748	34.4
Young's modulus (10^{-3}N/tex)	–	24.1	–	19.3	–	9.2
Hot water shrinkage** (%)	–	7.0	–	10.8	–	12.3
Hot air shrinkage** (%)	–	4.3	–	5.6	–	41.0

* BS = Butylene succinate, ** at 80 °C for 15 min

succinate (ES) contents of 0 to 20 mol.% were prepared by melt spinning (Mochizuki et al., 1997c). Tensile and thermal properties of as-spun and drawn PBS and PBES monofilaments (diameter 280 ± 5 μm) are shown in Table 4. The tensile tenacity of the fibers increased dramatically with drawing. Both tensile and Young's modulus of the drawn PBES monofilaments decreased when the ES content of the PBES copolymers was increased, while thermal shrink-age of the fibers was increased when the ES content was increased. This probably results from the less-ordered or lower crystallinity structure of the PBES with higher content of the ES.

WAXD data of PBS and PBES monofilaments are shown in Table 5. Crystallinity as well as crystallite size along the fiber axis would appear to decrease as the ES content of the PBES was increased. Table 6 shows small angle X-ray scattering (SAXS) data of

Tab. 5 Crystalline parameters of PBS and PBES monofilaments before and after enzymatic degradation, measured by wide-angle X-ray diffractometry (WAXD)

Drawing	Sample (BS/ES molar ratio)	Degrada-tion	Crystallite sizes D_{hkl} (nm) D_{h00} or D_{0k0} (22.3 °)*	D_{001} (24.9 °)*	Crystalline orientation Π (%) (22.3 °)*	Crystallinity index I_c (%)	Lattice distortion factor k
As-spun	PBS (100/0)	Before	65	–	–	30.4	1.54
		After	63	–	–	27.5	1.50
	PBES-1 (91/9)	Before	67	–	–	30.0	1.62
		After	68	–	–	26.6	1.57
	PBES-2 (78/22)	Before	67	–	–	28.8	1.69
		After	63	–	–	25.3	1.59
Drawn	PBS (100/0)	Before	61	78	93.5	61.5	0.76
		After	60	67	95.7	57.3	0.58
	PBES-1 (91/9)	Before	65	67	93.9	61.1	0.73
		After	64	63	95.6	56.3	0.58
	PBES-2 (78/22)	Before	65	56	93.3	55.6	0.83
		After	62	50	94.5	51.6	0.67

* Diffraction angle: 2θ

Tab. 6 Small-angle X-ray scattering (SAXS) of PBS and PBES monofilaments before and after enzymatic degradation

Drawing	Sample (BS/ES molar ratio)	Degradation	Maximum intensity (cps)	Difference of intensity	Long period (Å)	Scattering power index
As-spun	PBS (100/0)	Before	291	20	93	108,000
		After			130	
	PBES-1 (91/9)	Before	321	29	95	131,000
		After			96	
	PBES-2 (78/22)	Before	332	24	104	102,000
		After			118	
Drawn	PBS (100/0)	Before	410	180	114	220,000
		After			158	
	PBES-1 (91/9)	Before	510	315	106	358,000
		After			110	
	PBES-2 (78/22)	Before	455	272	92	356,000
		After			99	

cps, counts per second.

the fibers. The lower the ES content, the greater the long period along the fiber axis; this would suggest that there were more amorphous or less-ordered regions in the ES content-rich PBES copolymer.

3.2.2
Enzymatic Degradation

Enzymatic degradation of PBS and PBES monofilaments was performed in the same manner as the PCL monofilaments mentioned above. TOC values of as-spun monofilaments were found to be significantly higher than those of drawn filaments, and were shown to increase as the ES content of the PBES increased. This suggests that enzymatic degradation tends to proceed preferentially in the amorphous regions.

In order to investigate the structural effects on biodegradation, the fiber microstructure was examined using WAXD and SAXS both before and after enzymatic degradation. Following enzymatic degradation, crystallinity and crystallite size – particularly along the fiber axis direction – was

decreased in all PBS and PBES monofilaments; this was evidence that the crystalline regions can also be ultimately degraded. The decrease in lattice distortion factor in WAXD after enzymatic degradation means that the structurally disordered regions of crystals, and some disruption at the surface of lamella, were preferentially degraded, leading the residual crystal lattice to be more orderly. SAXS data indicated that the long period – the sum of the crystalline and amorphous layer thicknesses – of the fibers was increased after enzymatic degradation. This was probably due to loosening of amorphous regions caused by random scission of in-chain ester linkages of the polymer.

3.2.3
Environmental Degradation

PBS and PBES monofilaments were degraded to determine differences under various environmental conditions, including soil burial, seawater exposure, and aerobic biodegradation in activated sludge.

Tensile strength retention plotted against soil burial time is shown in Figure 7, and this is inversely related to the weight loss profile (Figure 8). The half-retention time of as-spun PBS and drawn PBES-2 monofilaments was approximately three months – much less than those of drawn PBS and PBES-1 monofilaments. It is interesting to note that the tendency of biodegradation, as monitored by weight loss in soil burial tests, correlates well with TOC values of enzymatic degradation. Thus, biodegradation in soil appears to occur by the fibers being colonized by microorganisms, which secrete an extracellular depolymerase (e.g., lipase) that degrades the fibers to soluble monomer and oligomer units. These soluble products are then absorbed through the microorganisms' cell walls and ultimately metabolized to carbon dioxide and water. This surface erosion mechanism is supported by SEM observations of a gradual reduction in fiber diameter with time under conditions of enzymatic degradation.

Marine exposure tests showed a similar (albeit slower) profile than soil burial tests. A slower rate of degradation in seawater may be caused by differences in the number of species of microorganisms that degrade the fibers, as well as the microorganism population.

The aerobic biodegradation of test and reference materials in activated sludge revealed that the drawn PBS monofilament showed almost the same level of degradability as drawn PCL and PHBV samples, whereas degradation of PBES-2 proceeded more readily.

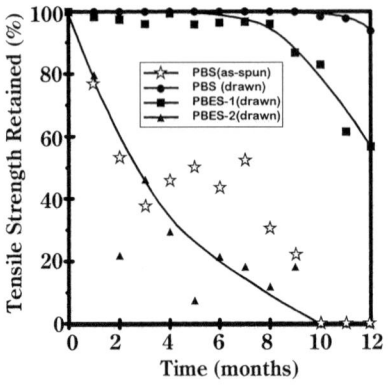

Fig. 7 Changes in tensile strength of PBS and PBES fibers during soil burial.

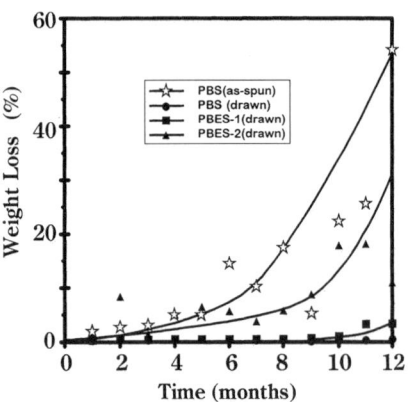

Fig. 8 Weight loss profiles of PBS and PBES fibers in soil burial.

4
Poly(lactic) Acid

4.1
Manufacture of PLA and its Products

PLA – one of the biodegradable aliphatic polyesters – has been highlighted as "green plastic" because of its availability from agricultural renewable resources such as corn. PLA is a biodegradable polymer with thermoplastics processability, and also has favorable mechanical properties and degradation characteristics. PLA can be processed on a conventional melt-extrusion equipment for PET, and the PLA products obtained provide the performance comparable to PET, with the added benefit of being fully degradable and compostable.

Recent advances in the fermentation of glucose, obtained from corn, has led to a dramatic reduction in the cost to manufacture the lactic acid necessary to produce PLA polymers. In addition, the technology to produce PLA polymers economically on a commercial scale has been developed worldwide by several companies. Notably, Cargill Dow LLC, a 50:50 joint venture between Cargill, Incorporated and The Dow Chemical Company, presently operates the world's largest (8000 tonnes per year) PLA (Nature-Works™) plant in Savage, Minneapolis, USA. In late 1999, they also began construction of a larger commercial plant with a world-scale PLA production capacity of 140,000 tonnes per year at Blair, Nebraska, and this should be on line in late 2001.

Unitika Ltd. (Osaka, Japan) has developed the world's first wide range of PLA products such as films, sheets, fibers, and nonwovens (e.g., spunbond fabrics) using their own polymer processing technology in collaboration with Cargill Dow LLC. These products are available commercially under the trademark TERRAMAC™ (terra means earth in Latin, while "mac" means son in Gaelic), giving the concept of "coming from nature and return to nature".

In the previous section, the biodegradable fibers made from aliphatic polyesters were reviewed, and the effects of biodegradation on their structure(s) discussed. Here, the performance features, biodegradation mechanism, and environmental degradation properties of the commercial PLA products including films, sheets, fibers, and non-wovens are described, and their potential applications are discussed.

4.2
Key Performance Features

TERRAMAC™ is a truly biodegradable product linked with carbon recycling of organic materials within the natural environment. The construction of a resource recycling system society is expected during the present century.

TERRAMAC™ products are not only biodegradable, but also are highly functional, notably when coupled with the intrinsic properties of PLA such as biocompatibility, compostability, bacteriostatic and antifungal characteristics, weather stability, dead-fold property, elastic recovery and good resilience, and thermal bonding capabilities.

PLA is a polymer produced by the polymerization of lactic acid by ring opening of its cyclic dimer, lactide. In accordance with FDA guidelines, PLA is "Generally Recognized As Safe" (GRAS). The limited migration observed during such studies represents no significant risk, since migrating species (lactic acid or its oligomers) are expected to be converted to lactic acid, as a safe food substance.

Under highly specific conditions of high temperature (above 60 °C) and humidity – as typified by a composting environment – PLA will disintegrate within one week to one month, followed by bacterial attack on the fragmented residues to produce carbon dioxide and water. In fact, TERRAMAC™ products degrade completely and safely when exposed to biologically active environments, such as aerobic composting or anaerobic digestion treatment plants, along with other organic materials such as grass, leaves, wood chips, craft paper, and food waste.

Under typical use and storage conditions, TERRAMAC™ products are extremely stable and safe because of the bacteriostatic and antifungal properties that distinguish PLA from other biodegradable materials. The bacteriostatic properties of TERRAMAC™ fibers under the standard method using *Staphylococcus aureus* ATCC 6538P as inoculum, are shown in Table 7. This unique

Tab. 7 Bacteriostatic properties of TERRAMAC™ fibers made from poly(lactic acid)

Material	Number of bacteria N (cells mL⁻¹)	log N	Bacteriostatic activity logB − logC	Bactericidal activity logA − logC	Washing
Inoculum*	$A = 2.5 \times 10^4$	4.4	–	–	
Nylon fabric (control)	$B = 1.7 \times 10^7$	7.2	–	–	
TERRAMAC	$C \leq 2.0 \times 10$	≤ 1.3	≤ 5.9	≤ 3.1	Before
TERRAMAC	$C \leq 2.0 \times 10$	≤ 1.3	≤ 5.9	≤ 3.1	After ten times
TERRAMAC/PET (3/7)	$C \leq 2.0 \times 10$	≤ 1.3	≤ 5.6	≤ 3.1	Before
TERRAMAC/PET (3/7)	$C = 1.1 \times 10^3$	3.0	3.9		After ten times
TERRAMAC/cotton (5/5)	$C \leq 2.0 \times 10$	≤ 1.3	≤ 5.6	≤ 3.1	Before

* *Staphylococcus aureus* ATCC 6538P. ⁺ A, no. of bacteria in inoculum; B, no. of bacteria in nylon fabric (control test); C, no. of bacteria in TERRAMAC tests.

feature offers distinct benefits in the areas of food production, sanitation, and agriculture.

4.3
Environmental Degradation

4.3.1
Soil Burial Test

The percentage decrease in tensile strength and relative viscosity of the PLA fibers following soil burial as a function of time are illustrated in Figure 9, these data indicating that degradation proceeds slowly, but steadily. After two years, the fibers have lost about 50% of their initial strength. A SEM photograph of PLA fibers subjected to 18 months of soil burial (Figure 10) shows that a number of fine cracks appear perpendic-

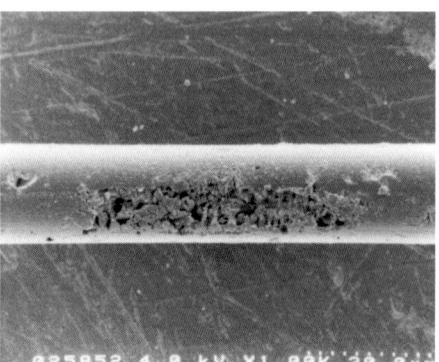

Fig. 10 SEM photograph of PLA fibers subjected to 18 months of soil burial.

ular to the fiber axis. It is supposed that hydrolytic degradation occurs preferably in the amorphous regions sandwiched between two crystalline zones, as described earlier.

4.3.2
Weathering Test

The weather stability of PLA fibers was determined using an accelerated weathering test (Sunshine Weather Meter), and compared with that of PET fiber. Figure 11 shows that the weather stability of PLA fibers is superior to that of PET fibers. Hence, the former materials can be significant players in the natural environment, such as agricul-

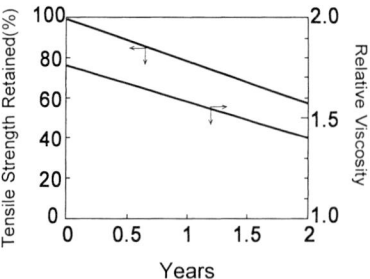

Fig. 9 Percentage decrease in tensile strength and relative viscosity of PLA fibers in soil burial.

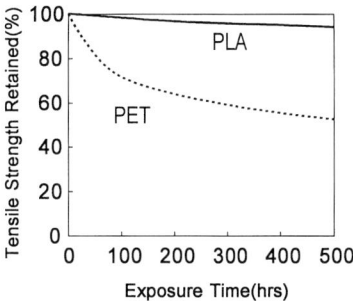

Fig. 11 Weather stability of PLA fibers examined by accelerated weathering test (Sunshine Weather Meter) in comparison with PET fiber.

tural and geotextiles – both areas where weather stability is essential for a product to be effective.

4.3.3
Aerobic and Anaerobic Composting Test

The compostability of PLA nonwovens (spunbond fabric, 25 g m^{-2}) was evaluated under simulated composting conditions. Evaluation of the ultimate aerobic biodegradability and disintegration was conducted by measurement of released biogas and disintegration at 60 ± 2 °C after 45 days, according to ISO 14855. This test method is designed to yield a percentage and rate of conversion of carbon of the substance to released carbon dioxide. Figure 12 shows the

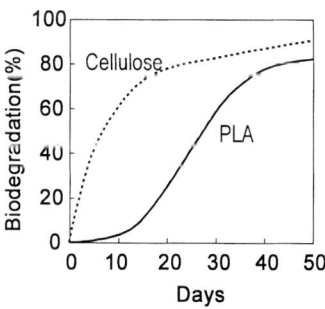

Fig. 12 Evaluation of ultimate aerobic biodegradability and disintegration of PLA nonwovens (spunbond fabric, 25 g m^{-2}), according to ISO 14855.

cumulative biodegradation of PLA spunbond fabric compared with cellulose powder as a positive reference, the PLA showing near-equal biodegradability with cellulose after 45 days.

Evaluation of the ultimate anaerobic biodegradability and disintegration under high-solids, anaerobic-digestion conditions was based on carbon conversion of the test substance to methane and carbon dioxide and disintegration after 45 days, according to FDIS 15985. The results represented in Figure 13 show that anaerobic decomposition of PLA nonwovens occurs at 52 ± 2 °C under high-solids and static, non-mixed conditions.

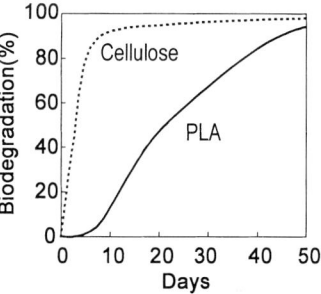

Fig. 13 Evaluation of ultimate anaerobic biodegradability and disintegration of PLA nonwovens (spunbond fabric, 25 g m^{-2}) under high-solids anaerobic-digestion conditions, according to FDIS 15985.

4.3.4
Biodegradation Mechanism

Few reports exist of the microbial degradation of high-molecular weight PLA. In the primary degradation phase, PLA undergoes chemical hydrolysis which is both temperature- and humidity dependent, and does not involve any microorganisms. As the Mn reaches approximately 10,000–20,000, microorganisms present in the soil begin to digest the lower-molecular-weight oligomer and lactic acid, producing carbon dioxide and water (Lunt, 1998). This two-stage mechanism (Table 8) differs distinctly from

Tab. 8 Two-stage degradation mechanism of poly(lactic acid)

Stage	Molecular weight	Rate of MW decrease	Weight loss	Hydrolysis reaction	Degradation mechanism
First stage	High	Slow (rate-determined)	None	Non enzymatic	Bulk
	Critical (Mn: 10,000–20,000)		Onset		
Second stage	Low	Rapid	Rapid	Non enzymatic and enzymtic	bulk and surface

that of many other biodegradable polymers which degrade by a single-step process, involving direct bacterial attack on the polymer.

In natural environments such as soil or water, therefore, the degradation of PLA proceeds slowly, but steadily, this being a convenient feature for the application of agriculture/horticulture and geotextiles. For biodegradable applications of PLA products, the preferred disposal route is by composting. In high-temperature (60–80 °C) and high-humidity environments under both aerobic and anaerobic conditions, PLA products will degrade rapidly (see Figures 12 and 13). However, under typical use and storage conditions, PLA products are extremely stable.

4.4
PLA Products and their Potential Applications

PLA is processed into various types of films, fibers, or spunbond fabrics on conventional melt-extrusion equipment. Here, a range of typical TERRAMAC™ PLA products is introduced, together with their potential applications, including agricultural/horticultural products (mulch, plant covers, plants pots, root covers, string), geotextiles (sand bags, erosion protection, slope stabilization, drainage), clothing and furnishings (shirts, uniforms, casual wear, towels, mats), personal hygiene products (wipes, disposable diapers, absorbent pad liners), packaging materials (wrapping, shrink label, food tray, blisters, strapping band, laminated paper, pouch), and industrial uses (filters, drainage bags, tapes, ropes, carpets).

A potential application of PLA products lies in environmental applications including geotextiles and agricultural/horticultural sectors, where they are recycled biologically into the soil when their service life is complete. Good weather stability and a slow rate of degradation of PLA in the natural environment will increase their service life. As an alternative, PLA products may also play a role in packaging and hygiene applications in which they are decomposed under aerobic or anaerobic composting conditions, along with other organic wastes. Such a disposal route is valuable where this type of composting infrastructure is in existence.

4.4.1
Films and Sheets

TERRAMAC™ products feature two types of PLA films, namely rigid and flexible, thus providing a wide variety of applications. The rate of degradation of the flexible-type film is greater than that of the rigid-type film, and can be controlled to some extent with regard to applications.

The biaxial orientation of PLA film offers good toughness and rigidity, as well as excellent gloss and transparency – similar to PET film. The high barrier property to flavor and aroma molecules avoids acalping concerns in flavor-sensitive packaging, although the gas barrier property of PLA film is poor. On the other hand, high gas permeability, such as oxygen permeability and water vapor transmission rates, in addition to high transparency properties of PLA films leads to modified-atmosphere packaging and "breathable" packaging of fresh vegetables and fruits so that freshness is maintained.

One interesting application is shrinkage packaging, and this is closely related to the level of the D-isomer of lactic acid in the PLA film formulation. Another application is that of paper composites which are coated or laminated with PLA films, rendering them more easily repulped than oriented polystyrene (OPS). Moreover, the dead-fold properties of PLA films lend themselves to twist-wrapping for items such as sweets and lettuce.

Recently, hand-soft blown films of PLA with plasticizer have been developed for use as an agricultural mulch, compost bags, and the backing sheets of personal hygiene products. The mechanical properties of such films resemble those of polyethylene or soft polyvinylchloride.

Unoriented sheet has extremely low haze and good moldability in thermoforming, and this appears promising for use as containers, trays, and blister-packing.

The use of PLA in films and sheets may be summarized as:

- biaxially oriented film (rigid type) – packaging, laminated pouch, laminated paper, twist wrapping.
- blown film (flexible type) – compost bags, agricultural mulch film, fertilizer bags, diapers.
- shrink film – packaging, labeling.
- sheet (cast) – thermoforming (container, tray, blister), card.
- uniaxially oriented products – slit-yarns, strings, ropes, strapping bands, sand bags.

4.4.2
Fibers and Nonwovens

An interesting feature of PLA fibers is that while processing temperatures are more typical of polyolefins (~220 °C), their yarn properties are similar to those of a polyester. PLA is an aliphatic polyester, and therefore contains no aromatic ring structures. The moisture regaining and wicking properties are superior to those of PET because PLA is more hydrophilic. Garments made from 100% PLA and blends of PLA with cotton and lyocell feel more comfortable to the wearer, while in addition the lower modulus leads to better properties of drape and feel. PLA fibers are also dyeable, using dispersion dyes at lower temperatures (100–120 °C). Fabrics containing PLA are also being considered for use as lightweight sports wear as well as more robust, outerwear clothing.

Although not a non-flammable polymer, the PLA fiber shows improved self-extinguishing characteristics, and also has low smoke generation compared with PET. In addition, the elastic recovery and crimp retention properties lead to excellent shape retention and crease resistance. As described earlier, the superior UV characteristic of PLA fabrics is attractive for exterior use.

By controlling the ratio of the D and L isomers in the PLA polymer chain, it is possible to induce different crystalline melting points. By incorporating polymers with high levels of the D-isomer (~10% D-monomer) into a sheath component, sheath/core-type bicomponent fibers with a sheath melting point of 120–130 °C can be produced. This type of conjugated fiber is used as binder fibers in nonwovens for thermal bonding.

Unitika Ltd. have successfully developed a high-performance PLA spunbond fabric with high strength and good dimensional stability. High air velocities, typical of those used in PET spunbond, result in a low thermal shrinkage (< 4% at 100 °C) fabric.

The use of PLA fibers may be summarized as follows:

- Monofilament and multifilament – knitted or woven fabrics, ropes, belts.
- Staple fiber (regular and sheath/core type for binder) – spun yarn, nonwovens, wipes, fiber cushion, plasters.
- Short-cut fiber – binder for paper making, nonwovens with wet process.
- Spunbond fabrics – plant covers, root covers, vertical drain sheets, disposable diapers, carpet backing, flower wrapping, bags, filters.

5
Conclusions

The high economic growth of the twentieth century has led to two major problems – namely, the exhaustion of fossil fuel resources, and the management of solid waste as a consequence of the "mass production, mass consumption, and mass disposal" of conventional synthetic polymer products with oil as their monomer source. It is likely that, during the twenty-first century, a socially oriented resource recycling system – from the viewpoint of both plastic waste management and resource conservation – should emerge.

Advances in the bacterial fermentation of D-glucose obtained from corn first led to an ability to produce lactic acid much more cheaply than could be obtained from petrochemical-derived products. The technology to produce PLAs economically on a commercial scale was also developed, and this will likely enable PLA polymers to enter the commodity plastics industry. It is possible that PLA might replace existing polymers where a renewable resource is beneficial, or where biodegradable performance is required. In addition, the unique properties of PLA polymers have attracted considerable attention across the spectrum of plastics applications.

One major potential application of PLA lies in packaging and disposable, personal care products, both of which are decomposed under either aerobic and/or anaerobic composting conditions, along with other organic wastes. In contrast, PLA products play a role in many environmental applications, including geotextiles, and in the agricultural and fishing industries where they are recycled biologically when their service life is complete. A slow rate of degradation, together with excellent weather stability in an outdoor environment, may combine to extend the service life of these PLA products.

6
References

Ajioka, M., Enomoto, K., Suzuki, K., Yamaguchi, A. (1995) Basic properties of polylactic acid produced by the direct polycondensation polymerization of lactic acid, *Bull. Chem. Soc. Jpn.* **68**, 2125–2131.

Bashir, Z., Hill, M. J., Keller, A. (1986) Comparative study of etching techniques for electron microscopy using melt processed polyethylene, *J. Mater. Sci. Lett.* **5**, 876–881.

Diamond, M. J., Freedman, B., Galibaldi, J. A. (1975) Biodegradable polyester films, *Int. Biodetn. Bull.* **11**, 127–132.

Doi, Y. (1990) *Microbial Polyesters.* New York: VCH Publishers.

Fields, R. D., Rodriguez, F., Finn, R. K. (1974) Microbial degradation of polyesters: polycaprolactone degraded by *P. pullulans*, *J. Appl. Polym. Sci.* **18**, 3571–3579.

Fujimaki, T. (1998) Processability and properties of aliphatic polyesters, 'BIONOLLE', synthesized by polycondensation reaction, *Polym. Degrad. Stab.* **59**, 209–214.

Lunt, J. (1998) Large-scale production, properties and commercial applications of polylactic acid polymers, *Polym. Degrad. Stab.* **59**, 145–152.

Mochizuki, M., Hirami, M. (1997a) Structural effects on the biodegradation of aliphatic polyesters, *Polymers for Advanced Technologies* **8**, 203–209.

Mochizuki, M., Hirami, M. (1997b) Biodegradable fibers made from truly-biodegradable thermoplastics, in: *Polymers and Other Advanced Materials* (Plasad, P. N., Mark, E., Fai, T. J., Eds.), New York: Plenum Press, 589–596.

Mochizuki, M., Hirano, M., Kanmuri, Y., Kudo, K., Tokiwa, Y. (1995) Hydrolysis of polycaprolactone fibers by lipase: effects of draw ratio on enzymatic degradation, *J. Appl. Polym. Sci.* **55**, 289–296.

Mochizuki, M., Mukai, K., Yamada, K., Ichise, N., Murase, S., Iwaya, Y. (1997a) Structural effects upon enzymatic hydrolysis of poly(butylene succinate-*co*-ethylene succinate)s, *Macromolecules* **30**, 7403–7407.

Mochizuki, M., Nakayama, K., Xian, R., Jiang, B. Z., Hirami, M., Hayashi, T., Masuda, T., Nakajima, A. (1997b) Studies on biodegradable poly(hexano-6-lactone) fibers 1. Structure and properties of drawn poly(hexano-6-lactone) fibers, *Pure and Appl. Chem.* **69**, 2567–2575.

Mochizuki, M., Murase, S., Inagaki, M., Kannmuri, Y., Kudo, K. (1997c) Structure and biodegradation of fibers made from poly(butylene succinate-*co*-ethylene succinate)s, *Sen'i Gakkaisi* **53**, 348–355.

Mochizuki, M., Hayashi, T., Nakayama, K., Masuda, T. (1999) Studies on biodegradable poly(hexano-6-lactone) fibers 2. Environmental degradation, *Pure Appl. Chem.* **71**, 2177–2188.

Potts, J. E., Clendinnings, R. A., Ackart, W. B., Niegisch, W. D. (1973) The biodegradability of synthetic polymers, *Polym. Sci. Technol.* **3**, 61–79.

Reeve, M. S., McCarthy, S. P., Downey, M. J., Gross, R. S. (1994) Polylactide stereochemistry: effect on enzymatic degradability, *Macromolecules* **27**, 825–831.

Tokiwa, Y., Suzuki, T. (1981) Hydrolysis of copolyesters containing aromatic and aliphatic ester blocks by lipase, *J. Appl. Polym. Sci.* **26**, 441–448.

Tokiwa, Y., Suzuki, T. (1986) Hydrolysis of polyesters by lipases, *Nature* **270**, 76–78.

Tomita, K., Kuroki, Y., Nagai, K. (1999) Isolation of thermophiles degrading poly(L-lactic acid, *J. Biosci. Bioeng.* **87**, 752–755.

Torres, A. Li, S. M., Roussos, S., Vert, M. (1996) Screening of microorganisms for biodegradation of poly(lactic acid) and lactic acid-containing polymers, *Appl. Environ. Microbiol.* **62**, 2393–2397.

Williams, D. F. (1981) Enzymic hydrolysis of polylactic acid, *Eng. Med.* **10**, 5–7.

2
Aliphatic Polyesters

Prof. Dr. Ann-Christine Albertsson[1], Prof. Dr. Indra K. Varma[2]
[1] Department of Polymer Technology, The Royal Institute of Technology, S-100 44
Stockholm, Sweden; Tel: +46-8-7908274; Fax: +46-8-100775;
E-mail: aila@polymer.kth.se
[2] Centre for Polymer Science and Engineering, Indian Institute of Technology, Hauz
Khas, New Delhi-110016, India; Tel: +91-11-6591425; Fax: +91-11-6591421;
E-mail: ikvarma@hotmail.com

$[I]_o$	initial concentration of initiator
$[M]_o$	initial monomer concentration
3-HB	3-hydroxybutyrate
bis-MPA	2,2-bis(hydroxymethyl) propionic acid

DPn	number average degree of polymerization
DXO	1,5-dioxepan-2-one
HA	hydroxyalkanoates
HEMA	2-hydroxyethyl methacrylate
LA	lactide
Mn	number average molecular weight
Mw	weight average molecular weight
PD,L-LA	poly(D,L-lactide)
PD-LA	poly(D-lactide)
PEG	poly(ethylene glycol)
PET	poly(ethylene terephthalate)
PHA	poly(hydroxyalkanoate)
PLA	polylactide
PL-LA	poly(L-lactide)
PS	*Pseudomonas* sp. lipase
ROP	ring-opening polymerization
Tg	glass transition
THF	tetrahydrofuran
β-PL	β-propiolactone
δ-VL	δ-valerolactone
ε-CL	ε-caprolactone

1

Introduction

Aliphatic polyesters are amongst the most important biocompatible and biodegradable materials that have received increasing attention during the past 15 years. Their applications in agriculture, coatings, medicine (drug delivery, body implants, sutures, etc.), adhesives and the packaging industry have grown significantly due to the availability of novel products with better performance characteristics. Current research activities in aliphatic polyesters are mainly focused on the synthesis, characterization, and development of polymers with better properties than the available state-of-the-art materials. Naturally occurring polyesters, such as plant cutin or shellac, have also been used in various fields as thermoplastics, adhesives, sealants, insulating materials, and coatings (for pharmaceutical products, food ingredients, etc.). This is primarily due to the non-toxic nature, insulating ability, oil resistibility, and thermoplasticity of these materials.

Aliphatic polyesters are synthesized either using chemical polymerization (polycondensation or ring-opening polymerization), or by bacterial fermentation. Enzyme-catalyzed synthesis of polyesters using lipase as a catalyst in anhydrous media is an emerging area of polyester synthesis (Nobes et al., 1996; Dong et al., 1999). Several review articles have been published dealing with the chemical or microbial synthesis of polyesters, their modification, and biodegradation (Albertsson, 1992; Albertsson and Karlsson, 1992, 1996a,b, 1998; Löfgren et al., 1995a; Arvanitoyannis, 1999; Albertsson and Varma, 2000). However, there is a need for a comprehensive review dealing with the *in vivo* and *in vitro* synthesis of aliphatic polyesters. The present article, which is more of an introduction to this fascinating interdisciplinary research, focuses on the current status of aliphatic polyesters, their structure and synthesis, biodegradation, applications, and future perspectives. A discussion on the naturally occurring polyester, shellac, has also been included to describe its multifunctional nature, as this is an important ingredient in the development of bio- and environmentally adaptable polymers (Albertsson and Karlsson, 1996a,b).

2
Historical Outline

Naturally occurring aliphatic polyesters, such as shellac, have been known since time immemorial, and their many applications have been referred to in several ancient scripts. In fact, it was the search for an alternative material to shellac that led to the development at the turn of this century of the first truly synthetic resin – Bakelite.

The initial studies of Carothers in the 1930s dealt with the synthesis of polyesters using condensation polymerization. However, high-molecular mass polymers with desirable mechanical properties could not be obtained using this technique because of certain inherent shortcomings. The polycondensation reaction must be carried out at high temperature for several hours in order to achieve high conversion. Since these reactions are reversible in nature, the product of condensation (usually water) must be removed continuously, and this leads to the depletion of low-boiling reactants and thereby creates an imbalance of functional groups. Hence, only oligomeric products with a molecular mass of the order of a few thousand could be obtained. Such polyesters, having appropriate functionality at the chain ends, found applications in the production of polyurethanes, or as plasticizers for thermoplastics.

In order to obtain high-molecular mass polymers, chain extension reactions of appropriate functionalized end-capped polyesters, with diacid chloride (Albertsson and Ljungqvist, 1986a,b, 1987), diiocyanate (Fujimaki, 1998) or bis-chloroformate (Ranucci et al., 2000) have been reported in the literature. High-molecular mass aliphatic polyester prepared through the polycondensation reaction of 1,4-butanediol, succinic acid and adipic acid or ethylene glycol and succinic acid, and the chain extended by further coupling reaction, has been available commercially since 1993 under the tradename of Bionolle™ (Showa Co., Japan).

Homopolymers and copolymers of α-hydroxy acids such as lactic acid or glycolic acid, prepared by ring-opening polymerization (ROP) of lactide (LA) or glycolide, have been known for almost 40 years. These have been the major constituents of degradable biomedical devices such as sutures, drug delivery systems, bone pins and plates, surgical adhesion, etc. Poly(lactic acid) was recognized in the 1960s for its potential use in temporary surgical implants and tissue repair. Synthetic resorbable sutures from poly(glycolic acid) were first developed in 1962 by American Cyanamid Co., and were available under the trade name of Dexon™ in 1970. Copolymers of glycolide and LA (Vicryl™) were introduced five years later as sutures with better performance. These materials have desirable mechanical properties, are biodegradable, and products of their degradation are biocompatible. Poly(ε-caprolactone) (poly(ε-CL)), is yet another biodegradable polyester which has attracted attention as a soil-degraded container material (Pitt et al., 1981). Blending or copolymerization has extended the application of these materials in different areas.

Many bacteria synthesize inclusions of poly(hydroalkanoate)s (PHAs) as energy reserve materials. Poly(3-hydroxybutyrate) (poly(3HB)), and poly(3-hydroxybyturate-co-3-hydroxyvalerate) belong to this category. (poly(3HB)) was mentioned in the microbiological literature as early as 1901 (Hocking and Marchessault, 1994). During the late 1950s and early 1960s, Baptist and Werber at W. R. Grace and Co. started producing this polymer for commercial evaluation as sutures and prosthetic devices, as well as in plastic laminates. These pioneering studies met with only limited success however, and remained dormant for several years until ICI made a major advance in the production of a copolymer of β-hydroxybutyrate and β-hydroxyvalerate. This family of materials, known as Biopol™, has much improved properties compared with the original homopolymer. In 1990, the first commercial product made from Biopol was launched in Germany as a biodegradable shampoo bottle. These polyesters are produced from renewable resources (i.e. by fermentation of glucose), are biodegradable, and are of commercial interest as a possible substitute of polyolefins.

3

Chemical Structure

Polyesters are characterized by the presence of ester groups in their backbone. They may be categorized in four groups on the basis of their occurrence or method of preparation as: (1) naturally occurring polyester; (2) microbial polyester; (3) condensation polyesters; and (4) polyesters obtained by ROP. The general route of synthesis can be described as shown in Figure 1.

3.1

Naturally Occurring Polyesters: Shellac

The resin portion of shellac is a complex mixture of monoesters (30%, soft resin) and polyesters (70%, hard resin) consisting of closely related sesquiterpenic acids of the cedrene skeleton, mainly jalaric acid, laccijalaric acid, and hydroxy fatty acids such as aleuritic acid (9,10,16-trihydroxy palmitic acid), 16-hydroxyhexadec-9-enoic acid, 9,10-dihydroxytetradecanoic acid (Singh et al., 1969, 1974a,b; Wadia et al., 1969; Khurana et al., 1970). Thus, isolated and conjugated double bonds, aldehydic and primary alcoholic groups (Prasad et al., 1991) are present in the backbone. As a consequence, shellac is a multifunctional resin. A marginal difference in the average ratios of terpinic acids, aleuritic acid and other fatty acids has recently been reported in the shellac samples that originated from India and Thailand (Wang et al., 1999). This ratio was 53:34:14 for Indian shellac, and 51:35:14 for Thailand shellac.

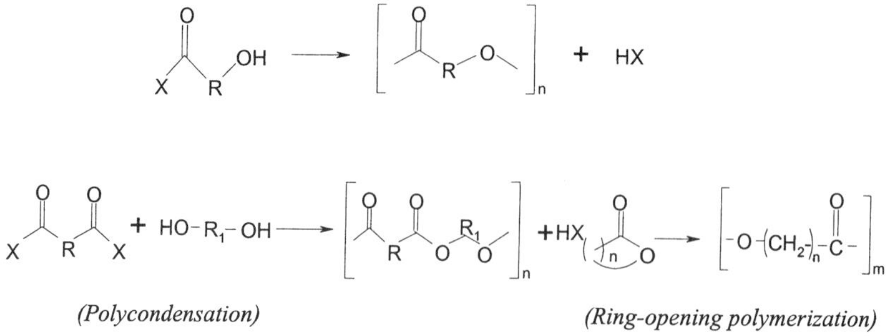

(Polycondensation)   *(Ring-opening polymerization)*

Fig. 1 General scheme of preparation of polyesters.

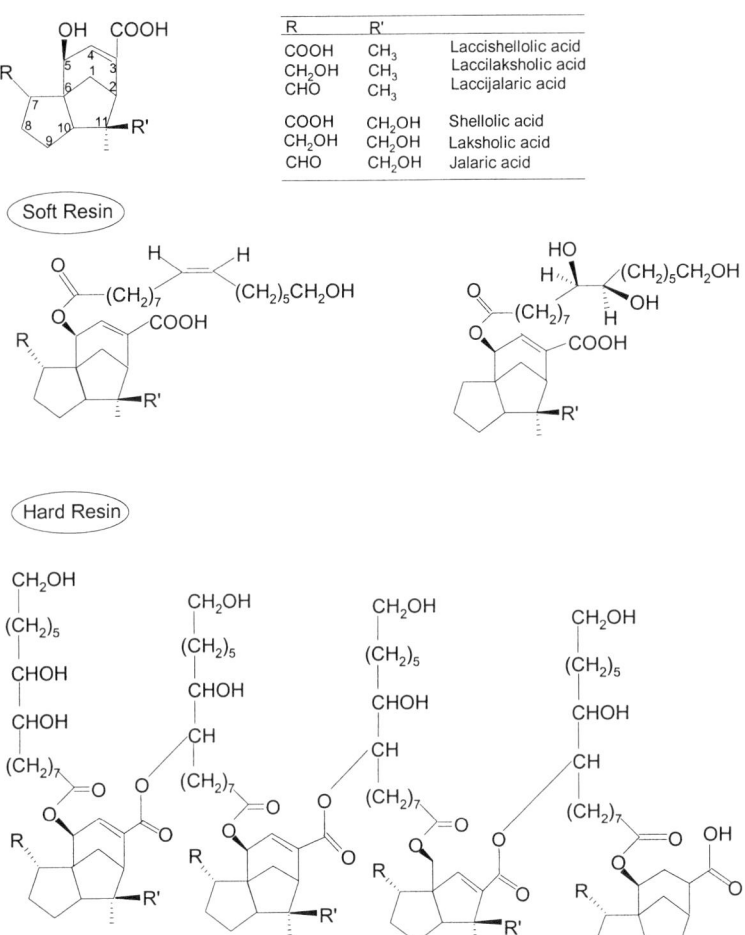

R	R'	
COOH	CH₃	Laccishellolic acid
CH₂OH	CH₃	Laccilaksholic acid
CHO	CH₃	Laccijalaric acid
COOH	CH₂OH	Shellolic acid
CH₂OH	CH₂OH	Laksholic acid
CHO	CH₂OH	Jalaric acid

Fig. 2 Main components of shellac (Wang *et al.*, 1999).

The chemical structure of the various constituents and representative structure of the polyester backbone of shellac may be depicted as shown in Figure 2.

3.2
Microbial Polyesters

PHAs are energy and carbon storage materials, and are accumulated intracellularly by numerous microorganisms under unfavorable growth conditions in the presence of excess of carbon source (Brandl et al., 1990; Doi, 1990; Steinbüchel and Valentin, 1995; Lee, 1996). The structure of polyesters derived from short chain length (3–5 carbon atoms) and medium chain length (6–16 carbon atoms), hydroxyalkanoic acids (PHA$_{SCL}$ and PHA$_{MCL}$ respectively) can be depicted as shown in Figure 3.

Among the more than 300 different microorganisms that are known to synthesize PHAs, only a few bacteria have been employed for the production because they

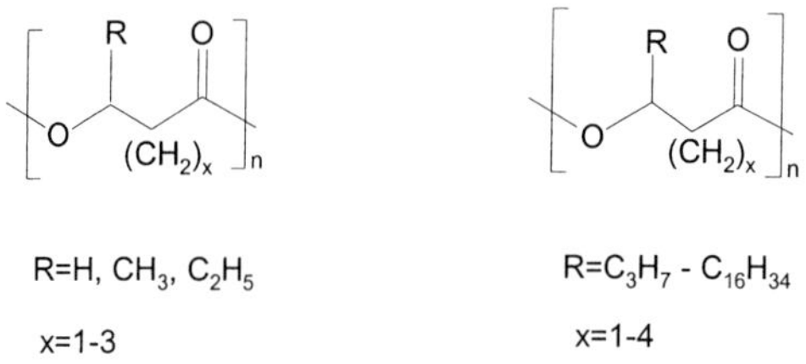

Short chain length (thermoplastics)   Medium chain length (elastomers)

R=H, CH$_3$, C$_2$H$_5$ R=C$_3$H$_7$ - C$_{16}$H$_{34}$

x=1-3   x=1-4

Fig. 3 General structural formula of bacterial poly(hydroxyalkanoate)s (Nobes et al., 1996).

can be cultivated efficiently to high cell densities with a high PHA content in a relatively short period of time (Lee and Choi, 1998). Bacteria that have been shown to produce poly(3HB) efficiently include *Ralstonia eutropha*, *Alcaligenes latus*, *Azotobacter vinelandii*, and recombinant *Escherichia coli*. Copolyesters of 3HB and 3-hydroxyvalerate have been produced by *A. eutrophus* from valeric and propionic acid under both air and oxygen:hydrogen (20:80) gas mixtures. Shifting the atmosphere changed the bacterial metabolism, and produced different monomer precursors for polymer synthesis. Such metabolic shifting may be used to produce PHAs with blockcopolymer within the polymer (Kelley et al., 1998).

3.3
Condensation Polyesters

Biodegradable aliphatic polyesters trademarked as Bionolle™ are produced through the polycondensation reaction of 1,4-butanediol, succinic acid (#1000 series) and adipic acid (#3000 series) or ethylene glycol and succinic acid (#6000 series) (Fujimaki, 1998). Chain extension by coupling reac-

tions has been carried out to obtain high-molecular weight materials. The structure of these copolymers can be depicted as shown in Figure 4.

Albertsson and Ljungqvist (1986a) synthesized ^{14}C-labeled poly(tetramethylene adipate) using adipic acid, tetramethylene glycol and tetraisopropyl titanate as a catalyst. The reaction was carried out in a nitrogen atmosphere for several hours at 190 °C. When water evolution ceased, the temperature was lowered to 140–150 °C. Further reaction was carried out at reduced pressure (1 torr) at 165 °C for several hours until the contents became highly viscous. The polymer was recovered by precipitating its acetone solution in methanol. Polycondensation of 3-hydroxyalkanoic acids and their derivatives has also been reported in the literature (Kobayashi et al., 1993).

A degradable aliphatic thermoplastic elastic block copolymer, poly(ethylene glycol)-(PEG)/poly(ethylene succinate) was synthesized by a two-step process with polycondensation of dimethyl succinate (0.6 mol), ethylene glycol (0.9 mol), and PEG (0.05 mol), followed by chain extension using adipolyl chloride (Albertsson and Ljungqvist, 1986b).

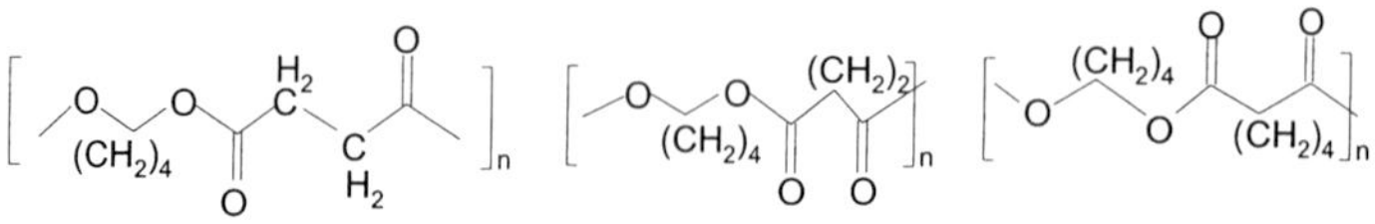

Poly(butylene succinate)   Poly(butylene succinate-adipate) copolymer

Fig. 4 Structural formula of condensation polyesters (Bionolle™) (Fujimaki, 1998)

The polycondensation was carried out at 240 °C for several hours. A series of block copolymers having poly(ethylene succinate)-b-poly(tetramethylene glycol) have also been synthesized by these authors (Albertsson and Ljungqvist, 1987) using polycondensation.

Recently α,ω-bis-hydroxy-terminated poly-(1,3-propylene succinate) has been chain-extended to yield high-molecular weight poly(ester-carbonates) (Ranucci et al., 2000) using the bis-chloroformate route. Thus, by using a molar ratio of 1,3-propanediol and succinic acid of 1.02, an oligomer having Mn of 2200 and Mw of 3000 was obtained. α,ω-Bis-chloroformate was then prepared by treating this oligomer with phosgene. Polycondensation of α,ω-bis-chloroformate-terminated poly(1,3-propylene succinate) with α,ω-bis-chloroformate yielded a poly(ester-carbonate) (Mn = 30,000 and Mw = 48,000).

Enzyme-catalyzed synthesis of polyesters, using lipase as a catalyst in anhydrous medium, has received increasing attention during the past few years (Knani, 1993; Knani et al., 1993; Athawale and Gaonkar, 1994; Brazwell, 1995; Mezoul et al., 1995; Suda et al., 1999). Low-molecular weight polyesters are obtained using this technique. For example, lipase-catalyzed polymerization of sebacic acid and 1,8-octanediol in aqueous medium yielded polyesters having a molecular weight of 1,600 g mol^{-1} (Kobayashi et al., 1998). In the polymerization of α,ω-dicarboxylic acid and glycol, the poly-merization behavior greatly depended on the chain length of monomers. The polymer was observed with 1,6-hexanediol (Suda et al., 1999). Polycondensation of 3HB acid or 12-hydroxydodecanoic acid using porcine pancreatic lipase yielded polyesters with Mw values ranging from a few hundred to several thousands (Shuai et al., 1999).

3.4
Polyesters Prepared by ROP

The monomers that have been used for the synthesis include glycolide, LA, β-propiolactone (β-PL), β-butyrolactone, γ-butyrolactone, δ-valerolactone (δ-VL), ε-CL, 1,5-dioxepan-2-one (DXO), pivalolactone, 1,4-dioxane-2-one, 2-methylene-1,3-dioxolane, 2-methylene-1,3-dioxepane etc. The structures of polyesters derived from some of these monomers are given in Figure 5.

The polymerizations are generally carried out in bulk or in solution (tetrahydrofuran (THF), dioxane, toluene, etc.). Dispersion polymerization of ε-CL using mixture of 1,4-dioxane and heptane and surface-active agents yields polymer in the form of microspheres and with a narrow molecular weight distribution (Slomkowski et al., 1998).

Few lactones polymerize spontaneously on standing or on heating (Cerrai et al., 1989), but most do so in the presence of a catalyst. ROP can be performed with a large number of initiators to form high-molecular mass products. Many organometallic com-

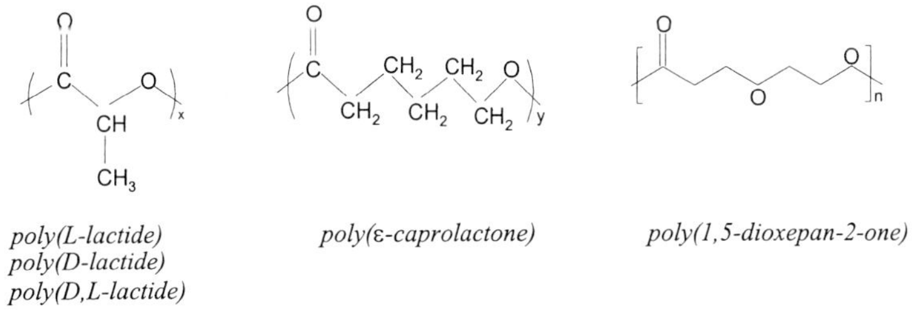

poly(L-lactide)
poly(D-lactide)
poly(D,L-lactide)

poly(ε-caprolactone)

poly(1,5-dioxepan-2-one)

Fig. 5 Aliphatic polyesters prepared by ring-opening polymerization.

pounds, such as oxides, carboxylates, and alkoxides are effective initiators for the controlled synthesis of polyesters using ROP of lactones (Löfgren et al., 1995a; Mecerreyes and Jérôme, 1999). Different initiation mechanisms can be divided into free radical, anionic, carbocationic, zwitterionic, co-ordinative, or based on active hydrogen species (Löfgren et al., 1995a). The highest yields and molecular weights have been obtained mainly by the anionic and co-ordinative ROP (Albertsson and Löfgren, 1996).

Enzymatic polymerization of lactones is a promising approach, and has been investigated by several workers (Knani and Kohn, 1993; Knani et al., 1993; Uyama and Kobayashi, 1993; Uyama et al., 1995; Henderson et al., 1996; Nobes et al., 1996; Bisht et al., 1997; Dong et al., 1999). Poly(ϵ-CL) of Mn 14,500, and with a molecular weight distribution of 1.23, has recently been reported using *Pseudomonas* sp. lipase (PS) as the catalyst (Dong et al., 1999). Water is consumed at the onset of polymerization, and released in parts during subsequent stages. A complex mechanism involving both ring-opening and linear condensation polymerization has been proposed for the enzymatic polymerization of lactones. The former was dominant at the early stage, while the latter was dominant in the later stage. Reaction media showed complex influences on the enzymatic polymerization. The proposed mechanism may be described as shown in Figure 6.

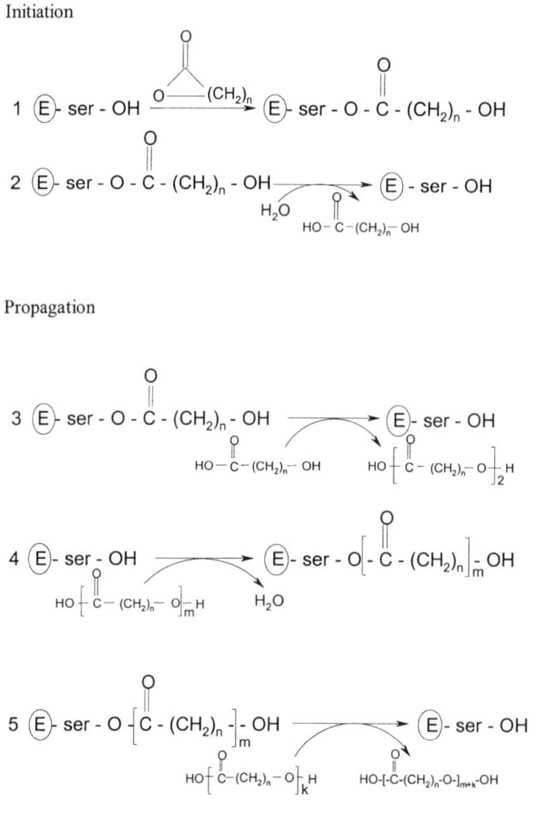

Initiation

Propagation

Fig. 6 Proposed mechanism for enzymatic polymerization of lactones (Dong et al., 1999).

High-molecular weight polylactides (PLAs) (Mw > 270,000) could be obtained by ROP of six-member D,L-, L,L- and D,D-LAs (Matsumura et al., 1998) by lipase in the temperature range of 80–130 °C. PS gave the highest molecular weight. Immobilization of lipase on celite significantly enhanced the polymerization with respect to low concentration of enzyme and Mw of the polymer. Polymerization of D,L-LA by lipase was better than that of L,L- or D,D-LAs.

Effective initiators for anionic polymerization of lactones include alkali metals, alkali metal alkoxides (such as lithium, sodium, or potassium), alkali metal naphthalenide complexes with crown ethers, and alkaline metals in graphite (e.g., potassium graphitide), etc. Depending upon the reaction conditions, type of initiators and monomer, the polymerization may proceed by living or non-living mechanisms. Chain growth (propagation) takes place by acyl-oxygen bond scission leading to the formation of alkoxide end-groups. Thus, alcoholate-initiated polymerization of lactones yields methyl ester and hydroxyl end-groups after termination with water. If alkyl oxygen cleavage had taken place, then methoxyl and carboxyl end-groups would have been formed.

Anionic polymerizations are sometimes associated with termination and transfer reactions. Huge amounts of cyclic oligomers, formed as a result of a backbiting reaction, were observed when ε-CL was polymerized using potassium *tert*-butoxide. However, in presence of lithium *tert*-butoxide in an apolar solvent (benzene), oligomer formation was significantly decreased. The polymerization proceeds by the acyl-oxygen scission. Smaller cyclic monomers such as β-PL do not form cyclic monomers to that extent.

β-PL, unlike other lactones, undergoes polymerization with weakly nucleophilic initiators such as metal carboxylates, tertiary amines, phosphines, and a variety of other initiators (Jedlinski et al., 1985). This is primarily due to high strain in the four-member ring. Pyridine and other tertiary amines initiate the anionic polymerization via a betaine that rapidly transforms into a pyridinium salt of acrylic acid. In order to minimize the chain transfer reactions, the polymerization is performed at a temperature between 0 and 10 °C.

Initiation by strong nucleophiles occurs by acyl-oxygen as well as alkyl-oxygen cleavage, whereas with weak nucleophiles, ring-opening occurs at the alkyl-oxygen bond. Similarly, polymerization of pivalolactone initiated with tertiary amines or phosphines is a fast process creating high-molecular weight products.

Block copolymers have been prepared by polymerizing 1,3-dioxan-2-one using stannous (II)-2-ethylhexanoate, $Sn(Oct)_2$, as initiator followed by addition of second monomer oxepane-2-one (Albertsson and Eklund, 1994).

Cationic polymerization is less useful than anionic polymerization for obtaining high-molecular weight polyesters. This is due to intramolecular transesterification (cyclization), proton and hydride transfer reactions. High-molecular weight polymers are obtained with strained monomers such as β-PL. Initiators for the ROP of lactones include protonic acids (HCl, RCOOH, RSO_3H, etc.), Lewis acids ($AlCl_3$, BF_3, $FeCl_3$, $ZnCl_2$, etc.), stabilized carbocations ($ET_3O^+BF_4^-$), and acylating agents ($CH_3CO^{(+)})^{(-)}OCl_4$).

The A-B diblock copolymer ε-CL and oxepan-2,7-dione has been synthesized using non-terminated cationic ends of poly-(ε-CL) (Lundmark et al., 1991). Cationic polymerization of DXO in bulk and solution using Lewis acids as initiators ($SnCl_4$, $FeCl_3$, $AlCl_3$, BCl_3 and BF_3OEt_2) has been reported by Albertsson and Palmgren (1996). BF_3OEt_2

was found to be a good initiator for both bulk and solution polymerization. The molecular weights of poly(DXO) increased with temperature up to 70 °C, but a further rise in temperature led to a decrease in molecular weight due to backbiting reactions. A Mw of 10,900 was obtained when the reaction was carried at 70 °C for 24 h using monomer to initiator ($[M]_o/[I]_o$) ratio of 500.

The coordinative initiation differs from the ionic polymerization in that the propagating species consists of a covalent bond species. This generally reduces the reactivity and polymerization rates. Decreased reactivity also leads to fewer side reactions, and often-controlled ROP of lactones may take place under these conditions. Tin alkoxides and phenoxides (Kricheldorf et al., 1991), aluminum alkoxides, mainly aluminum *iso*-prop-oxide and soluble bimetallic μ-oxo-alkoxides, essentially of zinc and aluminum have generated much interest because of their versatility as initiators (Heuschen et al., 1981; Hofman et al., 1987; Dubois et al., 1991b; Vanhoorne et al., 1992; Löfgren and Albertsson, 1994; Löfgren et al., 1994).

Block copolymers of ε-CL (Duda et al., 1998) with D,L-LA (Feng et al., 1983), styrene, or butadiene (Heuschen et al., 1981; Crivello et al., 1983) have been synthesized using these initiators. Efficient and versatile initiators based on α, β, γ, δ-derivatives of tetraphenylporphinato-aluminum for polymerization of ε-CL, β-lactones, δ-lactones, and LAs have been reported (Shimasaki et al., 1987; Inoue et al., 1992; Inoue and Aida, 1993).

Lanthanide compounds such as yttrium (Y) and lanthanum (Ln) alkoxide (McLain and Drysdale, 1992; McLain et al., 1992), samarium (Sm), and lutetium (Lu) complexes have been reported to yield high-molecular weight polyesters under mild conditions. Rapid polymerization of ε-CL at room temperature was reported when yttrium halides/alkoxide were used as initiators (Shen et al., 1996a,b,c). Stevels et al. (1996a,b) have used *in situ*-generated yttrium alkoxides as initiators for ε-CL and δ-VL. High-molecular mass poly(ε-CL) was obtained at room temperature when SmBr$_2$/Sm were used as initiators (Agarwal et al., 1999a). The use of bulky co-ordinated groups reduced the transesterification reactions, and polymers with narrow molecular weight distribution (MWD) could be obtained (Shen et al., 1996b,c). The performance of these initiators in terms of yield, molecular weight, polydispersity, and stereoregularity is affected by the ligands and by the oxidation state of the respective rare earth metals (Agarwal et al., 2000). High selectivity with minimal side reactions has been demonstrated in the ROP of cyclic esters with aluminum and lanthanide-based initiators (Dubois et al., 1991a,b; Chamberlain et al., 1999, 2000).

Aluminum alkoxide-initiated polymerization of lactones proceeds according to a "co-ordination-insertion" mechanism, which involves acyl-oxygen bond cleavage of the monomer and insertion into the aluminum–oxygen bond to the initiator. The co-ordination to the exocyclic oxygen to the metal results in polarization, thereby making the carbonyl carbon of the monomer more susceptible to nucleophilic attack. Transesterification reactions may also take place at elevated temperatures. Controlled synthesis of telechelic polymers, block and graft copolymers, and polymers of various architecture is possible by use of aluminum alkoxide of different structures and functionalities (Mecerreyes and Jérôme, 1999; Mecerreyes et al., 2000).

Aluminum isopropoxide co-ordinates to the exocyclic carbonyl oxygen, and the acyl-oxygen cleavage yields an isopropyl ester end-group. Termination of the growing chain with diluted HCl converted the propagating end to a hydroxyl group. A narrow molecular weight distribution and an in-crease in number average degree of polymer-ization (DPn) with increase in $[M]_o/[I]_o$ ratio indicated the living character of this poly-merization. A P(DXO) of Mn = 17,500 could be obtained by carrying out the polymer-ization in THF at 0 °C (Löfgren et al., 1994). The reaction follows first-order kinetics with respect to the monomer and initiator. The mechanism shown in Figure 7 has been suggested for the polymerization of DXO in bulk or in THF or toluene solution using aluminum isopropoxide as an initiator.

Controlled ROP of ε-CL is usually initiated by aluminum isopropoxide, $[Al(O^iPr)_3]$, in toluene at 0–25 °C. Under these conditions, this initiator exists as an aggregate of trimer and tetramer. However, freshly distilled $Al(O^iPr)_3$ consists mainly of trimer, which is a more reactive initiator for ROP. The initiation rate is high compared with prop-agation, so that narrow MWD is observed in the polymer. There is no termination reac-tion, and three chains grow per Al atom. Block polymers can be prepared by the addition of a second monomer (e.g. δ-VL).

In β-lactones, scission of either acyl-oxy-gen bond or the alkyl-oxygen bond may take place, thereby leading to the formation of alkoxide or carboxylate growing chain (In-oue et al., 1992; Inoue and Aida, 1993). A methylene chloride end-group was observed in $ZnCl_2$-initiated polymerization of ε-CL in xylene, thereby supporting the co-ordina-tion–insertion mechanism (Abraham et al., 2000).

Carboxylates are less nucleophilic than alkoxides, and are considered to behave more like a catalyst rather than as actual initiators. Metal carboxylates, such as stan-nous (II)-2-ethylhexanoate, $Sn(Oct)_2$, are used along with active hydrogen compounds (e.g. alcohols) as coinitiators (Mathisen and Albertsson, 1989; Mathisen et al., 1989; Rafler and Dahlmann, 1992). If no active hydrogen compound is added, then the actual initiating species may be hydroxyl-containing impurities (Nijenhuis et al., 1992; Löfgren et al., 1995a,b). $Sn(Oct)_2$ is a very effective and versatile catalyst which is commercially available, is easy to handle, and is soluble in common organic solvents

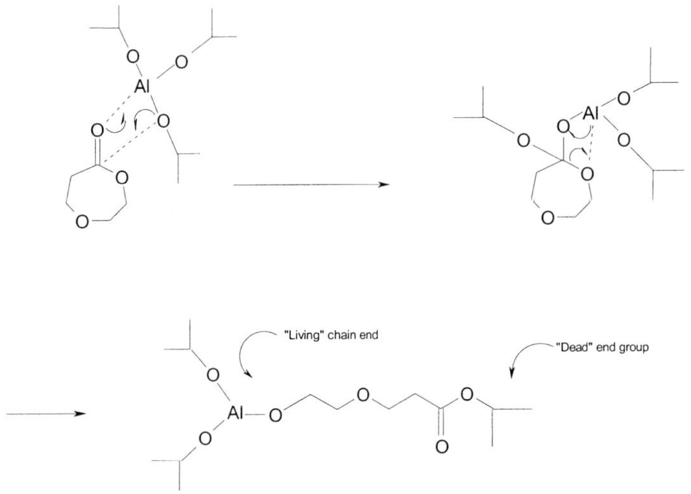

Fig. 7 Aluminum isopropoxide-initiated polymerization of 1,5-dioxepan-2-one.

and lactones. The mechanism of ROP in the presence of Sn(Oct)$_2$ has been recently examined by Kowalski et al. (1998, 2000) and Kricheldorf et al. (2000). The active species in ε-CL polymerization in THF was identified as OctSn[O(CH$_2$)$_5$C(O)]$_n$OR (Kowalski et al., 1998), suggesting thereby that polymerization proceeds on the tin(II)-alkoxide bond formed from Sn(Oct)$_2$.

Recent studies with tetrabutyl tin-initiated polymerization of ε-CL have indicated that Bu$_2$SnO dissolved in Bu$_4$Sn is the main initiator (Kricheldorf and Kreiser-Saunders, 2000). Almost 100% conversions are observed in living macrocyclic polymerization of ε-CL in bulk at 80 °C with 2,2-dibutyl-2-stanna-1,3-dioxepane as initiator (Kricheldorf and Eggerstedt, 1998).

Albertsson and coworkers have carried out extensive investigations in homo- and copolymerization of lactones in bulk as well as solutions using ROP. In bulk polymerization, temperatures in the range of 100–150 °C were used, while in solution polymerization temperature were kept low (0 to 60 °C) to minimize the side reactions such as intra- and intermolecular transesterification reactions. α-Methacryloyl-ω-hydroxyl-poly(ε-CL) macro monomer has been synthesized using Sn(Oct)$_2$ as catalyst and 2-hydroxyethyl methacrylate (HEMA) as an initiator. This macro monomer had a methacryloyl group at one end and a hydroxyl group at the other. The molecular weight and polydispersity of the macro monomer depended on the concentrations of Sn(Oct)$_2$ and HEMA, as well as the molar ratio of Sn/OH (Liu et al., 1998).

High-molecular weight poly(DXO) has been prepared at 110 °C using Sn(Oct)$_2$ as initiator (Albertsson and Palmgren, 1993); transesterification and degradation occurred above 130 °C. The initiation reaction in copolymerization of DXO and L-LA or D,L-LA in the presence of Sn(Oct)$_2$ is believed to take place as a result of water impurities. This leads to ring-opening, thereby generating a linear monomer molecule with an alcoholate ester end-group. This new catalytically active species, which is still attached to the tin reaction center, propagates the chain by ring-opening of the next co-ordinated cyclic monomer.

The Sn(Oct)$_2$ catalyst is generally active at elevated temperatures, leading to some intermolecular and intramolecular transesterification reactions (Penczek and Duda, 1996). Recently, a new catalyst system for ROP of lactones has been reported based on stannous (II) or scandium (III) trifluoromethane sulfonate (Möller et al., 2000; Nomura et al., 2000). These catalysts are very versatile, highly selective, and can be used under mild conditions.

3.5
Structural Modification by Copolymerization

The properties of aliphatic polyesters can be tailor-made by blending (Eastmond, 1999) and copolymerization (Albertsson and Gruvegård, 1995; Stridsberg et al., 1998), or by a change in the macromolecular architecture (e.g. hyper-branched polymers, star-shaped or dendrimers) (Grijpma and Pennings, 1994).

Homopolymers and random copolymers of ε-CL ($r_1 = 0.37$) and D,L-LA ($r_2 = 10.8$) have been reported using Ln halides as initiators (Shen et al., 1996a,b,c). Addition of epoxides or N-trimethylsilyltrialkyl (aryl) phosphinimines enhances the reactivity of Ln halides (Shen et al., 1996a). Graft copolymers, P(p-xylylene-g-ε-CL) have been prepared by *in situ* reductive coupling of aldehydes using SmI$_2$ (Agarwal et al., 1999b). Block copolymers have been prepared using halo-bridged Sm (III) complexes (Agarwal et al., 1999b).

The reactivity ratios in the copolymerization of DXO and L-LA were $r_{(DXO)} = 0.1$ and

$r_{(L-LA)} = 10$. Reactivity ratios in random co-polymerization of DXO and ε-CL have been reported as $r_{DXO} = 1.6$ and $r_{ε-CL} = 0.6$ (Albertsson and Gruvegård, 1995; Stridsberg et al., 1998). Reactivity ratios for the copolymerization of DXO and δ-VL were determined at 110 °C as $r_{VL} = 0.5$ and $r_{DXO} = 2.3$ (Gruvegård et al., 1998) At high conversion, depolymerization of poly(δ-VL) occurred, resulting in lower molecular weight and variations in copolymer composition.

Aluminum isopropoxide or $Sn(Oct)_2$ has been used for the preparation of block copolyesters (Pilati et al., 1999). Tri-block poly(ε-CL-b-DXO-ε-CL) was prepared by sequential addition of different monomers to a living polymerization system initiated with aluminum isopropoxide in THF or toluene solution (Löfgren et al., 1995b). Addition of a Lewis base (pyridine) did not significantly alter the polymerization rate in THF, though in toluene a small increase in rate was observed, especially at the later stages of polymerization. The optimal conditions were found to be polymerization in toluene solution at 25 °C with addition of 1 molar equivalent of pyridine (with respect to initiator). In order to increase the polymerization rate, the poly(ε-CL) block must first be synthesized at 0 °C, and the following two blocks at 25 °C.

Poly(L-LA-b-DXO-b-L-LA) has recently been reported using controlled ROP in a two-step process: polymerization of DXO using difunctional initiator-1,1,6,6-tetra-*n*-butyl-1,6-distanna-2,5,7,10-tetraoxacyclodecane, and addition and polymerization of L-LA in chloroform at 60 °C (Stridsberg and Albertsson, 2000) (Figure 8).

Poly(ε-CL-g-acrylamide) copolymers can be readily prepared using electron beam irradiation and Fe^{2+} ions provided by Mohr's salt as efficient terminators for grafted polyacrylamide chain radicals. The average molecular weight of grafted chains could be varied within wide limits by varying the Fe^{2+} ion concentration (Ohrlander et al., 1999a,b,c). Similar results were obtained when P(DXO) was grafted with acrylamide.

A new family of oligomers, with strongly branched structure (referred to as hyper-branched polyesters) is attracting considerable interest because of their unique mechanical and rheological properties, which can be easily tailored by changing the nature of the end-group. Albertsson (1992), using pentaerythritol, initiated studies in this field as tetra functional unit. Shi and Rånby (1996a,b,c) synthesized a series of methacrylated polyesters with different numbers of terminal double bonds per molecule, and investigated rheological properties, photopolymerization, and thermo-mechanical properties of cured films and glass fiber-reinforced laminates. The hyper-branched polyesters had much lower viscosity and much higher curing rates than the unsaturated polyester resins. The scheme for the synthesis of hyper-branched polyesters is shown in Figure 9.

The representative structure of the hyper-branched methacrylated polyester with 16 double bonds is shown in Figure 10.

Both dendrimers and hyper-branched macromolecules are prepared from AB_2 monomers, using different synthetic strategies. Dendrimers are produced by a stepwise series of reactions using two approaches:

- The divergent growth approach, which starts from a multifunctional core molecule and proceeds radially outwards.
- The convergent growth approach, in which the synthesis starts at the periphery and well-defined dendrons are prepared followed by final coupling to a multifunctional core (Hawker and Fréchet, 1990; Trollsås et al., 1998a,b; Hult et al., 1999).

Hyper-branched polymers are prepared in a single-step polymerization from ABx

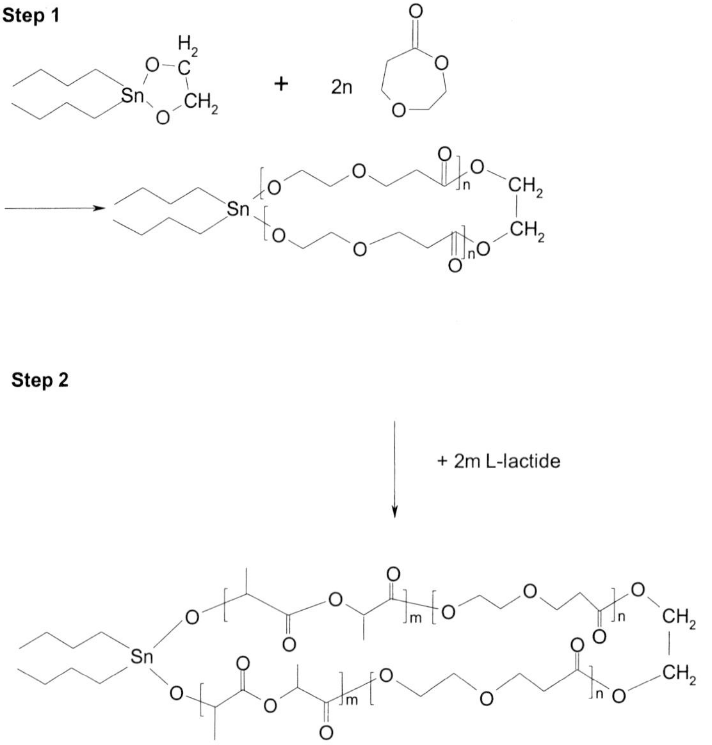

Step 1

Step 2

+ 2m L-lactide

Fig. 8 Reaction scheme for the ring-opening polymerization of poly(L-lactide-b-DXO-b-L-lactide) (Stridsberg and Albertsson, 2000).

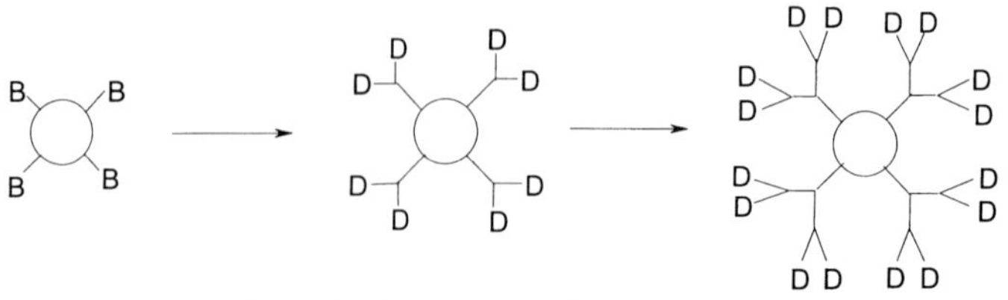

Fig. 9 Reaction scheme for the synthesis of hyper-branched polyester (Shi and Rånby, 1996a).

monomers. Thus, a perfectly branched structure is present in dendrimers, whereas irregular branching is present in hyper-branched polymers. Aluminum alkoxide-based initiators or tin-based catalysts have been successfully used for the preparation of hyper-branched (Trollsås et al., 1998a, 1999; Liu et al., 1999), dendrimer-like star polymers (Trollsås et al., 1998a) and star-shaped polymers (Joziasse et al., 2000). First- and second-generation versions of the benzyl ester of 2,2-bis-(hydroxymethyl) propionic acid (bis-MPA) are effective initiators for the ROP of lactones (ε-CL) in the presence of Sn(Oct)$_2$. The requisite AB$_2$ and AB$_4$ macro monomers and hyper-branched polyesters could be prepared by deprotection of the benzyl ester group from the initiator (Troll-

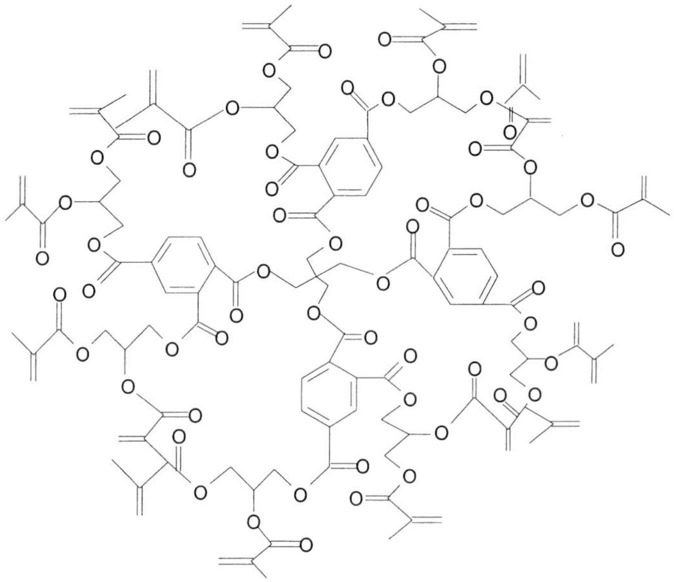

Fig. 10   Structure of the hyper-branched methacrylated polyester with 16 double bonds.

sås et al., 1998a,b, 1999). Additional functionality in these hyper-branched copolyesters could be imparted by using 1,4,9-trioxaspiro [4.6]-9-undecanone as a comonomer.

4

Occurrence

Shellac, the purified form of lac, is naturally occurring thermosetting aliphatic polyester resin that predates recorded history. It is of animal origin secreted by the lac insect *Kerria lacca* (*Laccifer lacca*) that grows parasitically on some trees such as ber, palas (*Butea frondosa*) or kusum. These trees grow mainly in India, Thailand, Mynammar, Sri Lanka, and the southern areas of China. Larvae of this insect attach to the twigs of the host tree and suck the tree's juices. They then secrete a protective coating of lac, and coalescence of the secretions of neighboring larvae forms a continuous covering on the twig. Harvesting is carried out twice a year

(*baisakhi* and *katiki*) by cutting the twigs from the trees, or twisting of branches to yield sticklac. The composition of sticklac may vary with the type of climate, nature of host plant, time of harvesting, and species of the insect. Seedlac is obtained from sticklac by washing and sieving with water or diluted soda to remove impurities, insect bodies, debris, etc. It consists of resin (70–80%), wax (6–7%), coloring matter (4–8%), and debris and moisture (15–25%).

Shellac is the refined form of seedlac prepared by: (1) hot filtration through woven cloth and stretching into sheets; or (2) an alcoholic process where seedlac is dissolved in methanol, filtered and recovered by solvent evaporation.

"Green" polyesters may be obtained by synthesis of monomers from renewable resources or bacterial fermentation. During the past decade, emphasis has been placed on the synthesis of diols (such as 1,3-propane diol) and dicarboxylic acids (succinic acid) using a microbial route instead of petro-based monomers. Glycerol may be converted to 1,3-propane diol by *Clostridium* species as well as Enterobacteriaceae (Deckwer, 1995). Fermentation of glucose by *Actinobacillus succinogenes* provides a good yield of succinic acid (Nghiem et al., 1997; Lee et al., 1999). Succinate fermentation is a novel process because the greenhouse gas CO_2 is fixed into succinate during glucose fermentation. Succinic acid can also be produced by *Anaerobiospirillum succinicipro-ducens* using glucose, lactose, sucrose, maltose, and fructose as carbon sources. Succinic acid can be easily converted to other chemicals useful in polyester synthesis such as 1,4-butane diol, tetrahydrofuran, γ-butyrolactone, and adipic acid, etc.

Many bacteria biosynthesize inclusions of high-molecular weight PHAs as a carbon reserve (Doi, 1990; Hocking and Marchessault, 1994). Bacteria can produce a wide variety of PHAs of which poly(3-HB) is the most frequently encountered, crystalline, thermoplastic material. More than 90 different repeat units have been identified in naturally occurring PHAs (Steinbüchel and Valentin, 1995). The main difference lies in number of methylene groups in the polymer backbone and the length and composition of the side chain at the stereocenter.

However, most of the monomers used in synthesis of polyester are prepared using petroleum as the feedstock. New synthetic routes have been devised for preparation of several of these monomers (Mathisen and Albertsson, 1989; Mathisen et al., 1989).

5
Degradation

Biodegradability of aliphatic polyesters depends on the backbone structure (e.g. the presence of hydrolyzable/oxidizable hydrophilic groups in the backbone, type of repetitive unit, composition, and sequence length), morphology (e.g., crystallinity, size of spherulites) orientation, and presence of additives (Mathisen and Albertsson, 1990; Albertsson, 1993; Karlsson and Albertsson, 1998a,b). Many publications are available on the degradation of polyesters, the majority of the studies having been conducted in order to simulate the conditions prevailing inside the human body, i.e., *in vitro* studies in buffer solution at 37 °C. It is believed that during degradation, chain scission occurs through simple hydrolysis (Kasuya et al., 1998), which is influenced by the presence of anions, cations, and enzymes (Holland et al., 1986; Park et al., 1992; Reeve et al., 1994). The process is autocatalytic, and products of hydrolysis such as carboxylic acid groups participate in the transition state. Water preferentially enters the amorphous parts (Albertsson and Ljungqvist, 1988a,b), but crystalline regions are also affected (Mathisen and Albertsson, 1990). Orientation reduces the rate of degradation (Albertsson and Ljungqvist, 1986b).

Bionolle™ is biodegradable in compost, in moist soil, in fresh water with activated sludge, and in sea water (Fujimaki, 1998). Electron beam-irradiated blends of Bionolle™ and rubber did not show any enzymatic degradation with lipase in the presence of $MgCl_2$ and phosphate buffer (pH 7.4) (Khan et al., 1999). Rates of *Rhizopus delemar* lipase or activated sludge-catalyzed biodegradation of cross-linked polyesters prepared by condensation of L- or D-malic acid and glycols having 2–6, 8–10 or 12 methylene groups depended on the methylene content of glycol and stereochemistry. The L-isomer was degraded more rapidly than the D-isomer (Nagata et al., 1999).

It was shown as early as the 1970s that poly(ε-CL) is easily biodegraded and utilized as carbon source by various microbial species (Fields et al., 1974; Tokiwa and Suzuki,

1977a,b). Biodegradability of poly(ϵ-CL) and biocompatibility of its degradation products is well documented in the literature. It is degradable in several biotic environments, such as river and lake waters, sewage sludge, farm soil, paddy soil, creek sediment, roadside sediment, pond sediment, and compost (Benedict et al., 1983; Albertsson et al., 1998). Abiotic degradation of poly(ϵ-CL) blends has also been studied at different pHs and different temperatures (Yasin and Tighe, 1992). The degradation mechanism proposed is hydrolysis of the polymer to 6-hydroxycaproic acid, an intermediate of ω-oxidation, and then β-oxidation to acetyl-S-CoA, which can then undergo further degradation in the citric acid cycle. Low- molecular mass poly(ϵ-CL) degraded more quickly than samples with high molecular mass (Fields et al., 1974). The molar mass decreases readily during biodegradation, and is accompanied by a broadening of molar mass distribution (Benedict et al., 1983; Tilstra and Johnsonbaugh, 1993). The appearance of parallel grooves on the polymer surface during biotic degradation has been reported (Albertsson et al., 1998; Eldsäter et al., 2000). No surface changes were observed in the samples exposed to an abiotic environment. Biotic hydrolysis in simulated compost gave rise to holes in the poly(ϵ-CL) films, in contrast to the situation in traditional garden waste (Karlsson and Albertsson, 1998a,b).

The degradation rate of poly(ϵ-CL) is not affected by its shape (i.e., as a film or as micro particles), and is enhanced by the presence of the enzyme lipase (Chen et al., 2000). The degree of crystallinity of poly-(ϵ-CL) increased with degradation, indicating thereby that preferential degradation takes place in amorphous regions (Albertsson et al., 1998; Chen et al., 2000). Recycling, and the addition of processing additives, reduced the degradation rates. Degra-

dation rates in biotic environments, composts and anaerobic sewage sludge were higher than in abiotic environments, due to synergism between high temperature and a richer fauna of microorganisms. The enzymatic degradation rates of radiation cross-linked poly(ϵ-CL) using lipase enzyme in a phosphate buffer solution at pH 7 were slower than that of uncross-linked poly(ϵ-CL), due to network structure (Darwis et al., 1998).

Microorganisms that degrade poly(ϵ-CL) are widely distributed in nature (Benedict et al., 1983; Nishida and Tokiwa, 1993). The enzyme that hydrolyzes naturally occurring hydrophobic (poly)esters such as cutin and lipids may also attack poly(ϵ-CL) (Mochizuki et al., 1995). In most environments, poly(ϵ-CL) has been found to biodegrade more slowly than other biodegradable polymers such as poly(hydroxybutyrate-co-valerate), regenerated cellulose, and starch, etc. Its half-life is about one year *in vivo* (Schindler et al., 1977).

Hydrolytic degradation of poly(ϵ-CL)/poly(L-LA) block copolyester at pH 7.4 and 37 °C over a 5-week period is controlled by the initial crystallinity of poly(ϵ-CL) and the overall composition. The rate of degradation increased with an increase in poly(L-LA) content. Microorganisms that secrete poly(ϵ-CL) depolymerase (cutinase), such as *Fusarium solani* and *Fusarium moniliforme*, were more effective with those polymers that had longer poly(ϵ-CL) sequence lengths (Lostocco et al., 1998). Hydrolytic degradation of a polyester–polyether block copolymer based on poly(ϵ-CL)/PEG/PLA increased with an increase in PEG content. This was attributed to increased hydrophilicity of the copolymer due to the increase in PEG content (Chen et al., 2000). Biodegradation rates of poly(ϵ-CL-b-PEG) also increased with increasing PEG content. Wei et al. (1997) have investigated the influence of composition, temperature, and morphology on *in vitro* degradation of poly(ε-CL), PLA and their block copolymers.

In vitro degradation of poly(trimethylene carbonate-co-caprolactone) containing 80% of ε-Cl was similar to that of poly(ε-CL) (Albertsson and Eklund, 1995). Electron beam irradiation and grafting of poly(ε-CL) with acrylamide affected its *in vitro* degradation at pH 7.4 and 37 °C in a phosphate buffer solution (Ohrlander et al., 1999a,b,c). Virgin poly(ε-CL) maintained Mn and Mw for up to 40 weeks, whereas a continuous decrease in molecular weight was observed in irradiated (5 Mrad) and grafted polymers.

Investigations of the degradation behavior of blends of polycaprolactone have been the subject of numerous publications (Yasin and Tighe, 1992; de Kesel et al., 1997; Chiellini et al., 1996). Blends of poly(ε-CL) with cornstarch granules are biodegraded by *Rhizopus arrhizus* lipase and *Bacillus subtilis* α-amylase. A reduction in poly(ε-CL) degradation was observed in its blends with poly(ethylene terephthalate) (PET) (Chiellini et al., 1996), whereas in films prepared from its blends with poly(vinyl alcohol) no sign of degradation was observed (de Kesel et al., 1997). Poly(ε-CL) blended with polystyrene, low-density polyethylene, Nylon-6, and PET showed different susceptibility towards enzymatic degradation compared with pure poly(ε-CL) (Albertsson and Karlsson, 1995b). Reactive compatibilized blends of PLA and poly(ε-CL), prepared in a twin-screw extruder using coupling agents, degraded at a faster rate on enzymatic degradation than the pure poly(ε-CL) or PLA, while the degradation rate of physical blends were in between those of poly(ε-CL) or PLA (Wang et al., 1998). The degradation behavior of solution cast poly(ε-CL) and poly(D,L-lactide), PD,L-LA, blend films was studied in phosphate buffer containing PS. The results indicated the selectivity of PS in promoting

degradation of poly(ε-CL) (Gan et al., 1999). Such behavior was not observed in the presence of porcine pancreatic lipase and *Candida cylindracea* lipase. Biodegradation studies of similar blends in soil have been investigated for 20 months, and indicate preferential enzymatic degradation of poly-(ε-CL) (Tsuji et al., 1998).

The homopolymer of DXO has good degradation properties. After 46 weeks of *in vitro* degradation, 30–50% of the initial molecular weight is retained (Mathisen et al., 1989; Albertsson and Palmgren, 1994). Degradable cross-linked polymers of DXO and a cross-linker, 2,2-bis-(ε-CL-4-yl) propane and bis-(ε-CL-4-yl exhibited a degradation time of 1 year (Palmgren et al., 1997). The amorphous D,L-LA/DXO copolymers degrade faster than the corresponding semi crystalline L-LA/DXO copolymers. An increase in LA content resulted in an increase in degradation rate (Albertsson and Löfgren, 1992). The main products of *in vitro* hydrolysis in phosphate buffer (pH 7.4) of copolymers of DXO and D,L- or L-LA were identified as the linear monomers, i.e., lactic acid and 2-hydroxyethoxypropionic acid (Karlsson et al., 1994). The amounts of products formed depend upon the copolymer composition and the degradation time. The *in vitro* degradation studies of poly(DXO), electron beam-irradiated poly-(DXO) (5 Mrad) and acrylamide-grafted P(DXO) showed complete weight loss after 55, 43, and 35 weeks, respectively (Ohrlander et al., 1999a,b,c).

6
Applications

Biodegradable aliphatic polyesters are being used in several fields ranging from agricultural implements to biomedical applications. Two major areas of their applications are: (1) as biomedical polymers; and (2) as ecological polymers (Ikada and Tsuji, 2000). Poly(ε-CL) has been promoted as a soil-degraded container material (Pitt et al., 1981), and has found applications in the growing and transplanting of trees, and as thin-walled tree seedling containers. It has also been used for implantable drug delivery devices (Capronol). In addition, it has been blended with starch and its derivatives for use as shopping bags (Darwis et al., 1998).

Potential applications of Bionolle™ injection-molded articles (such as cutlery, brushes, etc.), tubular films (composting bags, shopping bags), flexible packaging, food trays, cosmetic bottles, and beverage bottles have been projected, though for some of these applications there is a need to improve the performance of existing grades of material.

Aliphatic polyesters based on lactic acid have found use in a wide variety of medical applications, including bioresorbable surgical sutures, prosthetics, dental implants, pins, bone screws and plates for temporary internal fracture fixation, long-term delivery of antimalarial drugs, contraceptives, eye drugs, and controlled drug delivery systems. The US Food and Drug Administration have approved very few biodegradable polymers for use in biological systems (Davis et al., 1996), but one example is poly(LA-co-glycolide) containing leuprorelin acetate for the treatment of endometriosis and prostatic cancer.

Poly(ε-CL) is one of the most frequently used compounds in biodegradable drug delivery systems due to its biocompatibility, low glass transition (Tg), and its high permeability (Ali et al., 1993). Blends of poly(D,L-LA) and poly(DXO) have recently been investigated for sustained release of the non-steroidal anti-inflammatory drug, diclofenac sodium (Edlund and Albertsson, 2000a,b). The release rates were dependent

on the blend composition, thus providing a powerful means to control drug delivery. Sustained drug delivery has been achieved by using both copolymers as well as blends. Morphology is an important factor in determining the *in vitro* performance; poly(L-LA-co-DXO) exhibited higher moisture sensitivity than a blend of PL-LA-(DXO) of corresponding composition, due to its more crystallinic and dense morphology (Edlund and Albertsson, 1998, 2000a,b). A hydrophilic drug was released much faster from copolymer microspheres than from blend microspheres. Studies on bilayered biodegradable PEG/poly(butylene terephthalate) copolymers have recently been reported (van Dorp et al., 1999), the main focus of these being to identify an optimal polymer matrix for the development of a human skin substitute.

7
Outlook and Perspectives

Although many aliphatic polyesters are available commercially, few have achieved major sales at present. The main reason for their limited applications is the price of the product, or the short- and long-term performance of the material. Future research and developments to overcome these shortcomings are expected to increase the applications of these ecologically and environmentally friendly materials.

In biopolyesters, the cost of the raw materials, i.e., substrates, and the efficiency of the whole process is economically not very viable. Therefore, scientists and engineers are challenged to search for approaches to increase the efficiency and ultimately the productivity of the whole process. By mixing of substrates (e.g., acetate and glucose), a conversion efficiency of 82% has been reported (Ackermann and Babel, 1998).

The availability of the *R. eutropha* PHA biosynthesis genes has allowed the establishment of poly(3HB) biosynthesis in other organisms (e.g. recombinant strains of *E. coli*), and poly(3HB-co-3HV) and poly(4HB) have been prepared by constructing different strains of *E. coli* (Lee et al., 1994; Lee and Choi, 1998). Polyesters based on hydroxyalkanoic acids containing medium chain-length branches (6–16) or poly(4HB) have been prepared by cultivating specific *E. coli* strains on fatty acids (Hein et al., 1997; Langenbach et al., 1997). These examples illustrate that it would be possible to obtain biopolyesters in reasonable amounts, thereby increasing the efficiency of production from cells of recombinant *E. coli* strains if suitable pathways are designed by metabolic and genetic engineering. The establishment of the poly(3HB) biosynthesis pathway in plants opens up a promising perspective for the cheap production of large quantities of poly(3HB) by agriculture (Poirier et al., 1995; Leaf et al., 1996).

Studies in the area of polyesters prepared from renewable resources have already been initiated by Albertsson et al. (Ranucci et al., 2000). High-molecular weight polyesters were obtained using chain extension reactions to yield poly(ester-urethanes) and poly(ester-carbonate). The poly(ester-carbonate)s was found to be the most versatile material that could be tailor-made to match the different requirements. Future research into such polyesters is essential for developing materials with good mechanical properties and biodegradability.

Thermoplastic poly(ε-CL) is less versatile in terms of overall physical properties, and there is a need to improve its performance by structural modification, either by copolymerization or blending. Several such blends have been investigated in the past, but many unsolved problems remain with this approach (Eastmond, 1999).

A controllable biodegradability, desirable mechanical properties, suitable gas permeability and selectivity would improve the performance and extend the potential application areas of aliphatic polyesters, not only in agricultural, greenhouse or packaging industries, but also as a substitute for sensitive parts of the human body such as skin. There is a need for such focused studies in the future.

Acknowledgements

The financial assistance provided by The Swedish Research Council to one of the authors (I. K. Varma) is gratefully acknowledged.

8
References

Abraham, G. A., Gallardo, A., Lozano, A. E., Roman, J. S. (2000) ε-Caprolactone/ZnCl$_2$ complex formation: Characterization and ring-opening polymerization mechanism, *J. Polym. Sci., Part A: Polym. Chem.* **38**, 1355–1365.

Ackermann, J., Babel, W. (1998) Approaches to increase the economy of the PHB production, *Polym. Degrad. Stab.* **59**, 183–186.

Agarwal, S., Brandukova-Szmikowski, N. E., Greiner, A. (1999a) Reactivity of Sm(II) compounds as ring-opening polymerization initiators for lactones, *Macromol. Rapid Commun.* **20**, 274–278.

Agarwal, S., Brandukova-Szmikowski, N. E., Greiner, A. (1999b) Samarium(III)-mediated graft polymerization of ε-caprolactone and L-lactide on functionalized poly(p-xylylene)s: model studies and polymerizations, *Polym. Adv. Technol.* **10**, 528–534.

Agarwal, S., Mast, C., Dehnicke, K., Greiner, A. (2000) Rare earth metal initiated ring-opening polymerization of lactones, *Macromol. Rapid Commun.* **21**, 195–212.

Albertsson, A.-C. (1992) Biodegradation of polymers, in: *Handbook of Polymer Degradation* (Hamid, S. H., Maadhah, A. G., Amin, M. B., Eds.), New York: Marcel Dekker, 345–363.

Albertsson, A.-C. (1993) Degradable polymers, *J. Macromol. Sci., Pure Appl. Chem.* **A30**, 757–765.

Albertsson, A.-C., Edlund, U. (1998) Novel release systems from biodegradable polymers, *Polym. Preprints (Am. Chem. Soc., Div. Polym. Chem.)* **39**, 186–187.

Albertsson, A.-C., Eklund, M. (1994) Synthesis of copolymers of 1,3-dioxan-2-one and oxepan-2-one using coordination catalysts, *J. Polym. Sci.* **32**, 265–279

Albertsson, A.-C., Eklund, M. (1995) Influence of molecular structure on the degradation mechanism of degradable polymers. *In vitro* degradation of poly(trimethylene carbonate), poly(trimethylene carbonate-co-caprolactone) and poly(adipic anhydride), *J. Appl. Polym. Sci.* **57**, 87–103.

Albertsson, A.-C., Gruvegård, M. (1995) Degradable high-molecular weight random copolymers, based on ε-caprolactone and 1,5-dioxepan-2-one, with non-crystallizable units inserted in the crystalline structure, *Polymer* **36**, 1009–1016.

Albertsson, A.-C., Karlsson, S. (1992) Biodegradable polymers, in: *Comprehensive Polymer Science*, First Supplement (Allen, G., Aggarwal, S. L., Russo, S., Eds.), Oxford: Pergamon Press, 285–297.

Albertsson, A.-C., Karlsson, S. (1995) Degradable polymers for the future, *Acta Polym.* **46**, 114–123.

Albertsson, A.-C., Karlsson, S. (1996a) Degradable polymers, in: *Polymeric Materials Encyclopedia; Synthesis, Properties and Applications* (Salamone, J. C., Ed.), Boca Raton, FL: CRC Press, 150.

Albertsson, A.-C., Karlsson, S. (1996b) Macromolecular architecture – nature as a model for degradable polymers, *J. Macromol. Sci., Pure Appl. Chem.* **A33**, 1565–1575.

Albertsson, A.-C., Karlsson, S. (1998) Biodegradable polymers, in: *Encyclopedia of Environmental Analysis and Remediation* (Meyers, R. A., Ed.), London: John Wiley & Sons, 641–651.

Albertsson, A.-C., Ljungquist, O. (1986a) Degradable polymers. I. Synthesis, characterization and long term *in vitro* degradation of a ^{14}C-labeled aliphatic polyester, *J. Macromol. Sci., Chem.* **A23**, 393–409.

Albertsson, A.-C., Ljungquist, O. (1986b) Degradable polymers. II. Synthesis, characterization and degradation of an aliphatic thermoplastic block copolyester, *J. Macromol Sci., Chem.* **A23**, 411–422.

Albertsson A.-C., Ljungquist, O. (1987) Degradable Polymers. III. Synthesis and characterization of aliphatic thermoplastic block copolyesters, *J. Macromol. Sci., Chem.* **A24**, 977–990.

Albertsson, A.-C., Ljungquist, O. (1988a) Degradable polymers. IV. Degradation of aliphatic thermoplastic block copolyesters, *J. Macromol. Sci., Chem,* **A25**, 467–498.

Albertsson, A.-C., Ljungquist, O. (1988b) Degradable polyesters as biomaterials, *Acta Polym.* **39**, 95–104.

Albertsson, A.-C., Löfgren, A. (1992) Copolymers of 1,5-dioxepan-2-one and L- or D,L-lactide, synthesis and characterization, *Makromol. Chem., Macromol. Symp.* **53**, 221–231.

Albertsson, A.-C. Löfgren, A. (1995) Synthesis and characterization of poly(1,5-dioxepan-2-one-co-L-lactic acid) and poly(1,5-dioxepan-2-one-co-D,L-lactide), *J. Macromol. Sci., Pure Appl. Chem.* **A32**, 41–59.

Albertsson, A.-C., Löfgren, A. (1996) Biomedically degradable polymers, in: *Polymeric Materials Encyclopedia: Synthesis, Properties and Applications* (Salamone, J. C., Ed.), Boca Raton, FL: CRC Press Inc., 634–643.

Albertsson, A.-C., Palmgren, R. (1993) Polymerization and degradation of poly(1,5-dioxepan-2-one), *J. Macromol. Sci., Chem.* **A30**, 919–931.

Albertsson, A.-C., Palmgren, R. (1994) Synthesis of biodegradable elastomers based on 1,5-dioxepan-2-one, *Macromol. Rep.* **A31**, 1185–1189.

Albertsson, A.-C., Palmgren, R. (1996) Cationic polymerization of 1,5-dioxepan-2-one with Lewis acids in bulk and solution, *J. Macromol. Sci., Pure Appl. Chem.* **A33**, 747–758.

Albertsson, A.-C., Varma, I. K. (2000) Aliphatic polyesters: synthesis from lactones or by polycondensation, properties and applications, submitted.

Albertsson, A.-C., Löfgren, A., Zhang, Y., Bjursten, L.-M. (1994) Copolymers of 1,5-dioxepan-2-one and L- or D,L-dilactide: *in vivo* degradation behaviour, *J. Biomaterials Sci., Polym. Ed.* **6**, 411–423.

Albertsson, A.-C., Renstad, R., Erlandsson, B., Eldsäter, C., Karlsson, S. (1998) Effect of processing additives on (bio)degradability of film-blown poly(ε-caprolactone), *J. Appl. Polym. Sci.* **70**, 61–74.

Ali, S. A. M., Zhong, S.-P., Doherty, P. J., Williams, D. F. (1993) Mechanisms of polymer degradation in implantable devices. 1. Poly(ε-caprolactone), *Biomaterials* **14**, 648–656.

Arvanitoyannis, I. S. (1999) Totally and partially biodegradable polymer blends based on natural and synthetic macromolecules: preparation, physical properties, and potential as food packaging materials, *J. M. S.-Rev. Macromol. Chem. Phys.* **C39**, 205–271.

Athawale, V. D., Gaonkar, S. R. (1994) Enzymatic synthesis of polyesters by lipase catalyzed polytransesterification, *Biotechnol. Lett.* **16**, 149–154.

Benedict, C. V., Cameron, J. A., Huang, S. J. (1983) Polycaprolactone degradation by mixed and pure cultures of bacteria and yeast, *J. Appl. Polym. Sci.* **28**, 335–342.

Bisht, K. S., Henderson, L. A., Gross, R. A., Kaplan, D. L., Swift, G. (1997) Enzyme catalyzed ring-opening of ω-pentadecalactone, *Macromolecules* **30**, 2705–2711.

Brandl, H., Gross, R. A., Lenz, R. W., Fuller, R. C. (1990) Plastics from bacteria and for bacteria: poly(β-hydroxyalkanoate)s as natural, biocompatible and biodegradable polyesters, *Adv. Biochem. Eng. Biotechnol.* **41**, 77–93.

Brazwell, E. M., Filos, D. Y., Morrow, C. J. (1995) Biocatalytic synthesis of polymers. 3. Formation of a high molecular weight polyester through limitation of hydrolysis by enzyme-bound water and through equilibrium control, *J. Polym. Sci., Part A: Polym. Chem.* **33**, 89–95.

Cerrai, P., Tricoli, M., Andruzzi, F., Paci, M. (1989) Polyether polyester block copolymers by non-catalyzed polymerization of ε-caprolactone with poly(ethylene glycol), *Polymer* **30**, 338–343.

Chamberlain, B. M., Sun, Y., Hagadorn, J. R., Hemmesch, E. W., Young Jr., V. G., Pink, M., Hillmyer, M. A., Tolman, W. B. (1999) Discrete yttrium (III) complexes as lactide polymerization catalysts, *Macromolecules* **32**, 2400–2402.

Chamberlain, B. M., Jazdzewski, B. A., Pink, M., Hillmyer, M. A., Tolman, W. B. (2000) Controlled polymerization of D,L-lactide and ε-caprolactone by structurally well-defined alkoxo-bridged di- and triyttrium (III) complexes, *Macromolecules* **33**, 3970–3977.

Chen, D. R., Bei, J. Z., Wang, S. G. (2000) Polycaprolactone microparticles and their biodegradation, *Polym. Degrad. Stab.* **67**, 455–459.

Chiellini, E., Corti, A., Giovannini, A., Narducci, P., Paparella, A. M., Solaro, R. (1996) Evaluation of biodegradability of poly(ε-caprolactone)/poly(ethylene tererphthalate) blends, *J. Environ. Polym. Degrad.* **4**, 37–50.

Crivello, J. V., Lockhart, T. P., Lee, J. L. (1983) Diaryliodinium salts as thermal initiators of cationic polymerization, *J. Polym. Sci., Polym. Chem. Ed.* **21**, 97–109.

Darwis, D., Mitomo, H., Enjoji, T., Yoshii, F., Makuuchi, K. (1998) Enzymatic degradation of radiation crosslinked poly(ε-caprolactone), *Polym. Degrad. Stab.* **62**, 259–265.

Davis, S. S., Illum, L., Stolnik, S. (1996) Polymers in drug delivery, *Curr. Opin. Colloid Interface. Sci.* **1**, 660–666.

Deckwer, W. D. (1995) Microbial conversion of glycerol to 1,3-propanediol, *FEMS Microbiol. Rev.* **16**, 143–149.

de Kesel, C., Wauven, C. V., David, C. (1997) Biodegradation of polycaprolactone and its blends with poly(vinylalcohol) by micro-organisms from a compost of household refuse, *Polym. Degrad. Stab.* **55**, 107–113.

Doi, Y. (1990) *Microbial Polyesters.* New York: VCH.

Dong, H., Cao, S.-G., Li, Z.-Q., Han, S.-P., You, D.-L., Shen, J.-C. (1999) Study on the enzymatic polymerization mechanism of lactone and strategy for improving the degree of polymerization, *J. Polym. Sci., Part A: Polym. Chem.* **37**, 1265–1275.

Dubois, Ph., Jérôme, R., Teyssié, Ph. (1991a) Macromolecular engineering of polylactones and polylactides. 3. Synthesis, characterization, and applications of poly(ε-caprolactone) macromonomers, *Macromolecules* **24**, 977–981.

Dubois, Ph., Jérôme, R., Teyssié, Ph. (1991b) Aluminum alkoxides: A family of versatile initiators for the ring-opening polymerization of lactones and lactides, *Makromol. Chem., Macromol. Symp.* **42/43**, 103–116.

Duda, A., Biela, T., Libiszowski, J., Penczek, S., Dubois, Ph., Mecerreyes, D., Jérôme, R. (1998) Block and random copolymers of ε-caprolactone, *Polym. Degrad. Stab.* **59**, 215–222.

Eastmond, G. C. (1999) Poly(ε-caprolactone) blends, *Adv. Polym. Sci.* **149**, 59–223.

Edlund, U., Albertsson, A.-C. (1998) Tailored drug delivery from PTMC-PAA blends, *Proc. Int. Symp. Control. Rel. Bioact. Mater.* **25**, 725–726.

Edlund, U., Albertsson, A.-C. (2000a) Morphology engineering of a novel poly(L-lactide)/poly(1,5-dioxepan-2-one) microsphere system for controlled drug delivery, *J. Polym. Sci., Part A: Polym. Chem.* **38**, 786–796.

Edlund, U., Albertsson, A.-C. (2000b) Microspheres from poly(D,L-lactide)/poly(1,5-dioxepan-2-one) miscible blends for controlled drug delivery, *J. Bioact. Compat. Polym.* **15**, 1–16.

Eldsäter, C., Erlandsson, B., Renstad, R., Albertsson, A.-C., Karlsson, S. (2000) The biodegradation of amorphous and crystalline regions in film-blown poly(ε-caprolactone), *Polymer* **41**, 1297–1304.

Feng, X. D., Song, C. X., Chen, W. J. (1983) Synthesis and evaluation of biodegradable block copolymers of ε-caprolactone and D,L-lactide, *J. Polym. Sci., Polym. Lett. Ed.* **21**, 593–600.

Fields, R. D., Rodriguez, F., Finn, R. K. (1974) Microbial degradation of polyesters: Polycaprolactone degradation by *Pullularia pullulans*, *J. Appl. Polym. Sci.*, **18**, 3571–3580.

Fujimaki, T. (1998) Processability and properties of aliphatic polyester, BIONOLLE, synthesized by polycondensation reaction, *Polym. Degrad. Stab.* **59**, 209–214.

Gan, Z., Yu, D., Zhong, Z., Liang, Q., Jing, X. (1999) Enzymatic degradation of poly(ε-caprolactone)/poly(D,L-lactide) blends in phosphate buffer solution, *Polymer* **40**, 2859–2862.

Grijpma, D. W., Pennings, A. J. (1994) (Co)polymers of L-lactide 1. Synthesis, thermal properties and hydrolytic degradation, *Macromol. Chem. Phys.* **195**, 1633–1647.

Gruvegård, M., Lindberg, T., Albertsson, A.-C. (1998) Random copolymers of 1,5-dioxepan-2-one, *J. Macromol. Sci., Pure Appl. Chem.* **A35**, 885–902.

Hawker, C. J., Frechet, J. M. J. (1990) Preparation of polymers with controlled molecular architecture: A new convergent approach to dendritic macromolecules, *J. Am. Chem. Soc.* **112**, 7638–7647.

Hein, S., Söhling, B., Gottschalk, G., Steinbüchel, A. (1997) Biosynthesis of poly(4-hydroxybutyric acid) by recombinant strains of *Escherichia coli*, *FEMS Microbiol. Lett.* **153**, 411–418.

Henderson, L. A., Svirkin, Y. Y., Gross, R. A., Kaplan, D. L., Swift, G. (1996) Enzyme-catalyzed polymerizations of ε-caprolactone: Effects of initiator on product structure, propagation kinetics, and mechanism, *Macromolecules* **29**, 7759–7766.

Heuschen, J., Jérôme, R., Teyssié, Ph. (1981) Polycaprolactone-based block copolymers: Synthesis by anionic type catalysts, *Macromolecules* **14**, 242–246.

Hocking, P. J., Marchessault, R. H. (1994) Biopolyesters, in: *Chemistry and Technology of Biodegradable Polymers* (Griffin, G. J. L., Ed.), Glasgow: Blackie Academic & Professional, 48–96.

Hofman, A., Slomkowski, S., Penczek, S. (1987) Structure of active centers and mechanism of anionic and cationic polymerization of δ-valerolactone, *Makromol. Chem.* **188**, 2027–2040.

Holland, S. J., Tighe, B., Gould, P. (1986) Polymers for biodegradable medical devices 1. The potential of polyesters as controlled macromolecular release systems, *Controlled Release* **4**, 155–180.

Hult, A., Johansson, M., Malmström, E. (1999) Hyperbranched polymers, *Adv. Polym. Sci.* **143**, 1–34.

Ikada, Y., Tsuji, H. (2000) Biodegradable polyesters for medical and ecological applications, *Macromol. Rapid Commun.* **21**, 117–132.

Inoue, S., Aida, T. (1993) High-speed living ring-opening polymerization by a new catalyst based on metalloporphyrin, *Makromol. Chem., Macromol. Symp.* **73**, 27–36.

Inoue, S., Aida, T., Kuroki, M., Sugimoto, H. (1992) High-speed living polymerization with a new catalyst system based on metalloporphyrin, *Makromol. Chem., Macromol. Symp.* **64**, 151–158.

Jedlinski, Z., Kurock, P., Kowalczuk, M. (1985) Polymerization of β-lactones initiated by potassium solution, *Macromolecules* **18**, 2679–2683.

Joziasse, C. A. P., Grablowitz, H., Pennings, A. J. (2000) Star-shaped poly [(trimethylene carbonate)-co-(ε-caprolactone)] and its block copolymers with lactide/glycolide: synthesis, characterization and properties, *Macromol. Chem. Phys.* **201**, 107–112.

Karlsson, S., Albertsson, A.-C. (1998a) Abiotic and biotic degradation of aliphatic polyesters from "petro" versus "green" resources, *Macromol. Symp.* **127**, 219–225.

Karlsson, S., Albertsson, A.-C. (1998b) Biodegradable polymers and environmental interaction, *Polym. Eng. Sci.* **38**, 1251–1253.

Karlsson, S., Hakkarainen, M., Albertsson, A.-C. (1994) Identification by GC-MS and HS-GC-MS of *in vitro* degradation products of homo- and copolymers of L- and D,L-lactide and 1,5-dioxepan-2-one, *J. Chromatogr.* **A688**, 251–259.

Kasuya, K., Takagi, K., Ishiwatari, S., Yoshida Y., Doi, Y. (1998) Biodegradabilities of various aliphatic polyesters in natural waters, *Polym. Degrad. Stab.* **59**, 327–332.

Kelley, A. S., Jackson, D. E., Macosko, C., Srienc, F. (1998) Engineering the composition of co-polyesters synthesized by *Alcaligenes eutrophus*, *Polym. Degrad. Stab.* **59**, 187–190.

Khan, M. A., Idriss Ali, K. M., Yoshii, F., Makuuchi, K. (1999) Enzymatic degradation of Bionolle and Bionolle-rubber blends, *Polym. Degrad. Stab.* **63**, 261–264.

Khurana, R. G., Singh, A. N., Upadhye, A. B., Mhaskar, V. V., Dev, S. (1970) Chemistry of lac resin III. Lac acids. 3. Integrated procedure for their isolation from hard resin, chromatography, characteristics and quantitative determination, *Tetrahedron* **26**, 4167–4175.

Knani, D., Kohn, D. H. (1993) Enzymic polyesterification in organic media. 2. Enzyme-catalyzed synthesis of lateral-substituted aliphatic polyesters and copolyesters, *J. Polym. Sci., Part A: Polym. Chem.* **31**, 2887–2897.

Knani, D., Gutman, A. L., Kohn, D. H. (1993) Enzymic polyesterification in organic media. Enzyme-catalyzed synthesis of linear polyesters. 1. Condensation polymerization of linear hydroxyesters. 2. Ring-opening polymerization of (ε-caprolactone), *J. Polym. Sci., Part A: Polym. Chem.* **31**, 1221–1232.

Kobayashi, S., Uyama, H., Namekawa, S. (1998) In vitro biosynthesis of polyesters with isolated enzymes in aqueous systems and organic solvents, *Polym. Degrad. Stab.* **59**, 195–201.

Kobayashi, T., Hori, Y., Kakimoto, M. A., Imai, Y. (1993) Synthesis of biodegradable polyesters by polycondensation of methyl (R)-3-hydroxybutyrate and methyl (R) -3-hydroxyvalerate, *Makromol. Chem., Rapid Commun.* **14**, 785–790.

Kowalski, A., Duda, A., Penczek, S. (1998) Kinetics and mechanism of cyclic esters polymerization initiated with tin(II) octoate. 1. Polymerization of η-caprolactone, *Macromol. Rapid. Commun.* **19**, 567–572.

Kowalski, A., Duda, A., Penczek, S. (2000) Mechanism of cyclic ester polymerization initiated with tin(II) octoate. 2. Macromolecules fitted with tin(II) alkoxide species observed directly in MALDI-TOF spectra, *Macromolecules* **33**, 689–695.

Kricheldorf, H. R., Eggerstedt, S. (1998) Macrocycles 2. Living macrocyclic polymerization of ε-caprolactone with 2,2-dibutyl-2-stanna-1,3-dioxepane as initiator, *Macromol. Chem. Phys.* **199**, 283–290.

Kricheldorf, H. R., Kreiser-Saunders, I. (2000) Polylactones 49: Bu₄Sn-initiated polymerizations of ε-caprolactone, *Polymer* **41**, 3957–3963.

Kricheldorf, H. R., Sumbel, M. V., Kreiser-Saunders, I. (1991) Polylactones 20. Polymerization of ε-caprolactone with tributyltin derivatives: A mechanistic study, *Macromolecules* **24**, 1944–1949.

Kricheldorf, H. R., Kreiser-Saunders, I., Stricker, A. (2000) Polylactones 48. SnOct₂-initiated polymerizations of lactide: A mechanistic study, *Macromolecules* **33**, 702–709.

Langenbach, S., Rehm, B. H. A., Steinbüchel, A. (1997) Functional expression of the PHA synthase gene PhaCl from *Pseudomonas aeruginosa* in *Escherichia coli* results in poly(3-hydroxyalkanoate) synthesis, *FEMS Microbiol. Lett.* **150**, 303–309.

Leaf, T. A., Peterson, M. S., Stoup, S. K., Somers, D., Srienc, F. (1996) *Saccharomyces cerevisiae* expressing bacterial polyhydroxybutyrate synthase produces poly(3-hydroxybutyrate), *Microbiology* **142**, 1169–1180.

Lee, P. C., Lee, W. G., Lee, S. Y., Chang, H. N. (1999) Effects of medium components on the growth of *Anaerobiospirillum succiniciproducens* and succinic acid production, *Process Biochem.* **A35**, 49–55.

Lee, S. Y. (1996) Plastic bacteria? Progress and prospects for polyhydroxyalkanoates production in bacteria, Trends *Biotechnol.* **14**, 431–438.

Lee, S. Y., Choi, J.-I. (1998) Effect of fermentation performance on the economics of poly(3-hydroxybutyrate) production by *Alcaligenes latus*, *Polym. Degrad. Stab.* **59**, 387–393.

Lee, S. Y., Lee, K. M., Chang, H. N., Steinbüchel, A. (1994) Comparison of recombinant *Escherichia coli* strains for synthesis and accumulation of poly(3-hydroxybutyric acid) and morphological changes, *Biotechnol. Bioeng.* **44**, 1337–1347.

Liu, M., Vladimirov, N., Fréchet, J. M. J. (1999) A new approach to hyperbranched polymers by ring-opening polymerization of an AB monomer: 4-(2-hydroxyethyl)-ε-caprolactone, *Macromolecules* **32**, 6881–6884.

Liu, Y., Schulze, M., Albertsson, A.-C. (1998) [α-Methacryloyl-ω-hydroxyl-poly(ε-caprolactone)] macromonomer: Synthesis, characterization and copolymerization, *J. Macromol. Sci., Pure Appl. Chem.* **A35**, 207–232.

Löfgren, A., Albertsson, A.-C. (1994) Copolymers of 1,5-dioxepan-2-one and L- or D,L-dilactide. Hydrolytic degradation behaviour, *J. Appl. Polym. Sci.* **52**, 1327–1338.

Löfgren, A., Albertsson, A.-C., Dubois, Ph., Jérôme, R., Teyssié, Ph. (1994) Synthesis and characterization of biodegradable homopolymers and block copolymers based on 1,5-dioxepan-2-one, *Macromolecules* **27**, 5556–5562.

Löfgren, A., Albertsson, A.-C., Dubois, Ph., Jérôme, R. (1995a) Recent advances in ring-opening polymerization of lactones and related compounds, *J. M. S.-Rev. Macromol. Chem. Phys.* **C35**, 379–418.

Löfgren, A., Renstad, R., Albertsson, A.-C. (1995b) Synthesis and characterization of a new degradable thermoplastic elastomer based on 1,5-dioxepan-2-one and ε-caprolactone, *J. Appl. Polym. Sci.* **55**, 1589–1600.

Lostocco, M. R., Murphy, C. A., Cameron, J. A., Huang, S. J. (1998) The effects of primary structure on the degradation of poly(ε-caprolac-

tone)/poly(L-lactide) block copolymers, *Polym. Degrad. Stab.* **59**, 303–307.

Lundmark, S., Sjöling, M., Albertsson, A.-C. (1991) Polymerization of oxepane-2,7-dione in solution and synthesis of block copolymers of oxepane-2,7-dione and 2-oxepanone, *J. Macromol. Sci., Chem.* **42**, 2365–2370.

Mathisen, T., Albertsson, A.-C. (1989) Polymerization of 1,5-dioxepane-2-one. I. Synthesis and characterization of the monomer 1,5-dioxepan-2-one and its cyclic dimer 1,5,8,12-tetraoxacyclotetradecane-2,9-dione, *Macromolecules* **22**, 3838–3842.

Mathisen, T., Albertsson, A.-C. (1990) Hydrolytic degradation of melt extruded fibers from β-propiolactone, *J. Appl. Polym. Sci.* **38**, 591–601.

Mathisen, T., Masus, K., Albertsson, A.-C. (1989) Polymerization of 1,5-dioxepane-2-one. II. Polymerization of 1,5-dioxepane-2-one and its cyclic dimer also including a new procedure for the synthesis of 1,5-dioxepane-2-one, *Macromolecules* **22**, 3842–3846.

Matsumura, S., Mabuchi, K., Toshima, K. (1998) Novel ring-opening polymerization of lactide by lipase, *Macromol. Symp.* **130**, 285–304.

McLain, S. J., Drysdale, N. E. (1992) Living ring-opening polymerization of ε-caprolactone by yttrium and lanthanide alkoxides, *Polym. Prep. (Am. Chem. Soc., Div. Polym. Chem.)* **33**, 174–175.

McLain, S. J., Ford, T. M., Drysdale, N. E. (1992) Living ring-opening polymerization of (L,L)-lactide by yttrium and lanthanum alkoxides, *Polym. Prep. (Am. Chem. Soc., Div. Polym. Chem.)* **33**, 463–464.

Mecerreyes, D., Jérôme, R. (1999) From living to controlled aluminium alkoxide mediated ring-opening polymerization of di(lactones), a powerful tool for the macromolecular engineering of aliphatic polyesters, *Macromol. Chem. Phys.* **200**, 2581–2590.

Mecerreyes, D., Miller, R. D., Hedrick, J. L., Detrembleur, C., Jérôme, R. (2000) Ring-opening polymerization of 6-hydroxynon-8-enoic acid lactone: Novel biodegradable copolymers containing allyl pendent groups, *J. Polym. Sci., Part A: Polym. Chem.* **38**, 870–875.

Mezoul, G., Lalot, T., Brigodiot, M., Marechal, E. (1995) Enzyme catalyzed synthesis of aliphatic polyesters in organic media: Study of transesterification equilibrium shift and characterization of cyclic compounds, *J. Polym. Sci., Part A: Polym. Chem.* **33**, 2691–2698.

Mochizuki, M., Hirano, M., Kanmuri, Y., Kudo, K., Tokiwa, Y. (1995) Hydrolysis of polycaprolactone fibers by lipase: Effects of draw ratio on enzymic degradation, *J. Appl. Polym. Sci.* **55**, 289–296.

Möller, M., Kånge, R., Hedrick, J. L. (2000) Sn(OTf)₂ and Sc(OTf)₃: Efficient and versatile catalysts for the controlled polymerization of lactones, *J. Polym. Sci., Part A: Polym. Chem.* **38**, 2067–2074.

Nagata, M., Kono, Y., Sakai, W., Tsutsumi, N. (1999) Preparation and characterization of novel bio-degradable optically active network polyesters from malic acid, *Macromolecules* **32**, 7762–7767.

Nghiem, N. P., Davison, B. H., Suttle, B. E., Richardson, G. R. (1997) Production of succinic acid by *Anaerobiospirillum succiniciproducens*, *Appl. Biochem. Biotech.* **63–65**, 565–576.

Nijenhuis, A. J., Grijpma, D. W., Pennings, A. J. (1992) Lewis acid-catalyzed polymerization of L-lactide. Kinetics and mechanism of bulk poly-merization, *Macromolecules* **25**, 6419–6424.

Nishida, H., Tokiwa, Y. (1993) Distribution of poly(β-hydroxybutyrate) and poly(ε-caprolac-tone) aerobic degrading microorganisms in different environments, *J. Environ. Polym. De-grad.* **1**, 227–233.

Nobes, G. A. R., Kazlauskas, R. J., Marchessault, R. H. (1996) Lipase-catalyzed ring-opening poly-merization of lactones: A novel route to poly-(hydroxyalkanoate)s, *Macromolecules* **29**, 4829–4833.

Nomura, N., Taira, A., Tomioka, T., Okada, M. (2000) A catalytic approach for cationic living polymerization: Sc(OTf)₃-catalyzed ring-opening polymerization of lactones, *Macromolecules* **33**, 1497–1499.

Ohrlander, M., Wirsén, A., Albertsson, A.-C. (1999a) The grafting of acrylamide onto poly-(ε-caprolactone) and poly(1,5-dioxepan-2-one) using electron-beam pre-irradiation. I. *J. Polym. Sci., Part A: Polym. Chem.* **37**, 1643–1649.

Ohrlander, M., Lindberg, T., Wirsén, A., Albertsson, A.-C. (1999b) The grafting of acrylamide onto poly(ε-caprolactone) and poly(1,5-dioxepan-2-one) using electron-beam preirradiation. II. *J. Polym. Sci., Part A: Polym. Chem.* **37**, 1651–1657.

Ohrlander, M., Palmgren, R., Wirsén, A., Alberts-son, A.-C. (1999c) The grafting of acrylamide onto poly(ε-caprolactone) and poly(1,5-dioxepan-2-one) using electron-beam preirradiation. III. *J. Polym. Sci., Part A: Polym. Chem.* **37**, 1659–1663.

Palmgren, R., Karlsson, S., Albertsson, A.-C. (1997) Synthesis of degradable cross-linked polymers based on 1,5-dioxepan-2-one and crosslinker of bis-ε-caprolactone type, *J. Polym. Sci., Part A: Polym. Chem.* **35**, 1635–1649.

Park, T. G., Cohen, S., Langer, R. (1992) Poly(L-lactic acid)pluronic blends: Characterization of phase-separation behaviour, degradation and morphology and use as protein releasing ma-trices, *Macromolecules* **25**, 116–122.

Penczek, S., Duda, A. (1996) Selectivity as a measure of "livingness" of the polymerization of cyclic esters, *Macromol. Symp.* **107**, 1–15.

Pilati, F., Toselli, M., Messori, M., Priola, A., Bongiovanni, R., Malucelli, G., Tonelli, C. (1999) Poly(ε-caprolactone)-poly(fluoroalkylene oxide)-poly(ε-caprolactone) block copolymers. 1. Syn-thesis and molecular characterization, *Macro-molecules* **32**, 6969–6976.

Pitt, C. G., Chasalow, F. I., Hibionada, Y. M., Klimas, D. M., Schindler, A. (1981) Aliphatic polyesters. I. The degradation of poly(ε-capro-lactone) *in vivo*, *J. Appl. Polym. Sci.* **26**, 3779–3787.

Poirier, Y., Nawrath, C., Somerville, C. (1995) Production of polyhydroxyalkanoates, a family of biodegradable plastics and elastomers in bacteria and plants, *Bio/Technology* **13**, 142–150.

Prasad, N., Agarwal, S. C., Sengupta, S. C. (1991) Quantitative estimation of aleuritic acid in lac resin and its ether fractions, *Paint Resin* **61**, 28–29.

Rafler, G., Dahlmann, J. (1992) Biodegradable polymers. 6. Polymerization of ε-caprolactone, *Acta Polym.* **43**, 91–95.

Ranucci, E., Liu, Y., Söderqvist Lindblad, M., Albertsson, A.-C. (2000) New biodegradable polymers from renewable resources. High mo-lecular weight poly(ester carbonate)s from suc-cinic acid and 1,3-propanediol, *Macromol. Rapid Commun.* **21**, 680–684.

Reeve, M. S., McCarthy, S. P., Downey, M. J., Gross, R. A. (1994) Polylactide stereochemistry: Effect on enzymatic degradability, *Macromolecules* **27**, 825–831.

Schindler, A., Jeffcoat, R., Kimmel, G. L., Pitt, C. G., Wall, M. E., Zweidinger, R. (1977) in: *Contempo-rary Topics in Polymer Science* (Pearce, E. M., Schaefgen, J. R., Eds.), New York: Plenum Press, 251–286, vol. 2.

Shen, Y., Shen, Z., Shen, J., Zhang, Y., Yao, K. (1996a) Characteristics and mechanism of ε-caprolactone polymerization with rare earth halide systems, *Macromolecules* **29**, 3441–3446.

Shen, Y., Shen, Z., Zhang, Y., Yao, K. (1996b) Novel rare earth catalysts for the living polymerization and the block copolymerization of ε-caprolactone, *Macromolecules* **29**, 8289–8295.

Shen, Y., Zhu, K. J., Shen, Z., Yao, K.-M. (1996c) Synthesis and characterization of highly random copolymer of ε-caprolactone and D,L-lactide using rare earth catalyst, *J. Polym. Sci., Part A: Polym. Chem.* **34**, 1799–1805.

Shi, W., Rånby, B. (1996a) Photopolymerization of dendritic methacrylated polyesters. I. Synthesis and properties, *J. Appl. Polym. Sci.* **59**, 1937–1944.

Shi, W., Rånby, B., (1996b) Photopolymerization of dendritic methacrylated polyesters. II. Characteristics and kinetics, *J. Appl. Polym. Sci.* **59**, 1945–1950.

Shi, W., Rånby, B. (1996c) Photopolymerization of dendritic methacrylated polyesters. III. FRP composites, *J. Appl. Polym. Sci.* **59**, 1951–1956.

Shimasaki, K., Aida, T., Inoue, S. (1987) Living polymerization of δ-valerolactone with aluminum porphyrin: trimolecular mechanism by the participation of two aluminum-porphyrin molecules, *Macromolecules* **20**, 3076–3080.

Shuai, X. T., Jedlinski, Z., Kowalczuk, M., Rydz, J., Tan, H. M. (1999) Enzymatic synthesis of polyesters from hydroxyl acids, *Eur. Polym. J.* **35**, 721–725.

Singh, A. N., Upadhye, A. B., Wadia, M. S., Mhaskar, V. V., Dev, S. (1969) Lac resin II. Lac acids 2. Laccijalaric acid, *Tetrahedron* **25**, 3855–3867.

Singh, A. N., Upadhye, A. B., Mhaskar, V. V., Dev, S. (1974a) Chemistry of lac resin IV. Components of soft resin, *Tetrahedron* **30**, 867–874.

Singh, A. N., Upadhye, A. B., Mhaskar, V. V., Dev, S., Pol, A. V., Naik, V. G. (1974b) Chemistry of lac resin, 3. Structure, *Tetrahedron* **30**, 3689–3693.

Slomkowski, S., Sosnowski, S., Gadzinowski, M. (1998) Polyesters from lactides and ε-caprolactone. Dispersion polymerization versus polymerization in solution, *Polym. Degrad. Stab.* **59**, 153–160.

Steinbüchel, A., Valentin, H. E. (1995) Diversity of bacterial polyhyroxyalkanoic acids, *FEMS Microbiol. Lett.* **128**, 219–228.

Stevels, W. M., Ankoné, M. J. K., Dijkstra, P. J., Feijen, J. (1996a) Kinetics and mechanism of ε-caprolactone polymerization using yttrium alkoxides as initiators, *Macromolecules* **29**, 8296–8303.

Stevels, W. M., Ankoné, M. J. K., Dijkstra, P. J., Feijen, J. (1996b) A versatile and highly efficient catalyst system for the preparation of polyesters based on lanthanide tris(2,6-di-*tert*-butylphenolate)s and various alcohols, *Macromolecules* **29**, 3332–3333.

Stridsberg, K., Albertsson, A.-C. (2000) Controlled ring-opening polymerization of L-lactide and 1,5-dioxepan-2-one forming a triblock copolymer, *J. Polym. Sci., Part A: Polym. Chem.* **38**, 1774–1784.

Stridsberg, K., Gruvegård, M., Albertsson, A.-C. (1998) Ring-opening polymerization of degradable polyesters, *Macromol. Symp.* **130**, 367–378.

Suda, S., Uyama, H., Kobayashi, S. (1999) Dehydration polycondensation in water for synthesis of polyesters by lipase catalyst, *Proceed. Japan Academy Ser. B- Physical and Biol. Sci.* **75**, 201–206.

Tilstra, L., Johnsonbaugh, D. (1993) A test method to determine rapidly if polymers are biodegradable, *J. Environ. Polym. Degrad.* **1**, 247–255.

Tokiwa, Y., Suzuki, T. (1977a) Hydrolysis of polyesters by lipases, *Nature (London)* **270**, 76–78.

Tokiwa, Y., Suzuki, T. (1977b) Microbial degradation of polyesters. Part III. Purification and some properties of polyethylene adipate-degrading enzyme produced by *Penicillium* sp. strain 14–3, *Agric. Biol. Chem.* **41**, 265–274.

Trollsås, M., Hawker, C. J., Remenar, J. F., Hedrick, J. L., Johansson, M., Ihre, H., Hult, A. (1998a) Highly branched radial block copolymers via dendritic initiation of aliphatic polyesters, *J. Polym. Sci., Part A: Polym. Chem.* **36**, 2793–2798.

Trollsås, M., Hedrick, J. L., Mecerreyes, D., Jérôme, R., Dubois, Ph. (1998b) Internal functionalization in hyperbranched polyesters, *J. Polym. Sci., Part A: Polym. Chem.* **36**, 3187–3192.

Trollsås, M., Löwenhielm, P., Lee, V. Y., Möller, M., Miller, R. D., Hedrick, J. L. (1999) New approach to hyperbranched polyesters: Self-condensing cyclic ester polymerization of bis(hydroxymethyl)-substituted-ε-caprolactone, *Macromolecules* **32**, 9062–9066.

Tsuji, H., Mizuno, A., Ikada, Y. (1998) Blends of aliphatic polyesters. III. Biodegradation of solution-cast blends from poly(L-lactide) and poly(ε-caprolactone), *J. Appl. Polym. Sci.* **70**, 2259–2268.

Uyama, H., Kobayashi, S. (1993) Enzymatic ring-opening polymerization of lactones catalyzed by lipase, *Chem. Lett.* 1149–1150.

Uyama, H., Takeya, K., Hoshi, N., Kobayashi, S. (1995) Lipase catalyzed ring-opening polymerization of 12-dodecanolide, *Macromolecules* **28**, 7046–7050.

van Dorp, A. G. M., Verhoeven, M. C. H., Koerten, H. K., van Blitterswijk, C. A., Ponec, M. (1999) Bilayered biodegradable poly(ethylene glycol)/

poly(butylene terephthalate copolymer) (Polyactive™) as a substrate for human fibroblasts and keratinocytes, *J. Biomed. Mater. Res.* **47**, 292–300.

Vanhoorne, P., Dubois, Ph., Jérôme, R., Teyssié, Ph. (1992) Macromolecular engineering of polylactones and polylactides. 7. Structural analysis of copolyesters of ε-caprolactone and L- or D,L-lactide initiated by Al(O^{i}Pr)$_3$, *Macromolecules* **25**, 37–44.

Wadia, M. S., Khurana, R. G., Mhaskar, V. V., Dev, S. (1969) Chemistry of lac resin VI. Components of soft resin, *Tetrahedron* **25**, 3841–3854.

Wang, L., Ma, W., Gross, R. A., McCarthy, S. P. (1998) Reactive compatibilization of biodegradable blends of poly(lactic acid) and poly(ε-caprolactone), *Polym. Degrad. Stab.* **59**, 161–168.

Wang, L., Ishida, Y., Ohtani, H., Tsuge, S., Nakayama, T. (1999) Characterization of natural resin shellac by reactive pyrolysis-gas chromatography in the presence of organic alkali, *Anal. Chem.* **71**, 1316–1322.

Wei, P. Y., Fu, S. D., Wei, H. J., Ji, Y. Y., Yong, X. (1997) *In vitro* degradation of poly(ε-caprolactone), poly(lactide) and their block copolymers: Influence of composition, temperature and morphology, *React. Funct. Polym.* **32**, 161–168.

Yasin, M., Tighe, B. J. (1992) Polymers for biodegradable medical devices. 8. Hydroxybutyrate-hydroxyvalerate copolymers: Physical and degradation properties of blends with polycaprolactone, *Biomaterials* **13**, 9–16.

3
Biodegradable Polymer (Biopol®)

Dr. Jawed Asrar[1], Dr. Kenneth J. Gruys[2]
[1] Monsanto Company, 800 North Lindbergh Boulevard, St Louis, MO 63167, USA;
Tel: + 1-314-694-1291; Fax: + 1-314-694-6488; E-mail: jawed.asrar@monsanto.com
[2] Monsanto Company, 800 North Lindbergh Boulevard, St. Louis, MO 63167, USA;
Tel: + 1-636-737-7345; Fax: + 1-636-737-7015; E-mail: kenneth.j.gruys@monsanto.com

3HB	3-hydroxybutyrate
3HV	3-hydroyxvalerate
BN	boron nitride
CPA	cyclohexyl phosphonic acid
DBP	di-*n*-butyl phthalate
DSC	differential scanning calorimetry
GTA	glycerol triacetate
H_c	heat of crystallization
HEDP	1-hydroxyethylidene diphosphonic acid
IlvA	threonine deaminase
L/D	length/diameter
M_n	number average molecular weight
MCL	medium chain length
MVTR	moisture vapor transmission rate
NMR	nuclear magnetic resonance
OTR	oxygen transmission rate
PBA	polybutylene adipate
PCL	polycaprolactone
PDH	pyruvate dehydrogenase
PEE	pentaerythritol ethoxylate
PHA	poly(hydroxyalkanoate)
PhbA or BktB	β-ketothiolase
PhbB	acetoacetyl-CoA reductase
PhbC, PHB synthase; Poly(3HB)	poly(3-hydroxybutyrate)
Poly(3HB-*co*-3HV)	poly(3-hydroxybutyrate-*co*-3-hydroyxvalerate)

Poly(3HPE)	poly(3-hydroxy-4-pentenoic acid)
T_c	crystallization temperature
T_d	decomposition temperature
T_g	glass transition temperature
T_m	melting temperature
ZS	zinc stearate

1

Introduction

Poly(3-hydroxybutyrate) (Poly(3HB)) was first isolated and characterized in 1926 by Lemoigne at the Pasteur Institute in Paris (Lemoigne, 1926). Since then, this and other poly(3-hydroxyalkanoates) (PHAs) have been studied extensively by scientists who have come to the general conclusion that bacteria store Poly(3HB) as an energy reserve in much the same way that mammals accumulate fat. Although Poly(3HB) is the most common type of PHA, many different polymers and copolymers of this class are produced by a variety of organisms (Steinbüchel and Valentin, 1995). Beyond their biological curiosity, many PHAs have functional properties that are quite suitable for commercial applications. Whereas Poly(3HB) is somewhat brittle and thus limited in its commercial scope, a closely related copolymer, poly(3-hydroxybutyrate-co-3-hydroxyvalerate) (Poly(3HB-co-3HV)), is more flexible because of reduced crystallinity and is suitable for many commercial applications (Structure. Indeed, this copolymer was commercially developed by ICI and was sold under the tradename Biopol®. The focus of this chapter is Biopol®. In it we

will review all the pertinent literature on Biopol®, its properties, and those of other closely related PHAs. We also will discuss bacterial production of Biopol® and PHAs in general, recent learning on the Biopol® polymer science front, and efforts to produce this copolymer in transgenic plants.

2

Historical Outline

Poly(3HB) remained an academic curiosity until W. R. Grace in the US produced small quantities for commercial evaluation in the late 1950s and early 1960s. Several patents were issued as a result of that effort (Baptist, 1962a,b, 1965). Commercial interest lay dormant for over a decade until ICI began a research and development program. This project followed their single-cell protein animal feed project. Thus, ICI had the skills in place to run large-scale fermentation processes and polymer processing know-how was available in their plastics division (Holmes, 1984, 1985; Holmes et al., 1984). In the late 1980s, ICI began worldwide commercialization of a family of (Poly(3HB-co-3HV) copolymers with the tradename of Biopol®. In 1990, the agricultural and pharmaceutical businesses of ICI, including Biopol®, were spun-off as Zeneca Ltd. In 1996, Monsanto acquired the Biopol® business from Zeneca Ltd. Since this acquisition the emphasis at Monsanto was on producing Biopol® and related copolymers in plants, and improving their properties for different end-use applications. This continued until Monsanto stopped their research program and commercial Biopol® business at the end of 1998.

3
Microbial Biosynthesis of PHAs

The specific enzymes involved in the synthesis and later utilization of the energy reserve PHA polymers is known to vary between microorganisms. The well-studied Poly(3HB) biosynthetic pathway in *Ralstonia eutropha* is shown in Figure 1. The pathway involves the condensation of two molecules of acetyl-CoA by β-ketothiolase to form acetoacetyl-CoA which is subsequently reduced by acetoacetyl-CoA reductase to form D-(–)-3-hydroxybutyryl-CoA (3HB). Monomeric 3HB is then polymerized to form Poly(3HB) by PHB synthase. The three enzymes that catalyze these reactions are encoded by genes organized as an operon in this organism designate PhbA, PhbB, and PhbC for the ketothiolase, reductase, and synthase, respectively (Slater et al., 1988). Like other PHAs, this polymer accumulates in discrete, membrane-bound granules in the bacterial cell (Williamson and Wilkinson, 1958; Merrick and Douboroff, 1961).

The percentage of Poly(3HB) in bacterial cells is normally low, from 1 to 30%, but under controlled fermentation conditions of carbon excess and nitrogen limitation, overproduction of polymer can be encouraged to produce yields of up to 80% of the dry cell weight (Dawes and Senior, 1973; Ward et al., 1977). Numerous microbiological species are known to be suitable for the production of PHAs (Anderson and Dawes, 1990). The microorganisms may be wild-type or mutated, or may have the necessary genetic material introduced into them using recombinant DNA techniques.

In *R. eutropha*, glucose is the common carbon source for the production of Poly-(3HB) (Holmes, 1985). However, other substrates such as methanol, sucrose, ethanol, and acetic acid can be used by microorganisms to produce the homopolymer (Suzuki, 1986). It is through various intracellular pathways that PHA-producing organisms like *R. eutropha* are able to convert these substrates to the precursor acetyl-CoA. As an example, a two-stage culture method has been reported in which *R. eutropha* was grown in an organic medium under heterotrophic conditions for exponential growth where the cells were cultivated for Poly-(3HB) accumulation under autotrophic conditions. The O_2 concentration in the substrate gas (H_2 + CO_2) was below the explosion limit of 6.9%. Through this process Poly(3HB) was obtained at a high production rate and concentration (Tanaka et al., 1994).

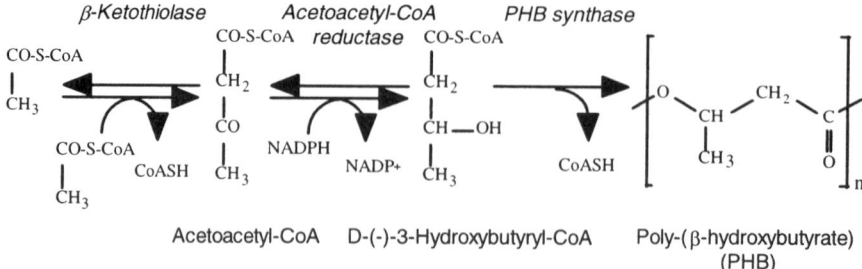

Fig. 1 Common three-step biochemical pathway to Poly(3HB) production as found in *R. eutropha*.

Interestingly, it is found that the Poly-(3HB) produced during exponential growth phase of *R. eutropha*, contains only carboxy chain ends; evidence of secondary hydroxyl is found only in polymers produced in the polymer accumulation phase (Madden et al., 1999). So, Poly(3HB) produced by fermentation always contains an excess of carboxy group in relation to the secondary hydroxyls. The number average molecular weight (M_n) of Poly(3HB) is frequently measured by end group analysis of the carboxy functionality. This is generally in excellent agreement with the true M_n measured by gel-permeation chromatography. However, M_n calculations based on the secondary hydroxyl are considerably higher due to the fact that not all the chains contain secondary hydroxyl end groups (Shah et al., 2000a). Therefore, M_n measured by secondary hydroxyls underestimates the actual number of chain ends and hence incorrectly leads to high M_n values.

By modifying the carbon source, *R. eutropha* can produce PHA copolymers. Pertinent to this chapter, Poly(3HB-*co*-3HV) or the Biopol® copolymer is typically produced using a combination of glucose and propionate in the growth media (Holmes, 1985). By adjusting the composition of the carbon sources, Poly(3HB-*co*-3HV) with up to 95 mol% 3HV content has been produced (Howells, 1982; Holmes, 1985; Doi, 1988). The fermentation process utilized by Zeneca and then Monsanto is based on descriptions written in various issued patents. In particular, Holmes et al. (1984) describe the use of the *R. eutropha* mutant NCIB 11599 grown under aerobic cultivation conditions to produce the Poly(3HB-*co*-3HV) copolymer when fed glucose and/or various amounts of propionic acid and other organic acids that lead to the 3HV component. The examples demonstrate that the copolymer can be produced up to 70% of the cell dry weight

and with 3HV levels up to 30% of the monomer fraction.

Alternate feeding of glucose and propionic acid to the phosphate-depleted batch cultures of *R. eutropha* produced Poly(3HB-*co*-3HV) with thermal properties markedly different from random copolymers of similar monomer content. These polymers exhibited a single glass transition and a single melting peak that was significantly higher than expected for a random copolymer. The polymers were found, by solvent fractionation and nuclear magnetic resonance (NMR), to be a mixture of Poly(3HB) and a random copolymer, Poly(3HB-*co*-3HV). When compared with random copolymers of similar monomer content, the polymers produced by alternate substrate feeding displayed no improvement in mechanical properties and possessed similar aging characteristics (Madden et al., 1998).

The pathway for Poly(3HB-*co*-3HV) biosynthesis is essentially identical to that shown in Figure 1, where the steps to the 3HV component starts with the condensation of acetyl-CoA with propionyl-CoA to form 3-ketovaleryl-CoA. Interestingly, it was recently found that this condensation step for 3-ketovaleryl-CoA production requires a different β-ketothiolase than the PhbA ketothiolase because of the narrow substrate specificity of this latter enzyme (Slater et al., 1998). This finding had important implications for recombinant systems as will be described below. Poly(3HB-*co*-4 hydroxybutyrate) copolymers can also be produced by *R. eutropha* when fed on nitrogen-free cultures of either butyrolactone and butyric acid or 4-hydroxybutyrate and 4-chlorobutyric acid (Doi et al., 1988).

Outside of *R. eutropha*, using *Pseudomonas oleovorans* and a range of *n*-alkanoic acids, PHAs containing up to 12 carbon atoms have been produced (Brandl et al., 1988; Gross et al., 1989). Although all the PHAs were heteropolymers containing up to six different monomers, the major monomer unit always had the same number of carbon atoms as the n-alkanoate substrate used. The introduction of functionalized chain-ends into PHA has also been reported (Hirt et al., 1996; Asrar et al., 1999; Madden et al., 1999; Shah et al., 2000a) and could facilitate subsequent modification, e.g. in the production of block copolymers. This has been accomplished through supplementation of the culture medium with various alcohols, diols, and polyols (Asrar et al., 1999; Madden et al., 1999; Shah et al., 2000a). A carboxyl group, belonging to the final 3HB monomer in a Poly(3HB) chain, is normally found at the chain terminus. However, this has been altered in vivo by the inclusion of hydroxy-containing compounds, which become incorporated at the polymer terminus via a chain transfer reaction in which the growing polymer chain is esterified to the hydroxy compound (Madden et al., 1999; Shah et al., 2000a). In addition to chain termination some glycols, e.g. ethylenglycol, are also found to be utilized as a carbon source for the production of Poly(3HB) (Shah, 2000b).

Escherichia coli does not normally synthesize PHAs since it lacks the PHA biosynthetic genes. However, it has been reported to accumulate Poly(3HB) levels approaching 80% of the dry cell weight when transformed with plasmids bearing the appropriate biosynthetic genes from R. eutropha. The major advantage of synthesizing Poly(3HB) in E. coli is that all the genetic engineering principles that apply to this organism can be utilized in optimizing the production of Poly(3HB) and other PHAs (Pouton and Akhtar, 1996). For example, high molecular weights, e.g. 4–5 MDa, can be achieved in this organism because E. coli does not contain PHA depolymerase enzymes (Sim et al., 1997). There are also potential advantages in terms of ease of extraction of the polymer and purity of the product (Pouton and Akhtar, 1996). Additionally, a Poly(3HB) synthesizing mutant of E. coli has been developed from which the polymer can be extracted by mild heat treatment rather than by chemical extraction techniques (Pool, 1989). This strain is thought to release the accumulated polymer through thermally induced cell lysis at a low temperature of 42 °C (Pouton and Akhtar, 1996).

4
Plant Production of Biopol®

The commercial production of Poly(3HB-co-3HV) has been done on a relatively small scale since it is not economically competitive with similar petrochemical-based polymers such as polyethylene and polypropylene. Therefore, efforts to produce PHAs with transgenic plants as the plastic factory has been pursued as an alternative (Poirier, 1999a; Poirier et al., 1995a). This route has the potential to be cost-competitive with traditional plastics because plant systems in theory only require CO_2, water, and sunlight for input. Of course, this is an oversimplification since other agriculture inputs are required to produce a crop (Gerngross and Slater, 2000). In addition, plant production of a copolymer is difficult to generate and control since the system relies on endogenous metabolic precursors in the plant. This is in contrast to the flexibility of supplementing the growth media with the required substrates in a microbial fermentation process. Nevertheless, efforts starting with the more simple Poly(3HB) homopolymer had shown promise that a plant production system could be viable. More specifically, Somerville and coworkers showed that Poly(3HB) levels of up to 14% of the plant dry weight could be produced in transgenic *Arabidopsis thaliana* leaves using the *R. eutropha* enzymes (Poirier et al., 1992; Nawrath et al., 1994) with little effects on plant growth. This was true only if the Poly(3HB) biosynthetic enzymes were targeted to the plant chloroplast (Nawrath et al., 1994). Expression of these genes in the plant cytosol caused severe stunting and sterility (Poirier et al., 1992). The reasons for the differences between cytosolic versus plastid production are not entirely clear, but it may very well be due to the availability of acetyl-CoA where in plastids the production of this metabolite is robust (Nawrath et al., 1994).

The unavailability of direct metabolic precursors for the 3HV component of Poly(3HB-*co*-3HV) in a plant led us at Monsanto to devise a pathway that might be amenable to this biosynthesis (Slater et al., 1999). This is illustrated in Figure 2. Since the plastid was shown to be the preferred site for PHA biosynthesis, metabolic pathways present in this organelle were the focus. In the plastid, threonine is one product of the aspartate family of amino acids and is also the starting metabolite in the biosynthesis of isoleucine.

The first committed step in isoleucine biosynthesis is the deamination of threonine to form α-ketobutyrate which is catalyzed by threonine deaminase. From there α-ketobutyrate is normally converted to 2-keto-2-hydroxybutyrate through a condensation reaction with pyruvate catalyzed by acetohydroxyacid synthase. This conversion step results in the maintenance of a very low steady-state concentration of α-ketobutyrate. It was hypothesized that significantly increasing the concentration of this 2-ketoacid in the plant plastid might allow a diversion of some of this carbon to form propionyl-CoA through the action of the plant plastid pyruvate dehydrogenase (PDH) complex. The normal function of this enzyme is in the oxidative decarboxylation of pyruvate to form acetyl-CoA, the direct precursor for fatty acid biosynthesis, and in the mitochondria the entry metabolite for the tricarboxylic acid cycle. α-Ketobutyrate is a one-carbon extended analogue of pyruvate, and we demonstrated that the plastid PDH complex can also catalyze the oxidative decarboxylation of

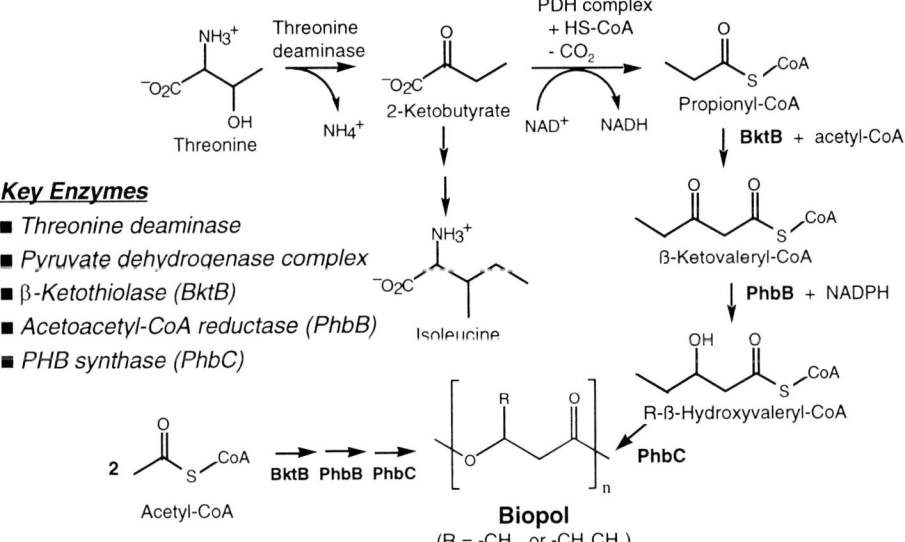

Fig. 2 Designed metabolic pathway for the production of Poly(3HB-*co*-3HV) in plant plastids.

this 2-ketoacid, although with less catalytic efficiency relative to pyruvate (Slater et al., 1999). The overexpression of threonine deaminase, either the wild-type or a deregulated form of the enzyme (biosynthetic threonine deaminase is normally feedback regulated by isoleucine in plant and microbial systems), was thought to be the means to significantly increase its concentration in the plant plastid. The PDH complex could then catalyze the formation of propionyl-CoA and following this step the normal PHA biosynthetic genes would carry reactions forward, in the same manner as in bacteria, to form the Biopol® copolymer as illustrated in Figure 2.

Transformation of plants, while now routine, nevertheless is time consuming and therefore not an efficient means to test a metabolic pathway in a recombinant system. Realizing this, we first used *E. coli* as the vehicle to see if the pathway shown in Figure 2 could be operative. The specific genes used in the initial transformation in *E. coli* were the *E. coli* threonine deaminase (*ilvA*), a wild-type or deregulated variant, and the three normal Poly(3HB) biosynthetic enzymes from *R. eutropha*, i.e. *phbA*, *phbB*, and *phbC*. The results from this experiment are given below in Table 1. As Table 1 shows, wild-type IlvA was expressed approximately 7-fold above the deregulated L481F mutant, but interestingly deregulation has a larger impact on the intracellular concentration of α-ketobutyrate. Unfortunately, no impact on

the percent 3HV component in the polymer (each recombinant strain produced from 30 to 40% PHA based on cell dry weight) was evident, i.e. as constructed, the pathway was not operational. While it was apparent that a block in the pathway was not at the first step (i.e. threonine deaminase produces elevated levels of α-ketobutyrate), a series of enzyme substrate specificity assays were required to determine that the PhbA β-ketothiolase would not catalyze the formation of β-ketovaleryl-CoA. Fully realizing that *R. eutropha* produces the copolymer, our group identified and characterized a second β-ketothiolase designated BktB that is primarily responsible in this organism for the production β-ketovaleryl-CoA during the biosynthesis of Poly(3HB-*co*-3HV) (Slater et al., 1998). Reconstruction of recombinant *E. coli* strains substituting BktB for PhbA made a functional pathway for the production of Poly(3HB-*co*-3HV). Specifically, an 8% fraction of 3HV in the copolymer was produced by simply growing the organism on glucose (Valentin et al., 1999a). This is well within the range of 3HV for the typical composition of commercial Biopol®.

Success in demonstrating the pathway in *E. coli* led to the transformation of two plant systems: *Arabidopsis thaliana* for leaf expression and *Brassica napus* (rapeseed) for seed production (Slater et al., 1999). To direct biosynthesis in the plastids of these tissues the *agrobacterium* transformation cassettes

Tab. 1 Results from a test of the pathway shown in Figure 2 in transformed *E. coli* (the cells overexpressed four proteins, a wild-type or deregulated threonine deaminase, and the three *R. eutropha* Poly(3HB) biosynthetic enzymes PhbA (β-ketothiolase), PhbB (acetoacetyl-CoA reductase), and PhbC (PHB synthase))

E. coli construct	IlvA (µg/mg) no Ile	IlvA (µg/mg) + 0.10 mM ille	Activity of no Ile (%)	α-Ketobuty-rate (µM)	P(3HB-co-3HV) (% 3HV)
phb ABC	< 0.04	<0.04	–	13	1.3
phb ABC + wild-type IlvA	4.2	0.16	3.8	42	1.1
phb ABC + L481F IlvA	0.59	0.56	95	580	0.9

were constructed with a plastid transit peptide fused to an open reading frame for all genes, and all gene expression was driven by either the 35S promoter or *Lesquerella* hydroxylase promoter (Broun et al., 1998) for *Arabidopsis* or rapeseed, respectively. The results from this work showed copolymer production in both plants (Slater et al., 1999). Levels of Poly(3HB-*co*-3HV) were modest, however, relative to production previously reported for Poly(3HB) biosynthesis in *Arabidopsis* (Nawrath et al., 1994), and less than that seen for Poly(3HB) production in *Brassica* using a similar construct and transformation strategy (Houmiel et al., 1999). Specifically, copolymer levels were less than 3% of the plant tissue dry weight (*Arabidopsis*) or seed weight (rapeseed). It was also clear that at least for *Brassica* oilseeds, increasing polymer production was inversely proportional to the 3HV content in the polymer. The reasons for this appear to be metabolic channeling of increased α-ketobutyrate (produced by threonine deaminase), to both isoleucine and 2-aminobutyrate. The latter metabolite is the direct transamination product of the 2-ketoacid. The consequence is that more carbon is directed towards these non-target metabolites and, because of this, the steady-state levels of α-ketobutyrate, though measurably increased (approximately 20-fold), is not sufficient for robust conversion to propionyl-CoA by the plant plastid PDH complex. As such, additional engineering will be required to make the pathway operative as a commercial production system. Nevertheless, the results obtained from this work represent a complex feet of plant metabolic engineering with the introduction of four transgenes (not counting the selection marker) and the modification of two independent plant pathways (i.e. fatty acid and amino acid biosynthesis) to make a non-related product (Poirier, 1999b). Additional

detail on this constructed pathway and the impact on plant metabolism is addressed in Volume 3a, Chapter 15 in this book series (Poirier and Gruys).

It is of interest to note that of the PHAs produced in transgenic plants from various research groups, that earlier reports noted a broad molecular weight distribution, ranging from 8 to 10, relative to bacterial PHA which typically produces narrower molecular weight distributions from 2 to 3 (Poirier et al., 1995b). However, Poly(3HB) and Poly(3HB-*co*-3HV) produced in *Arabidopsis* and canola at Monsanto resulted in a polymer that more typically resembles that seen in bacteria (Slater et al., 1999; Asrar et al., 2000a, 2001a). The reason for this difference in results from plant-produced PHA could well be due to the additional steps done at Monsanto to assure that the transgenic plants have a single loci (i.e. contain a single copy) for the individual PHA biosynthetic genes. This appears to be true for either a single loci that contains all the transgenes (a result from using a mutigene construct that has all the biosynthetic genes) or distinct loci for the individual transgenes (resulting from crossing homozygous plant lines that individually contain a portion of the pathway or cotransformation strategies using more than one plant vector).

If the technical hurdles towards the production of Biopol® in a plant system are overcome, a major challenge remains in the development of a process for the recovery of the copolymer from plant tissue. This is likely to be a major factor impacting the overall costs in a plant production system. As shown in the electron micrograph in Figure 3, the production of PHA, in this case Poly(3HB) in a *Brassica* seed plastid, will require separation from lipid, protein, and other cellular components. The characteristics of Poly(3HB-*co*-3HV), e.g. the crystalline nature and thermal properties (see

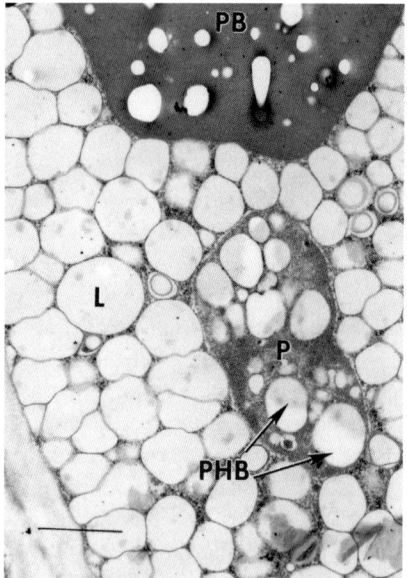

Fig. 3 Electron micrograph of a plastid from a *B. napus* seed producing Poly(3HB). Plastid (P), polymer (PHB), lipid bodies (L), and protein bodies (PB) are indicated. The bar indicates a 1 μm.

below), puts boundaries on the methods that can be utilized. In addition, the choice of crop and tissue type (i.e. seed or leaf) could have a large impact. With this in mind some possibilities include supercritical fluid extraction, selective non-halogenated solvent extraction of PHA from plant biomass, or routes that take advantage of density differences between PHA and biomass. It is worth noting that any type of solvent used in the process should already have regulatory approval as a synthetic food additive or flavoring agent. This is because a crop production system would likely be viable only if other byproducts such as oil, proteins, and carbohydrates can be recovered in addition to polymer. A current common laboratory method for PHA extraction which involves the use of chloroform is obviously not an option.

Recent studies by Mott et al. (Mott, 2000) took a preliminary look at processing of Poly(3HB) from wild-type (as an additive)

and transgenic Poly(3HB) rapeseed, matching to some degree the conditions used in commercial oil extraction using hexane in conjunction with mechanical crushing. The desirable outcome would be that slight modifications to existing commercial extraction methods could separate PHA from oil and that residual meal would still be suitable for animal feed. This was done on an ultra scale-down level and, while qualitative in nature, nevertheless showed that transgenic seed required a higher mechanical force than non-transgenic seed to yield the same level of oil. The same group also studied separation of Poly(3HB) from protein meal (Hughes et al., 2000). For general methods of PHA extraction from various plant materials, a recent patent application described the use of PHA-poor solvents in separation processes (Kurdikar et al., 2000a). These solvents include linear and branched alcohols and esters that dissolve less than 1% (w/v) of the PHA at a temperature below the solvent boiling point. However, under conditions of elevated temperature and pressure, significant solvation of PHAs occurs in these solvents. Even when processed and mixed with excess biomass material, a clean PHA precipitate could be obtained by the use of agitation while the solution mixture was cooled. This precipitation occurred 20–30 °C higher than the temperature at which other biomass components precipitated out of solution. Other similar methods include solvation in a PHA-good solvent followed by recovery through solvent evaporation, or addition of a PHA-poor solvent to cause selective precipitation of PHA (Kurdikar et al., 2000b). From this it can be seen that while promising methods for the recovery of PHAs from plant biomass are described, it is apparent that optimization will be required when higher levels of transgenic PHA-containing plant material become available for further experimentation.

5
Properties of Biopol®

5.1
Morphology of Biopol® and other PHAs

As mentioned earlier, Poly(3HB) and other PHAs accumulate in discrete spherical granules in the cell cytoplasm. Granules have a diameter that ranges from 100 to 800 nm (Dawes, 1973; Ellar, 1973) and are enclosed in a unit membrane about 2–4 nm thick (Ellar, 1973). The granules are typically composed of about 98% polymer and the rest is protein and phospholipid. ^{13}C-NMR studies (Barnard and Sanders, 1989; Bonthrone et al., 1992) on live cells suggest that Poly(3HB) is predominantly in a mobile state, but not in solution, in the granules. Water was shown to be an integral part of the granule and appears to act as a plasticizer for the polymer, although the exact conformation within which water is included to plasticize the polymer is not entirely clear. It is now proposed that the enzymes involved in the PHA biosynthetic pathway operate only on a mobile hydrated material and that the solid granules characteristic of dried cells are an artifact of the drying process. The crystallization of the polymer is thought to be under kinetic control and is inhibited by the submicron size of the particles, and their protein and phospholipid coat (Bonthrone et al., 1992). A method for extraction of amorphous granules has been reported (Fuller et al., 1992). It has also been possible to produce kinetically stable amorphous granules in vitro from crystalline Poly(3HB) and other PHAs, using surfactant to stabilize the submicron size granules (Fuller et al., 1992). These studies may well have some future impact on the processing of Poly(3HB) and other PHAs into articles of commercial use (Pouton and Akhtar, 1996).

5.2
Composition and Molecular Structure of PHAs

Poly(3HB) is isotactic and similar to isotactic polypropylene as both have pendant methyl groups attached to the main chain in a single conformation (Brandl et al., 1988). Poly(3HB) is a compact right-handed helix with a 2-fold screw axis and a fiber repeat of 5.95 Å. The helix conformation is stabilized by carbonyl–methyl group interactions and represents one of the few exceptions of a helix found in nature which does not depend on hydrogen bonding for its formation and stability. Poly(3HB) can have a weight average molecular weight of 0.1–3 MDa, although for processing the molecular weights are usually in the range of 200 to 800 kDa. The polydispersity is in the range of 2.2–3 (Brandl et al., 1988). Since Poly(3HB) is made biologically as a stored energy source in cell walls, it can be separated from cell material in very high purity. The major impurities in Poly(3HB) are inorganic nitrogen, phosphorous, and sulfur-containing compounds which are present at concentrations less than 200 p.p.m. (Barham, 1984). There are no catalyst residues to be concerned about in the polymer.

Poly(3HB-co-3HV) copolymers have been shown to exhibit isodimorphism (Bluhm et al., 1986; Scandola et al., 1992) with the 3HB and 3HV units cocrystallizing. As the 3HV content of Poly(3HB-co-HV) is increased, the melting point decreases [from around 178 °C for Poly(3HB)] and passes through a minimum of around 75 °C at around 40 mol% 3HV content and increases to around 110 °C for Poly(3HB-co-3HV) containing 97 mol% 3HV. A change from the Poly(3HB) lattice to the Poly(3HV) lattice takes place at around 40% 3HB. At less than 40 mol% 3HV, 3HV units can crystallize in the Poly(3HB) lattice and at greater than 40 mol% 3HV, 3HB units can crystallize in the Poly(3HV) lattice. Thus, isodimorphism makes it possible to achieve relatively high levels of crystallinity in these copolymers, retaining the useful hydrolysis and chemical resistance exhibited by Poly(3HB). Also the same nucleants can be used for Poly(3HB) and lower 3HV content Poly(3HB-co-3HV). Poly(3HB-co-3HV) satisfies the physical requirements for isodimorphism in that the two monomer units have approximately the same shape and occupy similar volumes, and the polymer chain conformations of both homopolymers are compatible with either crystalline lattice. Like Poly(3HB), the Poly(3HV) polymer chain has a 21 helix conformation, an orthorhombic unit cell and space group P_{212121} with a unit cell parameters $a = 9.32$ Å, $b = 10.02$ Å, and $c = 5.56$ Å (Yokouchi et al., 1974). NMR evidence indicates that at 3HV levels of 20–40 mol%, 3HV units are partially excluded from the Poly(3HB) lattice (Tokiwa et al., 1992). A recent study suggests that cocrystallization of the two monomer units certainly does occur but that the molar ratio of 3HV within crystals is approximately two-thirds of the total molar ratio present in the copolymer (Pouton and Akhtar, 1996).

5.3
Crystallization Behavior of Biopol®

The embrittlement of Poly(3HB) and its 3HV copolymers on storage at room temperature is not associated with relaxation of the amorphous regions but rather to the development of interlamellar secondary crystallization. The small crystallites produced underpin the amorphous regions and reduce the mobility of the chain segments thus raising the modulus and embrittling the material. Crystallizing the copolymer at high temperature leads to rejection of the 3HV units into the amorphous regions and reduces the extent of secondary crystallization which can develop at room temperature, such that these materials do not embrittle on storage. Dynamic mechanical and electrical thermal analysis were found to be useful techniques for measuring the mobility of the amorphous regions in partially crystalline samples, under conditions where differential scanning calorimetry (DSC) was unable to detect any glass transition (Biddlestone et al., 1996).

A patent (De Koning, 1994a) describes Poly(3HB) and high 3HB-containing PHAs in which aging had occurred and the original properties were restored by a heat treatment and subsequent aging is retarded. A patent has been filed on cooling the Polyester after preparation to less than 90 °C and then heating from 90 to 160 °C within 24 h of preparation to retard aging (Liggat and O'Brien, 1994a). Other patents were filed on PHA-plasticizer compositions (Liggat and O'Brien, 1994b) and Poly(3HB-co-3HV) copolymers (Liggat and O'Brien, 1994c) claiming to retard the aging process.

Tab. 2 Effect of 3HV content on the maximun crystalization rate

Polymer	Maximum crystallization rate (mm s⁻¹)	T_c (°C)	T_m (°C)
P(3HB)	4.5	88	197
P(3HB-co-6%3HV)	1.4	80	186
P(3HB-co-12%3HV)	0.43	78	173
P(3HB-co-16%3HV)	0.23	70	167

5.4
Thermal Properties of Biopol®

Thermal degradation of Poly(3HB) occurs quite rapidly at temperatures above the melting point (T_m) of the polymer (171 °C). The significant benefit of the Poly(3HB-*co*-3HV) copolymers is their lower melting points that enable them to be melt processed at lower temperatures than Poly(3HB), generally in the range of around 140–180 °C, significantly reducing thermal degradation (Cox, 1994).

Poly(3HB-*co*-*x*%3HV) copolymers are also thermoplastics with melting points, T_ms, in the range of 75–170 °C (Poirer, 1995) depending on the value of x (the mol% 3HV). Increasing the amount of 3HV in the copolymer reduces the Tm from around 180 °C for Poly(3HB) to around 75 °C for a copolymer containing 30–40% 3HV. The heat distortion temperature of the copolymers decreases from 140 to 92 °C as the 3HV content increases. The temperature for the onset of decomposition, T_d, with fast heat is in the 232–244 °C range (Marchessault et al., 1988), but is lower with longer exposure. The β-substituted aliphatic polyesters are unstable at greater than 170 °C.

The thermal degradation mechanism of Poly(3HB) has been studied and is considered to be primarily a random chain scission process by a β-elimination reaction via a six-membered cyclic transition state (Grassie et al., 1984a–c; Kunioka and Doi, 1990). Activation energies between 109 ± 13 kJ mol^{-1} for Poly(3HB) and 126 ± 13 kJ mol^{-1} for Poly(3HB-*co*-3HV) copolymers have been reported (Mitomo and Ota, 1991). The major pyrolysis product of Poly(3HB) is crotonic acid produced by the β-elimination reaction (Billingham et al., 1978; Grassie et al., 1984a–c). Significant quantities of Poly(3HB) oligomers and small quantities of isocrotonic acid were also evolved on heating

the polymer to around 300 °C. Secondary products such as propylene, ketene, CO_2, acetaldehyde, etc., were also formed by further decomposition of the primary products. Conventional thermal stabilizers have been shown to have little effect in preventing thermal degradation of Poly(3HB) (Billingham et al., 1978).

The thermal degradation mechanism of Poly(3HB-*co*-3HV) copolymers has been reported to be analogous to that for Poly(3HB), with a similar activation energy and degradation kinetics (Kunioka and Doi, 1990).

During thermal processing of Poly(3HB), both reactions, chain-end condensation leading to an increase in the molecular weight, as well as thermal degradation leading to a decrease in molecular weight, take place. Since each thermal degradation event produces a carboxyl and a crotonate chain end, the carboxyl concentration increases considerably with time and chain-end condensations are minimized leading to a drastic reduction in the molecular weight. It is found that an increase in the hydroxy-termination in Poly(3HB) leads to an increase in the thermal stability, most likely by prolonging the condensation and delaying the degradation reaction (Shah et al., 2000a).

5.5
Melt Rheology of Biopol® and its Impact on Processing

One of the key melt flow properties of a molten polymer is the extensional viscosity of the melt. 'Extensional thickening' and increase in the extensional viscosity above the linear viscoelastic limit is important to stabilize the polymer processing operations involving stretching of melts such as film blowing, fiber spinning, melt coating, etc. A polymer having low melt strength is unable to withstand the minimum strain required to draw the polymer to the desired dimension and will exhibit instabilities such as breakage, sagging, or draw resonance.

Rheological characterization of Poly(3HB-co-3HV) indicates the viscosity decreases dramatically with increasing shear rate (shear thinning). An increase in the molecular weight results in an increase in shear viscosity and extensional viscosity. In the deformation range studied, the effect of molecular weight on the extensional viscosity is larger than on shear viscosity. Thus the increased molecular weight results in increased melt strength and enhanced stability of a bubble during blown film processing (Asrar and Pierre, 2000a).

5.6
Effect of Branching in Biopol®

The melt strength can be improved by branching. A PHA melted in the presence of a free radical initiator at a temperature above the decomposition temperature of the free radical initiator for a sufficient length of time undergoes crosslinking. The thermally induced decomposition of the free radical initiator results in the production of reactive radicals which produce interchain crosslink formation between PHA chains. The branched PHA compositions can be made by reactive extrusion, where the extrusion temperatures and residence times are sufficient for melting the PHA and for causing the decomposition of the peroxide. When producing branched PHAs by this method, there are competing reactions occurring in the melt. Thermal decomposition of the PHA results in a decrease in its molecular weight while the branching reaction produces an increase in its molecular weight. The choice of extruder temperature, free radical initiator, and initiator concentration can be optimized to give control of resulting molecular weight and degree of branching.

The most effective concentrations of peroxide appear to be in the range of 0.05–0.1 up to 0.3 wt% (Asrar and D'Haene, 2000, 2001). At less than 0.05% there is no appreciable effect and at concentrations greater than 0.5 wt% the extruded materials become brittle. Branching also slows down the age-related embrittlement of articles produced from PHAs.

Figure 4 shows both the shear and the elongational viscosity of a linear and a branched Poly(3HB-co-3HV) prepared using 0.2 wt% DCP dicumyl peroxide as measured in a capillary rheometer. The data illustrate that the shear viscosities of both linear and branched polymers are almost equal, as these are mainly determined by the molecular weight of the polymer. Elongational viscosity is much larger for the branched material than for the linear polymer due to the existence of strain hardening in the elongational response of the branched product (D'Haene et al., 1999).

A novel approach for producing a branched bacterial polyester has recently been described. Poly(3HB) containing poly-hydroxy chain ends, produced by batch fermentation in presence of pentaerythritol ethoxylate (PEE), using *R. eutropha*, are used as precursors for producing branched Poly(3HB) during thermal processing. Chain-end condensation and transesterification reactions during thermal processing of PEE terminated Poly(3HB) results in hyper-branched Poly(3HB). Hydrodynamic volume of Poly(3HB) produced in presence or absence of the PEE are found to be the same. However, the hydrodynamic volume of PEE-terminated Poly(3HB) increases considerably after its thermal processing, indicating the formation of star-branched Poly(3HB) (Asrar et al., 2000c, 2001b).

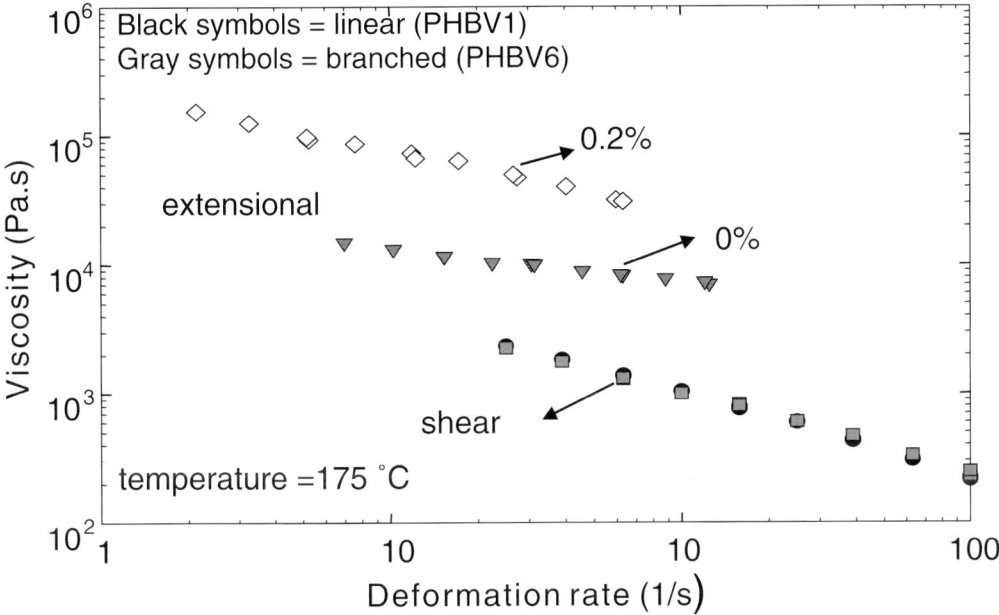

Fig. 4 Effect of branching on the shear viscosity (circles) and the elongational viscosity (triangles), as measured in a capillary rheometer. The elongational viscosity was determined from the entrance pressure drop using Cogswell's analysis (capillary data).

6
Processing Characteristics of PHAs

PHAs, as isolated from natural organisms, are difficult to process because of their relatively low decomposition temperatures (140–170 °C) near their melting points, low melt strength, their slow crystallization rates, etc. Because of thermal degradation in the melt, the melt flow index can change rapidly with time and volatile decomposition products must be handled safely. The slow crystallization rates lead to tacky products (e.g. fibers, films, molded articles) which adhere to themselves and the process equipment. Products produced by thermal processing undergo embrittlement with time (aging). Various approaches such as use of nucleating agents, plasticizers, branching, blends, etc., have been taken to overcome the processing and product difficulties and shortcomings.

6.1
Effect of Nucleating Agents

PHAs from biological sources can be of very high purity. As a result, Poly(3HB) melts may undergo homogeneous nucleation, but the nucleation density for Poly(3HB) is orders of magnitude less than for polyethylene and polypropylene. However, the growth rate, once nucleated, is similar to polypropylene, nylon 6 and PET. The low rate of nucleation in Poly(3HB) leads to development of large spherulites within the Poly(3HB). The large spherulites can significantly reduce the physical and mechanical properties of the polymers.

The overall rate of crystallization can be increased by the use of nucleating agents which promote crystallization of the molten or glassy resin by increasing the number of nuclei within the Poly(3HB). The increased number of nuclei leads to smaller diameter spherulites, a more rapid loss of tackiness, and concurrent increase in mechanical strength and reduced cycle times in thermal processes. Potential nucleating agents can be tested using DSC. As the molten polymer is cooled at a constant rate, an exotherm is produced as the polymer crystallizes. The temperature range over which crystallization occurs, the area of the peak, and the peak sharpness give an indication of the crystallization behavior of the material. Addition of a nucleating agent generally causes an increase in the crystallization peak and the peak area. Effective nucleating agents include saccharin and particulates such as talc, micronized mica, boron nitride (BN), chalk, calcium hydroxyapatite, and calcium carbonate (Webb, 1998). BN is commonly used as the nucleant for Poly(3HB) and other PHAs.

Saccharin ($T_m = 226\,°C$) can nucleate Poly(3HB) when added to the melt due to epitaxial growth on its surface (Barham, 1984). The nucleating effect of saccharin is due to adsorption of molecules onto the surface in what is close to their crystallographic form.

Technical grade NH_4Cl incorporated as an aqueous solution is found to be a more effective nucleating agent.

Poly(3HB) seed crystals were used to nucleate crystallization of Poly(3HB-co-12%3HV) ($T_m = 149.9\,°C$). At 5% of Poly-(3HB) seed crystals, the crystallization temperature (T_c) was 88.7 °C and heat of crystallization (H_c) was 45.7 J g^{-1}. At 1 p.h.r. of Poly(3HB) seed crystals, the T_c was 70.3 °C and H_c was 41.9 J g^{-1}. Without the Poly(3HB) seed crystals, no T_c was observed (Liggat, 1994d).

Cyclohexyl phosphonic acid (CPA) and zinc stearate (ZS) mixtures are more effective nucleating agents than BN for Poly(3HB-co-3HV) copolymers with high 3HV content. For example, when 0.10 p.h.r. CPA and 0.385 p.h.r. ZS were added to a Poly(3HB-co-22%3HV) polymer and compared with Poly(3HB-co-22%3HV) containing 1 p.h.r. BN at a T_c of 80 °C, the half-crystallization time and the half-heat of crystallization peak were similar for CPA/ZS and BN; 1.26 and 1.39 min and 19.85 and 19.26 J g^{-1}, respectively. At a T_c of 50 °C, the half-crystallization time for CPA/ZS was 0.40 versus 3.91 min for BN and the half-heat peak was 47.09 versus 11.73 J g^{-1}. Thus at lower T_cs, CPA/ZS yields significantly faster crystallization rates with improved energies of crystallization. With high 3HV containing polymers it should be possible to use lower mold temperatures and shorter cycle times with CPA/ZS. However, for low 3HV containing polymers, both types of nucleant yield similar crystallization rates (Herring and Webb, 1990).

More recently, organophosphorous compounds having at least two phosphonic acid groups such as 1-hydroxyethylidene diphosphonic acid (HEDP) have been shown to serve as nucleating agents and give products with excellent clarity. For example, a Poly(3HB-co-3%3HV) copolymer containing 0.2 p.h.r. HEDP showed a T_c of 78 °C and an H_c of 55 J g^{-1}. A combination of HEDP with Ca or Mg stearate has also been shown to be an effective nucleating agent. A Poly(3HB-co-8%3HV) copolymer containing 0.1 p.h.r. HEDP and 0.3 p.h.r. calcium or manganese stearate had a T_c of 72 °C and an H_c of 51 and 50 J g^{-1}, respectively, compared to a T_c of 56 °C and an H_c of 15 J g^{-1} for Poly(3HB-co-8%3HV) alone (Asrar and Pierre, 1999).

6.2
Effect of Plasticizers

The use of monomeric and polymeric plasticizers in PHAs to impart properties and impede loss of properties by secondary crystallization has been described. Plasticizers for Poly(3HB) and Biopol® include (1) high-boiling esters of polybasic acids such as phthalates, isophthalates, citrates, fumarates, glutamate, phosphates, or phosphites, (2) high-boiling esters and part esters of polyhydric alcohols, especially glycols, polyglycols, and glycerol, and (3) aromatic sulfonamides. For producing rigid, but not too brittle articles, 6–12 p.h.r. w/w is generally suitable.

Glycerol triacetate (GTA) (triacetin) and glycerol tributyrate (GTB) are effective plasticizers for Biopol® and show good compatibility. Solid-state ^{13}C-NMR showed the mobility of the Poly(3HB) chain is enhanced by the addition of GTA. Unfortunately, GTA is too volatile during melt processing or even storage at high ambient temperatures and the monomeric plasticizers are eluted with time with a corresponding decline in product properties. Acetyl tri-*n*-butyl citrate (Estaflex) is also used as a plasticizer for PHAs and is considered an improvement over GTA (Hammond et al., 1994).

Di-*n*-butyl phthalate (DBP) displays a plasticizing effect toward Poly(3HB) very similar to that found with polyvinylchloride (Gassner and Owen, 1994). A near linear relationship exists for the glass transition temperature (T_g) and DBP weight fraction from the T_g of Poly(3HB) of 8 °C to the T_g of DBP of –90 °C. The amount of DBP needed to lower the T_g of Poly(3HB) to –40 °C is about 30%. Poly(3HB) is able to crystallize closer to room temperature in the presence of DBP as the temperature of the crystallization exotherm decreases with increasing DBP content. This is not surprising and is attributed to a concomitant decrease of T_g, above which the macromolecules acquire enough mobility to rearrange and crystallize. The enthalpy (H_c) correlates with the decrease of Poly(3HB) in the polymer–DBP mixture, therefore the ability of Poly(3HB) to

crystallize does not decrease in the presence of plasticizer. A level of 39% DBP lowers the T_m of Poly(3HB) from 175 to 150 °C.

Use of high molecular weight polymers as non-exuding plasticizers has been reported. Polycaprolactone (PCL) and polybutylene adipate (PBA) have been used with Poly(3HB) but the mixtures are not compatible, and films made from the mixtures exhibit rapid aging and loss in tensile properties (Kumagai and Doi, 1992; Gassner and Owen, 1994)

Based on the mechanical properties and scanning electron microscopy of the immiscible blends, it was suggested that the Poly(3HB)/PCL blend has a macro-phase separated structure and the Poly(3HB)/PBA blend has a modulated structure with a micro-phase separation (Kumagai and Doi, 1992). Polyethylene oxide, which has a T_g of –59 °C, behaves like a high-molecular-weight plasticizer with Poly(3HB) (Bailey et al., 1990). Polymeric esters and epoxidized soybean oils have given very good results as plasticizers for PHAs (Hammond et al., 1994). Several oligoesters were recently reported to compatibilize blends of Biopol® with other biodegradable polyesters (Asrar and D'Haene, 2001a,b).

Biopol®, with molecular weight in excess of 400,000 and oligogmeric plasicizers are found suitable to produce highly oriented films (Asrar and Pierre, 2000b).

6.3
Physical and Mechanical Properties of Biopol®

Poly(3HB) can range from a predominantly amorphous material (less than 30% crystallinity) to a highly crystalline (greater than 80%) material, depending on its history. The density in the amorphous state is 1.177 g cm^{-3} and in the crystalline state it is in the range of 1.23–1.26 g cm^{-3} (Brandl et al., 1988; Waddington, 1994). The mechanical properties of Poly(3HB) decrease significantly below a molecular weight around 400 kDa and the material is quite brittle at around 200 kDa (Cox, 1994).

Poly(3HB-co-x%3HV) crystallinity is typically in the 39–69% range depending on the percentages of the respective monomers (Poirer et al., 1995). The density in the amorphous state is 1.16 g cm^{-3} (Waddington, 1994) and in the crystalline state is around 1.2 g cm^{-3} (Poirer et al., 1995). The mechanical properties of Poly(3HB-co-x%3HV) can also be varied by changing the percentages of the respective monomers. The higher the percentage of 3HV, the less crystalline and the more elastic the polymer becomes. Some thermal and mechanical properties are presented in Table 3. A study of the thermal characteristics in vivo has been published (Gogolewski et al., 1993), and a mechanical evaluation in vivo and in vitro has also been published (Miller and Williams, 1987).

Barrier properties are generally measured in terms of moisture vapor transmission rate (MVTR) and oxygen transmission rate (OTR). The MVTR appears to be fairly constant as a function of the 3HV content for Poly(3HB-co-3HV) copolymers and the OTR may increase somewhat with increasing 3HV level. Introducing orientation decreases OTR significantly, improving the barrier to oxygen. In general, the OTR for cast (unoriented) Poly(3HB-co-3HV) is less than (i.e. superior to) that for low-density polyethylene, polypropylene and polyvinylchloride with MVTR similar to polyvinylchloride and approaching low-density polyethylene.

7
Block Copolymers Using Hydroxy-terminated PHAs

Poly(3HB) can be degraded by acid-catalyzed methanolysis to form low-molecular-weight (average DP = 26), stereoisomerically pure Poly(3HB) chains. The hydroxyl terminus of these polymers was reacted with AlEt$_3$ to form a Poly(3HB)-O-AlEt$_2$ macroinitiator species. These macroinitiators were then used to carry out the ring-opening polymerizations of e-caprolactone, and lactide monomers to prepare Poly(3HB)–PCL, Poly(3HB)–D,L-PLA and Poly(3HB)-L-PLA A–B diblock copolymers of variable chain segment lengths (Reeve et al., 1993).

Poly(3HB-co-3HV)diol with M_n of 2300 and T_m of 140 °C could be prepared by transesterification reaction between baitenially produced Poly(3HB-co-3HV) and ethylene glycol using dibutyltin dillarate as a catalyst. The oligomers were considered well suited for the preparation of high-molecular-

Tab. 3 Effect of 3HV content on the mechanical properties of Biopol®

Polymer	Molecular weight (kDa)	T_g (°C)	T_m (°C)	T_d (°C)	H_f (J g^{-1})	T_s (MPa)	T_s (MPa)	Flexural modulus (MPa)	Elongation at yield (%)	Elongation at break (%)
Poly(3HB)	370	1	171	252	51	36	2500	28.50	2.2	2.5
Poly(3HB-co-7%3HV)	450	−1	160	243	32	24	1400	1600	2.3	2.8
Poly(3HB-co-11%HV)	529	2	145	235	12	20	1100	1300	5.5	17
Poly(3HB-co-22%HV)	227	−5	137	251	7	16	620	750	8.5	36

T_g: Glass transition temperature; T_m: Melting temperature; T_d: Decomposition temperature; H_f: Heat of fusion; T_s: Tensile strength.

weight block copolymers by chain extension (Hirt et al., 1996).

Poly(3HB-*co*-3HV) diol of higher molecular weight have also been chain extended to produce block copolymers (Asrar et al., 1999).

The Poly(3HB-*co*-4%3HV)-diol just described (Herring and Webb, 1990) was used as the 'hard segments' and PCL-diol or a Pad-*alt*-1,4-BD-DEG-EG-diol used as 'soft segments' with TMDI 2,2,4-trimethylhexamethylene diisocyanate or LDI (S)-2,6-diisocyanate methylcaproate (L-lysinemethylester diisocyanate) for synthesis of high-molecular-weight micro-phase-segregated block copolyesterurethanes (Hirt et al., 1996). The Young's modulus (between 40 MPa and 1.3 GPa) was found to depend directly on the fraction of crystallizable Poly(3HB-*co*-4%3HV)-diol in the block copolymer, while the type of non-crystallizable segment or diisocyanate had only a minor influence. Generally, the tensile strength increases and the elongation at break decreases with increasing content of the Poly(3HB-*co*-4%3HV)-diol. The chain length of the non-crystallizable segment indirectly influences the morphology and the mechanical properties of the polymers through changes in phase-segregation behavior. The block copolymers can be used for degradable implant materials that can be sterilized, and the time in use varied between several days and several years by selection of the 'soft segment' (Hirt et al., 1996). The random hydrolytic cleavage of the amorphous part of these polymers might result, *in vivo*, in the production of small crystalline particles of low-molecular- weight Poly(3HB) that could then undergo phagocytosis and biodegradation inside phagosomes. It was shown that macrophages are able, *in vitro*, to phagocytize and degrade the Poly(3HB)-diol segments, as degradation products were found in extracts of cell supernatants after 8 days incubation (Ciardelli et al., 1995).

8
Melt Processing of Biopol®

Biopol® can be melt processed on conventional equipment as their melting points cover a range similar to other thermoplastics including polyethylene and polypropylene. A range of fabrication processes have been demonstrated including extrusion, injection molding, extrusion blow molding, thermoforming, oriented and unoriented cast and blown films, mono- and multifilament spinning, coating, calendering, and foaming (Darnell et al., 1986).

Solid-state extrusion of Poly(3HB-*co*-3HV) has been carried out at temperatures below the T_m (e.g. 135–150 °C), depending on the 3HV content. The solid-state extrudates showed an extra melting endotherm about 15–20 °C above the normal T_m, which became increasingly dominant at lower extrusion temperatures. The solid-state extruded samples did not show significant chain orientation along the extrusion direction. When the extrusion temperature was raised closer to the T_m, the quality of the extrudates improved, as reflected in the mechanical properties (Wang et al., 1996).

8.1
Molding

Injection molding can be carried out using a conventional injection molding machine with a polyethylene-type screw of length/diameter (L/D) ratio = 20:1. The shot capacity should be close to the weight of the molding to avoid long residence time in the barrel and consequent thermal degradation. Mold heating of 60 °C is required for rapid crystallization and acceptable cycle time (Darnell et al., 1986).

Extrusion blow molding can be carried out on a conventional single or multi-head machine, with a polytheylene-type screw of L/D ratio=24:1 and mixing tip. A compression zone after the spider avoids problems such as weld lines. Mold and blow pin head temperatures of 60 °C are required for rapid crystallization and acceptable cycle time (Darnell et al., 1986).

8.2
Extrusion

Free-standing Poly(3HB-*co*-3HV) films have been obtained by coextruding Poly(3HB-*co*-3HV) between two layers of sacrificial polymer (e.g. polyolefins), stretching, and orienting the multilayer film, and then stripping away the polyolefin layers after the Poly(3HB-*co*-3HV) has had time to crystallize. The Poly(3HBV-*co*-3HV) film can be laminated to other films (e.g. water-soluble films) for applications (Martini et al., 1989).

In another approach, multilayer films are coextruded where Poly(3HB-*co*-3HV) is the internal layer surrounded by outer layers of biodegradable films for use as diaper films. The external layers are not stripped away, but remain as an integral part of the biodegradable multilayer film (Wnuk et al., 1994).

In another process, Poly(3HB-*co*-3HV) compounded with plasticizers, nucleating agents, and/or other additives is extruded on a preformed supportive bubble of EVA resin or low-density polyethylene (De Micheli et al., 1996)

It has recently been found that Poly(3HB-*co*-3HV) can produce stable bubble during the film blowing process which could be continuously stretched to produce oriented films with excellent properties provided the molecular weight of the polymer is in the range of 490–700 kDa (Asrar and Pierre, 2000a). Below 400 kDa the melt strength is too low creating an unstable bubble and above 700 kDa the melt viscosity is too high and the material is too stiff for the preparation of oriented film in a continuous process. Oriented films produced by such a process are found to have much better ductility and tensile properties compared to unoriented film. The increase in molecular weight does not effect the elongation at break of stretched films, but it increases the tensile strength at break. Monoriented films tested after 90 days did not show any sign of aging, while unstretched films are known to produce a large reduction in elongation with time due to secondary crystallization. All the polymers in the Table 4 contain 1 p.h.r. of BN and 10 p.h.r. of acetyltributylcitrate plasticizer.

8.3
Melt Coating of Biopol® on Paper and Paperboard

Paper and paperboard laminated with Poly(3HB-*co*-3HV) are manufactured by continuously casting a melt of Poly(3HB-*co*-3HV) to the surface of paper or board while attaching a release film to the polyester surface which can be removed after cooling (Kazuya et al., 1994).

It is important from a commercial standpoint to develop compositions and methods

Tab. 4 Effect of molecular weight on tensile properties of Biopol® (8% C5) films

Molecular weight (kDa)	Elongation (oriented/ unoriented) (%)	Tensile strength (oriented/ unoriented) (MPa)
796	90/34	110/27
625	80/15	108/26
525	80/11	62/22
390	unstable – bubble, no continuous orientation	

for cost-effective extrusion operations which to not depend on the use of a sacrificial non-PHA polymeric layer, and which provide PHA films and coatings having good physical and mechanical properties. Neck-in and draw resonance, periodic variation in cross-section of the polymer web eventually leading to web rupture, are the two major issues which need to be controlled in a high-speed coating process. It is also important to have good adhesion of the polymer to the substrate. Considerable effort is devoted to minimizing the neck-in and draw resonance while improving the adhesion to the substrate.

It is found that branched PHA, produced by reactive extrusion using small amounts of peroxide, are very suitable for coating on paper and board (Asrar and D'Haene, 2000). Branched PHAs produced from such a process also have improved physical, mechanical, and aging characteristics, relative to those produced from unmodified linear PHAs.

It was reported recently (Asrar et al., 2000b) that adhesion of the Biopol® to the substrate paper and the seal strength are dependent on the molecular weight of the Biopol® coating. Critical molecular weight for good heat sealing properties was in the range of 250–150 kDa (Asrar et al., 2000b). To produce such Biopol® coating, the molecular weight of the Biopol® pellets needs to be in the range of 350–250 kDa. Poly(3HB-co-3HV) with a molecular weight higher than 350 kDa generally produced coatings of considerably lower molecular weight and poor quality.

8.4
Melt Spinning of Fibers

Successful melt spinning of Poly(3HB) and Poly(3HB-co-3HV) copolymers requires careful control of crystallinity during the drawing step. The polymer and copolymers do not strain crystallize, so fibers cannot be drawn from the amorphous melt. Thus careful control of time and temperature to produce the correct level of crystallinity for successful orientation has enabled oriented fibers to be produced with increased tenacity and extension to break. The orientation achieved depends on the level of crystallinity (i.e. a window for orientation exists), the 3HV content (in general a higher 3HV content material which crystallizes more slowly will pass through the orientation window more slowly facilitating orientation), and draw speeds (Katsuhiko, 1994). Nucleating agents can be added to the Poly(3HB) to increase the crystallization rate and decrease residence time in processing steps (Holmes, 1984). Monofilament of 3 g d^{-1} tenacity (317 MPa tensile strength) has been reported (Mochizuki et al., 1994a,b).

Poly(3HB-co-8%3HV) was melt spun, to give a biodegradable fiber with breaking strength 310 MPa and knot strength 290 MPa (Yamamoto et al., 1995, 1997). Wide-angle X-ray spectroscopy studies showed the fibers possess bimodal chain orientation with two populations of crystals. In the first population the chain axes are oriented primarily along the fiber axis and in the second population the chain axes are oriented in the transverse direction to the fiber axis. At the annealing temperature of 100 °C, the relative population of the crystals oriented normal to the fiber axis was found to be large and their proportion decreases with an increase of annealing temperature. The tensile strength of these fibers improved by the increase of the proportion of the transversely oriented chains (Yamamoto et al., 1997).

PCL and Poly(3HB-co-3HV) were cospun in a 1:1 weight ratio, to give a 0.31 mm bicomponent monofilament with tensile strength 703 MPa, breakage temperature

112 °C, and good biodegradability (Mochizuki et al., 1993a).

Biodegradable composite fibers can be prepared from a core of Poly(3HB-*co*-3HV) and a sheath of polybutylene succinate or polyethylene succinate ($T_m = 70$ °C) (Yamada et al., 1995).

8.5
Multifilament

Biodegradable multifilaments of Poly(3HB-*co*-3HV) copolymers with tensile strength above 2.0 g d^{-1} have been prepared by melt spinning at 140–220 °C, cooling by air to 40–80 °C, and drawing greater than 1.2 times in one or more steps. For example, Poly(3HB-*co*-6%3HV) containing 1 part BN and 10 parts glycerol triacetate (optional) was melt spun at 180 °C though a 0.3 mm × 36 spinning nozzle, cooled to 60 °C, lubricant added, and drawn between rollers at 100 °C and unheated rollers to give multifilament with 200 denier/36 filaments at 10 to 500 m min^{-1} and good biodegradability. Biodegradability was determined by burying a sample in soil for 2 months, if the multifilament-filament lost its shape or 50% of its original tensile strength, biodegradability was considered good (Mochizuki et al., 1994b).

In a similar set of experiments, Poly(3HB-*co*-6%3HV) (molecular weight 750 kDa) blended varying amounts of PCL (molecular weight 80 kDa) was spun to give 200 denier/36 multifilaments with good biodegradability. At 20% PCL, 1 wt% BN, 10 wt% GTA, and a draw ratio of 7.0 the tenacity was 5.8 g d^{-1} (Mochizuki et al., 1994a).

8.6
Non-wovens

Heat-adherable biodegradable bicomponent fibers were prepared by cospinning a 1:1 mixture of PCL and Poly(3HB-*co*-3HV) at 260 °C, winding, drawing twice, crimping, lubricating, drying, and cutting to give 5 denier staple fibers with tensile strength 473 MPa. The fibers were formed into a web and heated at 70 °C to give a non-woven with tensile strength 2800 g cm^{-1} and good biodegradability (Mochizuki et al., 1993b).

UV-degradable, biodegradable composites have been made consisting of a non-woven blend fabric of 50% crimped composite fibers with Poly(3HB-*co*-3HV) as sheath and PET containing 3.5% TiO$_2$ as core at sheath/core at 50/50 and 50% rayon; the rayon decomposed extensively when left in the sun for 1 month and then buried in soil for 4 months (Sosa and Koizumi, 1994).

8.7
Coated Non-wovens

Biodegradable fiber webs can be produced from biodegradable fibers and biodegradable thermoplastic binders. For example, spreading 3% of a biodegradable Poly(3HB-*co*-3HV) powder ($T_m = 160$ °C) on a rayon fleece (10 g m^{-2}) and hot-roll pressing at 170 °C and 50 kg cm^{-1} gave a biodegradable non-woven (Takai, 1994).

Biodegradable paper-based laminates can be produced by coating a paper substrate with PCL, drying, extrusion coating with Poly(3HB-*co*-3HV), and aging for 1 week to give a laminate with good interlayer adhesion for making cups with good biodegradability (Taniguchi et al., 1994a). Laminates useful for cups, trays, and cartons may also be prepared by coating a paper substrate with a polyester adhesive (e.g. THW 3257), laminating with Poly(3HB-*co*-3HV), and aging for 1 week to give a laminate with good adhesion which is degradable in soil (Taniguchi et al., 1994b). See Figure 5.

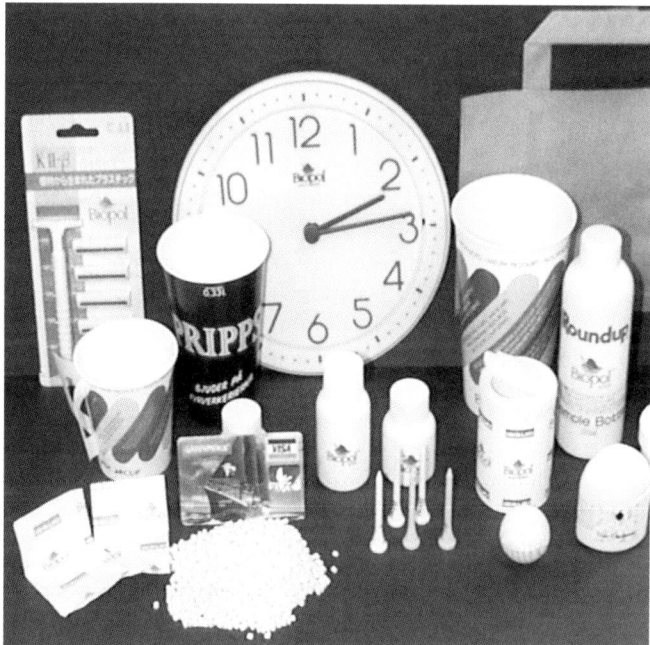

Fig. 5   Different products made from Biopol®

9
Biodegradation of Biopol® and Related PHAs

PHAs biodegrade on disposal into a microbially active environment. Biodegradation has been demonstrated in aerobic and anaerobic sewage, estuarine and marine sediment, lake, pond and sea water, soil compost, and simulated managed landfill. Many bacteria, streptomycetes, and molds have been isolated that degrade PHAs (Mergaert et al., 1993). The rate of biodegradation depends on environmental factors such as moisture level, nutrient supply, temperature, and pH, and on material parameters such as comonomer content, initial molecular weight, degree of crystallinity, surface, and formulation (e.g. plasticizers) (Cox et al., 1992). These material parameters can be controlled/modified within certain limits to optimize biodegradation rates for different applications (Cox et al., 1994).

The biodegradation rate of Poly(3HB-co-3HV) copolymers tends to be faster than the Poly(3HB) homopolymer (e.g. around 2 versus 6 weeks in anaerobic sewage), but the relative rate depends on the environment (Cox et al., 1992).

The rate of biodegradation of Poly(3HB-co-3HV) has been reported not to be affected significantly by the surface area or thickness of the samples, but is affected significantly by the crystallinity (Kim et al., 1995a). Crystalline domains are needed for binding of the depolymerase enzymes. However, highly crystalline PHAs biodegrade very slowly, while amorphous PHAs do not biodegrade at all.

Biodegradation appears to occur by colonization of the PHA surface by bacteria, which secrete an extracellular depolymerase that degrades the polymer to primarily monomer units in the vicinity of the cell. The soluble degradation products are then absorbed through the cell wall and metabolized. Apart from a small amount of biological material, the final products of biodegradation are CO_2 and water in aerobic conditions and methane and CO_2 in anaerobic conditions. The surface erosion mechanism is supported by observations of a gradual decrease in sample thickness with time and constant molecular weight of the residual sample during biodegradation (Cox et al., 1992).

The kinetics and mechanism of surface hydrolysis of Poly(3HB) film was studied at 25–37 °C and pH 6–8 with an extracellular PHB depolymerase purified from *Alcaligenes faecalis* T1. The primary product of enzymatic hydrolysis was the dimer of hydroxybutyric acid (HBA), which was subsequently hydrolyzed to monomer. The heat of enzyme adsorption on the Poly(3HB) surface was 43 kJ mol^{-1}, indicating a strong interaction between the enzyme binding domain and the hydrophobic surface of Poly(3HB). The activation energy of the enzyme-catalyzed hydrolysis was found to be 82 kJ mol^{-1} (Kasuya et al., 1995).

Enzymatic degradation of PHAs with different chain lengths (C4–C10) produced from various carbon substrates by *R. eutropha* or *P. oleovorans* were studied at 37 °C in a phosphate buffer (pH 7.4) containing the PHA depolymerase from *A. faecalis* T1. The rate of degradation, as determined by time-dependent changes in weight loss of films, was strongly dependent upon the composition of the polyesters and markedly decreased with an increase in the side chain length of the 3HA monomeric units. The polyester chains were finally degraded into the monomers and dimers of 3HA acids by the PHA depolymerase (Kanesawa et al., 1994).

10
Biocompatibility of PHAs

The degradation of Poly(3HB) produces D-3-hydroxybutyric acid which is a normal constituent of human blood. Work at the Middlesex Hospital in England, has shown that this material, which can be obtained by simple hydrolysis of the polymer, can be used as an intravenous or oral carbon supply and has a number of clinical advantages over the more commonly used glucose drip. It is reported that obese patients in this study, who were undergoing therapeutic starvation for 14 days, never complained of hunger (Holmes, 1985).

There is evidence for both enzymatic and hydrolytic degradation *in vivo* (Gogolewski et al., 1993). *In vitro* studies (Holland et al., 1987; Bailey et al., 1990) suggest that Poly(3HB) and Poly(3HB-*co*-3HV) copolymers degrade by hydrolysis in a multistage process where the majority of the molecular weight loss occurs before any significant mass loss. As 3HV content increases in Poly(3HB-*co*-3HV) copolymers, the percentage of amorphous regions which are readily attacked by hydrolytic degradation increases, thereby increasing degradation rates. In addition, elevated temperatures and alkaline conditions have been shown to increase degradation rates.

11
Applications of Biopol®

11.1
Fishing Lines and Nets

Monofilaments useful for fishing nets, ropes, and marine agriculture are produced from Poly(3HB-*co*-3HV) having a molecular weight around 400 kDa by a process in which resin is molten for less than 6 min at temperatures about 15 °C higher than the T_m of the polymer, spun at a shear rate above 2×10^{-2} s^{-1} above the filter layer in the spinneret pack and has a residence time in the pack less than 2 min to give monofilaments, which are solidified at T_c around 10 °C for 5–10 s and stretched (Inagaki et al., 1996).

In another example, drawn fibers of Poly(3HB-*co*-8%3HV) were twisted to obtain ropes and used in nets for crab cages. The nets showed good strength and biodegradability in the sea (Sim et al., 1997).

The tensile strength of Poly(3HB-*co*-3HV)s increases with decreasing 3HV content. Poly(3HB-*co*-3HV)s with (0–10%) 3HV exhibit tensile strength comparable to low-density polyethylene and polypropylene. Orientation further increases this property. Fishing nets require good tensile strength. The relatively slow biodegradation rate of Poly(3HB-*co*-3HV)s in sea water permits a useful lifetime of articles such as fishing nets. On disposal, or loss, the Poly(3HB-*co*-3HV) net (Poly(3HB-*co*-3HV) density around 1.2) sinks to the sea bed where more rapid biodegradation is expected to occur in the more microbially active silt (Cox, 1994).

Studies of PHA films, both in sea and freshwater, by scanning electron microscopy show rapid colonization by bacteria and microalgae, including several species of diatoms. Very few hyphae were seen on the films even after more than 30 days in water, which contrasts with previous studies in soil where fungal hyphae were far more numerous. Signs of decomposition, such as etched areas associated with bacteria, were found. Microalgae, although frequently observed, appeared to use the PHA films as a growth support rather than degrading it. Degradation was faster in aquatic medium than in soil (Lopez-Llorca et al., 1994).

To take advantage of their respective properties, high-density polyethylene as the core and Poly(3HB-*co*-3HV) as the sheath

were melt spun at a 60:40 v/v ratio to give fibers with tenacity 5.4 g d^{-1} and exhibiting very small amount of adhesion of marine substances on immersion of a net of the fibers in seawater for 6 months (Brandl et al., 1988).

11.2
Other Marine Applications

A biodegradable net of spun monofilaments of Biopol® coated with PVOH was used as a host for growing seaweed from seedling. Seaweed survival and tensile strength retention were 60 and 70%, respectively, after 3 months outdoors versus 15 and 32%, respectively, for the uncoated Biopol® (Hamada, 1997).

Blends of PCL and Poly(3HB-*co*-3HV) have good anti-algae properties and are useful as nets for seafood cultivation (Suzuki et al., 1996).

11.3
Biomedical Applications

Dry and melt spun non-woven Poly(3HB) fibers are suitable for medical applications. Poly(3HB) and Poly(3HB-*co*-3HV) can be steam sterilized without losing properties and are resistant to alcohol (Kledzki et al., 1994). If sterilized, they may be left in place to aid blood clotting without the rejection problems associated with cotton materials. Swabs, pads, or other articles made from Poly(3HB) and left in the body by design or by accident will not cause toxemia and are slowly absorbed by the body. Being hydrophilic they will take up aqueous liquids. They differ from cotton wool in having little or no tendency to break off small fibers, but even if small pieces were to enter a wound, they would be safe. There is no need to enclose them in a retaining gauze and they can be readily tailored to size at the point of use.

Poly(3HB) is very biocompatible, producing an exceptionally mild foreign body response. A monofilament surgical suture degrades very slowly (Miller and Williams, 1987) and could require several years to be totally resorbed by the body. The time required for biodegradation is related to surface area and multifilament sutures resorb much more quickly (Holmes, 1985).

Monofilaments of Poly(3HB) and Poly(3HB-co-3HV) were studied *in vivo* and *in vitro*, and assessed for changes in mechanical properties and topography. *In vivo* biodegradation was observed only with Poly(3HB) when pre-degraded by 10 Mrad of γ-irradiation before implantation. High-temperature *in vitro* hydrolysis suggested that 3HV copolymer addition retarded the rate of degradation of Poly(3HB). Hydration reactions had the most effect on the ultimate properties of the materials. In contrast, the elastic properties appeared to be relatively unaffected (Miller and Williams, 1987). Thus it was concluded that neither Poly(3HB) nor Poly(3HB-co-3HV) in monofilament form is very biodegradable, although susceptibility to degradation may be increased by exposure to gamma radiation (Williams and Miller, 1987).

Levels of cell attachment on gel-spun Poly(3HB) fibers were extremely low. Cell viability pre- and post-testing was 90%. On alkali-treated fibers, cells exhibited full spread morphologies, and high-level attachment and subsequent adhesion possibly due to polymer degradation and creation of new surface (Davies and Tighe, 1995).

Poly(3HB) has been gel-spun into a novel form with one possible application as a wound scaffolding device, designed to support and protect a wound against further damage, while promoting healing by encouraging cellular growth on and within the device from the wound surface. The nonwoven combines a large volume with a low mass and is called 'wool' because of its similarity in appearance to 'cotton wool'. The hydrolytic degradation of the wool was investigated in an accelerated model of pH 10.6 and 70 °C. The Poly(3HB) wool gradually collapsed during degradation which was characterized by a reduction in T_g, T_m, and a fusion enthalpy peak of maximum crystallinity (88%) which coincided with the point of matrix collapse (Foster and Tighe, 1994). An enzymatic assay of HBA monomers (HBA dehydrogenase converts NAD to NADH which is associated with an increase in light absorption at 340 nm) was used to monitor hydrolysis of gel spun Poly(3HB) fibers (Foster and Tighe, 1995).

Adsorbable polymeric scaffolds for cell culture and transplantation, tissular reconstruction, and *in vivo* drug or protein release have been proposed by salt leaching/solvent casting a Poly(3HB-co-3HV). Electron microscopy and mercury porosimetry show highly porous, well-interconnected microstructures with a porosity level of 0.85 and a mean pore diameter of 122 μm. The weight loss of the porous structures is about 50% for a 140-day period of hydrolytic degradation in phosphate-buffered solutions, pH 7.4 at 70 °C. Incubations from 1 to 35 days of canine anterior cruciate ligament fibroblasts in scaffolds have shown a limited proliferation rate (150×10^6 cells (g maximal d)$^{-1}$) but high protein synthesis ($2.4 \pm 0.1 \times 10^{-2}$ ng cell^{-1} day^{-1} at day 28). Additional work is underway on potential applications for orthopedic reconstructions (Chaput et al., 1995a).

The hemocompatibility of a material can be evaluated through the study of adsorption of proteins stimulating the thrombus formation. Films of Poly(3HB-co-3HV) were treated with perfluorohexane and H_2 plasmas. It was shown that hydrophobic polymer films adsorb more albumin than fibrogen. Moreover, the amounts of protein adsorbed seem to be mainly a function of the surface roughness of the films – the highest amounts being adsorbed on the rougher plasma-treated films, while the smoother perfluorohexane-treated surfaces adsorbed less proteins (Coussot-Rico et al., 1994).

Poly(3HB-*co*-3HV) films (7, 14, and 22% 3HV) were analyzed for *in vitro* cytotoxicity and aqueous accelerated degradation, *in vivo* degradation, and tissue reactions. The Poly(3HB-*co*-3HV) materials and extracts were found to elicit few or mild toxic responses, did not lead to *in vivo* tissue necrosis or abscess formation, but did provoke acute inflammatory reactions slightly decreasing with time. Poly(3HB-*co*-3HV) shows low rates of degradation *in vitro* as well as *in vivo*. The weight loss rate generally increases with the copolymer (3HV) content and ranges from 0.15 to 0.30 (*in vitro*) to 0.25% day^{-1} (*in vivo*). Compositional and physicochemical changes in Poly(3HB-*co*-3HV) materials were rapidly detected during the accelerated hydrolysis, but were much slower to appear *in vivo*. The structural and mechanical integrity on the Poly(3HB-*co*-3HV) materials tend to disappear early *in vitro* as well as *in vivo*. After 90 weeks in dorsal muscular tissue of adult sheep, 50–60% of the initial weight of the Poly(3HB-*co*-3HV) polymer still remained (Chaput et al., 1995b)

Biocompatibility and degradation mechanisms of block copolyurethanes containing telechelic Poly(3HB) segments have found growing interest for possible biomedical applications for these materials. The random hydrolytic cleavage of the amorphous part of these polymers might result, *in vivo*, in the production of small crystalline particles of low-molecular-weight Poly(3HB) that could undergo phagocytosis and biodegradation inside phagosomes. Studies suggested the macrophages could degrade the low molecular weight Poly(3HB) (Ciardelli et al., 1995).

Resorption of Poly(3HB), PLA, PGA/C, and PDS sutures in sheep and rats was studied versus implantation site and species (Sim, 1997). Poly(3HB-*co*-3HV) resorbs in 6–12 months versus 2–3 weeks for PLA. Sutures of Poly(3HB-*co*-4HB) have been prepared and their properties studied (Kimura et al., 1994; Oota et al., 1994).

A review on synthesis and uses of Poly(3HB) and Poly(3HB-*co*-3HV) as biodegradable thermoplastics and biodegradable fibers and textiles as well as their potential use in medicine has been published (Babel et al., 1990).

12
Copolymers with Long Alkyl Chain and other Functionalities

PHAs can be produced as homopolymers or copolymers having the general structure shown in Structure.

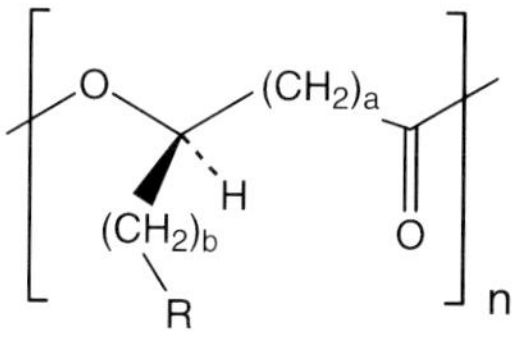

The non-specificity of the biosynthetic enzymes in polymer-accumulating organisms has proven to be an advantage offering the potential of synthesizing unusual polymers not readily synthesized chemically (Pouton and Akhtar, 1996). Thus the microbial PHA synthase enzymes have broad substrate ranges and are capable of incorporating a large number of hydroxyalkanoic acid monomers as constituents of biosynthetic PHA depending upon growth conditions, precursor substrate availability, and the source of the PHA synthase enzyme. Examples (Shi et al., 1995) of the diversity of repeat units incorporated into PHAs include β-hydroxypropionates with β-substituents that contain vinyl (Doi and Abe, 1990), nitrile (Kim et al., 1990, 1995; Kim, 1991b), phenyl (Fritzsche and Lenz, 1990), cyanophenoxy and nitrophenoxy (Kim et al., 1995b), and halogen (Abe et al., 1990; Lenz et al., 1990) functionalities. Also, 3-hydroxypropionate and 4-hydroxybutyrate repeat units have been found in PHAs produced using the bacterium *R. eutropha* (Kunioka et al., 1988; Nakamura et al., 1991). The diversity in composition of biosynthetic PHA polymers is underscored by the fact that at least 91 hydroxyalkanoic acid monomers have been identified as substrates for PHA synthases (Steinbüchel and Valentin, 1995).

Considerable effort has been directed to the production of PHAs with lower melting point, lower crystallinity, and high ductility by copolymerization with monomers containing longer alkyl chains both as a side group as well as in the main chain. Incorporation of monomers containing long alkyl side chain, e.g. 3-hydroxy hexanoate, lead to considerable change in bulk and solution properties. Increase in the hexanoate group in the copolymer leads to an increase in solubility, ductility and a decrease in the melting point, crystallinity, rate of crystallization, and the spherulite size. Interestingly, an increase in the hexanoate content in the copolymer leads to an increase in the activation energy for thermal degradation probably due to the difficulty in the formation of six-membered ring, the transition state for the βe-elimination process involved in random chain scission.

Although ductile polymers could be produced by copolymerizing with monomers containing long alkyl side chains, a good balance of properties was achieved when the long alkyl chain was part of the main chain (Kunioka et al., 1988; Nakamura et al.,

1991). Loss in ductility with time was not observed when 4-hydroxybutyrate was copolymerized with 3HB and 3HV (Madden et al., 2000).

Medium-chain-length (MCL)-PHAs have been classified as thermoplastic elastomers (Gagnon et al., 1992; Marchessault and Morin, 1992). However, their low T_m and low crystallization rate limit their applicability. A MCL-PHA grown on 75 vol% *n*-octane and 25 vol% 1-octene contained some unsaturated bonds corresponding to 3-hydroxyoctenoic acid and 3-hydroxyhexenoic acid. The PHA, after crosslinking by irradiation, became a true rubber with constant properties over a wide temperature range from T_g (around −15 to 27 °C depending on dose) up to the degradation temperature. The polymer is completely biodegradable (Molitoris et al., 1996) and is thought to be the first microbially produced biodegradable rubber (de Koning et al., 1994b). Production of a homopolyester with unsaturation, poly(3-hydroxy-4-pentenoic acid) (Poly(3HPE)), from a structurally unrelated carbon source, sucrose, was also reported recently (de Rodrigues et al., 2000). *Burkholderia* sp. DSMZ # 9243 was found to simultaneously accumulate two homopolyesters, Poly(3HB) and Poly(3HPE), when sucrose was used as the carbon source. Poly(3HPE) was easily separated from Poly(3HB) due to its high solubility in THF and other organic solvents compared to Poly(3HB). The presence of unsaturated pendent groups in Poly(3HPE) is found to considerably lower the degree of crystallinity and the melting point compared to the fully saturated analogue Poly(3HV) without significantly effecting the T_g. Presence of unsaturation in the polymer also provides opportunity for crosslinking and further polymer modification. Purified Poly-(3HPE) was crosslinked by UV radiation and subjected to epoxidation using 3-chloroperoxybenzoic acid. Introduction of epoxides into the 3HPE homopolyester was found to increase the glass transition temperature of the resulting polymer (Valentin et al., 1999b).

13
Patents Describing Production and Properties of Biopol® and Related PHAs

As has been demonstrated in this chapter there is a large volume of scientific literature available on Biopol® that cover its biosynthesis, properties, and applications. In view of this it is not surprising that there are also a number of patents on this copolymer and related PHAs. Primary among these are patents on composition of matter, properties and applications, and production in microbial or plant systems. Table 5 is a listing of various patents related to Biopol®. It is by no means extensive, but attempts to highlight those patents that are key to technologies important to commercial production and applications.

Tab. 5 Major patents regarding Biopol®

Patent number	Holder	Inventor	Title	Date of publication
WO 94/11519	Monsanto/Zeneca	P. A. Fentem	Production of polyhydroxyalkanoates in plants	26/5/94
US 5502273	Monsanto/Zeneca	P. A. Fentem	Production of polyhydroxyalkanoates in plants	26/3/96 16/1/01
US 6175061				
US 5650555	Michigan State University	C. R. Somerville et al.	Transgenic plants producing polyhydroxyalkanoates	22/7/97
US 5610041	Michigan State University	C. R. Somerville et al.	Processes for producing polyhydroxybutyrate and related poly-hydroxyalkanoates in the plastids of higher plants	11/3/97
WO 98/00557	Monsanto	K.J. Gruys et al.	Methods of optimizing substrate pools and biosynthesis of poly-β-hydroxybutyrate-co-poly-β-hydroxyalerate in bacteria and plants	8/1/98 24/8/99
US 5942660				
WO 00/52183	Monsanto	T. A. Mitsky et al.	Multigene expression vectors for the biosynthesis of products via multienzyme biological pathways	8/9/00
US 6043063	Monsanto	D. L. Kurdikar et al.	Methods of PHA extraction and recovery using non-halogenated solvents	3/28/00
US 5750848	Monsanto	N. Kruger et al.	DNA sequence useful for the production of polyhydroxyalkanoates	5/12/98
US 6087471	Monsanto	D. L. Kurdikar et al.	High temperature PHA extraction using PHA-poor solvents	7/11/00
US 5627276	Monsanto/Zeneca	W. Greer	Macromolecular degradation of nucleic acids	5/6/97
US 5550173	Monsanto/Zeneca	T. Hammond et al.	Polyester composition and oligomers useful as components thereof-polymer	8/27/96
US 6111006	Monsanto/Zeneca	D. S. Waddington	Process for preparing films and coatings	7/7/97
US 4477654	Monsanto/Zeneca	P. A. Holmes et at.	3-Hydroxybutyrate polymers	10/16/84
US 5061743	Monsanto/Zeneca	J. M. Herring et al	3-Hydroxybutyrate polymer composition	10/29/91
US 5364778	Monsanto/Zeneca	D. Byrom	Copolymer production	11/15/94
US 5516825	Monsanto/Zeneca	H. S. Montador et al.	Biodegradable polyester plasticizer	5/14/96
US 5599891	Monsanto/Zeneca	D. M. Horowitz et al.	Polymer composition comprising a crystallisable	2/4/97
US 5622847	Monsanto/Zeneca	R. H. Cumming et al.	Process for separation of solid materials	4/22/97
US 5534616	Monsanto	D. S. Waddington	Polyhydroxyalkanoates and film formation there from	7/19/96
US 5578382	Monsanto/Zeneca	D. S. Waddington	Polyhydroxyalkanoates and film formation there from	11/26/96
US 5753782	Monsanto/Zeneca	T. Hammond et al.	Polyester composition	5/19/98
US 4910145	Monsanto/Zeneca	P. A. Holmes et al.	Separation process	3/20/90

Tab. 5 (cont.)

Patent number	Holder	Inventor	Title	Date of publication
US 6080562	Monsanto/Zeneca	D. Byrom et al.	HV/HB copolymer production	6/27/00
US 5894062	Monsanto	J. M. Liddell	Process for the recovery of polyhydroxyalkanoic acid	4/13/99
US 5891936	Monsanto	J. M. Liddell et al.	Production of polymer composition	4/6/99
US 5977250	Monsanto	N. George et al.	Latex of polyhydroxyalkanoate	11/2/99
WO 9631347	Monsanto/Zeneca	S. D. Waddington et al.	A method for preparing a compostible laminate with polycaprolactone and a poly 3-hydroxyalkanoic acid	10/10/96
WO 9721762	Monsanto	N. George	Dispersions of polyhydroxyalkanoates in water	6/19/97
US 5798440	Monsanto	J. M. Liddell et al.	Increasing the particle size of polymers	8/25/98
WO 9515260	Monsanto/Zeneca	J. D. Kemmish et al.	Multilayer biodegradable film	6/8/95
US 6091002	Monsanto	J. Asrar et al.	Polyhydroxyalkanoates of narrow molecular weight distribution prepared in transgenic plants	7/18/00 8/15/01
US 6228263				
EP 0996670A1	Monsanto	J. Asrar et al.	PHA compositions and methods for their use in the production of PHA films	5/3/00
US 6127512	Monsanto	J. Asrar et al.	Plasticized polyhydroxyalkanoate compositions and methods for their use in the production of shaped polymeric articles	10/3/00
US 6191203	Monsanto	J. Asrar et al.	Polymer blends containing polyhydroxyalkanoates and compositions with good retention of elongation	2/20/01
US 5994478	Monsanto	J. Asrar et al.	Hydroxy-terminated polyhydroxyalkanoates	11/30/99 12/5/00 3/20/01
US 6156852				
US 6204341				
US 5502273	Monsanto/Zeneca	S. W. J. Bright et al.	Production of polyhydroxyalkanoates in plants	3/26/96
US 6025028	Monsanto	J. Asrar et al.	Polyhydroxyalkanoate coatings	2/15/00
US 6096810	Monsanto	J. Asrar et al.	Modified polyhydroxyalkanoates for production of coatings and films	8/01/00 3/13/01
US 6201083				
US 5973100	Monsanto	J. Asrar et al.	Nucleating agents for polyhydroxyalkaroates and other thermoplastic polyesters and their production	2/4/99
EP 0304293A2	Mitsubishi Kasei Corporation	Y. Doi	Copolyester and process for producing the same	2/22/88

14
References

Abe, C., Taima, Y., Nakamura, Y., Doi, Y. (1990) New bacterial copolyester of 3-hydroxyalkanoates and 3-hydroxy-ω-fluoroalkanoates produced by *Pseudomonas oleovorans, Polym. Commun.* **31**, 404–406.

Akhtar, S., Pouton, C., Notarianni, L. (1992) Crystallization behavior and drug release from bacterial poly (hydroxyalkanoates), *Polymer* **33**, 117–126.

Anderson, A. J., Dawes, E. A. (1990) Occurrence, metabolism, metabolic role, and industrial uses of bacterial polyhydroxyalkanoates, *Macro. Rev.* **54**, 450–472.

Asrar, J., Pierre, J. R. (1999b) Nucleating agents for polyhydroxyalkanoates and other thermoplastic polyesters and methods for their production and use (Monsanto Co.), US patent 5 973 100.

Asrar, J., Shah, D. T., Tran, M. (1999) Hydroxy-terminated polyhydroxyalkanoates (Monsanto Co.), US patent 5 994 478.

Asrar, J., D'Haene, P. (2000) Modified poly-hydroxyalkanoates for production of coatings and films (Monsanto Co.), US patent 6 096 810.

Asrar, J., Pierre, J. R. (2000a) Poly(hydroxyalka-noate) compositions and method of their use in the production of films (Monsanto Co.), European patent 0996670A1.

Asrar, J., Pierre, J. R. (2000b) Plasticized poly-hydroxyalkanoate pellet compositions and methods for their use in the production of shaped polymeric articles (Monsanto Co.), US patent 6127512.

Asrar, J., Mitsky, T. A., Shah, D. T. (2000a) Polyhydroxyalkanoates of narrow molecular weight distribution prepared in transgenic plants (Monsanto Co.), US patent 6 091 002.

Asrar, J., Pierre, J. R., D'Haene, P. (2000b) Poly-hydroxyalkanoate coatings (Monsanto Co.), US patent 6 025 028.

Asrar, J., Shah, D. T., Tran, M. (2000c) Hydroxy-terminated polyhydroxyalkanoates (Monsanto Co.), US patent 6 156 852.

Asrar, J., D'Haene, P. (2001a) Polymer blends containing polyhydroxyalkanoates and composi-tions with good retention of elongation (Mon-santo Co.), US patent 6 191 203.

Asrar, J., D'Haene, P. (2001b) Modified poly-hydroxyalkanoates for production of coatings and films (Monsanto Co.), US patent 6 201 083.

Asrar, J., Mitsky, T. A., Shah, D. T. (2001a) PHAs of narrow molecular weight distribution in transgenic plants (Monsanto Co.), US patent 6 228 263.

Asrar, J., Shah, D. T., Tran, M. (2001b) Hydroxy-terminated polyhydroxyalkanoates (Monsanto Co.), US patent 6 204 341.

Babel, W., Riis, V., Hainich, E. (1990) Microbial thermoplastics: biosynthesis, properties, and application, *Plaste Kautsch* **37**, 109–115.

Bailey, W. J., Kuruganti, V., Angle, J. S. (1990) Biodegradable polymers produced by free-radical ring-opening polymerization, *ACS Symp. Ser.* **433**, 149–160.

Baptist, J. N. (1962a) Poly(3-hydroxybutyric acid) (W. R. Grace & Co.), US patent 3 036 959.

Baptist, J. N. (1962b) Poly(3-hydroxybutyric acid) (W. R. Grace & Co.), US patent 3 044 942.

Baptist, J. N. (1965) Method of making absorbable surgical sutures from poly beta hydroxy acids (W. R. Grace & Co.), US patent 3 225 766.

Barham, P. J. (1984) Nucleation behavior of poly-3-hydroxy-butyrate, *J. Mater. Sci.* **19**, 3826–2834.

Barnard, G. N., Sanders, J. K. M. (1989) The poly-beta-hydroxybutyrate granule *in vivo*. A new insight based on NMR spectroscopy of whole cells, *J. Biol. Chem.* **264**, 3286–3291.

Biddlestone, F., Harris, A., Jay, J. N. (1996) The physical aging of amorphous Poly(hydroxybuty-rate), *Polym. Int.* **39**, 221–229.

Billingham, N. C., Henman, T. J., Holmes, P. A. (1978) Degradation and stabilization of polyesters of biological and synthetic origin, in: *Developments in Polymer Degradation* (Grassie, N., Ed.), Amsterdam: Elsevier, 81–121, Vol. 7.

Bluhm, T. L., Hamer, G. K., Marchessault, R. H., Fyfe, C. A., Veregin, R. P. (1986) Isodimorphism in bacterial Poly(β-hydroxybutyrate-*co*-β-hydroxyvalerate), *Macromolecules* **19**, 2871–2876.

Bonthrone, K. M., Clauss, J., Horowitz, D. M., Hunter, B. K., Sanders, J. K. M. (1992) The biological and physical chemistry of polyhydroxyalkanoates as seen by NMR spectroscopy, *FEMS Microbiol. Rev.* **103**, 269–277.

Brandl, H., Gross, R. A., Lenz, R. W., Fuller, C. W. (1988) *Pseudomonas oleovorans* as a source of Poly(β-hydroxyalkanoates) for potential applications as biodegradable polyesters, *Appl. Environ, Microbiol.* **54**, 1977–1982.

Broun, P., Boddupalli, S., Somerville, C. (1998) A bifunctional oleata 12-hydroxylase: desaturase from *Lesquerella fendleri*, *Plant J.* **13**, 201–210.

Chaput, C., DesRosiers, E. A., Assad, M., Brochu, M., Yahia, L., Selmani, A., Rivard, C. (1995a) Processing biodegradable natural polyesters for porous soft materials, *NATO ASI Ser. E.* **294**, 229–245.

Chaput, C., Yahia, L., Selmani, A., Rivard, Mater, C. (1995b) Natural Poly(hydroxybutyrate-hydroxyvalerate) polymers as degradable biomaterials, *Res. Soc. Symp. Proc.* **394**, 111–116.

Ciardelli, G., Saad, B., Hirt, T., Keiser, O., Neuenschwander, P., Suter, W. U., Uhlschmid, G. K., Mater, J. (1995) Phagocytosis and biodegradation of short-chain poly[(*R*)-3-hydroxybutyric acid] particles in macrophage cell line, *Sci. Mater. Med.* **6**, 725–730.

Coussot-Rico, P., Clarotti, G., Ait Ben Aoumar, A., Najimi, A. (1994) Relation between surface energy of polyhydroxyalkanoate-based membrane and the adsorption of proteins on the same membranes, *Eur. Polym. J.* **30**, 1327–1333.

Cox, M. K. (1992) The effect of material parameters on the properties and biodegradation of 'Biopol', in: *Biodegradable Polymers and Plastics* (Vert, M., Feijen, J., Albertsson, A., Scott, G., Chiellini, E., Eds), Cambridge: Royal Society of Chemistry, 95–100.

Cox, M. K. (1994) Properties and applications of polyhydroxyalkanoates, *Biodegradable Plastics and Polymers* (Doi, Y., Fukuda, K., Eds), Amsterdam: Elsevier, 120–135.

D'Haene, P., Remsen, E. E., Asrar, J. (1999) Preparation and characterization of a branched bacterial polyester, *Macromolecule* **32**, 5229–5235.

Darnell, J., Lodish, H., Baltimore, D. (1986) Cell walls in plants are constructed of multiple layers of cellulose, in: *Molecular Cell Biology* (Scientific American Books, Ed.), Washington, DC: Chapter 5, 179–187.

Davies, S., Tighe, B. (1995) Cell attachment to gel-spun polyhydroxybutyrate fibers, *Polym. Prepr. (Div. Polym. Chem.)* **36**, 103–104.

Dawes, E. A., Senior, P. J. (1973) The role and regulation of energy reserve polymers in microorganisms, *Adv. Microbial. Physiol.* **10**, 10135–10266.

De Koning, G. J. M. (1994a) Poly((*R*)-3-hydroxybutyrate) embrittled by ageing and ageing reversed by heat treatment process (Zeneca Ltd), WO 94/17121.

De Koning, G. J. M., van Bilsen, H. M. M., Lemstra P. J., Hazenberg, W., Witholt, B., Preusting, H., Zan der Galien, J. G., Schirmer, A., Jendrossek, D. (1994b) A biodegradable rubber by crosslinking Poly(hydroxyalkanoate) from *Pseudomonas oleovorans*, *Polymer* **35**, 2090–2097.

De Micheli, C., Navarini, F., Roncoroni, V. (1996) Process for the manufacture of totally biodecomposable films with high mechanical characteristics and relevant products and applications, European patent 736563.

de Rodrigues, M. F., Valentin, H. E., Berger, P. A., Tran, M., Asrar, J., Gruys, K. J., Steinbüchel, A. (2000) Polyhydroxyalkanoate accumulation in *Burkholderia sp.*: a molecular approach to elucidate the genes involved in the formation of two homopolymers consisting of short-chain-length 3-hydroxyalkanoic acids, *Appl. Microbol. Biotechnol.* **53**, 453–460.

Doi, Y. (1988) Random copolymer comprising D-(-)-3-hydroxybutyrate and D-(-)-3-hydroxyvalerate, and microbial process for production thereof, European patent 288908.

Doi, Y., Abe, C. (1990) Biosynthesis and characterization of a new bacterial copolyester of 3-hydroxyalkanoates and 3-hydroxy β chloroalkanoates, *Macromolecules* **23**, 3705–3707.

Doi, Y., Kunioka, M., Nakamura, Y., Soga, Y. (1988) Nuclear magnetic resonance studies on unusual bacterial copolyesters of 3-hydroxybutyrate and 4-hydroxybutyrate, *Macromolecules* **21**, 2722–2727.

Ellar, D., Lundgren, D., Okamura, G. K., Marchessult, R. H. (1973) Morphology of poly-β-hydroxybutyrate granules, *J. Mol. Biol.* **35**, 489–502.

Foster, L. J. R., Tighe, B. J. (1994) The degradation of gel-spun Poly(β-hydroxybutyrate) 'wool', *J. Environ. Polym. Degrad.* **2**, 185–194.

Foster, L. J. R., Tighe, B. J. (1995) Enzymatic assay of hydroxybutyric acid monomer formation in Poly(beta-hydroxybutyrate) degradation studies, *Biomaterials*, **16**, 341–3.

Fritzsche, K., Lenz, R. W. (1990) An unusual bacterial polyester with a phenyl pendant group, *Makromol. Chem.* **191**, 1957–1965.

Fuller, R. C., O'Donnell, J. P., Saulnier, J., Redlinger, T. E., Foster, J., Lenz, R. W. (1992) The supramolecular architecture of the polyhydroxyalkanoate inclusions in *Pseudomonas oleovorans*, *FEMS Microbiol. Rev.* **103**, 279–288.

Gagnon, K. D., Lenz, R. W., Farris, R. J., Fuller, R. C. (1992) Crystallization behavior and its influence on the mechanical properties of a thermoplastic elastomer produced by *Pseudomonas oleovorans*, *Macromolecules* **25**, 3723–3728.

Gassner, F., Owen, A. J. (1994) Physical properties of Poly(β-hydroxybutyrate)–Poly(β-caprolactone) blends, *Polymer* **35**, 2233–2236.

Gerngross, T. U., Slater, S. C. (2000) How green are green plastics? *Sci. Am.* **August**, 36–41.

Gogolewski S., Jovanovic, M., Perren, S. M., Dillon, J. G., Hughes, M. K. (1993) Tissue response and *in vivo* degradation of selected polyhydroxyacids: polylactides (PLA), poly(3-hydroxybutyrate) (PHB), and poly(3-hydroxybutyrate-*co*-3-hydroxyvalerate) (PHB/VA), *J. Biomed. Mater. Res.* **27**, 1135–1148.

Grassie, N., Murray, E. J., Holmes, P. A. (1984a) The thermal degradation of Poly(-(D)-β-hydroxybutyric acid). Part I. Identification and quantitative analysis of products, *Polym. Degr. Stab.* **6**, 47–61.

Grassie, N., Murray, E. J., Holmes, P. A. (1984b) The thermal degradation of Poly(-(D)-β-hydroxybutyric acid). Part 2. Changes in molecular weight, *Polym. Degr. Stab.* **6**, 95–103.

Grassie, N., Murray, E. J., Holmes, P. A. (1984c) The thermal degradation of Poly(-(D)-β-hydroxybutyric acid). Part 3. The reaction mechanism, *Polym. Degr. Stab.* **6**, 127–134.

Gross, R. A., DeMello, C., Lenz, R. W., Fuller, R. C. (1989) The biosynthesis and characterization of Poly(β-hydroxyalkanoates) produced by *Pseudomonas oleovorans*, *Macromolecules* **22**, 1106–1115.

Hamada, T. (1997) Biodegradable fiber nets for growing seaweed with long life (Kuraray Co.), Japanese patent 09000096.

Hammond, T., Liggat, J. J., Montador, J. H., Webb, A. (1994) Plasticizers for biodegradable polyesters, especially polyhydroxyalkanoate (Zeneca Ltd), WO 94/28061.

Herring, J. M., Webb, A. (1990) Nucleating agents for 3-hydroxybutyrate polymer compositions (ICI), European patent 400855.

Hirt, T. D., Neuenschwander, P., Suter, U. W. (1996) Telechelic diols from poly[(R)-3-hydroxybutyric acid] and poly{[(R)-3-hydroxybutyric acid]-*co*-[(R)-3-hydroxyvaleric acid]}, *Macromol. Chem. Phys.* **197**, 1609–1614.

Holland, S. J., Jolly, A. M., Yasin, M., Tighe, B. J. (1987) Polymers for biodegradable medical devices. II. Hydroxybutyrate–hydroxyvalerate copolymers: hydrolytic degradation studies, *Biomaterials* **8**, 289–295.

Holmes, P. A. (1984) 3-Hydroxybutyrate polymer fibers (ICI), European patent 104731.

Holmes, P. A. (1985) Applications of PHB-a microbially produced biodegradable thermoplastic, *Phys. Technol.* **16**, 32–36.

Holmes, P. A., Collins, S. H., Wright, L. F. (1984) 3-Hydroxybutyrate polymers (ICI), US patent 4 447 654.

Houmiel, K. L., Slater, S., Broyles, D., Casagrande, L., Colburn, S., Gonzalez, K., Mitsky, T. A., Reiser, S. E., Shah, D., Taylor, N. B., Tran, M., Valentin, H. E., Gruys, K. J. (1999) Poly(β-hydroxybutyrate) production in oilseed leukoplasts of *Brassica napus*, *Planta* **209**, 547–550.

Howells, E. R. (1982) Single-cell protein and related technology, *Chem. Ind.* **7**, 508–511.

Hughes, A., Mott, I. E. C., Dunnill, P. (2000) Studies with natural rapeseed and microbially derived polyhydroxybutyrate to simulate extraction of plastic from transgenic material, *Bioproc. Eng.* **23**, 257–263.

Inagaki, K., Kan, Y., Takahashi, S. (1996) Biodegradable polyester monofilaments and their manufacture (Chikyu Kankyo Sangyo Gijutsu KK), Japanese patent 08218216.

Kanesawa, Y., Tanahashi, N., Doi, Y., Saito, T. (1994) Enzymic degradation of microbial Poly(3-hydroxyalkanoates), *Polym. Degrad. Stab.* **45**, 179–185.

Kasuya, K., Inoue, Y., Yamada, K., Doi, Y. (1995) Kinetics of surface hydrolysis of poly[(R)-3-hydroxybutyrate] film by PHB depolymerase from *Alcaligenes faecalis* T1, *Polym. Degrad. Stab.* **48**, 167–174.

Katsuhiko, T. (1994) Biopol properties and processing, in: *Biodegradable Plastics and Polymer* (Doi, Y., Fukuda, K., Eds.), Amsterdam: Elsevier, 362.

Kazuya, H., Tetsuo, A., Motoko, Y., Masayuki, T. (1994) Laminate and its production (Toppan Printing Co. Ltd), Japanese patent 06270368.

Kim, I., Lee, M., Seo, I., Shin, P. (1995a) The effects of properties on the biodegradation of bio-degradable polymers, *Pollimo* **19**, 727–733.

Kim, O., Gross, R. A., Rutherford, D. R. (1995b) Bioengineering of Poly(β-hydroxyalkanoates) for advanced material applications: incorporation of cyano and nitrophenoxy side chain substituents, *Can. J. Microbiol.* **41** (Suppl. 1) 32–43.

Kim, Y. B. (1991) Preparation, characterization, and modification of poly-beta-hydroxyalkanoates from *Pseudomonas oleovorans*, PhD Thesis, University of Massachusetts, Amherst, MA.

Kim, Y. B., Lenz, R. W., Fuller, R. C. (1990) Preparation and characterization of Poly(β-hydroxyalkanoates) obtained from *Pseudomonas oleovorans* grown with mixtures of 5-phenylvaleric acid and *n*-alkanoic acids, *Macromolecules* **24**, 5256–5260.

Kimura, Y., Yamane, H., Komatsuzaki, S. (1994) Biodegradable hydroxybutyric acid-based polyesters for medical uses (Nippon Zeon Co.), Japanese patent 06336523.

Kledzki, A. K., Gassan, J., Heyne, A. M. (1994) Effect of the sterilization process on the biodegradable plastics, *Makromol. Chem.* **219**, 11–26.

Kunioka, M., Doi, Y. (1990) Thermal degradation of microbial copolyesters: Poly(3-hydroxybutyrate-*co*-3-hydroxyvalerate) and Poly(3-hydroxybutyrate-*co*-4-hydroxybutyrate), *Macromolecules* **23**, 1933–1936.

Kumagai, Y., Doi, Y. (1992) Enzymic degradation and morphologies of binary blends of microbial Poly(3-hydroxybutyrate) with Poly(ε-caprolactone), Poly(1,4-butylene adipate) and Poly(vinyl acetate), *Polym. Degrad. Stab.* **36**, 241–248.

Kunioka, M., Nakamura, Y., Doi, Y. (1988) New bacterial copolyesters produced in *Alcaligenes eutrophus* from organic acids, *Polym. Commun.* **29**, 174–176.

Kurdikar, D. L., Strauser, F. E., Solodar, A. J., Paster, M. D. (2000a) High temperature PHA extraction using PHA-poor solvents (Monsanto Co.), US patent 6 087 471.

Kurdikar, D. L., Strauser, F. E., Solodar, J. A., Paster, M. D., Asrar, J. (2000b) Methods of PHA extraction and recovery using non-halogenated solvents (Monsanto Co.), US patent 6 043 063.

Lemoigne, M. (1926) Products of dehydration and of polymerization of β-hydroxybutyric acid, *Bull. Soc. Chem. Biol.* **8**, 770–782.

Lenz, R. W., Kim, B. W., Ulmer, H., Fritzsche K., Fuller, R. C. (1990) Functionalized Poly(β-hydroxyalkanoates) produced by bacterial, *Polym. Prep.* **31**, 408–409.

Liggat, J. J., O'Brien, G. (1994a) Retarding age embrittlement of polyesters, polyester compositions and polyester moldings by heat treatment (Zeneca Ltd), WO 94/28047.

Liggat, J. J., O'Brien, G. (1994b) Restoring mechanical properties of polyhydroxyalkanoates (Zeneca Ltd), WO 94/28048.

Liggat, J. J., O'Brien, G. (1994c) Heat treatment of polyesters for reduction of mechanical property losses in aging (Zeneca Ltd), WO 94/28049.

Liggat, J. J. (1994d) Biodegradable polyester composition for blow molded bottles (Zeneca Ltd), WO 94/28070.

Lopez-Llorca, L. V., Colom-Valiente, M. F. M., Carcases, J. (1994) Study of biofouling of poly-hydroxyalkanoate (PHA) films in water by scanning electron microscopy, *Micron* **25**, 45–51.

Madden, L. A., Anderson, A. J., Asrar, J. (1998) Synthesis and characterization of Poly(3-hydroxybutyrate) and Poly(3-hydroxybutyrate-*co*-3-hydroxyvalerate) polymer mixtures produced in high-density fed-batch cultures of *Ralstonia eutropha* (*Alcaligenes eutrophus*), *Macromolecules* **31**, 5660–5667.

Madden, L. A., Anderson, A. J., Shah, D. T., Asrar, J. (1999) Chain termination in polyhydroxyalkanoate synthesis: involvement of exogenous hydroxy-compounds as chain transfer agents, *Int. J. Biol. Macromol.* **25**, 43–53.

Madden, L. A., Anderson, A. J., Asrar, J., Berger, P., Garrett, P. (2000) Production and characterization of Poly(3-hydroxybutyrate-*co*-3-hydroxyvalerate-*co*-4-hydroxybutyrate) synthesized by *Ralstonia eutropha* in fed-batch cultures, *Polymer* **41**, 3499–3505.

Marchessault, R. H., Bleuhm, T. L., Deslandes, Y., Hamer, G. K., Orts, W. J., Sundararajan, P. R., Taylor, M. G., Bloembergen, S., Holden, D. A. (1988) Poly(β-hydroxyalkanoates): biorefinery polymers in search of applications, *Macromol. Symp.* **19**, 235–254.

Marchessault, R. H., Morin, F. G. (1992) Solid-state carbon-13 NMR study of the molecular dynamics in amorphous and crystalline Poly(β-hydroxyalkanoates), *Macromolecules* **25**, 576–581.

Martini, F., Perazzo, L., Vietto, P. (1989) Manufacture of Polymeric products, US patent 4 880 592.

Mergaert, J., Webb, A., Anderson, C., Wouters, A., Swings, J. (1993) Microbial degradation of Poly(3-hydroxybutyrate) and Poly(3-hydroxybutyrate-*co*-3-hydroxyvalerate) in soils, *Appl. Environ. Microbiol.* **59**, 3233–3238.

Merrick, J. M., Douboroff, M. (1961) Enzymic synthesis of Poly(β-hydroxybutyric acid) in bacteria, *Nature* **189**, 890–892.

Miller, N. D., Williams, D. (1987) On the biodegradation of poly-beta-hydroxybutyrate (PHB) homopolymer and poly-beta-hydroxybutyrate–hydroxyvalerate copolymers, *Biomaterials* **8**, 129–137.

Mitomo, H., Ota, E. (1991) Thermal decomposition of Poly(β-hydroxybutyrate) and its copolymer, *Sen 'I Gakkaishi* **47**, 89–94.

Mochizuki, M., Kan, Y., Takahashi, S., Kanemoto, N. (1993a) Biodegradable bicomponent fibers (Unitika Ltd), Japanese patent 05093316.

Mochizuki, M., Kan, Y., Takahashi, S., Kanemoto, N. (1993b) Biodegradable bicomponent fibers and nonwoven fabrics (Unitika Ltd), Japanese patent 05093318.

Mochizuki, M., Kan, Y., Takahashi, S., Kanemoto, N., Muta, Y. (1994a) Biodegradable polyester fibers including hydroxyalkanoate-units and their manufacture (Unitika Ltd, Zeneca KK), Japanese patent 06264305.

Mochizuki, M., Kan, Y., Takahashi, S., Kanemoto, N., Muta, Y. (1994b) Biodegradable multifilaments of Poly(β-hydroxyalkanoates) and their manufacture (Unitika Ltd, Zeneca Ltd), Japanese patent 06264306.

Molitoris, H. P., Moss, S. T., de Koning, G. J. M., Jendrossek, D. (1996) Scanning electron microscopy of polyhydroxyalkanoate degradation by bacteria, *Appl. Microbiol. Biotechnol.* **46**, 570–579.

Mott, I. E. C., Hughes, A., Dunnill, P. (2000) An ultra scale-down process study for the production of polyhydroxybutyrate from transgenic rapeseed, *Bioprocess Eng.* **22**, 451–459.

Nakamura, S., Kunioka, M., Doi, Y. (1991) Biosynthesis and characterization of bacterial Poly(3-hydroxybutyrate-*co*-3-hydroxypropionate), *Macromol. Rep.* **A28** (Suppl. 1), 15–24.

Nawrath, C., Poirier Y., Somerville, C. (1994) Targeting of the polyhydroxybutyrate biosynthetic pathway to the plastids of *Arabidopsis thaliana* results in high levels of polymer accumulation, *Proc. Natl Acad. Sci. USA* **91**, 12760–12764.

Oota, T., Noguchi, H., Ishii, Y. (1994) Biodegradable hydroxybutyric acid polyester compositions with good moldability (Mitsubishi Chem. Ind.), Japanese patent 06157878.

Poirier, Y., Dennis, D. E., Klomparens, K., Somerville, C. (1992) Polyhydroxybutyrate, a biodegradable thermoplastic, produced in transgenic plants, *Science* **256**, 520–523.

Poirier, Y., Nawrath, C., Somerville, C. (1995a) Production of polyhydroxyalkanoates, a family of biodegradable plastics and elastomers, in bacteria and plants, *Bio/Technology* **13**, 142–150.

Poirier, Y., Schechtman, L. A., Satkowski, M. M., Noda, I., Somerville, C. (1995b) Synthesis of high molecular weight Poly([R]-(–)-3-hydroxybutyrate) in transgenic *Arabidopsis thaliana* plant cells, *Int. J. Biol. Macromol.* **17**, 7–12.

Poirier, Y., Gruys, K. J. (2002) Production of PHAs and transgenic plants, in: *Biodegradable Polymers* (Steinbüchel, A., Doi, Y., Eds.), Volume 3, Chapter 13 (in press).

Poirier, Y. (1999a) Production of new polymeric compounds in plants, *Curr. Opin. Biotechnol.* **10**, 181–185.

Poirier, Y. (1999b) Green chemistry yields a better plastic, *Nat. Biotechol.* **17**, 960–961.

Pool R. (1989) In search of the plastic potato, *Science* **245**, 1187–1189.

Pouton, C. W., Akhtar, S. (1996) Biosynthetic polyhydroxyalkanoates and their potential in drug delivery, *Adv. Drug Del. Rev.* **18**, 133–162.

Reeve, M. S., McCarthy, S. P., Gross, R. A. (1993) Preparation and characterization of (R)-Poly-(β-hydroxybutyrate)–Poly(ε-caprolactone) and (R)-Poly(β-hydroxybutyrate)–Poly(lactide) degradable diblock copolymers, *Macromolecules* **26**, 888–894.

Scandola, M., Ceccorulli, G., Pizzoli, M., Gazzano, M. (1992) Study of the crystal phase and crystallization rate of bacterial Poly(3-hydroxy-butyrate-*co*-3-hydroxyvalerate), *Macromolecules* **25**, 1405–1410.

Shah, D. T., Tran, M., Berger, P. A., Asrar, J., Madden, L. Z., Anderson, A. J. (2000a) Synthesis and properties of hydroxy-terminated Poly(hydroxyalkanoate)s, *Macromolecules* **33**, 2875–2880.

Shah, D. T., Tran, M., Berger, P. A., Agarwal, P., Asrar, J., Madden, L. A., Anderson, A. J. (2000b) Synthesis of Poly(3-hydroxybutyrate) by *Ralstonia eutropha* in the presence of ¹³C-labeled ethylene glycol, *Macromolecules* **33**, 6624–6626.

Sharma, R., Ray, A. R. (1995) Polyhydroxybutyrate, its copolymers and blends, *Rev. Macromol. Chem. Phys.* **C35**, 327–359.

Shi, F., Rutherford, D. R., Gross, R. A. (1995) A new strategy to control microbial polyester structure, *Polym. Prepr.* **36**, 430–432.

Sim, S. J., Snell, K. D., Hogan, S. A., Stubbe, J., Rha, C., Sinskey, A. J. (1997) PHA synthase activity controls the molecular weight and polydispersity of polyhydroxybutyrate *in vivo*, *Nat. Biotechnol.* **15**, 63–67.

Slater, S. C., Voige, W. H., Dennis, D. E. (1988) Cloning and expression in *Escherichia coli* of the *Alcaligenes eutrophus* H16 poly-β-hydroxybutyrate biosynthetic pathway, *J. Bacteriol.* **170**, 4431–4436.

Slater, S. C., Houmiel, K. L., Tran, M., Mitsky, T. A., Taylor, N. B., Padgette, S. T., Gruys, K. J. (1998) Multiple β-ketothiolases mediate Poly(β-hydroxyalkanoate) copolymer synthesis in *Ralstonia eutropha*, *J. Bacteriol.* **180**, 1979–1987.

Slater, S., Mitsky, T. A., Houmiel, K. L., Hao, M., Reiser, S. E., Taylor, N. B., Tran, M., Valentin, H. E., Rodriguez, D. J., Stone, D. A., Padgette, S. R., Kishore, G., Gruys, K. J. (1999) Metabolic engineering of *Arabidopsis* and *Brassica* for poly(3-hydroxybutyrate-co-3-hydroxyvalerate) copolymer production, *Nat. Biotechol.* **17**, 1011–1016.

Sosa, K., Koizumi, T. (1994) Biodegradable and UV-degradable polyester composite fibers (Kuraray Co.), Japanese patent 06093516.

Steinbüchel A., Valentin, H. E. (1995) Diversity of bacterial polyhydroxyalkanoic acids, *FEMS Microbiol. Lett.* **128**, 219–228.

Suzuki, M., Sasaki, I., Okuno, M. (1996) Polymer blends for use in nets for seafood cultivation (Gunze KK), Japanese patent 08228640.

Suzuki, T., Yamane, T., Shimizu, S. (1986) Mass production of poly-β-hydroxybutyric acid by fully automatic fed-batch culture of methylotroph, *Appl. Microbiol. Biotechnol.* **23**, 322–329.

Takai, Y. (1994) Biodegradable fiber webs and their manufacture (Daiwa Spinning Co. Ltd), Japanese patent 06248547.

Tanaka, K., Ishizake, A., Ferment, J. (1994) Production of poly-D-3-hydroxybutyric acid from carbon dioxide by a two-stage culture method employing *Alcaligenes eutrophus* ATCC 1769T, *Bioengineering* **77**, 425–427.

Taniguchi, M., Nakagawa, Y., Hachifusa, K., Aizawa, T., Yoshikawa, M. (1994a) Biodegradable paper-based laminates (Toppan Printing Co. Ltd), Japanese patent 06293113.

Taniguchi, M., Nakagawa, Y., Hachifusa, K., Aizawa, T., Yoshikawa, M. (1994b) Biodegradable polyester-paper laminates (Toppan Printing Co. Ltd), Japanese patent 06316042.

Tokiwa, Y., Iwamoto, A., Takeda, K. (1992) Biodegradable plastic composition, biodegradable plastic shaped body and method of producing same (AIST, JSP Corp.), US patent 5 124 371.

Valentin, H. E., Broyles, D. L., Casagrande, L. A., Colburn, S. M., Creely, W. L., DeLaquil, P. A.,

Felton, H. M., Gonzalez, K. A., Houmiel, K. L., Lutke, K., Magadeo, D. A., Mitsky, T. A., Padgette, S. R., Reiser, S. E., Slater S., Stark, D. M., Stock, R. T., Stone, D. A., Taylor, N. B., Thorne, G. M., Tran, M., Gruys, K. J. (1999a) PHA production, from bacteria to plants, *Int. J. Biol. Macromol.* **25**, 303–306.

Valentin, H. E., Berger, P. A., Gruys, K. J., de Rodrigues, M. F., Steinbüchel, A, Tran, M., Asrar, J. (1999b) Biosynthesis and characterization of Poly(3-hydroxy-4-pentenoic acid), *Macromolecules* **32**, 7389–7395.

Waddington, D. S. (1994) Polyhydtroxyalkanoates and film formation therefrom (Zeneca Ltd), WO 94/16000.

Wang, D., Yamamoto, T., Cakmak, M. (1996) Processing characteristics and structure development in solid-state extrusion of bacterial copolyesters: poly(3-hydroxybutyrate-*co*-3-hydroxyvalerate), *J. Appl. Polym. Sci.* **61**, 1957–1970.

Ward, A. C., Rowley, B. I., Dawes, E. A. (1977) Effect of oxygen and nitrogen limitation on poly-ß-hydroxybutyrate biosynthesis in ammonium-grown *Azotobacter beijerinckii*, *J. Gen. Microbiol.* **102**, 61–68.

Webb, A. (1998) Tampon applicators and compositions for making same, European patent 0291024.

Williams, D. F., Miller, N. D. (1987) The degradation of polyhydroxybutyrate (PHB), *Adv. Biomater.* **7**, 471–476.

Williamson, D. H., Wilkinson, J. F. (1958) The isolation and estimation of the poly-ß-hydroxybutyrate inclusions of *Bacillus* species, *J. Gen. Microbiol.* **19**, 198–209.

Wnuk, A. J., Koger II, T. J., Young, T. A. (1994) Biodegradable, liquid-impervious multilayer film compositions, WO 94/00293.

Yamada, K., Kan, Y., Murase, S. (1995) Biodegradable composite fibers with Poly(hydroxyalkanoic acid) core and Poly(alkylene succinate) sheath (Unitika Ltd, Chikyu Kankyo Sangyo Gijutsu KK), Japanese patent 07324227.

Yamamoto, T., Kimizu, M., Kikutani, T., Furuhashi, Y., Cakmak, M. (1997) The effect of drawing and annealing conditions on the structure and properties of bacterial Poly(3-hydroxybutyrate-*co*-3-hydroxyvalerate) fibers, *Int. Polym. Process.* **12**, 29–37.

Yamamoto, T., Kimizu, M., Maekawa, Y., Shinkawa, T. (1995) High-strength biodegradable fibers of *c*-axis oriented crystals combined with other crystals and manufacture thereof (Ishikawa Prefecture, Chukoh Chem. Ind.), Japanese patent 07300720.

Yokouchi, M., Chatani, Y., Tadokoro, H., Tani, H. (1974) Structural studies of polyesters. VII. Molecular and crystal structures of racemic Poly(β-ethyl-β-propiolactone), *Polym. J.* **6**, 248–255.

4
Applications of PHAs in
Medicine and Pharmacy

Dr. Simon F. Williams[1], Dr. David P. Martin[2]
[1] Tepha, Inc. 303 Third Street, Cambridge, MA 02142, USA;
Tel.: +01-617-492-0505; Fax: +01-617-492-1996; E-mail: williams@metabolix.com
[2] Tepha, Inc. 303 Third Street, Cambridge, MA 02142, USA;
Tel.: +01-617-492-0505; Fax: +01-617-492-1996; E-mail: martin@metabolix.com

HA	hydroxyapatite
mcl-PHA	medium chain-length PHA
M_w	molecular weight
PCL	polycaprolactone
PGA	polyglycolic acid
PHA	polyhydroxyalkanoate
PLA	polylactic acid
poly(3HB)	poly-R-3-hydroxybutyrate
poly(3HB-co-3HV)	poly-R-3-hydroxybutyrate-co-R-3-hydroxyvalerate
poly(3HB-co-4HB)	poly-R-3-hydroxybutyrate-co-4-hydroxybutyrate
poly(3HO-co-3HH)	poly-R-3-hydroxyoctanoate-co-R-3-hydroxyhexanoate
poly(3HP)	poly-3-hydroxypropionate
poly(4HB)	poly-4-hydroxybutyrate
Poly(5HV)	poly-5-hydroxyvalerate
poly(6HH)	poly-6-hydroxyhexanoate
T_g	glass transition temperature
T_m	melting temperature

1
Introduction

Polyhydroxyalkanoates (PHAs) are a class of naturally occurring polyesters that are produced by a wide variety of different microorganisms (Steinbüchel, 1991). Although they are derived biologically, the structures of these polymers bear a fairly close resemblance to some of the synthetic absorbable polymers currently used in medical applications. Owing to their limited availability, the PHAs have remained largely unexplored, yet these polymers offer an extensive range of properties that extend far beyond those currently offered by their synthetic counterparts.

At the last count there were well over 100 different types of hydroxy acid monomers that had been incorporated into PHA polymers, and the list is continuing to grow (Steinbüchel and Valentin, 1995). These monomers include hydroxyalkanoate units ranging from 2- to 6-hydroxy acids substituted with a wide range of groups including alkyl, aryl, alkenyl, halogen, cyano, epoxy, ether, acyl, ester, and acid groups (see Figure 1). By no means will all of these monomers be useful or suitable for medical use; however, they provide a set of materials with properties that range from rigid and stiff to flexible and elastomeric, including polymers that degrade relatively quickly *in vivo* and others that are slow to degrade. In

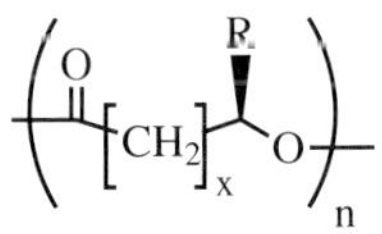

Typical values of x, n and R
x = 1 to 4
n = 1,000 to 10,000
R = alkyl group (C_mH_{2m+1})
  or functionalized alkyl group

Fig. 1 General chemical structure of the PHAs.

addition, the PHA polymers are thermoplastic in nature, with a wide range of thermal properties, and can be processed using conventional techniques (Holmes, 1988).

2
Historical Outline

As a class of polymers, the PHAs are relative newcomers, with many of the different types having been discovered during only the past 20 years. One of the simplest members of the class, poly-*R*-3-hydroxybutyrate, poly(3HB), is an exception as it was first identified in 1925 and is the most well-known PHA polymer. It should be noted however, that the properties of poly(3HB) are not representative of the polymer class as a whole.

During the 1980s, the British company, Imperial Chemical Industries (ICI), developed a commercial process to produce poly(3HB), and a related copolymer known as poly-*R*-3-hydroxybutyrate-*co-R*-3-hydroxyvalerate, poly(3HB-*co*-3HV). These polymers were sold under the tradename of Biopol®, and were developed primarily as renewable and biodegradable replacements for petroleum-derived plastics. As a result of these activities and others (Lafferty et al., 1988), both polymers became widely available, which in turn provided opportunities for their evaluation as medical biomaterials. While these efforts have resulted in several promising clinical trials, and development efforts continue, products containing these materials have yet to be approved for *in vivo* medical use.

In 1993, ICI transferred its biological division to Zeneca, which continued to develop PHAs for commodity applications under the tradename Biopol. Zeneca, however, sold its Biopol assets to Monsanto in the mid-1990s. In 2001, an American company,

Metabolix, Inc. acquired the Biopol assets from Monsanto, and is developing transgenic approaches to the large-scale manufacture of PHAs through fermentation and agricultural biotechnology.

More recent interest in the use of PHA polymers for medical applications has arisen primarily in response to the needs of the emerging field of tissue engineering, where a much wider range of absorbable polymers are being sought for use as tissue scaffolds. In fact, in the past two years PHAs have become one of the leading classes of biomaterials under investigation for the development of tissue-engineered cardiovascular products because they can offer properties not available in existing synthetic absorbable polymers.

An American company, Tepha, Inc., is currently engaged in the development of a range of tissue-engineered products based on PHA polymers, and is expanding the number available for medical research to meet both the needs of tissue engineering and the development of more traditional medical devices. As a result of these efforts during the past two years, the number of materials currently under evaluation has expanded and now includes three additional PHA polymers, namely, poly-*R*-3-hydroxy-octanoate-*co*-*R*-3-hydroxyhexanoate (poly(3-HO-*co*-3HH)), poly-4-hydroxybutyrate (poly-(4HB)), and poly-*R*-3-hydroxybutyrate-*co*-4-hydroxybutyrate (poly(3HB-*co*-4HB)). This brings the total number of PHA polymers currently under investigation for medical application to five (Figure 2).

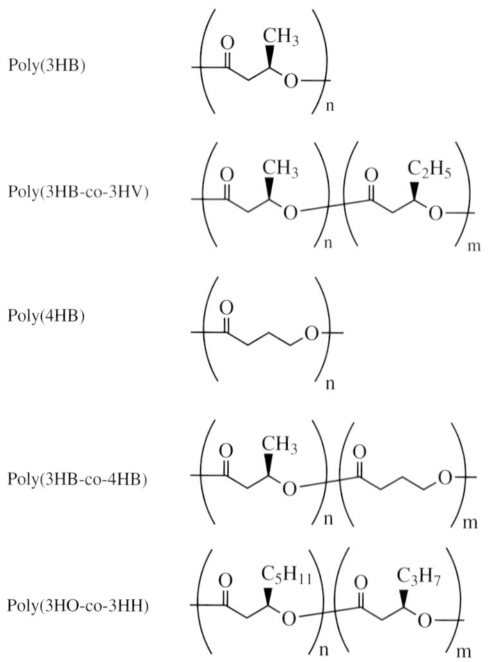

Poly(3HB)

Poly(3HB-co-3HV)

Poly(4HB)

Poly(3HB-co-4HB)

Poly(3HO-co-3HH)

Fig. 2 Chemical structures of PHAs currently under medical investigation.

3

PHA Preparation and Properties: A Primer

3.1

Production

The PHA polymers are accumulated as discrete granules within certain microorganisms at levels reaching 90% of the dry cell mass, and can be isolated fairly readily by breaking open the cells and using either an aqueous-based or solvent-based extraction process to remove cell debris, lipid, nucleic acids, and proteins. Traditionally, these polymers have been produced by fermentation from sugars or oils, often with co-feeds, and the majority of medical studies on poly(3HB), poly(3HB-*co*-3HV), and poly(3HO-*co*-3HH), have been based on polymers derived via this route.

During the late 1980s, the genes responsible for PHA production were isolated, and this has led more recently to the development of transgenic methods for PHA production (see Williams and Peoples, 1996, and references therein). This breakthrough has provided a new means of tailoring the properties of PHA polymers to particular applications, and represents a potentially important advance in the development of technologies to produce designer biomaterials for medical use. Poly-4-hydroxybutyrate (poly(4HB)), for example, is produced using this technology. Transgenic PHA production may also prove to be important in the medical field from a regulatory standpoint, since this technology allows the production host to be selected. For example, PHAs may now be produced by fermentation in *Escherichia coli* K12, a well-characterized host used extensively by the biotechnology industry.

In general, PHA polymers are produced with relatively high molecular weights (M_w) *in vivo*. Commercial grades of poly(3HB) copolymers typically have M_w that are at least 500,000, although PHAs with much longer pendant groups (known as medium chain-length PHAs, mcl-PHAs), such as poly(3HO-*co*-3HH), typically have M_w that are closer to 100,000. Polydispersity is typically around 2.0. By isolating the enzymes responsible for PHA production, namely PHA synthases, researchers have also been able to produce PHA polymer *in vitro* with ultra-high M_w exceeding several million (Gerngross and Martin, 1995), and also *in vivo* in transgenic organisms (Kusaka et al., 1997; Sim et al., 1997).

3.2

Mechanical and Thermal Properties

As a class of polymers, the PHAs offer an extensive design space with properties spanning a large range, and usefully extending the relatively narrow property range offered by existing absorbable synthetics (Engelberg and Kohn, 1991). The mechanical properties of the five PHAs currently being investigated for medical use are shown in Table 1. The homopolymer, poly(3HB), is a relatively stiff, rigid material that has a tensile strength comparable with that of polypropylene. The introduction of a comonomer into this polymer backbone, however, significantly increases the flexibility and toughness of the polymer (extension to break and impact strength), and this is accompanied by a reduction in polymer stiffness (Young's modulus). This is evident in the poly(3HB) copolymers, poly(3HB-*co*-3HV) and poly(3HB-*co*-4HB) (Doi, 1990; Sudesh et al., 2000).

A progressive and substantial change in the mechanical properties of poly(3HB) also occurs when the pendant groups are extended from the polymer backbone. The mcl-PHA, poly(3HO-*co*-3HH), for example, shares the same backbone as poly(3HB), but in contrast is a highly flexible thermoplastic elastomer with properties comparable with those of commercially produced materials (Gagnon et al., 1992).

Extending the distance between the ester groups in the PHA backbone can also have a dramatic impact on mechanical properties. The homopolymer poly(4HB), for example, is a highly ductile, flexible polymer with an extension to break of around 1000%, compared with poly(3HB), which has an extension to break of less than 10%. Combining these different monomers to form copolymers, as in poly(3HB-co-4HB), produces one series of materials with a wide range of useful mechanical properties that can be tailored to specific needs. Interestingly, at levels of around 20–40% 4HB, the poly(3HB-co-4HB) copolymers actually behave like elastic rubbers.

The thermal properties of PHAs also span wide ranges (see Table 1). Typical melting temperatures (T_m) range from around 55 °C for poly(3HO-co-3HH) to about 180 °C for the poly(3HB) homopolymer. Glass transition temperatures (T_g) span the range from about –55 °C to about 5 °C. In general, T_m values decrease as the pendant groups become longer. This is particularly important in the melt processing of poly(3HB), which is unstable at temperatures just above its melting point. Incorporation of other monomers into the poly(3HB) polymer backbone yields lower-melting poly(3HB) copolymers that can be more readily processed. T_g values are also depressed by the incorporation of monomers with longer pendant groups, the depression being relatively modest in the poly(3HB) copolymers, but pronounced for the mcl-PHAs.

3.3
Sterilization of PHA Polymers

For medical use, most PHAs have been sterilized using ethylene oxide, without causing any significant changes to the physico-chemical properties of the polymers. However, low-melting PHAs such as poly(4HB) and poly(3HO-co-3HH) are generally sterilized using a cold cycle, particularly if the polymer has been fabricated ready for use. Residual ethylene oxide levels in poly(3HO-co-3HH) after cold sterilization with ethylene oxide for 8 h at 38 °C with 65% humidity have been reported to be < 1 ppm after one week (Marois et al., 1999a).

Several studies have described the effects of γ-irradiation on PHA polymers derived from 3-hydroxy acids, such as poly(3HB), poly(3HB-co-3HV), and poly(3HO-co-3HH). It has been reported that poly(3HB), unlike polyglycolic acid (PGA), can be sterilized by γ-irradiation doses on the order of 2.5 Mrad (Holmes, 1985), although it is likely that some reduction in molecular weight results from this treatment. At higher doses (10–20 Mrad) the mechanical integrity of both

Tab. 1 Mechanical and thermal properties of some representative PHAs

PHA	Poly-(3HB)	Poly(3HB-co-20%3HV)	Poly(4HB)	Poly(3HB-co-16%4HB)	Poly(3HO-co-12%3HH)
Melting temperature [T_m, °C]	177	145	60	152	61
Glass transition temperature [T_g, °C]	4	−1	−50	−8	−35
Tensile strength [MPa]	40	32	104	26	9
Tensile modulus [GPa]	3.5	1.2	0.149	n.d.	0.008
Elongation at break [%]	6	50	1000	444	380

poly(3HB) and poly(3HB-*co*-3HV) are significantly compromised (Miller and Williams, 1987). Luo and Netravali (1999) have also reported significant changes in the mechanical properties and M_w of poly(3HB-*co*-3HV) after exposure to γ-irradiation at doses of 10–25 Mrad.

Exposure of poly(3HO-*co*-3HH) to γ-irradiation at a dose of 2.5 Mrad at room temperature has also been reported to result in a loss of molecular weight on the order of 17%; this is caused by random chain scission, accompanied by some degree of physical cross-linking (Marois et al., 1999a). Thus, while γ-irradiation is generally recognized as a desirable alternative to ethylene oxide for sterilization, care must be exercised in its use on PHA polymers, and the procedures carefully validated.

A few PHA polymers may also be sterilized by steam (Baptist and Ziegler, 1965), particularly if they have T_m over 140 °C, and are thermally stable at this temperature. Holmes (1985) has reported that poly(3HB) powders can be sterilized in this manner.

4
Biocompatibility

Without doubt, the biological response to PHA polymers *in vivo* represents the most important property of these biomaterials if a medical application is being contemplated. Most of the information currently available relates to poly(3HB) and poly(3HB-*co*-3HV), and has been recently reviewed (Hasirci, 2000). A small amount of information on poly(3HO-*co*-3HH), poly(3HB-*co*-4HB), and poly(4HB) has also been published. Care should be exercised in interpreting these data however, since most studies have been based on the use of industrial rather than medical grades of PHA polymers. Notably, Garrido (1999) has described the presence of

cellular debris in industrial samples of poly(3HB-*co*-3HV), Rouxhet et al. (1998) detected a number of contaminants on the surface of these samples by X-ray photo-electron spectroscopy, and Williams et al. (1999) reported that an industrial sample of poly(3HB) contained more than 120 endotoxin units per gram. Two methods to remove endotoxin have been reported recently, one being based primarily on the use of peroxide (Williams et al., 1999) and the other by use of sodium hydroxide (Lee et al., 1999).

4.1
Natural Occurrence

Some of the monomers incorporated into PHA polymers are known to be present *in vivo*, and both their metabolism and excretion are well understood. The monomeric component of poly(3HB), *R*-3-hydroxybutanoic acid, for example, is a normal metabolite found in human blood. This hydroxy acid is a ketone body, and is present at concentrations of 3–10 mg per 100 mL blood in healthy adults (Hocking and Marchessault, 1994). This monomer has been administered to obese patients undergoing therapeutic starvation to reduce protein loss (Pawan and Semple, 1983), and also evaluated as an intravenously administered energy source in both humans (Hiraide and Katayama, 1990) and piglets (Tetrick et al., 1995). There is also interest in the use of this monomer in ocular surgery as an irrigation solution to maintain the tissues (Chen and Chen, 1992).

The monomeric component of poly(4HB), 4-hydroxybutanoic acid, is also a naturally occurring substance that is widely distributed in the mammalian body, being present in the brain, kidney, heart, liver, lung, and muscle (Nelson et al., 1981). This 4-hydroxy acid has been used for over 35 years as an intravenous agent for the induction of anesthesia and for long-term sedation (Entholzner et al., 1995). It is also one of the most promising treatments for narcolepsy (Scharf et al., 1998), although unfortunately as with many hypnotics there has been some illegitimate use of this compound. However, since the half-life of the acid is short (35 min), and relatively high doses (several grams) are required to obtain any hypnotic effect, small implants of poly(4HB) could not induce general sedation, for example.

In addition to the known presence of certain PHA monomers in humans, low molecular-weight forms of poly(3HB) have also been detected in human tissues. Reusch and colleagues first identified poly(3HB) in blood serum (0.6 – 18.2 mg L^{-1}) complexed with low-density lipoproteins, and with the carrier protein albumin (Reusch et al., 1992). The oligomers have also been detected in human aorta (Seebach et al., 1994), and are known to form ion channels *in vivo* when complexed with polyphosphate (Reusch et al., 1997).

4.2

In vitro Cell Culture Testing

Relatively few studies have attempted to characterize the tissue response of PHAs caused by leachables such as impurities, additives, monomers, and degradation products. Chaput et al. (1995) evaluated the cytotoxic responses of three poly(3HB-co-3HV) compositions (7, 14, and 22% hydroxyvalerate) using direct contact and agar diffusion cell culture tests, and reported that the solid polymers elicited mild to moderate cellular reactions *in vitro*. However, the cytotoxicity of extracts from these polymers varied with the medium, surface-to-volume ratio, time and temperature. Dang et al. (1996) also evaluated an extract from an industrial sample of poly(3HB-co-3HV) in

an *in vitro* cell culture test method with mouse fibroblasts, and reported that the extract appeared slightly to suppress cellular activity.

In other *in vitro* testing, Rivard et al. (1995) showed that porous poly(3HB-co-9%3HV) substrates (Selmani et al., 1995), when seeded with canine anterior cruciate ligament (ACL) fibroblasts, sustained a cell proliferation rate similar to that observed in collagen sponges for around 35 days, with maximal cell density occurring after 28 days. Interestingly, the poly(3HB-co-9%3HV) substrates maintained their structural integrity during the culturing, whereas the collagen foams contracted substantially and produced significantly less protein. In evaluating poly(3HB) as a potential drug delivery matrix, Korsatko et al. (1983a) also reported no significant differences in cellular growth with mice fibroblasts.

Saito et al. (1991) evaluated poly(3HB) sheets in an inflammatory test using the chorioallantoic membrane of the developing egg, and reported that the polymer did not cause any inflammation.

Several reports have described the effects of small, low molecular-weight, crystalline particles of poly(3HB) on the viability of cultured macrophages, fibroblasts, co-cultures of Kupffer cells and hepatocytes, and osteoblasts (Ciardelli et al., 1995; Saad et al., 1996a,b,c). These particles represent one of the degradation products expected to arise *in vivo* from the absorption of poly(3HB) and DegraPol®, a phase-segregated multiblock polyesterurethane copolymer. At low concentrations, the small poly(3HB) particles were found to be well tolerated by macrophages, fibroblasts, Kupffer cells and hepatocytes. Macrophages, Kupffer cells, and to a lesser extent fibroblasts and osteoblasts, were all found to take up (phagocytose) the small particles of poly(3HB) (1 – 20 μm), and evidence of biodegradation by macrophages was also found (Ciardelli et al., 1995). Hepatocytes, in contrast, demonstrated no signs of poly(3HB) phagocytosis. At high concentrations (>10 µg mL^{-1}), phagocytosis of poly(3HB) particles was found to cause cell damage and cell activation in macrophages and to a lesser degree in osteoblasts, but not in fibroblasts (Saad et al., 1996a,b,c). Separately, the chondrocyte compatibility of a DegraPol foam was also evaluated *in vitro*. Rat chondrocytes were found to attach to about 60% of the foam compared with a polystyrene control, and proliferated at comparable rates (Saad et al., 1999), leading to the conclusion that the DegraPol foam had acceptable chondrocyte compatibility.

Of particular interest in the evolving field of tissue engineering was a report by Rouxhet et al. (1998) on the effect of adhesion and proliferation of monocytes-macrophages to a poly(3HB-*co*-8%3HV) film when modified by hydrolysis or coated with different proteins. As anticipated, the cells were found to have a greater affinity for the polymer surface after it had been hydrolyzed to liberate additional carboxylate and hydroxyl functions. However, it was also found that adhesion of this cell type increased significantly when fibronectin was adsorbed to the polymer surface, but not when collagen or albumin were pre-absorbed.

Cellular attachment to porous tubes made from poly(3HO-*co*-3HH) under different seeding conditions has been evaluated (Stock et al., 1998). Although dynamic cell seeding techniques were found initially to result in a higher rate of ovine smooth muscle cellular attachment compared with static seeding, higher attachment was not sustained under simulated blood flow conditions. Cell attachment to a composite material of PGA and poly(3HO-*co*-3HH) has also been reported (Sodian et al., 1999). After seeding with myofibroblasts and endothelial cells, these composites were incubated in a bioreactor under pulsatile flow. After eight days, near-confluent layers of cells were observed with the formation of extracellular matrix. Sodian et al. (2000a) also studied cellular attachment to porous samples of poly(4HB), and compared the results to those obtained with a porous poly(3HO-*co*-3HH) material and a PGA mesh. After seeding and incubating these materials with ovine vascular cells for eight days, there were significantly more cells on the PGA, although after exposure to flow no significant differences were found. A considerable amount of collagen development was noted for each sample, with the highest amounts present in the PGA meshes. Cellular attachment to a composite of poly(4HB) with a PGA mesh has also been evaluated *in vitro* recently, and compared with the mesh alone and a poly(4HB) foam (Nasseri et al., 2000). Better cell migration into the composite, and better shape retention were observed.

4.3
In vivo Tissue Responses

Some of the earliest investigations of the *in vivo* tissue responses to PHA polymers were made by W. R. Grace and Co. in the mid-1960s (Baptist and Ziegler, 1965). In these early studies, film strips of poly(3HB) were implanted subcutaneously and intramuscularly in rabbits, and removed after eight weeks. Examination of the implant sites revealed granulomatous foreign body reactions, but these did not affect the underlying area.

Since these early investigations, many reports have been made describing the *in vivo* tissue responses of poly(3HB) and poly(3HB-*co*-3HV) in both biocompatibility and application-directed studies. Chaput et al. (1995) described one of the longest *in vivo* studies, in which poly(3HB-*co*-3HV) films (containing 7, 14, and 22% valerate) were sterilized by ethylene oxide and implanted intramuscularly in sheep for up to 90 weeks. No abscess formation or tissue necrosis was seen in the vicinity of the implants. However, after 1 week *in vivo*, acute inflammatory reactions with numerous macrophages, neutrophils, lymphocytes and fibrocytes were observed in a capsule at the interface between the polymers and the muscular tissues. After 11 weeks, the observed reaction was less intense with a lower density of inflammatory cells present, though lymphocytes were still observed in the capsule and muscular tissues. At this stage, the capsules were reported to consist primarily of connective tissue cells, and were dense and well-vascularized with highly organized oriented fibers and fibroblastic cells aligned in parallel with the polymer surfaces. A large number of fatty cells were also observed in the capsule, as well as at the interface and in adjacent muscles after long-term implantation (at 70 and 90 weeks). Interestingly, few differences were observed between the capsules, tissue characteristics or cellular activity in terms of the compositions of the three poly(3HB-*co*-3HV) polymers.

Similar results were also observed by Gogolewski et al. (1993) when poly(3HB) and poly(3HB-*co*-3HV) samples were implanted subcutaneously in mice. Fibrous capsules of around 100 µm thickness developed after one month, and these increased to 200 µm by three months, but then thinned to 100 µm at six months. However, the number of inflammatory cells was found to increase with valerate content, and a few granulocytes were still present around blood vessels near encapsulated implants containing 22% valerate at six months. Separately, Tang et al. (1999) suggested that leachable impurities and low molecular-weight poly(3HB) are at least partly responsible for increased colla-

gen deposition following an *in vivo* study of subcutaneous poly(3HB) implants in rats.

Williams et al. (1999) reported a 40-week subcutaneous implant study of poly(3HO-*co*-3HH) in mice. At two weeks, there was minimal reaction to the implants which had been encapsulated by a thin layer of fibroblasts, four to six cell layers thick, surrounded by collagen. There was no evidence of macrophages, and the tissue response continued to be very mild at 4, 8, 12 and 40 weeks, with the amount of connective tissue surrounding the implants remaining fairly constant. The polymer proved to be particularly inert, and could be readily removed with little tissue adherent to the implants. An extract from poly(3HO-*co*-3HH) was also tested in a standard skin sensitization test (ASTM F270), but no discernable erythema/eschar formation was observed.

Subcutaneous implants of poly(4HB) have also been reported to be well tolerated *in vivo* during the course of their degradation (Martin et al., 1999), with minimal inflammatory responses occurring.

It is worth noting that, *in vivo*, as most PHA polymers break down they release hydroxy acids that are significantly less acidic and less inflammatory than many currently used synthetic absorbable polymers (Taylor et al., 1994). For example, poly(3HB) and poly(4HB), are derived from 3- and 4-hydroxybutanoic acids (pK_a 4.70 and 4.72, respectively), that are significantly less acidic than the 2-hydroxy acids (glycolic acid, pK_a 3.83; lactic acid, pK_a 3.08) found in PGA and poly-lactic acid (PLA). Furthermore, significant differences in the mechanism of degradation of these synthetic polymers, which can degrade autocatalytically from the inside outward, can result in substantial amounts of acidic degradation products being released. In one clinical study, for example, around 5% of the patients receiving PGA screws had an inflammatory reaction to the implants that was sufficient to warrant operative drainage (Böstman, 1991).

Finally, Holmes (1988) has reported that poly(3HB) shows negligible oral toxicity, the LD_{50} being greater than 5 g kg^{-1}.

5
Biodistribution

The biodistribution of poly(3HB) microspheres in mice (Bissery et al., 1984a), and poly(3HB) granules in rats (Saito et al., 1991) has been investigated using ^{14}C-labeling and, as anticipated, results have been found to depend upon particle size. In the first study, microspheres of 1–12 µm diameter were injected intravenously into mice, and traced at 0.5, 1, and 24 h, and every seven days thereafter. After 30 min, 47% of the radioactivity was found in the lungs, 14% in the liver, and 2.1% in the spleen. After 1 h, concentrations in the lungs and liver had increased to 62 and 16%, respectively, and by 24 h there was still 60% in the lungs and 24% in the liver. Thereafter, the amounts remained fairly constant, but fell somewhat in the lungs. In the rat study, granules of 500–800 nm diameter were injected through a tail vein into rats and traced at intervals of 2.5 h, 1 day, 13 days, and 2 months. After 2.5 h, approximately 86% of the radioactivity had accumulated in the liver, with 2.5% and 2.4% of the total distributed in the spleen and lungs, respectively. During the following two months, radioactivity levels in most of the tissues decreased slowly, but steadily.

6
Bioabsorption

The rates of bioabsorption of PHA polymers *in vivo* vary considerably, and depend pri-

marily upon their chemical compositions. Other factors such as their location, surface area, physical shape and form, crystallinity, species, and molecular weight can also be very important. While useful information can be derived from *in vitro* studies, results of *in vitro* studies with PHA polymers are not always good indicators of *in vivo* behavior.

6.1
In vitro Degradation

In order to investigate the mechanism of degradation of poly(3HB) and poly(3HB-*co*-3HV) *in vivo*, a number of studies have been conducted to determine their rates of hydrolysis *in vitro* (see Holland et al., 1987, 1990; Yasin et al., 1990; Knowles and Hastings, 1992; Chaput et al., 1995). These studies have used complementary techniques such as gravimetric and molecular weight analysis, as well as measurements of surface and tensile properties to monitor different aspects of degradation and develop a concept of the overall degradation process. This has led to the following general scheme of *in vitro* behavior for poly(3HB) and poly(3HB-*co*-3HV). Initially, some surface modification is observed, with water diffusing into the polymer and porosity increasing. Crystallinity also increases, but there is relatively little change in molecular weight in the first few months, and tensile properties remain fairly constant. As the porosity increases, hydrolysis of the polymer chains releases degradation products that can diffuse away more easily. The molecular weight decreases, erosion increases, and both weight and tensile strength begin to decrease more rapidly. At about one 1 year, the initial resistance to degradation is followed by an accelerated degradation with the material becoming more brittle, but not losing its physical integrity. After one year, the most apparent change in the physical appearance of the polymer is loss of surface gloss and the development of surface rugosity.

Other *in vitro* studies have examined the action of additives such as polysaccharides (Yasin et al., 1989), polycaprolactone (PCL) (Yasin and Tighe, 1992), as well as lipases, PHA depolymerases, and several extracts on PHA degradation. Although PHA depolymerases are abundant in the environment and are responsible for PHA biodegradation in soil, there is currently no evidence that these enzymes are present *in vivo*. Mukai et al. (1993) investigated the action of 16 lipases on five different PHA polymers prepared either by fermentation or synthetically, and found that none of these enzymes catalyzed the hydrolysis of poly(3HB). However, the other four PHA polymers were hydrolyzed by lipases, with the number of lipases capable of hydrolyzing the PHA polymer chains decreasing in the following order: poly-3-hydroxypropionate (poly-(3HP)) > poly(4HB) > poly-5-hydroxyvalerate (poly(5HV)) > poly-6-hydroxyhexanoate (poly(6HH)). Interestingly, two lipases have been detected recently in tissue adjacent to poly(3HB) implants in rats, raising the possibility of their involvement in poly(3HB) bioabsorption (Löbler et al., 1999). The copolymer, poly(3HB-*co*-3HV), formulated as microspheres with PCL, and loaded with bovine serum albumin, has also been incubated with four different extracts *in vitro* (Atkins and Peacock, 1996a). The percentage weight loss decreased in the order newborn calf serum > pancreatin > synthetic gastric juice > Hanks' buffer, and it was speculated that the enhanced biodegradation in newborn calf serum, and surface erosion in pancreatin, must be due to enzymatic activity in these extracts.

The *in vitro* degradation of poly(3HO-*co*-3HH) has also been examined for up to 60 days (Marois et al., 1999b). When exposed to acid phosphatase and β-glucuronidase for

this time period, no significant surface or chemical modifications were observed, and no significant weight loss was detected. It was concluded that this polymer, which shares a common backbone with poly(3HB), degrades slowly by chemical hydrolysis.

Degradation of poly(4HB) *in vitro* has recently been reported (Martin et al., 1999). The homopolymer is fairly resistant to hydrolysis at pH 7.4, and over a 10-week period very little degradation was observed, although a 20–40% reduction in average molecular mass did occur during this time period.

6.2
In vivo Bioabsorption

In early studies, some confusion arose around the stability of poly(3HB) and poly(3HB-*co*-3HV) *in vivo*. Korsatko et al. (1983a, 1984) and Wabnegg and Korsatko (1983) evaluated poly(3HB) for use as matrix retard tablets and reported that the polymer was degraded *in vivo* at a rate directly proportional to the elapsed time (a zero-order reaction). However, it was reported later that monofilaments derived from poly(3HB) and poly(3HB-*co*-3HV) (8 and 17% valerate) showed little, if any, loss of strength when implanted subcutaneously in rats for up to six months (Miller and Williams, 1987), except after γ-irradiation. Many subsequent studies have confirmed that poly(3HB) and poly(3HB-*co*-3HV) do degrade *in vivo*, albeit slowly (Hasirci, 2000). Typically, poly(3HB) is completely absorbed *in vivo* in 24–30 months (Malm et al., 1992b; Hazari et al., 1999a). During the first four weeks *in vivo*, the degree of crystallinity of a sample of poly(3HB) implanted in the peritoneal cavity was reported to have increased, presumably as a result of the amorphous regions of the polymer degrading more rapidly than the crystalline domains (Behrend et al., 2000a). After four weeks, crystallinity, Young's modulus, and microhardness were each shown to have decreased fairly steadily, this being consistent with a surface process.

Kishida et al. (1989) attempted to develop a method to accelerate the bioabsorption of poly(3HB) and poly(3HB-*co*-3HV) *in vivo* by adding basic compounds to the polymers. Although *in vitro* the rate of hydrolysis was found to increase, the effect *in vivo* was minimal, presumably because the basic accelerators had leached out.

The mcl-PHA, poly(3HO-*co*-3HH), also degrades slowly *in vivo*. Williams et al. (1999) reported that the molecular weight (M_w) of subcutaneous implants of poly(3HO-*co*-3HH) in mice decreased from 137,000 at implantation to around 65,000 over 40 weeks, and that there were no significant differences between the molecular weights of samples taken from the surfaces and interiors of the implants. The latter finding suggests that slow, homogeneous hydrolytic breakdown of the polymer occurs.

While poly(3HB), poly(3HB-*co*-3HV), and poly(3HO-*co*-3HH) are generally degraded slowly *in vivo*, consistent with *in vitro* observations, the homopolymer, poly(4HB) is an exception. Martin et al. (1999) found the *in vivo* degradation of this polymer to be relatively rapid, and to vary with porosity. Over a 10-week period it was reported that film, 50%, and 80% porous samples implanted subcutaneously in rats, lost 20%, 50%, and nearly 100% of their mass, respectively. The average molecular mass of the polymer also decreased significantly, but independently of sample configuration. These data suggest that the degradation of poly(4HB) *in vivo* depends in part on surface area, and that the mechanical properties of poly(4HB) implants are likely to undergo a gradual change rather than the more abrupt changes seen with other synthetic absorb-

ables, such as PGA. This might be advantageous, for example, in tissue regeneration applications where a sudden loss of a mechanical property is undesirable, or more gradual loss of implant mass and steady in growth of new tissue are beneficial.

7
Applications

Until recently, only poly(3HB) and poly(3HB-*co*-3HV) were available commercially, and consequently the majority of investigations into applications have focused on a relatively narrow set of polymer properties within the PHA design space. This situation is beginning to change however, with more recent studies involving poly(3HO-*co*-3HH), poly(4HB) and poly(3HB-*co*-4HB).

7.1
Cardiovascular

Without doubt, the major medical use of PHAs has been in the development of cardiovascular products.

7.1.1
Pericardial Patch

One of the most advanced applications of PHA polymers in cardiovascular products has been the development of a regenerative poly(3HB) patch that can be used to close the pericardium after heart surgery, without formation of adhesions between the heart and sternum (Bowald and Johansson, 1990; Malm et al., 1992a,b; Bowald and Johansson-Ruden, 1997). These adhesions represent a significant complication if a second operation is necessary, thereby increasing the risk of rupturing the heart or a major vessel, and prolonging the overall duration of the operation. In an initial study, native pericardium was excised from 18 sheep, and replaced with a nonwoven poly(3HB) patch (Malm et al., 1992a). Patches were then harvested between 2 and 30 months after the operation, examined for adhesions, infection, and inflammatory response, and compared with controls where native pericardium had been removed and left open. Moderate adhesions were present in all the controls, whereas no adhesions developed in 14 of the animals receiving poly(3HB) patches. Interestingly, the pericardium was regenerated in all animals receiving a patch, with the surface of the patches being completely covered with mesothelium-like cells after two months, and a dense underlying collagen layer developing over 12 months. A pronounced tissue response to the patch was observed, with the polymer being slowly phagocytosed by polynuclear macrophages – a finding which has led others to question the biocompatibility of the patch (Tomizawa et al., 1994). Polymer remnants were still present after 24 months, and some macrophages were still found at 30 months, but no platelet aggregates were detected.

Following animal studies with the poly(3HB) patch, a randomized clinical study of 50 human patients admitted for bypass surgery and/or valvular replacement was undertaken (Duvernoy et al., 1995). Using computed tomography (CT), 39 of these patients (19 with the patch and 20 without) were examined for the presence of adhesions at 6 and 24 months, and a lower incidence of postoperative adhesions was reported for the group receiving PHA patches, based on the presence of fat located between the patch and the cardiac surface.

In contrast to these studies, Nkere et al (1998) found no significant difference in a short-term study of adhesion formation among calves undergoing bypass surgery with and without the poly(3HB) patch, as well as calves not undergoing bypass surgery but with their pericardium left open. It was noted however that there might be a species variation, and also that the duration of the studies was different. Also, in comparison with Malm's sheep study, the calves in this study had been subjected to bypass, which was considered more clinically relevant.

7.1.2
Artery Augmentation
Non-woven patches of poly(3HB) have been evaluated in the augmentation of the pulmonary artery as scaffolds for the regeneration of arterial tissue in low-pressure systems (Malm et al., 1994). A total of 19 lambs was used for the trial, with 13 receiving poly(3HB) patches, and six receiving Dacron patches as a control group. No aneurysms were observed in either group, and the pores of the non-woven poly(3HB) were small enough to prevent bleeding. All the patches were harvested between 3 and 24 months, and endothelial layers were found on both patch materials. Beneath the endothelium-like surface, the configuration closely resembled native artery, with smooth muscle cells, collagen and elastic fibers in the poly(3HB) explants. By contrast, a thin collagenous layer had formed under the endothelium lining of the Dacron implants due to the well-known inflammatory reaction to Dacron fiber, and a dense infiltration of lymphocytes was present. As in the case of the pericardial studies, the nonwoven poly(3HB) patch was phagocytosed by polynucleated macrophages, and macrophages persisted, even at 24 months. However, no platelet aggregates were found on the luminal surface.

Highly porous foam patches of poly(4HB) seeded with endothelial, smooth muscle cells, and fibroblasts have also been evaluated in artery augmentation, and with good results (Stock et al., 2000a). A total of six cell-seeded patches, and one unseeded control patch were implanted into the pulmonary artery of sheep. Echocardiography and examination of the cell-seeded explants at 4, 7 and 20 weeks indicated that progressive tissue regeneration had occurred, but with no evidence of thrombus, dilation, or stenosis. Examination of the control patch at 20 weeks revealed slight bulging at the site of implantation, and less tissue regeneration.

7.1.3

Atrial Septal Defect Repair

Malm et al. (1992c) also tested the efficacy of the nonwoven poly(3HB) patch in the repair of atrial septal defects created in six calves. Implants were evaluated between 3 and 12 months, and complete endothelial layers facing the right and left atrium were observed, with a subendothelial layer of collagen and some smooth-muscle cells. As before, the patch was degraded by polynuclear macrophages, with small particles of polymer still present at 12 months, and a foreign body reaction persisting. Nonetheless, the patches prompted the formation of regenerated tissue that resembled native atrial septal wall, and had sufficient strength to prevent the development of shunts in the atrial septal position.

7.1.4

Cardiovascular Stents

One of the main problems with the use of metallic stents in cardiovascular applications is the subsequent restenosis that can result from excessive growth of the blood vessel wall. This is believed to be due (at least in part) to irritation caused by the metallic stent on the vessel wall. A potential solution to this problem may lie in the development of an absorbable stent that can prevent reocclusion of the vessel in the short term, but then be absorbed so that it does not cause any persistent irritation of the vessel wall.

Attention is beginning to focus on the use of PHAs in absorbable stents as well as coatings, often in combination with drug delivery systems. Van der Giessen et al. (1996) evaluated several bioabsorbable polymers, including poly(3HB-*co*-3HV), as candidate biomaterials for cardiovascular stents. Strips of polymer were deployed on the surface of coil wire stents, and implanted in porcine coronary arteries of 2.5–3.0 mm diameter. After four weeks, most of the materials tested, including poly(3HB-*co*-3HV), had provoked extensive inflammatory responses and fibrocellular proliferation. However, these results were shown to be inconsistent with *in vitro* results, and that other factors such as implant geometry, implant design, and degradation products may have been responsible for some of the observed response. It was also noted that the polymers were not sterilized prior to implantation.

The homopolymer, poly(3HB), has also been fabricated into a cardiovascular stent (Schmitz and Behrend, 1997), and tested in a rabbit model (Unverdorben et al., 1998). It was also reported that poly(3HB) stents plasticized with triethyl citrate (Behrend et al., 2000b) and fabricated by laser cutting of a molded construct had an average elastic recoil of about 20–24% immediately after dilation, and of 27–29% after 120 h *in vitro* (Behrend et al., 1998). After implantation into the arteries of rabbits, the poly(3HB) stents instigated a temporary intimal proliferation, and were observed to degrade fairly rapidly *in vivo*.

7.1.5

Vascular Grafts

Vascular grafts are currently inserted to repair or replace compromised blood vessels in arterial or venous systems that have been subjected to damage or disease, for example atherosclerosis, aneurysmal disease or traumatic injury. For the larger-diameter vessels, synthetic grafts are frequently employed, and these can be impregnated with protein to make them completely impervious to blood, and thereby ready for anastomotic procedures. Several studies have shown, however, that when protein substrates are impregnated into grafts they may promote undesirable immunological reactions. In order to try and develop an improved sealant for a synthetic graft, Noisshiki and Komatsuzaki (1995) investigated the possible use of poly(3HB-co-4HB) as a graft coating, the coated grafts being implanted into dogs and examined at 2 and 10 weeks. Subsequently, it was noted that degradation of the polymer had already started after two weeks.

Marois et al. (1999c, 2000) investigated the use of poly(3HO-co-3HH) as an impregnation substrate. Polyester grafts impregnated with poly(3HO-co-3HH) were implanted in rats, and compared with both protein- and fluoropolymer-impregnated grafts for periods ranging from 2 to 180 days. No infiltration of tissue into the poly(3HO-co-3HH)-impregnated graft occurred because of the presence of the polymer; moreover, polymer degradation was found to be very slow with just a 30% reduction in molecular weight after six months. Tissue response, after an initial acute phase seen in all grafts, was reported as generally mild, and additional investigations with this biomaterial were recommended.

The elastomeric polymer, poly(3HO-co-3HH), has also been evaluated as a component of an autologous cell-seeded tissue-engineered vascular graft in lambs (Shum-Tim et al., 1999). Tubular conduits (7 mm diameter) comprising a nonwoven PGA mesh on the inside and layers of poly(3HO-co-3HH) outside were prepared, and seeded with a mixed cell population of endothelial cells, smooth muscle cells, and fibroblasts, obtained by the expansion of

sections of carotid artery harvested from lambs. After seven days, the seeded conduits were used to replace 3–4-cm abdominal aortic segments in lambs. The lambs were sacrificed at 10 days and 5 months after surgery, and the conduits compared with unseeded controls. All control conduits became occluded during the study, but the cell-seeded tissue-engineered grafts remained patent (open to blood flow), except for one stricture, and no aneurysms were observed. This contrasts sharply with results obtained using a polyglactin-PGA composite that was limited by high porosity, stiffness, and a relatively short degradation time, and developed aneurysms within a few weeks. Histologic analysis of the poly(3HO-co-3HH)-PGA grafts was reported to reveal an insignificant inflammatory response, with increased cell density, collagen formation and mechanical properties that were approaching those of native aorta.

It has also been proposed that poly(3HB) might be used to repair severed blood vessels by the insertion of a tube of this material (Baptist and Ziegler, 1965).

7.1.6
Heart Valves

Perhaps the most remarkable results with PHA polymers have been obtained in the development of cell-seeded tissue-engineered heart valves. These valves promise to provide unique solutions to the deficiencies of mechanical and animal valves currently in clinical use, such as the need for anticoagulant therapy and repeat surgery to replace defective or even outgrown valves in young children. In initial studies in this area, scaffolds seeded with autologous vascular cells and based on porous PGA and PLA had been used to replace a single pulmonary valve leaflet in lambs, but attempts to replace all three pulmonary valve leaflets had failed due to the relatively high stiffness and rigidity of the PGA-PLA biomaterials. However, when the leaflets were replaced with porous poly(3HO-*co*-3HH), and sutured to a conduit composed of a poly(3HO-*co*-3HH) film sandwiched between layers of PGA mesh, these difficulties were overcome (Stock et al., 2000b). Echocardiography of the seeded constructs implanted in lambs indicated no thrombus formation with only mild, nonprogressive, valvular regurgitation up to 24 weeks after implantation. Histologic examination revealed organized and viable tissue, without thrombus formation. Notably, thrombus developed on all leaflets in the unseeded control scaffolds after four weeks. After six weeks *in vivo*, no PGA remained, but poly(3HO-*co*-3HH) was still substantially present with a slight reduction in molecular weight (26%). A variant of this scaffold, with leaflets derived from poly(3HO-*co*-3HH) sandwiched between PGA mesh, has also been evaluated under pulsatile flow *in vitro* (Sodian et al., 1999). Under these conditions, it was shown that vascular cells attached to the scaffold, proliferated, and oriented in the direction of flow after four days.

The design of the heart valve scaffolds have been further refined in subsequent studies, and other PHA polymers have also been evaluated (Sodian et al., 2000a). Sodian et al. (2000b,c,d) also described the fabrication of a functional tissue-engineered heart valve entirely from porous poly(3HO-*co*-3HH), and seeding of the scaffold with vascular cells from ovine carotid artery which resulted in cellular in-growth into the pores and formation of a confluent layer under pulsatile flow conditions.

Recently, in one of the most astonishing results of tissue engineering described to date, Hoerstrup et al. (2000) succeeded in developing a PHA-based heart valve scaffold in lambs that was completely replaced at eight weeks by a functional trileaflet heart valve. Two components contributed to this success. First, the use of poly(4HB), which was coated on a PGA mesh to provide a rapidly degrading yet flexible scaffold; and second, the use of an *in vitro* pulse duplicator system. After 20 weeks *in vivo*, the mechanical properties of the valve were reported to resemble those of the native valve, and histologic analysis showed uniform, layered cuspal tissue with endothelium (see Figure 8 of Schoen and Levy, 1999). Echocardiography demonstrated mobile, functioning leaflets without stenosis, thrombus, or aneurysm to 20 weeks, and importantly, the inner diameter of the valve construct was found to have increased from 19 mm at implant to 23 mm at 20 weeks. The latter finding is particularly exciting for the development of a tissue-engineered heart valve that can be used in young children, and will grow with the child.

7.2
Dental and Maxillofacial

7.2.1
Guided Tissue Regeneration

In guided tissue regeneration, barrier membranes are used to encourage regeneration of new periodontal ligament by creating a space or pocket that excludes gingival connective tissue from the healing periodontal wound, and also by preventing the downgrowth of epithelial tissue into the wound. Galgut et al. (1991) evaluated the histologic response of rats to poly(3HB-*co*-3HV) membranes that could be used in this application, and found that the membranes were well tolerated. Compared with Gore-Tex™ (polytetrafluoroethylene, PTFE) membranes, little downgrowth of epithelial tissue was observed with poly(3HB-*co*-3HV).

During the closure of palatal defects, mucoperiosteal flaps are frequently moved to the midline of the palate, leaving two areas of denuded bone adjacent to the dentition. These wounds heal by migration of kerati-nocytes and fibroblasts, as well as by wound contraction. Later, the formation of scar tissue occurs, and it is believed that attach-ment of this tissue to areas of the denuded bone can disturb subsequent maxillary growth, or might affect the development of dentition. Using beagle dogs, Leenstra et al. (1995) investigated the use of nonporous poly(3HB-*co*-3HV) films to keep the muco-periosteum and bone separated until the transition of teeth was completed (at about 24 weeks). At two weeks, it was reported that one of the poly(3HB-*co*-3HV) films had deformed; however, this was attributed to the method of insertion. At 8 and 12 weeks, the films were unimpaired and surrounded by fibrous capsules, and it was concluded that poly(3HB-*co*-3HV) films were more suitable for this procedure in terms of mechanical properties and tissue response than was PLA or PCL.

7.2.2
Guided Bone Regeneration

In addition to using barrier membranes to create new periodontal ligament, mem-branes can also be used to generate new bone in jaw bone defects, as well as to increase the width and height of the alveolar ridge. Kostopoulos and Karring (1994a) reported successful bone regeneration in jaw bone defects in rats using poly(3HB-*co*-3HV) membranes to create spaces for bone fill. Mandibular defects were produced in rats and either covered with poly(3HB-*co*-3HV) membranes, or left uncovered. During the following 15 to 180 days, increasing bone fill was achieved with poly(3HB-*co*-3HV) membranes, whereas in the uncovered con-trol group ingrowth of other tissues oc-curred, and only 35–40% of the defect area was filled with bone after 3–6 months. Kostopoulos and Karring (1994b) also used

poly(3HB-*co*-3HV) membranes successfully to increase the height of the rat mandible. In two-thirds of the rats, the space created by the poly(3HB-*co*-3HV) membrane was com-pletely filled with bone by six months, although in some cases soft tissue migrated through ruptures in the membranes, there-by inhibiting bone formation. In contrast, bone formation was negligible when mem-branes were not used. It was noted that while this biomaterial might form the basis of a barrier membrane, some modifications to the physical properties would be required.

Barrier membranes of poly(3HB-*co*-3HV) have been reinforced with polyglactin fibers and used to cover dental implants placed in fresh extraction sockets in dogs (Gotfredsen et al., 1994). However, very poor results were obtained. After 12 weeks, inflammatory infiltrates were seen adjacent to the po-ly(3HB-*co*-3HV)-polyglactin membrane that interfered with bone healing, and less bone fill was observed compared with control sites with no membrane. Given the results of Kostopoulos and Karring (1994a,b), and the known inflammatory response of tissues to PGA degradation products, the observed result might be attributed to the polyglactin component of the membrane.

7.3
Drug Delivery

The potential use of poly(3HB) and po-ly(3HB-*co*-3HV) in drug delivery has been evaluated in a number of studies, and the field has been reviewed several times (Hol-land et al., 1986; Juni and Nakano, 1987; Koosha et al., 1989; Pouton and Akhtar, 1996; Nobes et al., 1998, Scholz, 2000). Studies have included investigations of these polymers as subcutaneous implants, com-pressed tablets for oral administration, and microparticulate carriers for intravenous use.

7.3.1
Implants and Tablets

In the early 1980s, Korsatko et al. (1983a,b) studied the release of a model drug, 7-hydroxyethyltheophylline (HET), from poly(3HB) tablets prepared by homogeneously compounding and compressing the polymer and the drug. In both *in vitro* and *in vivo* experiments, the drug was released in a linear manner, but the rate of release from subcutaneous implants in mice was found to be about two- to three-fold slower than *in vitro*. At a drug loading of 10%, sustained release was observed for approximately 10 weeks *in vitro*, and up to 20 weeks *in vivo*. At loadings up to 30%, release could be sustained for up to 50 days; however, at substantially higher levels (60–80%) all the drug was released within 24 h. Changes in tablet compaction pressures did not alter the release rates. In subsequent studies, Korsatko et al. (1987) evaluated the influence of poly(3HB) molecular weight on drug release, and found the release of the antihypertensive drug midodrin-HCl from compressed tablets to be increased as the polymer's molecular weight increased from 3000 to 600,000. In comparison, when poly(3HB)-coated granules of the beta-blocker, Celiprolol-HCl were prepared using a fluid bed dryer system, it was found that smaller amounts of the higher molecular-weight polymers were required to slow the release of Celiprolol-HCl, presumably because of improved film formation at higher molecular weight.

Could et al. (1987) investigated the release of fluorescein, and dextrans labeled with fluorescein, from tablets of poly(3HB-*co*-3HV) prepared by direct compression. Consistent with the observations of Korsatko et al. (1983a,b), the rate of release was found to increase significantly at higher loadings. Faster rates of release were also observed as the percentage of valerate in the copolymer was decreased, leading to reduced compressibility, and more rapid influx of fluid into the tablet. The higher molecular-weight dextrans also released more rapidly than fluorescein, but this was attributed to the creation of more porous and hydrophilic matrices and led to the testing of porosigen (pore-forming) additives as a means of controlling the rate of drug release. Two additives, microcrystalline cellulose and lactose, were tested, and both were found to enhance the rate of drug release from the modified tablets.

Release profiles of 5-fluorouracil (5-FU) from melt-pressed poly(3HB) disks have been examined *in vitro* over a range of drug concentrations (10–50%) by Juni and Nakano (1987). The rate of release was found to increase with drug loading, and complete release was observed in five days at a 50% loading. Gangrade and Price (1992) also used melt compression to prepare PHA drug delivery systems with lower porosity, but chose the lower-melting poly(3HB-*co*-3HV) copolymer, and incorporated progesterone at loadings of 5 to 50%. Consistent with the findings of others, increased loadings resulted in faster rates of release, with 85% of the drug being released in three days at a 50% loading compared with 50% at a 5% loading.

The homopolymer, poly(3HB), has been evaluated as a potential candidate for the development of a gastric retention drug delivery device that could extend the absorption period of a drug from the stomach (Cargill et al., 1989). However, further development was not pursued because no degradation of the device was observed in a 12-h period during an *in vitro* dissolution assay. Jones et al. (1994) developed an auricular poly(3HB-*co*-3HV) implant for cattle containing metoclopramide at a loading of 50% as a prophylactic treatment for fescue toxicosis (which is caused by cattle grazing on endophyte-infected fescue). The implant was prepared by melt compression, and released metoclopramide at an effective rate of 12 mg per day *in vivo*.

Compression-molded compacts of poly(3HB) loaded with tetracycline have been evaluated in the treatment of periodontal disease (Collins et al., 1989; Deasy et al., 1989). Using *in vitro* studies, poly(3HB) compacts loaded with 50% tetracycline were developed that could deliver therapeutic levels of the antibiotic for eight to nine days. Six patients with gingivitis were treated with these compacts, and their saliva was monitored for the release of tetracycline. Therapeutic levels of tetracycline were detected over the 10-day study period, and an improvement in the gingival condition from moderate to mild inflammation was reported. However, when the treatment was stopped the improvement was not maintained.

Subcutaneous implants of poly(3HB) containing gonadotropin-releasing hormone (GnRH) have been tested for their ability to release this hormone, and stimulate luteinizing hormone (LH) secretion, promote preovulatory follicle growth, and induce ovulation in acylic sheep (McLeod et al., 1988). In comparison with oil-based formulations, the poly(3HB) implants containing 40–50 µg of GnRH consistently produced elevated plasma levels of LH for periods of two to four days, and a high incidence of ovulation was obtained, particularly when two implants per animal were used.

Akhtar et al. (1989, 1991, 1992) investigated the release of a model drug, methyl red, from both solution-cast and melt-processed films of poly(3HB) and poly(3HB-*co*-3HV), and examined the effect of varying the temperature during polymer crystallization. It was found that the faster-crystallizing homopolymer, poly(3HB), was better able to trap methyl red than the slower-crystallizing

poly(3HB-*co*-3HV) copolymers, contributing to a slower release of the drug from the homopolymer *in vitro*.

Hasirci et al. (1998) described the development of poly(3HB-*co*-3HV) rods containing the antibiotic Sulperazone for the treatment of osteomyelitis. Rods loaded with 20 and 50% of the drug were prepared by adding granules of the antibiotic to poly(3HB-*co*-3HV) solvent solutions, and molding the resulting pastes into rods. These rods were then introduced into rabbit tibias containing metal implants infected with *Staphylococcus aureus* (obtained from chronic osteomyelitis patients). After 15 days the infection had been eradicated. A similar approach to the treatment of osteomyelitis using poly(3HB-*co*-3HV) rods containing sulbactam-cefoperazone has also been reported (Yagmurlu et al., 1999). Recently, Korkusuz et al. (2001) evaluated the use of poly(3HB-*co*-4HB) and poly(3HB-*co*-3HV) rods as antibiotic carriers for the treatment of osteomyelitis. A bone infection, experimentally induced with *S. aureus*, was effectively treated with solvent-blended rods containing Sulperazone or Duocid. It was noted that the poly(3HB-*co*-4HB) rods were preferred as they were less rigid and easier to handle than the poly(3HB-*co*-3HV) rods.

Kharenko and Iordanskii (1999) prepared poly(3HB) tablets containing the vasodilator, diltiazem, with drug loadings up to about 45%, and monitored release rates of these delivery systems *in vitro*. Near-complete release of the drug was observed at the highest loading concentrations, with slower release being observed at lower loadings.

7.3.2
Microparticulate Carriers

A number of groups have examined the potential use of poly(3HB) and poly(3HB-*co*-3HV) polymers as microparticulate carriers for drug delivery. In general, these systems have been produced using solvent evaporation techniques, and variables such as drug loading, polymer composition, molecular weight, crystallization rate, particle size, and the use of additives have been investigated. Although there are exceptions, the following observations are fairly typical: (1) Increased valerate content in copolymers of poly(3HB-co-3HV) usually slows the rate of drug release, presumably because the copolymers are less crystalline than poly(3HB). Incorporation of valerate into poly(3HB) also tends to yield microparticles that are less susceptible to physical damage than poly(3HB); (2) Smaller particle sizes decrease loading capacity, but increase the rate of drug release; (3) Small changes in polymer molecular weight have little impact on release rates; however, large changes can increase crystallinity and lead to enhanced release rates; (4) Drug release from poly(3HB) and poly(3HB-co-3HV) is completed well before any significant degradation of the polymers has begun, so drug release is entirely diffusion controlled; and (5) Lower drug loadings reduce the release rate.

One of the earliest studies of poly(3HB) microspheres in drug delivery was carried out by Bissery et al. (1983, 1984a). Using a solvent evaporation technique, ^{14}C-labeled poly(3HB) microspheres (1–12 µm) were prepared and found, as expected, to concentrate primarily in the lungs of mice upon intravenous administration. However, when the microspheres were loaded with an anticancer agent, lomustine [N-(2-chloroethyl)-N'-cyclohexyl-N-nitrosourea; CCNU], and administered to Lewis lung carcinoma-bearing mice, little effect was observed (Bissery et al., 1985). From in vitro studies it was found that the drug was completely released from the microspheres in 24 h at a loading of 7.4%, compared with a release time of over 90 h when PLA microspheres were used (Bissery et al., 1984b). Notably, the poly(3HB) microspheres were found to be somewhat irregular in shape, which was attributed to the highly crystalline nature of the homopolymer.

Brophy and Deasy (1986) examined the release profiles of sulphamethizole from poly(3HB) and poly(3HB-co-3HV) microparticles (53–2000 µm) formed by grinding a solvent-evaporated matrix of these components, and reported that increased rates of release were observed when the molecular weight of poly(3HB) was increased. This observation was attributed to poor distribution of the drug in the highly crystalline poly(3HB) polymer. As anticipated, faster rates of release were also observed as the poly(3HB) particle size was decreased, and as drug loading was increased. Incorporation of valerate into the poly(3HB) polymer chain decreased the release rate, presumably on account of the improved distribution of the drug. Overcoating of the poly(3HB) microparticles with PLA reduced the burst effect (the initial, rapid release of the drug from the delivery device surface), as well as significantly reducing the overall release rate. A sustained release of sulphamethizole was demonstrated in vivo when poly(3HB) microparticles (425–600 µm) loaded at 50% were administered intravenously to dogs. Release was complete in about 24 h, and correlated well with in vitro results.

Slower release of several anticancer antibiotics, including doxorubicin, aclarubicin, 4'-O-tetrahydropyranyl doxorubicin, bleomycin, and prodrugs of 5-fluoro-2'-deoxyuridine from poly(3HB) microspheres has been reported (Juni et al., 1985, 1986; Juni and Nakano, 1987; Kawaguchi et al., 1992). For example, at a 13% loading of aclarubicin-HCl, poly(3HB) microspheres with a mean diameter of 170 µm were reported to release only 10% of the drug over five days in vitro. The observed release rates could, however, be increased by the incorporating fatty acid esters into the poly(3HB) microspheres. These esters were believed to facilitate drug release by forming channels in the poly(3HB) matrix (Kubota et al., 1988). Abe et al. (1992a,b) also reported faster release of an anticancer agent, lastet, from poly(3HB) microspheres when acylglycerols were incorporated into the microspheres.

In vivo, poly(3HB) microspheres containing two prodrugs of 5-fluoro-2'-deoxyuridine were reported to induce higher antitumor effects against P388 leukemia in mice when compared with administration of the free prodrugs over five consecutive days (Kawaguchi et al., 1992).

Koosha et al. (1987, 1988, 1989) reported the use of high-pressure homogenization to produce poly(3HB) nanoparticles containing prednisolone. At drug loadings up to 50%, a biphasic release pattern was observed *in vitro*, with an initial burst effect followed by a slow release of the drug that was complete in one to two days. Similar results were observed with tetracaine. Akhtar et al. (1989) has also prepared smaller poly(3HB) particles (20–40 µm) by spray-drying, and reported fairly rapid release of a model drug (methyl red) from this matrix.

Microspheres of poly(3HB) have been evaluated *in vivo* as controlled release systems for the oral delivery of vaccines that could potentially protect vaccine antigens from digestion in the gut, and target delivery to Peyer's patches (Eldridge et al., 1990). When a single oral dose of poly(3HB) microspheres containing coumarin was administered to mice, very good absorption of microspheres (of diameter <10 µm) was observed in the Peyer's patches 48 h later.

Conway et al. (1996, 1997) have evaluated the adjuvant properties of poly(3HB) microspheres *in vivo*, and observed a potent antibody response to encapsulated bovine serum albumin (BSA), with the highest titer being generated to preparations containing particles with the smallest sizes. Linhardt et al. (1990) also considered using poly(3HB) and poly(3HB-*co*-3HV) as vaccine delivery vehicles.

The effect of several different parameters on the release of progesterone (a low molecular-weight model drug) from poly(3HB) and poly(3HB-*co*-3HV) microspheres prepared with an emulsion solvent evaporation technique has been evaluated by Gangrade and Price (1991). Using scanning electron microscopy (SEM), the use of gelatin as an emulsifier was shown to provide more spherical microspheres than when using polyvinyl alcohol, sodium lauryl sulfate or methyl cellulose, and that smoother surfaces were obtained when the solvent was switched from chloroform to methylene chloride. Generally, poly(3HB) microspheres had very rough surfaces which became smoother as valerate was incorporated. Interestingly, release of the drug from poly(3HB-*co*-9%3HV) was slower than from either poly(3HB) or poly(3HB-*co*-24%3HV). Examination of the internal surfaces of these microspheres by SEM revealed fewer cavities and less porosity for the poly(3HB-*co*-9%3HV) microspheres, and this was consistent with the slower release observed. Porosity was also found to decrease when the microspheres were prepared at increasing temperatures between 25 and 40 °C. Typically, 60–100% of the drug was released from these microspheres in 12 h at drug loadings up to 12%.

Embleton and Tighe (1992a,b) also investigated the effects of increasing the valerate content of poly(3HB-*co*-3HV), temperature, and molecular weight on microsphere formation, and obtained results consistent with those of Gangrade and Price (1991). When 10–50% PCL was incorporated into these poly(3HB-*co*-3HV) microspheres, a systematic increase in porosity was observed with increasing PCL content, that was attributed to the elution of PCL from the hardened microcapsule wall once the poly(3HB-co-3HV) had precipitated during formation (Embleton and Tighe, 1993). Porosity was also found to increase significantly when polyphosphate-Ca^{2+} complexes were introduced into poly(3HB-co-3HV) microspheres (Gürsel and Hasirci, 1995).

Atkins and Peacock (1996b) also used PCL to prepare microcapsules (21–200 μm) from a blend of poly(3HB-co-3HV)/PCL(20%) with an inner reservoir of BSA loaded in an agarose core. The protein encapsulation was, however, low (<12 %) and only slightly influenced by loading. Release from the microcapsules loaded with up to 50% protein was observed in vivo for about 24 days.

Microspheres of poly(3HB) have been evaluated as potential embolization agents in renal arteries (Kassab et al., 1999), and with rifampicin as a chemoembolization agent (Kassab et al., 1997). Release of rifampicin during in vitro studies was consistent with other controlled release studies, with near-complete release of the drug being observed at high loadings (40%) in 24 h, and a somewhat prolonged release at lower loadings. Renal angiograms obtained before and after embolization with poly(3HB) microspheres (120–200 μm) using a contrast agent showed that 10 mg of the microspheres was sufficient to slow renal arterial blood flow, with subsequent partial occlusion of the pre-capillaries in two adult dogs. When a second injection was given, complete embolization was achieved. Histopathologic examination of the kidneys revealed changes consistent with renal artery obstruction and blockade of the blood supply to the kidneys.

In addition to using poly(3HB) compacts, microspheres of poly(3HB-co-3HV) have been used to deliver tetracycline and its hydrochloride salt for the treatment of periodontitis (Sendil et al., 1998, 1999).

Depending upon valerate content, between 42 and 90% of the tetracycline was released at 100 h, with loadings up to about 11%. Interestingly, it was found that molecular weight variation in the poly(3HB-co-3HV) compositions tested might have influenced the observed release of tetracycline. Encapsulation efficiency of the tetracycline hydrochloride salt was significantly less than the neutral form of the antibiotic.

A novel approach for preparing PHA drug delivery systems (Nobes et al., 1998) involves encapsulating a given drug during the in vitro enzymatic formation of poly(3HB) granules (Gerngross and Martin, 1995). Using this technique, it was possible to encapsulate approximately 4.5% of a model drug compound, Netilmicin, compared with between 4.1 and 17% using solvent evaporation.

Wang and Lehmann (1999) recently prepared poly(3HB) microspheres containing levonorgestrel with an average particle size of 64 μm, and found that this system prolongs the release of the drug by 1.8-fold compared with administration of the drug alone. The microspheres were also found to be effective in vivo, inducing a contraceptive effect in mice. Chen et al. (2000) recently described microspheres (30–40 μm) loaded with diazepam, and reported the characteristic biphasic release pattern with an initial burst effect.

Andersson et al. (1999) disclosed the use of supercritical fluid technology to incorporate the water-insoluble Helicobacter pylori adhesion protein A (HpaA) into poly(3HB) particles. This was achieved by preparing an emulsion with the protein, polymer, and methylene chloride, and then extracting the organic solvent from the emulsion with supercritical carbon dioxide to induce particle formation. Particles containing 0.6% protein were produced, with sizes of 1–3 μm.

7.4
Prodrugs

In 1999, it was discovered that poly(4HB) could be used as a prodrug of 4-hydroxy-butyrate (Williams and Martin, 2001). In this study, rats were dosed (by gavage) with low molecular-weight polymers of poly(4HB) (138 mg kg^{-1}), and the serum was assayed for the presence of monomer. The serum concentration of 4-HB increased to ~86 µM within 30 min, and remained elevated at approximately three- to five-fold the baseline value (9 µM) for about 8 h. In contrast, administration of the monomer resulted in a characteristic rapid increase to 182 µM within 30 min, followed by a rapid decrease to baseline within 2 h. The prolonged release of the monomer from poly(4HB) might potentially be beneficial in the treatment of narcolepsy, alcohol withdrawal, and several other indications. The therapeutic potential of poly(4HB) was also alluded to recently (Sudesh et al., 2000).

7.5
Nerve Repair

Partly on account of the piezoelectric properties of poly(3HB), interest has arisen in the use of this polymer for nerve repair (Aebischer et al., 1988). The basic approach involves aligning severed nerve ends within a small tube of poly(3HB), thus avoiding the need for sutures. Hazari et al. (1999a) and Ljungberg et al. (1999) evaluated the use of a nonwoven poly(3HB) sheet as a wrap to repair transected superficial radial nerves in cats for up to 12 months. Axonal regeneration was shown to be comparable with closure with an epineural suture for a nerve gap of 2–3 mm, and that the inflammatory response created by poly(3HB) was also similar to that found in primary epineural repair. In a subsequent study, the same

material was used to bridge an irreducible gap of 10 mm in rat sciatic nerve, and the results were compared to an autologous nerve graft (Hazari et al., 1999b). Good axonal regeneration in the poly(3HB) conduits with a low level of inflammatory infiltration was observed over 30 days, although the rate and amount of regeneration in the poly(3HB) conduit did not fully match that of the nerve graft.

7.6
Nutritional Uses

7.6.1
Human Nutrition

Several groups have evaluated oligomeric forms of the ketone body, R-3-hydroxybutanoic acid, as an alternative to the sodium salt of the monomer, for potential nutritional and therapeutic uses. Use of these polymeric forms might provide controlled release systems for the monomer and, importantly, overcome the problems associated with administering large amounts of sodium ion *in vivo*. Tasaki et al. (1998) reported the results of infusing dimers and trimers of R-3-hydroxybutyrate into rats, as well as exposure of these compounds to human serum samples and liver homogenate. Although mixtures of these compounds were not hydrolyzed by human serum, the monomer was liberated upon exposure to the liver homogenate as well as after infusion in rats. *In vitro*, the monomer was also liberated after incubation with the enzyme carboxylesterase.

Veech (1998, 2000) and Martin et al. (2000) have evaluated oligomers and oligolides of R-3-hydroxybutyrate *in vivo*, and observed release of the ketone body over prolonged periods. Potential uses of these delivery systems might include seizure control, reduction of protein catabolism, appetite suppression, control of metabolic disease, use in parenteral nutrition, increased cardiac efficiency, treatment of diabetes and insulin-resistant states, control of damage to brain cells in conditions such as Alzheimer's, and treatment of neurodegenerative disorders and epilepsy.

7.6.2
Animal Nutrition

The potential use of poly(3HB) and poly(3HB-co-3HV) polymers as a source of animal nutrition has been evaluated in vivo. Brune and Niemann (1977a) initially reported studies in rats, and subsequently described the digestion of poly(3HB) in pigs (Brune and Niemann, 1977b). In the latter work, when whole cells containing poly(3HB) were used as a nutrient source, approximately 65% of the poly(3HB) was excreted. Forni et al. (1999a,b) found the digestibility of poly(3HB-co-3HV) treated with sodium hydroxide to be increased when compared with the untreated polymer, in both sheep and pigs. Holmes (1988) also reported that poly(3HB) is degraded in the bovine rumen, while Peoples et al. (1999) studied the digestion of poly(3HB) and poly(3HO-co-3HH) in broiler chicks, concluding that the available energy from these polymers lies between that provided by carbohydrates and oils.

7.7
Orthopedic

Several studies of the use of PHA polymers for internal fixation have been undertaken. Vainionpää et al. (1986) described the use of compression-molded T-plates, prepared from poly(3HB) reinforced with carbon fiber (7%), to fix osteotomies of the tibial diaphysis in rabbits. The implants were fixed to the tibia with absorbable PGA sutures, and compared with implants prepared from

reinforced Vicryl®. After 12 weeks, better results were obtained with the reinforced poly(3HB) plates relative to the Vicryl plates, with the latter frequently leading to nonunion of the osteotomies, breakage, and angulation.

Doyle et al. (1991) reported that poly(3HB) can be reinforced with hydroxyapatite (HA) to increase its stiffness to a level approaching that of cortical bone (7–25 GPa). At a HA loading of 40% wt., a poly(3HB)/HA composite had a Young's modulus value of 11 GPa, although strength decreased from ~40 MPa for poly(3HB) to ~20 MPa for the composite. Under in vitro conditions in buffered saline at 37 °C it was noted that the modulus of the filled poly(3HB) sample decreased more rapidly, falling from 9 GPa to 4 GPa over four months at a loading of 20% wt. HA. Bending strength of the same filled sample also fell by about 50% over the same time period, from 55 MPa to around 25 MPa. The behavior of poly(3HB) filled with HA was also studied in vivo. No significant differences between implants derived from poly(3HB) or poly(3HB) filled with HA were observed when rivets were prepared from these materials and inserted into predrilled holes in rabbit femurs. Over time, increased amounts of new bone were found on the implant surfaces, and by six months the implants were closely encased in new cortical bone. The overall tissue responses were considered favorable, and some indication of osteogenic activity for poly(3HB) was noted. Boeree et al. (1993) and Galego et al. (2000) have also studied the mechanical properties of poly(3HB) and poly(3HB-co-3HV)/HA composites, and concluded that they could serve as alternatives to corticocancellous bone grafts.

Since the homopolymer poly(3HB) is piezoelectric, it might help to induce new local bone formation if used as an implant. The addition of other additives might further enhance this property, for example bioactive glasses that upon dissolution and deposition at an implant surface may encourage new bone formation. In this light, Knowles et al. (1991) studied the piezoelectric characteristics of poly(3HB-*co*-3HV) composites with glass fiber at 20, 30 and 40% wt., and found the piezoelectric potential output of these composites to be fairly close to that of bone. In subsequent studies, these composites were evaluated *in vitro* and *in vivo* (Knowles and Hastings, 1993a,b), and have also been studied with the inclusion of HA (Knowles et al., 1992). During *in vitro* studies of the poly(3HB-*co*-3HV)-glass composites, the glass component was found to be highly soluble in the polymer, and the observed weight loss was attributed to dissolution of the glass from the composites. These studies correlated well with observations *in vivo*. The poly(3HB-*co*-3HV)-glass composites were implanted subcutaneously, and as nonload-bearing femoral implants in rats. Initially, relatively high cellular activity was observed that was attributed to ions being released from the glass (and causing a soft tissue reaction), though this activity decreased with time. At four weeks in the femoral implants, cells could be seen entering the surface porosity formed by the solubilizing glass, and over time new bone was seen developing on the implant surface (Knowles and Hastings, 1993a,b). However, when composites of poly(3HB-*co*-3HV)/HA and poly(3HB-*co*-3HV)/HA-glass were implanted in rabbit femurs, and evaluated using a mechanical push-out test during the first eight weeks *in vivo*, it was found that the former bonded better than the latter. One explanation for this observation was attributed to the release of ions from the poly(3HB-*co*-3HV)/HA-glass composite inducing a soft tissue reaction that inhibited the formation of hard tissue at the implant surface (Knowles et al., 1992; Knowles, 1993).

In addition to composites with hydroxy-apatite and glass, Jones et al. (2000) tested composites of poly(3HB-*co*-3HV) with tri-calcium phosphate filler in subcutaneous and femoral implants. These implants were compared with composites derived from PLA and tri-calcium phosphate, and found to degrade about four times more slowly *in vivo*.

In an *in vitro* study, Rivard et al. (1996) prepared highly porous foams of poly(3HB-*co*-3HV) and seeded these scaffolds with chondrocytes and osteoblasts. Maximal cell densities were achieved after 21 days, with cellular diffusion taking place throughout the porous foams.

7.8
Urology

In the mid-1960s it was proposed that poly(3HB) could be used to repair a ureter by inserting a short tube of this material (Baptist and Ziegler, 1965). More recently, Bowald and Johansson (1990) described the use of poly(3HB-*co*-3HV) in the development of a tube for urethral reconstruction. A solution of the poly(3HB-*co*-3HV) copolymer was used to coat thin, knitted tubes of Vicryl, and the tubes were then implanted into four dogs to replace the urethra. After six to nine months, it was claimed that a fully functional urethra tissue had been reconstructed in all animals.

7.9
Wound Management

7.9.1
Sutures
As early as the mid-1960s it was suggested that poly(3HB) could be used as an absorbable suture (Baptist and Ziegler, 1965). Others have also proposed that poly(3HB-*co*-3HV) could be used as a suture coating,

and have coated braided sutures of PGA with solutions of this polymer (Wang and Lehmann, 1991).

7.9.2
Dusting Powders

Holmes (1985) has produced powders of poly(3HB) with small particle sizes, and proposed that these can be used as medical dusting powders, particularly with surgical gloves.

7.9.3
Dressings

Webb and Adsetts (1986) described wound dressings based on volatile solutions of poly(3HB) and poly(3HB-co-3HV) that could form thin films over wounds; these would be especially useful for emergency treatments. These films could potentially prevent airborne bacterial contamination of the wound, but would still be permeable to water vapor. As the preferred solvents are chlorinated hydrocarbons, however, these solutions might present other health hazards both to the patient and administrator. Steel and Norton-Berry (1986) also described wound dressings based on poly(3HB), their method involving the preparation of nonwoven fibrous materials of poly(3HB) and PHBV that could be used like swabs, gauze, lint or fleece.

Davies and Tighe (1995) evaluated the use of poly(3HB) fibers as a potential wound scaffold that could provide a framework for the laying down of a permanent dermal architecture. Using in vitro cell attachment assays with human epithelial cells, it was found that pre-treatment of the poly(3HB) fibers with either base or strong acid improved cell attachment and spreading.

Ishikawa (1996) described the implantation of poly(3HB-co-4HB) films in the abdominal cavity of rats between incisions in the skin and intestine to prevent coales-

cence. After one month, the incisions had substantially healed, and no coalescence had occurred. However, no degradation of the film in vivo was observed at one year.

7.9.4
Soft Tissue Repair

In early studies of the potential uses of poly(3HB) it was proposed that this polymer could be used in the surgical repair of hernias (Baptist and Ziegler, 1965).

Patches of poly(3HB) with a smooth surface on one side and a porous surface on the other have been evaluated as resorbable scaffolds for the repair of soft-tissue defects, and specifically for the closure of lesions in the gastrointestinal tract (Behrend et al., 1999). Using in vitro cell culture, moderate adhesion of mouse and rat intestine fibroblasts to the patch material was observed. When the patches were sutured over incisions in the stomachs of rats that had been closed with two surgical knots, adhesion of the patch to the gastric wall was found to be better than that seen with Vicryl patches, and good ingrowth and tissue regeneration was evident on the porous side of the implanted poly(3HB) patch.

8
Future Directions

With technology now in place to allow the properties of PHA polymers to be tailored to specific applications, coupled with a significant increase in the need for new absorbable biomaterials, this class of polymers currently appears to have a bright future in medicine and pharmacy. Indeed, a wide new range of applications for PHA polymers has recently been described which includes their use as suture anchors, meniscus repair devices, interference screws, bone plates and bone plating systems, meniscus regeneration devices, ligament and tendon grafts, spinal fusion cages and bone dowels, bone graft substitutes, surgical mesh and repair patches, slings, adhesion prevention barriers, skin substitutes, dural substitutes, bulking and filling agents, ureteric and urethral stents, vein valves, ocular cell implants, and hemostats (Williams, 2000; Williams et al., 2000).

9
Patents

Patents cited in studies described in this chapter are listed in Table 2.

Tab. 2 Patents referenced in this work

Patent Number	Assignee	Inventor(s)	Title	Date of Publication
WO 88/06866	Brown University Research Found.	Aebischer, P., Valentini, R.F., Galletti, P.M.	Piezoelectric nerve guidance channels	September 22, 1988
WO 99/52507	Astra Aktiebolag	Andersson, M.-L., Boissier, C., Juppo, A.M., Larsson, A.	Incorporation of active substances in carrier matrixes	October 21, 1999
US 3,225,766	W.R. Grace and Co.	Baptist, J.N., Ziegler, J.B.	Method of making absorbable surgical sutures from poly beta hydroxy acids	December 28, 1965
EP 0 349 505 A2	Astra Meditec AB	Bowald, S.F., Johansson, E.G.	A novel surgical material	January 3, 1990
EP 0 754 467 A1	Astra Aktiebolag	Bowald S.F., Johansson-Ruden, G.	A novel surgical material.	January 22, 1997
US 5,116,868	John Hopkins University	Chen, C.-H., Chen, S.C.	Effective ophthalmic irrigation solution	May 26, 1992
EP 355,453 A2.	Kanegafuchi Kagaku Kogyo Kabushiki Kaisha	Hiraide, A., Katayama, M.	Use of 3-hydroxybutyric acid as an energy source	February 28, 1990
GB 2 160 208A.	Imperial Chemical Industries, Plc.	Holmes, P.A.	Sterilised powders of poly(3-hydroxybutyrate)	December 18, 1985
US 5,480,394 WO93/05824	Terumo Kabushiki Kaisha	Ishikawa, K.	Flexible member for use as a medical bag	January 2, 1996 Apr. 1, 1993
WO 99/32536	Metabolix, Inc.	Martin, D.P., Skraly, F.A., Williams, S.F.	Polyhydroxyalkanoate compositions having controlled degradation rates	July 1, 1999
WO 00/04895	Metabolix, Inc.	Martin, D.P., Peoples, O.P., Williams, S.F.	Nutritional and therapeutic uses of 3-hydroxyalkanoate oligomers	February 3, 2000

Tab. 2 (cont.)

Patent Number	Assignee	Inventor(s)	Title	Date of Publication
JP7275344A2	Nippon Zeon Co. Ltd.	Noisshiki, Y., Komatsuzaki, S.	Medical materials for soft tissue use	October 24, 1995
WO 99/34687	Metabolix, Inc.	Peoples, O.P., Saunders, C., Nichols, S., Beach, L	Animal nutrition compositions	July 15, 1999
EP 0 770 401 A2.	Biotronik Mess- und Therapiegeräte GmbH & Co.	Schmitz, K.-P., Behrend, D.	Method of manufacturing intraluminal stents made of polymer material	February 5, 1997
US 4,603,070	Imperial Chemical Industries, Plc.	Steel, M.L., Norton-Berry, P.	Non-woven fibrous material	July 29, 1986
WO 98/41200 WO 98/41201	British Tehnology Group Ltd.	Veech, R.L.	Therapeutic compositions	September 24, 1998
WO 00/15216	BTG Int. Ltd.	Veech, R.L.	Therapeutic compositions	March 23, 2000
US 5,032,638	American Cyanamid Co.	Wang, D.W., Lehmann, L.T.	Bioabsorbable coating for a surgical device	July 16, 1991
GB 2,166,354	Imperial Chemical Industries, Plc.	Webb, A., Adsetts, J.R.	Wound dressings	May 8, 1986
WO 00/51662	Tepha, Inc.	Williams, S.F.	Bioabsorbable, biocompatible polymers for tissue engineering	September 8, 2000
WO 00/56376	Metabolix, Inc.	Williams, S.F., Martin, D.P., Skraly, F.	Medical devices and applications of polyhydroxyalkanoate polymers,	September 28, 2000
WO 01/19361A2	Tepha, Inc.	Williams, S.F., Martin, D.P.	Therapeutic uses of polymers and oligomers comprising gamma-hydroxybutyrate	March 22, 2001

10
References

Abe, H., Doi, Y., Yamamoto, Y. (1992a) Controlled release of lastet, an anticancer drug, from poly(3-hydroxybutyrate) microspheres containing acylglycerols, *Macromol. Rep.* **A29**, 229–235.

Abe, H., Yamamoto, Y., Doi, Y. (1992b) Preparation of poly(3-hydroxybutyrate) microspheres containing lastet an anticancer drug and its application to drug delivery system, *Kobunshi Ronbunshu* **49**, 61–67.

Aebischer, P., Valentini, R. F., Galletti, P. M. (1988) Piezoelectric nerve guidance channels, PCT Patent Application No. WO 88/06866.

Akhtar, S., Pouton, C. W., Notarianni, L. J., Gould, P. L. (1989) A study of the mechanism of drug release from solvent evaporated carrier systems of biodegradable P(HB-HV) polyesters, *J. Pharm. Pharmacol.* **41**, 5.

Akhtar, S., Pouton, C. W., Notarianni, L. J. (1991) The influence of crystalline morphology and copolymer composition of drug release from solution cast and melt-processed P(HB-HV) copolymer matrices, *J. Controlled Release* **17**, 225–234.

Akhtar, S., Pouton, C. W., Notarianni, L. J. (1992) Crystallization behaviour and drug release from bacterial polyhydroxyalkanoates, *Polymer* **33**, 117–126.

Andersson, M.-L., Boissier, C., Juppo, A. M., Larsson, A. (1999) Incorporation of active substances in carrier matrixes, PCT Patent Application No. WO 99/52507.

Atkins, T. W., Peacock, S. J. (1996a) *In vitro* biodegradation of poly(β-hydroxybutyrate-hydroxyvalerate) microspheres exposed to Hanks buffer, newborn calf serum, pancreatin and synthetic gastric juice, *J. Biomater. Sci. Polymer Edn.* **7**, 1075–1084.

Atkins, T. W., Peacock, S. J. (1996b) The incorporation and release of bovine serum albumin from poly-hydroxybutyrate-hydroxyvalerate microcapsules, *J. Microencapsulation* **13**, 709–717.

Baptist, J. N., Ziegler, J. B. (1965) Method of making absorbable surgical sutures from poly beta hydroxy acids, US Patent No. 3,225,766.

Behrend, D., Lootz, D., Schmitz, K. P., Schywalsky, M., Labahn, D., Hartwig, S., Schaldach, M., Unverdorben, M., Vallbracht, C., Laenger, F. (1998) PHB as a bioresorbable material for intravascular stents, *Am. J. Cardiol., Tenth Annual Symposium Transcatheter Cardiovascular Therapeutics, Abstract TCT-8*, 4S.

Behrend, D., Nischan, C., Kunze, C., Sass, M., Schmitz, K.-P. (1999) Resorbable scaffolds for tissue engineering, Proc. European Medical and Biological Engineering Conference, *Med. Biol. Eng. Comput.* **37** (Suppl.), 1510–1511.

Behrend, D., Schmitz, K.-P., Haubold, A. (2000a) Bioresorbable polymer materials for implant technology, *Adv. Eng. Mater.* **2**, 123–125.

Behrend, D., Kramer, S., Schmitz, K.-P. (2000b) Biodegradation and biocompatibility of resorbable polyester, *Zell. Interakt. Biomater.* **28**–32.

Bissery, M.-C., Puisieux, F., Thies, C. (1983) A study of process parameters in the making of microspheres by the solvent evaporation procedure, *Third Expo. Congr. Int. Technol. Pharm.* **3**, 233–239.

Bissery, M.-C., Valeriote, F., Thies, C. (1984a) *In vitro* and *in vivo* evaluation of CCNU-loaded microspheres prepared from poly(L)-lactide and poly(β-hydroxybutyrate), in: *Microspheres and Drug Therapy, Pharmaceutical, Immunological and Medical Aspects* (Davis, S.S., Illum, L., McVie, J.G., Tomlinson, E., Eds.), Amsterdam: Elsevier, 217–227.

Bissery, M.-C., Valeriote, F., Thies, C. (1984b) *In vitro* lomustine release from small poly(β-hydroxybutyrate) and poly(D,L-lactide) microspheres, *Proc. Int. Symp. Controlled Release Bioact. Mater.* **11**, 25–26.

Bissery, M.-C., Valeriote, F., Thies, C. (1985) Fate and effect of CCNU-loaded microspheres made of poly(D,L)lactide (PLA) or poly-β-hydroxybutyrate (PHB) in mice. *Proc. Int. Symp. Controlled Release Bioact. Mater.* **12**, 181–182.

Boeree, N. R., Dove, J., Cooper, J. J., Knowles, J., Hastings, G. W. (1993) Development of a degradable composite for orthopedic use: mechanical evaluation of an hydroxyapatite-polyhydroxybutyrate composite material, *Biomaterials* **14**, 793–796.

Böstman, O. M. (1991) Absorbable implants for the fixation of fractures, *J. Bone Joint Surg. Am.* **73**, 148–153.

Bowald, S. F., Johansson, E. G. (1990) A novel surgical material, European Patent Application No. 0 349 505 A2.

Bowald, S. F., Johansson-Ruden, G. (1997). A novel surgical material. European Patent Application No. 0 754 467 A1.

Brophy, M. R., Deasy, P. B. (1986) *In vitro* and *in vivo* studies on biodegradable polyester microparticles containing sulphamethizole, *Int. J. Pharm.* **29**, 223–231.

Brune, H., Niemann, E. (1977a) Use and compatibility of bacterial protein (*Hydrogenomonas*) with various contents of poly-beta-hydroxybutyric acid in animal nutrition. 1. Weight development and N balance in growing rats, *Z. Tierphysiol. Tierernähr. Futtermittelkd* **38**, 13–22.

Brune, H., Niemann, E. (1977b) Value and compatibility in animal nutrition of bacterial protein (*Hydrogenomonas*) containing various amounts of poly-beta-hydroxybutyric acid. 2. Body weight, nitrogen balance and fatty acid pattern of liver, muscle and kidney depot fat of growing swine, *Z. Tierphysiol. Tierernähr. Futtermittelkd* **38**, 81–93.

Cargill, R., Engle, K., Gardner, C. R., Porter, P., Sparer, R. V., Fix, J. A. (1989) Controlled gastric emptying. II. *In vitro* erosion and gastric residence times of an erodible device in Beagle dogs, *Pharm. Res.* **6**, 506–509.

Chaput, C., Yahia, L'H., Selmani, A., Rivard, C.-H. (1995) Natural poly(hydroxybutyrate-hydroxyvalerate) polymers as degradable biomaterials, *Mater. Res. Soc. Symp. Proc.* **385**, 49–54.

Chen, C.-H., Chen, S. C. (1992) Effective ophthalmic irrigation solution, US Patent No. 5,116,868.

Chen, J.-H., Chen, Z.-L., Hou, L.-B., Liu, S.-T. (2000) Preparation and characterization of di-

azepam-polyhydroxybutyrate microspheres, *Gongneng Gaofenzi Xuebao* **13**, 61–64.

Ciardelli, G., Saad, B., Hirt, T., Keiser, O., Neuenschwander, P., Suter, U. W., Uhlschmid, G. K. (1995) Phagocytosis and biodegradation of shortchain poly[β-3-hydroxybutyric acid] particles in macrophage cell line, *J. Mater. Sci. Mater. Med.* **6**, 725–730.

Collins, A. E. M., Deasy, P. B., MacCarthy, D., Shanley, D. B. (1989) Evaluation of a controlledrelease compact containing tetracycline hydrochloride bonded to tooth for the treatment of periodontal disease, *Int. J. Pharm.* **51**, 103–114.

Conway, B. R., Eyles, J. E., Alpar, H. O. (1996) Immune response to antigen in microspheres of different polymers, *Proc. Int. Symp. Controlled Release Bioact. Mater.* **23**, 335–336.

Conway, B. R., Eyles, J. E., Alpar, H. O. (1997) A comparative study on the immune responses to antigens in PLA and PHB microspheres, *J. Controlled Release* **49**, 1–9.

Dang, M.-H., Birchler, F., Ruffieux, K., Wintermantel, E. (1996) Toxicity screening of biodegradable polymers. I. Selection and evaluation of cell culture test methods, *J. Environ. Poly. Degrad.* **4**, 197–203.

Davies, S., Tighe, B. (1995) Cell attachment to gelspun polyhydroxybutyrate fibers, *Poly. Prepr. (Am. Chem. Soc., Div. Polym. Chem.)* **36**, 103–104.

Deasy, P. B., Collins, A. E. M., MacCarthy, D. J., Russell, R. J. (1989) Use of strips containing tetracycline hydrochloride or metronidazole for the treatment of advanced periodontal disease, *J. Pharm. Pharmacol.* **41**, 694–699.

Doi, Y. (1990) *Microbial Polyesters*, New York: VCH.

Doyle, C., Tanner, E. T., Bonfield, W. (1991) *In vitro* and *in vivo* evaluation of polyhydroxybutyrate and of polyhydroxybutyrate reinforced with hydroxyapatite, *Biomaterials* **12**, 841–847.

Duvernoy, O., Malm, T., Ramström, J., Bowald, S. (1995) A biodegradable patch used as a pericardial substitute after cardiac surgery: 6- and 24-month evaluation with CT, *Thorac. Cardiovasc. Surg.* **43**, 271–274.

Eldridge, J. H., Hammond, C. J., Meulbroek, J. A., Staas, J. K., Gilley, R. M., Tice, T. R. (1990) Controlled vaccine release in the gut-associated lymphoid tissues. I. Orally administered biodegradable microspheres target the Peyer's patches, *J. Controlled Release* **11**, 205–214.

Embleton, J. K., Tighe, B. J. (1992a) 5. Regulation of polyester microcapsule morphology, *Drug Tar-*

geting Delivery, 1 (Microencapsulation of Drugs) 45–54.

Embleton, J. K, Tighe, B. J. (1992b) Polymers for biodegradable medical devices. IX: Microencapsulation studies; effects of polymer composition and process parameters on poly-hydroxybutyrate-hydroxyvalerate microcapsule morphology, *J. Microencapsulation* **9**, 73–87.

Embleton, J. K., Tighe, B. J. (1993) Polymers for biodegradable medical devices. X. Microencapsulation studies: control of poly-hydroxybutyrate-hydroxyvalerate microcapsules porosity via poly-caprolactone blending, *J. Microencapsulation* **10**, 341–352.

Engelberg, I., Kohn, J. (1991) Physico-mechanical properties of degradable polymers used in medical applications: a comparative study, *Biomaterials* **12**, 292–304.

Entholzner, E., Mielke, L., Pichlmeier, R., Weber, F., Schneck, H. (1995) EEG changes during sedation with gamma-hydroxybutyric acid, *Anesthetist* **44**, 345–350.

Forni, D., Bee, G., Kreuzer, M., Wenk, C. (1999a) Novel biodegradable plastics in sheep nutrition. 2. Effects of NaOH pretreatment of poly(3-hydroxybutyrate-*co*-3-hydroxyvalerate) on *in vivo* digestibility and on *in vitro* disappearance, *J. Anim. Physiol. Anim. Nutr.* **81**, 41–50.

Forni, D., Wenk, C., Bee, G. (1999b) Digestive utilization of novel biodegradable plastic in growing pigs, *Ann. Zootech.* **48**, 163–171.

Gagnon, K. D., Fuller, R. C., Lenz, R. W., Farris, R. J. (1992) A thermoplastic elastomer produced by the bacterium *Pseudomonas oleovorans*, *Rubber World* **207**, 32–38.

Galego, N., Rozsa, C., Sánchez, R., Fung, J., Vázquez, A., Tomás, J. S. (2000) Characterization and application of poly(β-hydroxyalkanoates) family as composite biomaterials, *Polym. Test.* **19**, 485–492.

Galgut, P., Pitrola, R., Waite, I., Doyle, C., Smith, R. (1991) Histological evaluation of biodegradable and non-degradable membranes placed transcutaneously in rats, *J. Clin. Periodontol.* **18**, 581–586.

Gangrade, N., Price, J. C. (1991) Poly(hydroxybutyrate-hydroxyvalerate) microspheres containing progesterone: preparation, morphology and release properties, *J. Microencapsulation* **8**, 185–202.

Gangrade, N., Price, J. C. (1992) Properties of implantable pellets prepared from a biodegradable polyester, *Drug Dev. Ind. Pharm.* **18**, 1633–1648.

Garrido, L. (1999) Nondestructive evaluation of biodegradable porous matrices for tissue engineering, in *Methods in Molecular Medicine, Vol. 18: Tissue Engineering Methods and Protocols* (Morgan, J. R., Yarmush, M. L., Eds.), Totowa, NJ: Humana Press, 35–45.

Gerngross, T. U., Martin, D. P. (1995) Enzyme-catalyzed synthesis of poly[β-(-)-3-hydroxybutyrate]: formation of macroscopic granules *in vitro*, *Proc. Natl. Acad. Sci. USA* **92**, 6279–6283.

Gogolewski, S., Jovanovic, M., Perren, S. M., Dillon, J. G., Hughes, M. K. (1993) Tissue response and *in vivo* degradation of selected polyhydroxyacids: Polylactides (PLA), poly(3-hydroxybutyrate) (PHB), and poly(3-hydroxybutyrate-*co*-3-hydroxyvalerate) (PHB/VA), *J. Biomed. Mater. Res.* **27**, 1135–1148.

Gotfredsen, K., Nimb, L., Hjørting-Hansen, E. (1994) Immediate implant placement using a biodegradable barrier, polyhydroxybutyrate-hydroxyvalerate reinforced with polyglactin 910. An experimental study in dogs, *Clin. Oral Impl. Res.* **5**, 83–91.

Gould, P. L., Holland, S. J., Tighe, B. J. (1987) Polymers for biodegradable medical devices IV. Hydroxybutyrate-valerate copolymers as non-disintegrating matrices for controlled-release oral dosage forms, *Int. J. Pharm.* **38**, 231–237.

Gürsel, I., Hasirci, V. (1995) Properties and drug release behaviour of poly(3-hydroxybutyric acid) and various poly(3-hydroxybutyrate-hydroxyvalerate) copolymer microcapsules, *J. Microencapsulation* **12**, 185–193.

Hasirci, V. (2000) Biodegradable biomedical polymers, in: *Biomaterials and Bioengineering Handbook* (Wise, D.L., Ed.), New York: Marcel Dekker, 141–155.

Hasirci, V., Gürsel, I., Türesin, F., Yigitel, G., Korkusuz, F., Alaeddinoglu, G. (1998) Microbial polyhydroxyalkanoates as biodegradable drug release materials, in: *Biomedical Science and Technology* (Hincal, A. A. and Kas, H. S., Eds.), New York: Plenum Press, 183–187.

Hazari, A., Johansson-Rudén, G., Junemo-Bostrom, K., Ljungberg, C., Terenghi, G., Green, C., Wiberg, M. (1999a) A new resorbable wrap-around implant as an alternative nerve repair technique, *J. Hand Surg.* **24B**, 291–295.

Hazari, A., Wiberg, M., Johansson-Rudén, G., Green, C., Terenghi, G. (1999b) A resorbable nerve conduit as an alternative to nerve autograft in nerve gap repair, *Br. J. Plast. Surg.* **52**, 653–657.

Hiraide, A., Katayama, M. (1990) Use of 3-hydroxybutyric acid as an energy source, European Patent Application No. 355,453 A2.

Hocking, P. J., Marchessault, R. H. (1994) Biopolyesters, in: *Chemistry and Technology of Biodegradable Polymers* (Griffin, G.J.L., Ed.), Glasgow: Blackie, 48–96.

Hoerstrup, S. P., Sodian, R., Daebritz, S., Wang, J., Bacha, E. A., Martin, D. P., Moran, A. M., Guleserian, K. J., Sperling, J. S., Kaushal, S., Vacanti, J. P., Schoen, F. J., Mayer, J. E. (2000) Functional trileaflet heart valves grown *in vitro*, *Circulation* **102** (suppl. III), III-44–III-49.

Holland, S. J., Tighe, B. J., Gould, P. L. (1986) Polymers for biodegradable medical devices. I. The potential of polyesters as controlled macromolecular release systems, *J. Controlled Release* **4**, 155–180.

Holland, S. J., Jolly, A. M., Yasin, M., Tighe, B. J. (1987) Polymers for biodegradable medical devices II. Hydroxybutyrate-hydroxyvalerate copolymers: hydrolytic degradation studies, *Biomaterials* **8**, 289–295.

Holland, S. J., Yasin, M., Tighe, B. J. (1990) Polymers for biodegradable medical devices VII. Hydroxybutyrate-hydroxyvalerate copolymers: degradation of copolymers and their blends with polysaccharides under *in vitro* physiological conditions, *Biomaterials* **11**, 206–215.

Holmes, P. A. (1985) Sterilised powders of poly(3-hydroxybutyrate), UK Patent Application No. 2 160 208A.

Holmes, P. A. (1988) Biologically produced β-3-hydroxyalkanoate polymers and copolymers, in: *Developments in Crystalline Polymers* (Bassett, D.C., Ed.), London: Elsevier, 1–65, Vol. 2.

Ishikawa, K. (1996) Flexible member for use as a medical bag, US Patent No. 5,480,394.

Jones, R. D., Price, J. C., Stuedemann, J. A., Bowen, J. M. (1994) *In vitro* and *in vivo* release of metoclopramide from a subdermal diffusion matrix with potential in preventing fescue toxicosis in cattle, *J. Controlled Release* **30**, 35–44.

Jones, N. L., Cooper, J. J., Waters, R. D., Williams, D. F. (2000) Resorption profile and biological response of calcium phosphate filled PLLA and PHB7V, in: *Synthetic Bioabsorbable Polymers for Implants*, (Agrawal, C.M., Parr, J.E., Lin, S.T. Eds.), Scranton: ASTM, 69–82.

Juni, K., Nakano, M. (1987) Poly(hydroxy acids) in drug delivery, *CRC Crit. Rev. Ther. Drug Carrier Syst.* **3**, 209–232.

Juni, K., Nakano, M., Kubota, M., Matsui, N., Ichihara, T., Beppu, T., Mori, K., Akagi, M. (1985) Controlled release of aclarubicin from poly(beta-hydroxybutyric acid) microspheres, *Igaku no Ayumi* **132**, 735–736.

Juni, K., Nakano, M., Kubota, M. (1986) Controlled release of aclarubicin, an anticancer antibiotic, from poly-β-hydroxybutyric acid microspheres, *J. Controlled Release* **4**, 25–32.

Kassab, A. Ch., Xu, K., Denkbas, E. B., Dou, Y., Zhao, S., Piskin, E. (1997) Rifampicin carrying polyhydroxybutyrate microspheres as a potential chemoembolization agent, *J. Biomater. Sci., Polym. Edn.* **8**, 947–961.

Kassab, A. Ch., Piskin, E., Bilgic, S., Denkbas, E. B., Xu, K. (1999) Embolization with polyhydroxybutyrate (PHB) microspheres: *In-vivo* studies, *J. Bioact. Compat. Polym.* **14**, 291–303.

Kawaguchi, T., Tsugane, A., Higashide, K., Endoh, H., Hasegawa, T., Kanno, H., Seki, T., Juni, K., Fukushima, S., Nakano, M. (1992) Control of drug release with a combination of prodrug and polymer matrix: Antitumor activity and release profiles of 2′,3′-diacyl-5-fluoro-2′-deoxyuridine from poly(3-hydroxybutyrate) microspheres, *J. Pharm. Sci.* **81**, 508–512.

Kharenko, A. V., Iordanskii, A. L. (1999) Diltiazem release from matrices based on polyhydroxybutyrate, *Proc. Int. Symp. Controlled Release Bioact. Mater.* **26**, 919–920.

Kishida, A., Yoshioka, S., Takeda, Y., Uchiyama, M. (1989) Formulation-assisted biodegradable polymer matrices, *Chem. Pharm. Bull.* **37**, 1954–1956.

Knowles, J. C. (1993) Development of a natural degradable polymer for orthopaedic use, *J. Med. Eng. Tech.* **17**, 129–137.

Knowles, J. C., Hastings, G. W. (1992) *In vitro* degradation of a polyhydroxybutyrate/polyhydroxyvalerate copolymer, *J. Mater. Sci. Mater. Med.* **3**, 352–358.

Knowles, J. C., Hastings, G. W. (1993a) *In vitro* and *in vivo* investigation of a range of phosphate glass-reinforced polyhydroxybutyrate-based degradable composites, *J. Mater. Sci. Mater. Med.* **4**, 102–106.

Knowles, J. C., Hastings, G. W. (1993b) Physical properties of a degradable composite for orthopaedic use which closely matches bone, *Proceedings, First International Conference on Intelligent Materials* (Takagi, T., Ed.), Lancaster: Technomic, 493–504.

Knowles, J. C., Mahmud, F. A., Hastings, G. W. (1991) Piezoelectric characteristics of a polyhydroxybutyrate-based composite, *Clin. Mater.* **8**, 155–158.

Knowles, J. C., Hastings, G. W., Ohta, H., Niwa, S., Boeree, N. (1992) Development of a degradable composite for orthopedic use: *in vivo* biomechanical and histological evaluation of two bioactive degradable composites based on the polyhydroxybutyrate polymer, *Biomaterials* **13**, 491–496.

Koosha, F. Muller, R. H., Washington, C. (1987) Production of polyhydroxybutyrate (PHB) nanoparticles for drug targeting, *J. Pharm. Pharmacol.* **39**, 136.

Koosha, F., Muller, R. H., Davis, S. S. (1988) A continuous flow system for *in-vitro* evaluation of drug-loaded biodegradable colloidal carriers, *J. Pharm. Pharmacol.* **40**, 131.

Koosha, F., Muller, R. H., Davis, S. S. (1989) Polyhydroxybutyrate as a drug carrier, *CRC Crit. Rev. Ther. Drug Carrier Syst.* **6**, 117–130.

Korkusuz, F., Korkusuz, P., Eksioglu, F., Gursel, I., Hasirci, V. (2001) *J. Biomed. Mater. Res.* **55**, 217–228.

Korsatko, W., Wabnegg, B., Tillian, H. M., Braunegg, G., Lafferty, R. M. (1983a) Poly-D-(-)-3-hydroxybutyric acid – a biodegradable carrier for long term medication dosage. 2. The biodegradation in animals and *in vitro* – *in vivo* correlation of the liberation of pharmaceuticals from parenteral matrix retard tablets, *Pharm. Ind.* **45**, 1004–1007.

Korsatko, W., Wabnegg, B., Braunegg, G., Lafferty, R. M., Strempfl, F. (1983b) Poly-D-(-)-3-hydroxybutyric acid (PHBA) – a biodegradable carrier for long term medication dosage. 1. Development of parenteral matrix tablets for long term application of pharmaceuticals, *Pharm. Ind.* **45**, 525–527.

Korsatko, W., Wabnegg, B., Tillian, H. M., Egger, G., Pfragner, R., Walser, V. (1984) Poly-D-(-)-3-hydroxybutyric acid (poly-HBA) – a biodegradable former for long term medication dosage. 3. Studies on compatibility of poly-HBA – implantation tablets in tissue culture and animals, *Pharm. Ind.* **46**, 952–954.

Korsatko, W., Korsatko, B., Lafferty, R. M., Weidmann, V. (1987) The influence of the molecular weight of poly-D-(-)-3-hydroxybutyric acid on its use as a retard matrix for sustained drug release, *Proceedings, Third European Congress of Biopharmacology and Pharmokinetics* 234–242, Vol. 1.

Kostopoulos, L., Karring, T. (1994a) Guided bone regeneration in mandibular defects in rats using a bioresorbable polymer, *Clin. Oral Impl. Res.* **5**, 66–74.

Kostopoulos, L., Karring, T. (1994b) Augmentation of the rat mandible using guided tissue regeneration, *Clin. Oral Impl. Res.* **5**, 75–82.

Kubota, M., Nakano, M., Juni, K. (1988) Mechanism of enhancement of the release rate of aclarubicin from poly-β-hydroxybutyric acid microspheres by fatty acid esters, *Chem. Pharm. Bull.* **36**, 333–337.

Kusaka, S., Abe, H., Lee, S. Y., Doi, Y. (1997) Molecular mass of poly[β-3-hydroxybutyric acid] produced in a recombinant *Escherichia coli*, *Appl. Microb. Biotechnol.* **47**, 140–143.

Lafferty, R. M., Korsatko, B., Korsatko, W. (1988) Microbial production of poly-β-hydroxybutyric acid, in: *Biotechnology* Vol. 6b (Rehm, H.-J., Reed, G., Eds.), Weinheim: VCH, 135–176.

Lee, S. Y., Choi, J.-I., Han, K., Song, J. Y. (1999) Removal of endotoxin during purification of poly(3-hydroxybutyrate) from gram-negative bacteria, *Appl. Environ. Microbiol.* **65**, 2762–2764.

Leenstra, T. S., Maltha, J. C., Kuijpers-Jagtman, A. M. (1995) Biodegradation of non-porous films after submucoperiosteal implantation on the palate of Beagle dogs, *J. Mater. Sci. Mater. Med.* **6**, 445–450.

Linhardt, R. J., Flanagan, D. R., Schmitt, E., Wang, H. T. (1990) Biodegradable poly(esters) and the delivery of bioactive agents, *Poly. Prepr. (Am. Chem. Soc., Div. Polym. Chem.)* **31**, 249–250.

Ljungberg, C., Johansson-Ruden, G., Boström, K. J., Novikov, L., Wiberg, M. (1999) Neuronal survival using a resorbable synthetic conduit as an alternative to primary nerve repair, *Microsurgery* **19**, 259–264.

Löbler, M., Sass, M., Michel, P., Hopt, U. T., Kunze, C., Schmitz, K.-P. (1999) Differential gene expression after implantation of biomaterials into rat gastrointestine, *J. Mater. Sci. Mater. Med.* **10**, 797–799.

Luo, S., Netravali, A. N. (1999) Effect of ^{60}Co γ-radiation on the properties of poly(hydroxybutyrate-*co*-hydroxyvalerate), *Appl. Polym. Sci.* **73**, 1059–1067.

Malm, T., Bowald, S., Bylock, A., Saldeen, T., Busch, C. (1992a) Regeneration of pericardial tissue on absorbable polymer patches implanted into the pericardial sac. An immunohistochemical, ultrasound and biochemical study in sheep. *Scand. J. Thorac. Cardiovasc. Surg.* **26**, 15–21.

Malm, T., Bowald, S., Bylock, A., Busch, C. (1992b) Prevention of postoperative pericardial adhesions by closure of the pericardium with absorbable polymer patches. An experimental study. *J. Thorac. Cardiovasc. Surg.* **104**, 600–607.

Malm, T., Bowald, S., Karacagil, S., Bylock, A., Busch, C. (1992c) A new biodegradable patch for closure of atrial septal defect, *Scand. J. Thor. Cardiovasc. Surg.* **26**, 9–14.

Malm, T., Bowald, S., Bylock, A., Busch, C., Saldeen, T. (1994) Enlargement of the right ventricular outflow tract and the pulmonary

artery with a new biodegradable patch in trans-
annular position, *Eur. Surg. Res.* **26**, 298–308.

Marois, Y., Zhang, Z., Vert. M., Deng, X., Lenz, R.,
Guidoin, R. (1999a) Effect of sterilization on the
physical and structural characteristics of polyhydroxyoctanoate (PHO), *J. Biomater. Sci. Polymer
Edn.* **10**, 469–482.

Marois, Y., Zhang, Z., Vert. M., Deng, X., Lenz, R.,
Guidoin, R. (1999b) Hydrolytic and enzymatic
incubation of polyhydroxyoctanoate (PHO): a short-
term *in vitro* study of a degradable bacterial polyester, *J. Biomater. Sci. Polymer Edn.* **10**, 483–499.

Marois, Y., Zhang, Z., Vert, M., Beaulieu, L., Lenz,
R. W., Guidoin, R. (1999c) *In vivo* biocompatibility and degradation studies of polyhydroxyoctanoate in the rat: A new sealant for the polyester
arterial prosthesis, *Tissue Eng.* **5**, 369–386.

Marois, Y., Zhang, Z., Vert, M., Deng, X., Lenz, R.
W., Guidoin, R. (2000) Bacterial polyesters for
biomedical applications: *In vitro* and *in vivo*
assessments of sterilization, degradation rate and
biocompatibility of poly(β-hydroxyoctanoate)
(PHO), in: *Synthetic Bioabsorbable Polymers for
Implants* (Agrawal, C.M., Parr, J.E., Lin, S.T.,
Eds.), Scranton: ASTM, 12–38.

Martin, D. P., Skraly, F. A., Williams, S. F. (1999)
Polyhydroxyalkanoate compositions having con-
trolled degradation rates, PCT Patent Application
No. WO 99/32536.

Martin, D. P., Peoples, O. P., Williams, S. F. (2000)
Nutritional and therapeutic uses of 3-hydroxy-
alkanoate oligomers, PCT Patent Application No.
WO 00/04895.

McLeod, B. J., Haresign, W., Peters, A. R., Humke,
R., Lamming, G. E. (1988) The development of
subcutaneous-delivery preparations of GnRH for
the induction of ovulation in acyclic sheep and
cattle, *Anim. Reprod. Sci.* **17**, 33–50.

Miller, N. D., Williams, D. F. (1987) On the
biodegradation of poly-β-hydroxybutyrate (PHB)
homopolymer and poly-β-hydroxybutyrate-
hydroxyvalerate copolymers, *Biomaterials* **8**,
129–137.

Mukai, K., Doi, Y., Sema, Y., Tomita, K. (1993)
Substrate specificities in hydrolysis of polyhydroxyalkanoates by microbial esterases, *Biotechnol. Lett*, **15**, 601–604.

Nasseri, B. A., Pomerantseva, I., Ochoa, E. R.,
Martin, D. P., Oesterle, S. N., Vacanti, J. P. (2000)
Comparison of three polymer scaffolds for
vascular tube formation, *Tissue Eng.* **6**, 667.

Nelson, T., Kaufman, E., Kline, J., Sokoloff, L. (1981)
The extraneural distribution of γ-hydroxybuty-
rate, *J. Neurochem.* **37**, 1345–1348.

Nkere, U. U., Whawell, S. A., Sarraf, C. E.,
Schofield, J. B., O'Keefe, P. A. (1998) Pericardial
substitution after cardiopulmonary bypass surgery: A trial of an absorbable patch, *Thorac.
Cardiovasc. Surg.* **46**, 77–83.

Nobes, G. A. R., Marchessault, R. H., Maysinger, D.
(1998) Polyhydroxyalkanoates: materials for delivery systems, *Drug Delivery*, **5**, 167–177.

Noisshiki, Y., Komatsuzaki, S. (1995) Medical
materials for soft tissue use, Japanese Patent
Application No. JP7275344A2.

Pawan, G. L. S., Semple, S. J. G. (1983) Effect of 3-
hydroxybutyrate in obese subjects on very-low-
energy diets and during therapeutic starvation,
Lancet **1**, 15–17.

Peoples, O. P., Saunders, C., Nichols, S., Beach, L.
(1999) Animal nutrition compositions, PCT
Patent Applications No. WO 99/34687.

Pouton, C. W., Akhtar, S. (1996) Biosynthetic
polyhydroxyalkanoates and their potential in
drug delivery, *Adv. Drug Del. Rev.* **18**, 133–162.

Reusch, R. N., Sparrow, A. W., Gardiner, J. (1992)
Transport of poly-β-hydroxybutyrate in human
plasma, *Biochim. Biophys. Acta* **1123**, 33–40.

Reusch, R. N., Huang, R., Kosk-Kosicka, D. (1997)
Novel components and enzymatic activities of the
human erythrocyte plasma membrane calcium
pump, *FEBS Lett.* **412**, 592–596.

Rivard, C. H., Chaput, C. J., DesRosiers, E. A.,
Yahia, L'H., Selmani, A. (1995) Fibroblast seeding
and culture in biodegradable porous substrates,
J. Appl. Biomater. **6**, 65–68.

Rivard, C. H., Chaput, C., Rhalmi, S., Selmani, A.
(1996) Bio-absorbable synthetic polyesters and
tissue regeneration: a study of three-dimensional
proliferation of ovine chondrocytes and osteoblasts, *Ann. Chir.* **50**, 651–658.

Rouxhet, L., Legras, R., Schneider, Y.-J. (1998)
Interactions between biodegradable polymer,
poly(hydroxybutyrate-hydroxyvalerate), proteins
and macrophages, *Macromol. Symp.* **130**, 347–366.

Saad, B., Ciardelli, G., Matter, S., Welti, M.,
Uhlschmid, G. K., Neuenschwander, P., Suter, U.
W. (1996a) Characterization of the cell response
of cultured macrophages and fibroblasts to
particles of short-chain poly[β-3-hydroxybutyric
acid], *J. Biomed. Mater. Res.* **30**, 429–439.

Saad, B., Ciardelli, G., Matter, S., Welti, M.,
Uhlschmid, G. K., Neuenschwander, P., Suter, U.
W. (1996b) Cell response of cultured macrophages, fibroblasts, and co-cultures of Kupffer
cells and hepatocytes to particles of short-chain
poly[β-3-hydroxybutyric acid], *J. Mater. Sci. Mater.
Med.* **7**, 56–61.

Saad, B., Matter, S., Ciardelli, G., Uhlschmid, G. K., Welti, M., Neuenschwander, P., Suter, U. W. (1996c) Interactions of osteoblasts and macrophages with biodegradable and highly porous polyesterurethane foam and its biodegradation products, *J. Biomed. Mater. Res.* **32**, 355–366.

Saad, B. Neuenschwander, P., Uhlschmid, G. K., Suter, U. W. (1999) New versatile, elastomeric, degradable polymeric materials for medicine, *Int. J. Biol. Macromol.* **25**, 293–301.

Saito, T., Tomita, K., Juni, K., Ooba, K. (1991) *In vivo* and *in vitro* degradation of poly(3-hydroxybutyrate) in rat, *Biomaterials* **12**, 309–312.

Scharf, M. B., Lai, A. A., Branigan, B., Stover, R., Berkowitz, D. B. (1998) Pharmacokinetics of gammahydroxybutyrate (GHB) in narcoleptic patients, *SLEEP* **21**, 507–514.

Schmitz, K.-P., Behrend, D. (1997) Method of manufacturing intraluminal stents made of polymer material, European Patent Application No. 0 770 401 A2.

Schoen, F. J. Levy, R. J. (1999) Tissue heart valves: current challenges and future research perspectives, *J. Biomed. Mater. Res.* **47**, 439–465.

Scholz, C. (2000) Poly(beta-hydroxyalkanoates) as potential biomedical materials: an overview, *ACS Symp. Ser. (Polymers from renewable resources: biopolyesters and biocatalysis)*, **764**, 328–334.

Seebach, D., Brunner, A., Bürger, H. M., Schneider, J., Reusch, R. N. (1994) Isolation and ¹H-NMR spectroscopic identification of poly(3-hydroxybutanoate) from prokaryotic and eukaryotic organisms, *Eur. J. Biochem.* **224**, 317–328.

Selmani, A., Chaput, C., Paradis, V., Yahia, L'H., Rivard, C.-H. (1995) Microscopical analysis of 2D/3D porous poly(hydroxybutyrate-co-hydroxyvalerate) for tissue engineering, *Adv. Sci. Technol.* **12**, 165–172.

Sendil, D., Gürsel, I., Wise, D. L., Hasirci, V. (1998) Antibiotic release from biodegradable PHBV microparticles, in: *Biomed. Sci. Technol. [Proceedings, Fourth International Symposium (1998), Meeting Date 1997]*, (Hincal, A. A., Kas, H. S., Eds.), New York: Plenum Press, 89–96.

Sendil, D., Gürsel, I., Wise, D. L., Hasirci, V. (1999) Antibiotic release from biodegradable PHBV microparticles, *J. Controlled Release* **59**, 207–217.

Shum-Tim, D., Stock, U., Hrkach, J., Shinoka, T., Lien, J., Moses, M. A., Stamp, A., Taylor, G., Moran, A. M., Landis, W., Langer, R., Vacanti, J. P., Mayer, J. E. (1999) Tissue engineering of autologous aorta using a new biodegradable polymer, *Ann. Thorac. Surg.* **68**, 2298–2305.

Sim, S. J., Snell, K. D., Hogan, S. A., Stubbe, J., Rha, C., Sinskey, A. J. (1997) PHA synthase activity controls the molecular weight and polydispersity of polyhydroxybutyrate *in vivo*, *Nature Biotechnol.* **15**, 63–67.

Sodian, R., Sperling, J. S., Martin, D. P., Stock, U., Mayer, J. E., Vacanti, J. P. (1999) Tissue engineering of a trileaflet heart valve – Early *in vitro* experiences with a combined polymer, *Tissue Eng.* **5**, 489–493.

Sodian, R., Hoerstrup, S. P., Sperling, J. S., Martin, D. P., Daebritz, S., Mayer, J. E., Vacanti, J. P. (2000a) Evaluation of biodegradable, three-dimensional matrices for tissue engineering of heart valves, *ASAIO J.* **46**, 107–110.

Sodian, R., Sperling, J. S., Martin, D. P., Egozy, A., Stock, U., Mayer, J. E., Vacanti, J. P. (2000b) Fabrication of a trileaflet heart valve scaffold from a polyhydroxyalkanoate biopolyester for use in tissue engineering, *Tissue Eng.* **6**, 183–188.

Sodian, R., Hoerstrup, S. P., Sperling, J. S., Daebritz, S. H., Martin, D. P., Schoen, F. J., Vacanti, J. P., Mayer, J. E. (2000c) Tissue engineering of heart valves: *in vitro* experiences, *Ann. Thorac. Surg.* **70**, 140–144.

Sodian, R., Hoerstrup, S. P., Sperling, J. S., Daebritz, S., Martin, D. P., Moran, A. M., Kim, B. S., Schoen, F. J., Vacanti, J. P., Mayer, J. E. (2000d) Early *in vivo* experience with tissue-engineered trileaflet heart valves, *Circulation* **102** (suppl. III), III-22–III-29.

Steel, M. L., Norton-Berry, P. (1986) Non-woven fibrous material, US Patent 4,603,070.

Steinbüchel, A. (1991) Polyhydroxyalkanoic acids, in: *Biomaterials* (Byrom, D., Ed.), New York: Stockton Press, 123–213.

Steinbüchel, A., Valentin, H. E. (1995) Diversity of bacterial polyhydroxyalkanoic acids, *FEMS Microbiol. Lett.* **128**, 219–228.

Stock, U. A., Sodian, R., Herden, T., Martin, D. P., Egozy, A., Vacanti, J. P., Mayer, J. E. (1998) Influence of seeding techniques on cell-polymer attachment of biodegradable polymers – polyhydroxyalkanoates (PHA), *Abstract from the 2nd Bi-Annual Meeting of the Tissue Engineering Society*, December 4–6, Orlando, Florida.

Stock, U. A, Sakamoto, T., Hatsuoka, S., Martin, D. P., Nagashima, M., Moran, A. M., Moses, M. A., Khalil, P. N., Schoen, F. J., Vacanti, J. P., Mayer, J. E. (2000a) Patch augmentation of the pulmonary artery with bioabsorbable polymers and autologous cell seeding, *J. Thorac. Cardiovasc. Surg.* **120**, 1158–1168.

Stock, U. A., Nagashima, M., Khalil, P. N., Nollert, G. D., Herden, T., Sperling, J. S., Moran, A., Lien, J., Martin, D. P., Schoen, F. J., Vacanti, J. P., Mayer, J. E. (2000b) Tissue-engineered valved conduits in the pulmonary circulation, *J. Thorac. Cardiovasc. Surg.* **119**, 732−740.

Sudesh, K., Abe, H., Doi, Y. (2000) Synthesis, structure and properties of polyhydroxyalkanoates: biological polyesters, *Prog. Polym. Sci.* **25**, 1503−1555.

Tang, S., Ai, Y., Dong, Z., Yang, Q. (1999) Tissue response to subcutaneous implanting poly-3-hydroxybutyrate in rats, *Disi Junyi Daxue Xuebao* **20**, 87−89.

Tasaki, O., Hiraide, A., Shiozaki, T., Yamamura, H., Ninomiya, N., Sugimoto, H. (1998) The dimer and trimer of 3-hydroxybutyrate oligomer as a precursor of ketone bodies for nutritional care, *JPEN J Parenter. Enteral Nutr.* **23**, 321−325.

Taylor, M. S., Daniels, A. U., Andriano, K. P., Heller, J. (1994) Six bioabsorbable polymers: *In vitro* acute toxicity of accumulated degradation products, *J. Appl. Biomater.* **5**, 151−157.

Tetrick, M. A., Adams, S. H., Odle, J., Benevenga, N. J. (1995) Contribution of D-(-)-3-hydroxybutyrate to the energy expenditure of neonatal pigs, *J. Nutr.* **125**, 264−272.

Tomizawa, Y., Moon, M. R., Noishiki, Y. (1994) Antiadhesive membranes for cardiac operations, *J. Thorac. Cardiovasc. Surg.* **107**, 627−629.

Unverdorben, M., Schywalsky, M., Labahn, D., Hartwig, S., Laenger, F., Lootz, D., Behrend, D., Schmitz, K., Schaldach, M., Vallbracht, C. (1998) Polyhydroxybutyrate (PHB) stent-experience in the rabbit, *Am. J. Card. Transcatheter Cardiovascular Therapeutics, Abstract TCT-11,* 5S.

Vainionpää, S., Vihtonen, K., Mero, M., Pätiälä, H., Rokkanen, P., Kilpikari, J., Törmälä, P. (1986) Biodegradable fixation of rabbit osteotomies, *Acta Orthopaed. Scand.* **57**, 237−239.

Van der Giessen, W. J., Lincoff, A. M., Schwartz, R. S., van Beusekom, H. M. M., Serruys, P. W., Holmes, D. R., Ellis, S. G., Topol, E. J. (1996) Marked inflammatory sequelae to implantation of biodegradable and nonbiodegradable polymers in porcine coronary arteries, *Circulation* **94**, 1690−1697.

Veech, R. L. (1998) Therapeutic compositions, PCT Patent Application Nos. WO 98/41200 and WO 98/41201.

Veech, R. L. (2000) Therapeutic compositions, PCT Patent Application No. WO 00/15216.

Wabnegg, B., Korsatko, W. (1983) About the compatibility of retard tablets consisting of poly-D-(-)-3-hydroxybutyric acid as a carrier of active agents delivered applied parenterally, *Sci. Pharm.* **51**, 372.

Wang, D. W., Lehmann, L. T. (1991) Bioabsorbable coating for a surgical device, US Patent No. 5,032,638.

Webb, A., Adsetts, J. R. (1986) Wound dressings, UK Patent Application No. 2,166,354.

Williams, S. F. (2000) Bioabsorbable, biocompatible polymers for tissue engineering, PCT Patent Application No. WO 00/51662.

Williams, S. F., Martin, D. P. (2001) Therapeutic uses of polymers and oligomers comprising gamma-hydroxybutyrate. PCT Patent Application No. WO 01/19361A2.

Williams, S. F., Peoples, O. P. (1996) Biodegradable plastics from plants, *CHEMTECH* **26**, 38−44.

Williams, S. F., Martin, D. P., Horowitz, D. H. Peoples, O. P. (1999) PHA applications: addressing the price performance issue I. Tissue engineering, *Int. J. Biol. Macromol.* **25**, 111−121.

Williams, S. F., Martin, D. P., Skraly, F. (2000) Medical devices and applications of polyhydroxyalkanoate polymers, PCT Patent Application No. WO 00/56376.

Yagmurlu, M. F., Korkusuz, F., Gursel, I., Korkusuz, P., Ors, U., Hasirci, V. (1999) Sulbactam-cefoperazone polyhydroxybutyrate-*co*-hydroxyvalerate (PHBV) local antibiotic delivery system: *in vivo* effectiveness and biocompatibility in the treatment of implant-related experimental osteomyelitis, *J. Biomed. Mater. Res.* **46**, 494−503.

Yasin, M., Tighe, B. J. (1992) Polymers for biodegradable medical devices VIII. Hydroxybutyrate-hydroxyvalerate copolymers: physical and degradative properties of blends with polycaprolactone, *Biomaterials* **13**, 9−16.

Yasin, M., Holland, S. J., Jolly, A. M., Tighe, B. J. (1989) Polymers for biodegradable medical devices VI. Hydroxybutyrate-hydroxyvalerate copolymers: accelerated degradation of blends with polysaccharides, *Biomaterials* **10**, 400−412.

Yasin, M., Holland, S. J., Tighe, B. J. (1990) Polymers for biodegradable medical devices V. Hydroxybutyrate-hydroxyvalerate copolymers: effects of polymer processing on hydrolytic degradation, *Biomaterials* **11**, 451−454.

5
Polylactides

Prof. Dr. Hideto Tsuji
Department of Ecological Engineering, Faculty of Engineering, Toyohashi
University of Technology, Tempaku-cho, Toyohashi, Aichi 441–8580, Japan;
Tel.: +81-532-44-6922; Fax: +81-532-44-6929; E-mail: tsuji@eco.tut.ac.jp

CLL	ε-caprolactone
[COOH]	concentration of terminal carboxyl group
DDS	drug delivery system
DLA	D-lactide
DLLA	DL-lactide
DR	draw ratio
E	Young's modulus
[ester]	ester group concentration
GPC	gel permeation chromatography
HMB	hexamethylbenzene
HMW	high-molecular weight
$[H_2O]$	water concentration
J^0_e	equilibrium compliance
k	hydrolysis rate constant
LA	lactide
Lc	crystalline thickness
LLA	L-lactide
LMW	low-molecular weight
MLA	meso-lactide
Mn	number-average molecular weight
$M_{n,t}$	M_n at hydrolysis time t
$M_{n,0}$	M_0 at hydrolysis time zero
$M_{n,s}$	peak molecular weight of the lowest molecular weight specific peak
M_v	viscosity-average molecular weight
M_w	weight-average molecular weight
Ny6	poly(ε-amino caproic acid)
P	physical property
PBS	poly(butylene succinate)
PCL	poly(ε-caprolactone)
PCL-b-PEO	poly(ε-caprolactone)-b-poly(ethylene oxide)
PDLA	poly(D-lactide), poly(D-lactic acid)
poly(DLA-GA)	poly(D-lactide-*co*-glycolide)
PDLLA	poly(DL-lactide)
PDLLA-b-PEO	poly(DL-lactide)-b-poly(ethylene oxide)
poly(DLLA-CL)	poly(DL-lactide-*co*-ε-caprolactone)
PEBS	poly(ethylene/butylene succinate)
PEO	poly(ethylene oxide)
PGA	polyglycolide
poly(HB-HV)	poly[(*R*)-3-hydroxybutyrate-*co*-3-hydroxyvalerate]
PHMS	poly(hexamethylene succinate)
PHO	poly(3-hydroxyoctanoate)
PLA	polylactide, poly(lactic acid)
poly(LA-GA)	poly(lactide-*co*-glycolide)
PLLA	poly(L-lactide), poly(L-lactic acid)
poly(LLA-DLA)	poly(L-lactide-*co*-D-lactide)

poly(LLA-CL)	poly(L-lactide-*co*-ε-caprolactone)
poly(LLA-GA)	poly(L-lactide-*co*-glycolide)
poly(LLA-Ly)	poly(L-lactide-*co*-lysine)
Pluronic	ABA triblockcopolymers of poly(ethylene oxide) (A) and poly(propylene oxide) (B)
PML	poly(meso-lactide)
PPHOS	polyphosphazene
PSA	poly(sebacic anhydride)
PVA	poly(vinyl alcohol)
PVAc	poly(vinyl acetate)
PDXO	poly(1,5-dioxepan-2-one)
P_0	physical property of the polymer with infinite M_n
ROP	ring-opening polymerization
R-PHB	poly[(R)-3-hydroxybutyrate]
R,S-PHB	poly[(R,S)-3-hydroxybutyrate]
SEM	scanning electron microscopy
SR	self-reinforced
syn-PLA	syndiotactic poly(lactide), syndiotactic poly(lactic acid)
T_a	annealing temperature
T_g	glass transition temperature
T_m	melting temperature
T_m^0	equilibrium melting temperature
T_{1C}	spin lattice relaxation time
x_c	crystallinity
$[a]^{25}_{589}$	specific optical rotation at 25 °C and a wavelength of 589 nm
ΔH_m	enthalpy of melting
ΔH_c	enthalpy of cold crystallization
Δh^0	heat of fusion (per unit mass)
Δn	birefringence
δ_p	polymer solubility parameter
δ_s	solvent solubility parameter
ε_B	elongation-at break
$[\eta]$	intrinsic viscosity
η_0	terminal viscosity
ρ_c	crystal density
σ	specific fold surface free energy
σ_B	tensile strength
σ_γ	yield stress
3D	three-dimensional

1

Introduction

During the past few decades, polylactides, poly(lactic acid)s (PLAs) and their copolymers have attracted much attention in terms of their ecological, biomedical, and pharmaceutical applications, for the following reasons.

- They can be produced from renewable resources such as starch.
- They have mechanical properties comparable with those of commercial polymers such as polyethylene, polypropylene, and polystyrene.
- They are degradable in the human body, as well as in natural environments.
- The toxicity of their degradation products (lactic acid and its oligomers) in the human body and in natural environments is very low.

PLAs can be synthesized either by the polycondensation of lactic acid, or by ring-opening polymerization (ROP) of lactide (LA) (IUPAC name: 3,6-dimethyl-1,4-dioxane-2,5-dione). Lactic acid (2-hydroxy propionic acid) is optically active, and has two enantiomeric L- and D- (S- and R-) forms. In accordance with the two enantiomeric forms, their homopolymers have respective stereoisomers, and their crystallizablility changes depending on their tacticity and optical purity. On the other hand, Ikada et al. (1987) found that stereocomplexation occurs between the enantiomeric poly(L-lactide), poly(L-lactic acid) (PLLA) and poly(D-lactide), poly(D-lactic acid) (PDLA) due to their peculiar strong interaction. The physical properties, hydrolysis and biodegradation behavior of PLAs can be controlled by altering their molecular characteristics such as molecular weight and monomer sequence distribution, and their highly ordered structures such as crystallinity and crystalline thickness, by polymer blending, additives,

material shapes, etc. The recent developments to produce PLAs at low costs (US $1–2 per kg) will accelerate their use as the commodity plastics. This chapter outlines the basic aspects of synthesis, processing, structures, physical properties, hydrolysis, and biodegradation of PLAs. The commercial aspects of PLAs are described in detail in Chapters 8 and 9 of this volume.

2

Historical Outline

Pelouze (1845) first discovered the formation of a linear dimer of lactic acid, i.e., lactoyl-lactic acid, through the esterification reaction of lactic acid by removal of water at high temperature (130 °C), whilst Nef (1914) later confirmed the presence of lactic acid oligomers from trimer to heptamer when lactic acid was dehydrolyzed at an elevated temperature (90 °C) at reduced pressure (15 mmHg). A two-step polymerization procedure using the cyclic dimer of lactic acid, i.e. lactide (LA), for the synthesis of high-molecular weight (HMW) poly(lactide)s, i.e., poly(lactic acid)s (PLAs), was suggested by Carothers et al. (1932) and developed by Lowe (1954). Since the late 1960s, PLAs and their copolymers have been investigated for their biomedical applications such as sutures and prostheses (Wise et al., 1979), and in the 1970s the commercial biodegradable sutures Vicryl and Glactin 910 were produced from the copolymers of lactide and glycolide (Shneider, 1972). On the other hand, De Santis and Kovacs (1968) analyzed the crystal structure (α-form) of the isotactic poly(L-lactide) (PLLA). Since the late 1970s, PLAs and their copolymers have been studied for the pharmaceutical applications as the matrices for drug delivery systems (DDSs) (Wise et al., 1979).

Additional basic information concerning the synthesis, physical properties, crystallization behavior, and medical and pharmaceutical applications of PLAs and their copolymers was obtained throughout the 1980s and 1990s. Kricheldorf and colleagues (Kricheldorf et al., 1990; Kricheldorf and Kreiser-Saunders, 1996) and Dubois et al. (1997) investigated the synthesis and molecular characterization of LA homo- and copolymers, whilst Pennings and coworkers studied the crystallization (Kalb and Pennings, 1980; Vasanthakumari and Pennings, 1983) and spinning of PLLA and the physical properties of its spun and drawn fibers (Leenslag and Pennings, 1987a). Ikada and coworkers (Idada, 1998; Ikada and Tsuji, 2000) investigated the medical and pharmaceutical applications of PLAs and their copolymers, whilst Tsuji and colleagues (Tsuji and Ikada, 1999a; Tsuji, 2000a,b) studied the crystallization, physical properties, hydrolysis, and biodegradation of PLA-based materials, including a PLA stereocomplex. This stereocomplex is formed by the peculiarly strong interaction between enantiomeric PLLA and poly(D-lactic acid), poly(D-lactide) (PDLA) (Ikada et al., 1987), and it was Okihara et al. (1991) who first proposed the crystal structure of the PLA stereocomplex, where PLLA and PDLA are packed in side-by-side fashion. Li and Vert (1995) investigated the hydrolysis behavior of massive PLA-based materials, and demonstrated that the core-accelerated hydrolysis occurs by the catalytic oligomers formed by hydrolysis. During the latter 1990s, in addition to medical applications as scaffolds for tissue regeneration, PLAs have been attracting much attention in terms of ecological applications, and have been recognized as being substitutes of the commercial "petro" polymers because they can be produced from renewable resources such as starch, and at low cost (ca. US$ 1–2 per kg).

3
Synthesis and Purification of Monomers

Lactic acid is a well-known, simple hydroxy acid with an asymmetric carbon atom, and is synthesized by the bacterial fermentation of carbohydrates such as sugar from starch. As mentioned earlier, lactic acids are optically active compounds, including L- and D-forms. The melting temperature (T_m) of a 1:1 racemic mixture (compound) of L- and D-lactic acids is 52.8 °C, which is higher than that of the respective enantiomers (16.8 °C). The cyclic dimers of lactic acids, LAs, have three different forms; L-lactide (LLA) with two L-lactyl units, D-lactide (DLA) with two D-lactyl units, and meso-lactide (MLA) with one L-lactyl unit and one D-lactyl unit. Racemic lactide or DL-lactide (DLLA) is a 1:1 physical mixture or a 1:1 racemic compound (stereocomplex) of LLA and DLA, and has a T_m (124 °C) higher than that of LLA (95–99 °C), DLA (95–99 °C), and MLA (53–54 °C). The structures and T_m values of lactic acids and LAs are shown in Figure 1 (Tsuji and Ikada, 1999a). The laboratory method for LA synthesis is thermal depolymerization of low-molecular weight (LMW) PLAs, followed by distillation of LAs under reduced pressure. The crude LAs thus obtained are normally purified by recrystallization using ethyl acetate as a solvent. Purification of LAs is indispensable to increase the molecular weight of the resulting polymers. The industrial procedures for synthesis and purification of lactic acids and LAs have been described in detail (Kharas et al., 1994; Hartmann, 1998).

4
Synthesis of Polymers

A variety of LA homo- and copolymers having different molecular structures and, therefore, different physical properties, hy-

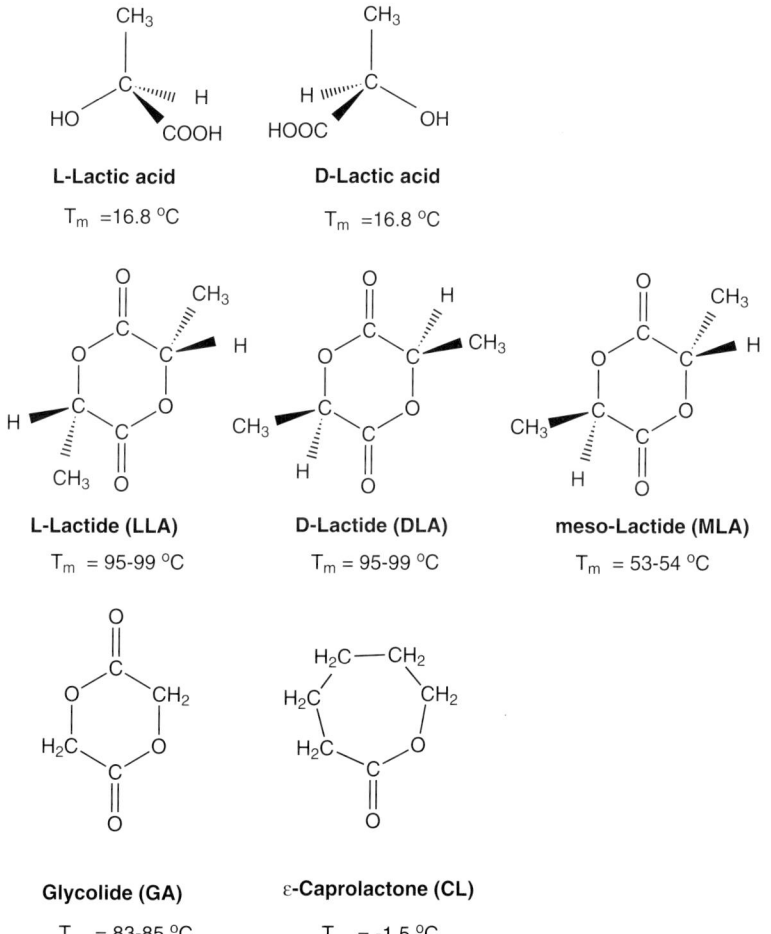

Fig. 1 Structure and melting point (T_m) of L- and D-lactic acids and lactides (LAs), glycolide (GA), and ε-caprolactone (CL).

drolyzability, and biodegradability have been synthesized using LAs, lactic acids, and comonomers.

4.1
Homopolymers

PLAs can be synthesized by two methods, as illustrated in Figure 2: (1) Polycondensation of lactic acids, and (2) ROP of LAs (Tsuji and Ikada, 1999a). The catalysts reported to be effective in increasing the molecular weight of the resulting PLAs up to the order of 10^5 and 10^6 for polycondensation (Ajioka et al., 1995; Moon et al., 2000) and for ROP (Lillie and Schulz, 1975; Kohn et al., 1984; Leenslag and Pennings, 1987b; Kricheldorf and Sumbél, 1989), respectively, are tin-based catalysts such as tin (II or IV) chloride and tin (II) bis-2-ethylhexanoic acid (stannous octoate or tin octoate). These tin-based catalysts are generally used because of their high solubility in LAs and PLA oligomers and low toxicity; moreover, they have re-

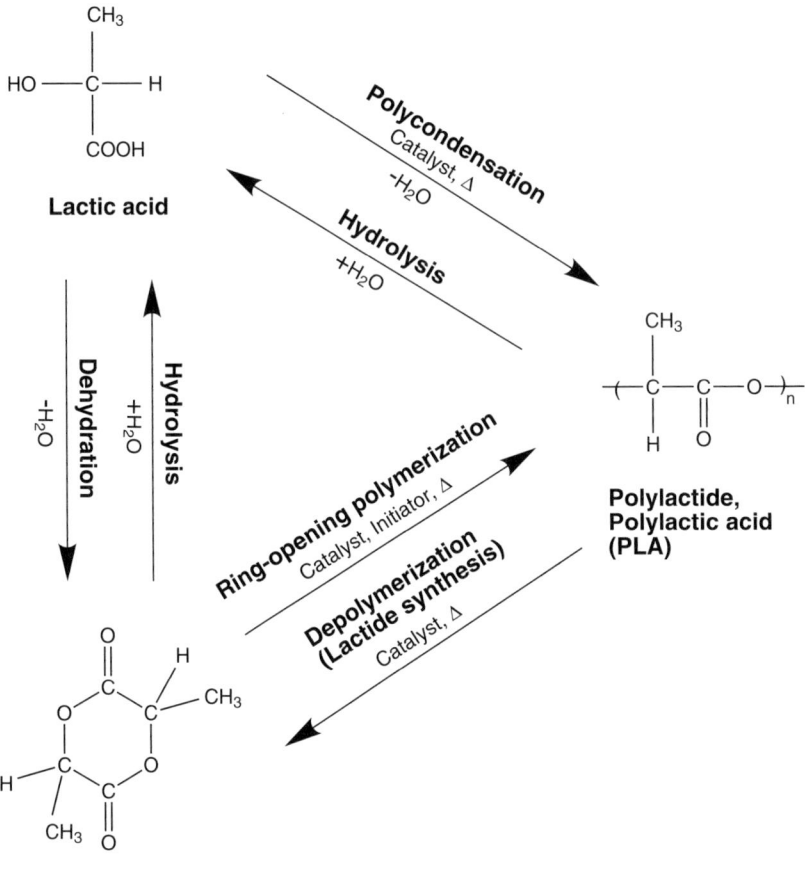

Lactide (LA)

Fig. 2 Schematic representation of polycondensation and dehydration of lactic acid, synthesis and ring-opening polymerization of lactide (LA), and hydrolysis of polylactide, polylactic acid (PLA) and LA.

ceived approval from the FDA. Bulk polymerization in the presence of these catalysts is favorable to avoid the problem of racemization and transesterification during the ROP of lactide in solution. In contrast, aluminum alkoxides are known to be effective catalysts to synthesize mono-dispersed PLAs (Trofimoff et al., 1987) and their block-copolymers, utilizing the "living" nature of polymerized chains (Song and Feng, 1984). Recently, PLAs with weight-average molecular weight (M_w) as high as 2.7×10^5 and oligomeric lactic acids have been synthe-

sized by lipase-catalyzed ROP of LAs (Matsumura et al., 1998) and polycondensation of lactic acids (Ohya et al., 1995), respectively.

Hyon et al. (1997) synthesized poly(DL-lactic acid) with a molecular weight range of 2×10^3 to 2×10^4 by polycondensation varying the reaction temperature, time, and pressure. PLAs with relatively high molecular weights are obtained by polycondensation only when water is efficiently removed from the polymerization mixture of lactic acid, PLA, and water. Such conditions involve: (1) a temperature range of 180–200 °C; (2) low pressure (<5 mmHg); (3) long reaction time; and (4) the addition of an appropriate catalyst and azeotropic solvent of water. Ajioka et al. (1995) and Moon et al. (2000) synthesized poly(L-lactic acid) having a M_w of more than 1×10^5 using tin-based catalysts and with diphenylether as an azeotropic solvent, and using those activated by proton acids, respectively. Too high a reaction temperature results in the reduction of optical purity of the PLAs, even when optically pure L- or D-lactic acid is used as a monomer. In contrast, PLAs with very high molecular weights ($M_w > 1 \times 10^6$) can be synthesized using the ROP method under the following conditions (Leenslag and Pennings, 1987b): (1) low concentration of initiator; (2) relatively low temperature (<130 °C); and (3) long reaction time. The molecular weight of PLAs synthesized by ROP of LAs can be varied by altering the concentration of initiator having a hydroxyl group, such as alcohols (Schindler et al., 1982; Tsuji and Ikada, 1999a).

PLA composed of equimolar L- and D-lactyl (a half of lactide) units, can be synthesized by polycondensation of DL-lactic acid and ROP either of DLLA or MLA (Schindler and Gaetano, 1988; Kricheldorf and Boettcher, 1993). The monomer distribution and tacticity of this stereocopolymer varies depending on the monomer, the nature of the polymerization catalyst, and the polymerization conditions such as temperature and time (Kasperczyk, 1995). Recently, Ovitt and Coates (1999) successfully synthesized syndiotactic PLA (syn-PLA) from MLA using an aluminum alkoxide, whilst Chisholm et al. (1997) proposed new assignments of resonance lines to the monomer sequences in ^{13}C- and ^{1}H-NMR spectra of poly(DL-lactide) (PDLLA) and poly(meso-lactide) (PML) in the region of the methine group (Table 1).

PLAs with different L-lactyl unit contents can be prepared by altering the enantiomeric

excess of the monomers: L- and D-lactic acids (Fukuzaki et al., 1989), LLA and DLA (Tsuji and Ikada, 1992), and LLA and MLA (MacDonald et al., 1996). The stereoblock-copolymers were synthesized by two-step living ROP of LLA with DLA or DLLA (Yui et al., 1990; Spinu et al., 1996), and also by single-step ROP of DLLA using Shiff's base/aluminum methoxide as initiator (Sarasua et al., 1998; Spassky et al., 1998).

4.2
Linear Copolymers

Since the copolymers from LAs and glycolide (GA) are described in detail in Chapter 6 of this volume, the synthesis of LA copolymers is briefly outlined here. Typical comonomers of LAs often used for ROP are GA and CL (see Figure 1). The copolymers of LA with GA or CL, as well as the LA homopolymers and stereocopolymers, have been intensively studied as matrices for drug delivery systems (DDS) (Lewis, 1990; Pitt, 1990) and biodegradable scaffolds for tissue regeneration (Kharas et al., 1994; Agrawal et al., 1997). The other comonomers utilized for copolymer preparation have been summarized in reviews (for example, Kharas et al., 1994; Hartmann, 1998).

Similarly to PLA stereocopolymers, the monomer distribution in the copolymers depends strongly on the monomer pairs, the nature of catalysts, and the polymerization conditions. The assignments of resonance lines to monomer sequences in ^{13}C- and ^{1}H-NMR spectra of poly(L-lactide-*co*-glycolide) [poly(LLA GA)] (Kasperczyk, 1996) or poly (L-lactide-*co*-ε-caprolactone) [poly(LLA-CL)] (Kasperczyk and Bero, 1991) are summarized in Table 1. The proton resonance lines that are useful in determining the respective monomer contents of the copolymers are those at around 5.2 ppm for methine group of the lactyl unit, at 4.8 ppm for the meth-

Tab. 1 Assignments of resonance lines to monomer sequences in ¹³C- and ¹H-NMR spectra of stereo-copolymers from DLLA or MLA (Chisholm et al., 1997), poly(LLA-GA) (Kasperczyk, 1996), and poly(LLA-CL) (Kasperczyk and Bero, 1991)

Polymer	Nuclei	Monomer sequence[a–d]	Chemical shift [ppm]
Stereocopolymers	Methine carbon	iss/ssi	69.22
from DLLA or MLA		sss	69.17
		sii/iis, sis	69.10
		ssi/iss	68.99
		iii,isi,iis/sii	68.91
	Methine proton	sis,iss/ssi	5.100
		sii,iis	5.089
		iis/sii	5.044
		iii	5.035
		sss, isi, ssi, iss	5.027
Poly(LLA-GA)	Lactyl carbon	LLGG	168.89
		LLLL	168.82
		GLG	168.76
	Glycolyl carbon	GGGG	166.35
		GGLL	166.21
	Methylene protone	GLGGG, GGGLG	4.96
		LGGLG, GLGGL	4.94
		GGGG	4.90
		LLGGL, LGGLL	4.88
		GGGGL, LGGGG	4.87
		LLGGG, GGGLL	4.86
		LLGLL, GLGLL, LLGLG, GLGLG	4.83
		GGGLG, GLGGG	4.80
		LGGGL, GLGGL, LGGLG	4.78
Poly(LLA-CL)	Lactyl carbon	CL̲C	170.83
		LLLL̲C, CLLL̲C	170.32
		CLL̲C	170.26
		CL̲LC	170.22
		CL̲LLC	170.11
		CL̲LLL	170.08
		LLL̲LC	169.72
		CLL̲LC	169.67
		CLL̲LL	169.57
		LLL̲LLL	169.57
	Caprolactone carbon	CC̲C	173.49
		CLC̲C	173.44
		LLC̲C	173.42
		CC̲LL	172.84
		LLC̲LL	172.77
		CLC̲LC, LLC̲LC	172.72
		CC̲LC	172.53

[a]Sequences are given along the chain direction of CO to O; [b]s and i represent syndiotactic and isotactic sequences, respectively; [c]G and L represent glycolyl (a half of GA) and L-lactyl (a half of LLA) units, respectively; [d]C and L represent CL and L-lactyl (a half of LLA) units, respectively.

ylene group of the glycolyl unit, and at 4.1 ppm for the γ-methylene group of the CL unit (Tsuji and Ikada, 1994, 2000a). The monomer sequences of poly(LLA-GA) (Gilding and Reed, 1979; Kricheldorf and Kreiser, 1987a) and poly(LLA-CL) (Kricheldorf and Kreiser, 1987b; Choi et al., 1994) are known to be blocky and difficult to be randomized (Shen et al., 1996). Two-step block copolymerization was performed for LLA or DLLA with GA (Rafler and Jobman, 1994) or with CL (Song and Feng, 1984) to obtain their block copolymers. Recently, (R)-β-butyrolactone(B) was used as comonomer of LLA(A), and their relatively random (Hori et al., 1993) and A-B-A block (Hiki et al., 2000) copolymers were synthesized, while A-B-A block copolymers was prepared from LLA(A) and 1,5-dioxepan-2-one(B) by Stridsberg and Albertsson (2000).

Hydrophobic LA sequences have been copolymerized with hydrophilic sequences such as poly(ethylene oxide) (PEO) (Younes and Cohn, 1987), poly(propylene oxide) (Kimura et al., 1989), and their block copolymer (Pluronic™) (Yamaoka et al., 1999) to increase the hydrophilicity, flexibility, and biodegradability of PLA-based materials.

4.3
Graft Copolymers

Graft copolymers can be divided into two types: (1) PLAs or their copolymers grafted polymers; and (2) other polymers grafted PLAs, or their copolymers. The former examples involve PLLA, PDLLA, poly(DLL-GA), or poly(LLA-CL) grafted hydrophilic dextrans (Li et al., 1997; Youxin et al., 1998), PVA (Onyari and Huang, 1996; Breitenbach and Kissel, 1998), pullulan (Ohya et al., 1998a), amylose (Ohya et al., 1998b), dextran (de Jong, 2000), hydrophobic vinyl polymers (Eguiburu, 1996), polyurethane (Hsu and Chen, 2000) and poly(p-xylylene)s

(Agarwal et al., 1999), whilst the latter examples include hydrophilic poly(acrylamide) (Tsuchiya et al., 1993), poly(acrylic acid) (Södergård, 1998), and PEO (Cho et al., 1999) grafted PLLA. The proposed methods of preparing the graft copolymers containing PLAs and their copolymers chains were: (1) copolymerization of macromonomers of PLAs or their copolymers having a vinyl group with vinyl monomers; (2) ROP of LAs initiated by macromolecular polyols such as PVA and polysaccharides; and (3) irradiation-initiated polymerization of LAs from a polymer, and that of vinyl monomers or alkene oxide from PLAs or their copolymers. In some of the above-mentioned studies, graft polymerizations were performed on the surfaces of bulk polymeric materials.

4.4
Branched Polymers

Branched PLAs can be prepared by the ROP of LAs initiated with branched polyols, which was proposed by Schindler et al. (1982). The polyols utilized are glycerin (Arvanitoyannis et al., 1995), pentaerythritol (Kim and Kim, 1994), and sorbitol (Arvanitoyannis et al., 1996). Atthoff et al. (1999) synthesized well-defined 2-, 4-, 6-, and 12-arm PLLAs and PDLLAs from L- or DL-lactic acid using 2,2'-bis(hydroxymethyl)propanoic acid derivatives (hydroxy functional dendrimers) as initiators.

4.5
Cross-linked Polymers

The cross-linked PLAs were prepared by copolymerization of LAs with bicyclolactones or carbonates (Grijpma et al., 1993; Nijenhuis et al., 1996a), whilst Storey and coworkers (Storey et al., 1993; Storey and Hickey, 1994) synthesized a network structure containing PLA chains by copolymerization of methyl methacrylate and styrene with LMW three-arm methacrylate-endcapped PDLLA and by cross-linking of LMW three-arm PDLLA using tolylene-2,6-diisocyanate. Nijenhuis et al. (1996a) also investigated two methods to introduce cross-links into PLA molecules, and found that dicumyl peroxide and electron beam irradiation were effective and non-effective, respectively, for cross-linking.

5
Polymer Blending

The blending of PLAs with other polymers is commercially advantageous in order to obtain biodegradable materials with a wide variety of physical properties and hydrolysis, biodegradation, and drug-release behaviors. For this purpose, numerous polymer blends have been prepared, and these PLAs – or their copolymer-based polymer blends – can be divided into six groups, as shown in Table 2. Among these polymer blends, stereocomplexationable polymer blends have been intensively studied because of their specific properties in solution as well as in bulk (see below; Tsuji, 2000a). The miscibility of PLA-based blends has attracted much attention, and some miscible polymer blends of PLAs and their copolymers with other polymers may even become immiscible as the polymer molecular weights are increased because of decreased polarity of polymer chains and the entropy of mixing. The reported miscible polymer blends are those of PLLA with either poly[(R)-3-hydroxybutyrate] (R-PHB) (Koyama and Doi, 1997) or poly[(R,S)-3-hydroxybutyrate] (R,S-PHB) (Ohkoshi et al., 2000) and those from PDLLA and poly(1,5-dioxepan-2-one) (PDXO) (Edlund and Albertsson, 2000) where the number average molecular weight (M_n) of PLLA or R,S-PHB is as low as 9×10^3, and that of PDXO as low as 2.1×10^4.

6
Additives

Of the reported additives, hydroxyapatite – which is a component of mammalian bones – has been the most frequently and widely utilized for PLAs and copolymers to increase the biocompatibility and bending modulus (Hyon et al., 1985) and to accelerate bone regeneration (Higashi et al., 1986) when utilized as a scaffold for bone regeneration.

Other additives are plasticizers such as glycerin to decrease Young's modulus and increase elongation-at-break and drug-release rate (Pitt et al., 1979), cyclic monomer LAs to enhance hydrolysis (Nakamura et al., 1989), catalyst deactivators to reduce lactide formation during thermal processing (Hartmann, 1999), and nucleation reagents such as talc to accelerate crystallization (Kolstad, 1996). Renstad et al. (1998) found that the addition of SiO_2 had no significant effects on thermal degradation during processing and biodegradation, while the addition of $CaCO_3$ reduced the rates of thermal degradation during processing and biodegradation.

7
Fiber-reinforced Plastics

PGA, PLLA, and PDLLA fibers have been utilized to reinforce PLLA matrices (Vainionpää et al., 1989; Törmälä, 1992). Among these reinforced PLLA materials, those reinforced by PLLA, i.e., self-reinforced (SR) PLLA, can avoid interfacial problems occurring at the interface between the two different materials. SR-PLLA has a bending modulus of about 250–270 MPa, which is much higher than the 57 to 145 MPa of non-reinforced PLLA and this decreases to 10 MPa after 1 year of *in vivo* degradation (Majola et al., 1992).

Tab. 2 Polymer blends based on PLAs and their copolymers

Polymer pairs Biodegradable polymers		PLAs and copolymers	References
PLAs (Self blends)	HMW PDLLA	LMW PDLLA	Bodmeier et al. (1989); Asano et al. (1991); Mauduit et al. (1996)
	HMW PLLA	LMW PLLA	Von Recum et al. (1995)
PLAs (Stereocomplexationable enantiomeric blends)	PDLA (or copolymers with DLA monomer sequences)	PLLA (or copolymers with LLA monomer sequences)	Ikada et al. (1987); Murdoch and Loomis (1988a,b, 1989; Yui et al. (1990); Loomis and Murdoch (1990); Loomis et al. (1990); Tsuji et al. (1991a–c, 1992a,b, 1994, 2000a); Tsuji and Ikada (1992, 1993, 1994, 1996a, 1999a,b); Tsuji (2000a,c); Tsuji and Miyauchi (in press); Tsuji and Suzuki (2000); Okihara et al. (1991); Brochu et al. (1995); Stevels et al. (1995); Brizzolara et al. (1996a,b); Spinu et al. (1996); Cartier et al. (1997)
PLAs (Non-stereocomplexationable diastereoisomeric blends)	PDLLA	PLLA	Jorda and Wilkes (1988) Tsuji and Ikada (1992, 1995a, 1996b, 1997); Tsuji and Miyauchi (2000c); Serizawa et al. (2001)
		PDLA	Tsuji and Ikada (1992, 1997); Tsuji and Miyauchi (2000c)
		PLAs with different optical purities	Tsuji and Ikada (1992)
LA or lactic acid Copolymers	LLA-rich PLAs	PLLA	Tsuji and Ikada (1992)
	DLA-rich PLAs	PDLA	Tsuji and Ikada (1992)
	poly(LLA-GA)	PLLA	Cha and Pitt (1990); Tsuji and Ikada (1994)
	poly(DLA-GA)	PDLA	Tsuji and Ikada (1994); Matsumoto et al. (1997)
	poly(LLA-CL)	PLAs with different optical purities	Grijpma et al. (1994)
Aliphatic polyesters	poly(LLA-Ly)	PLLA	Hiljanen-Vainio et al. (1996)
	PCl	PLLA	Cook et al. (1997) Hiljanen-Vainio et al. (1996); Yang et al. (1997); Lostocco et al. (1998); Tsuji and Ikada (1998a); Tsuji et al. (1998); Wang et al. (1998); Kim et al. (2000)
		PDLLA	Domb (1993); Zhang et al. (1995); Tsuji and Ikada (1996c) Gan et al. (1999); Meredith and Amis (2000); Aslan et al. (2000)
		PLLA	Blümm and Owen (1995); Koyama and Doi (1997); Yoon et al. (2000)
	R-P-3B	PDLLA	Domb (1993); Zhang et al. (1996); Koyama and Doi (1995)
		poly(DLLA-CL)	Koyama and Doi (1996)
		PDLLA-b-PEO	Zhang et al. (1995, 1997)
	R,S-PHB	PLLA	Ohkoshi et al. (2000)
	poly(HB-HV)	PLLA	Iannace et al. (1994, 1995)
	PHO	PDLLA	Mallardé et al. (1998)
	PHMS	PLLA	Lostocco and Huang (1997)
	PBS	PLLA	Inoue et al. (1998)
	PEBS	PLLA	Liu et al. (1997)
	PCL-b-PEO	PDLLA-b-PEO	Zhang et al. (1995)
Miscellaneous Biodegradable polymers	Cellulose	PLLA	Nagata et al. (1998)
	N-6	PLLA	Hu et al. (1994, 1995)
	PDXO	PDLLA	Edlund and Albertsson (2000)
	PEO	PLLA	Younes and Cohn (1988); Domb (1993) Nakafuku and Sakoda (1993); Nakafuku (1994, 1996); Nijenhuis et al. (1996b) Sheth et al. (1997); Yang et al. (1997); Tsuji et al. (2000a)
		PDLLA	Domb (1993)
		poly(LA-GA)	Domb (1993)
	P uronic	PLLA	Park et al. (1992)
	PPHOS	poly(LA-GA)	Ibim et al. (1997)

Tab. 2 (cont.)

Polymer pairs Biodegradable polymers	PLAs and copolymers	References
PSA	PDLLA	Domb (1993); Shakesheff et al. (1994); Davies et al. (1996)
		Chen et al. (1998)
PVA	PLA-GA	Domb (1993)
	PLLA	Tsuji and Muramatsu (2001a,b)
	poly(DLLA-GA)	Pitt et al. (1992, 1993)
	PLLA	
PVAc		Gajria et al. (1996)

HMW: high-molecular weight; LMW: low-molecular weight; Ny6: poly(ε-aminocaproic acid); PBS: poly[butylene succinate]; PCL: poly(ε-caprolactone); PCL-b-PEO: poly(ε-caprolactone)-b-poly(ethylene oxide); poly(DLA-GA): poly(DLA-GA): poly(DL-lactide-co-glycolide); PDLLA: poly(DL-lactide); PDLLA-b-PEO: poly(DL-lactide)-b-poly(ethylene oxide); PDXO: poly(1,5-dioxepan-2-one); PEBS: poly(ethylene/butylene succinate); PEO: poly(ethylene oxide); poly(HB-HV): poly((R)-3-hydroxybutyrate-co-3-hydroxyvalerate]; PHMS: poly(hexamethylene succinate); PHO: poly(3-hydroxyoctanoate); PLA: poly(lactide), poly(lactic acid); poly(LA-GA): poly(lactide-co-glycolide); PLLA: poly(L-lactic acid); poly(LLA-CL): poly(L-lactide-co-ε-caprolactone); poly(LLA-GA): poly(L-lactide-co-glycolide); poly(LLA-Ly): poly(L-lactide-co-lysine); Pluronic: ABA triblockcopolymers of polyethylene oxide (A) and poly(propylene oxide) (B); PPHOS: poly[phosphazene]; PSA: poly(sebacic anhydride); PVA: poly(vinyl alcohol); PVAc: poly(vinyl acetate); R-PHB: poly[(R)-3-hydroxybutyrate]; R.S-PHB: poly[(R,S)-3-hydroxybutyrate]

8
Molding

Solidification methods of polymers include: (1) melt-molding; (2) solution-casting or solvent evaporation; and (3) non-solvent precipitation. Of these methods, the melt molding approach is most favored for the preparation of biodegradable materials for ecological, medical, and pharmaceutical applications, as it does not require any solvents such as chloroform and methylene chloride, which are toxic and harmful to the environment and to human beings.

8.1
Thermal Treatments

The parameters for thermal treatment or crystallization from the melt include: (1) pretreatments prior to crystallization, such as melting and quenching from the melt; (2) melting temperature and time; (3) crystallization or annealing temperature and time (T_a and t_a, respectively); and (4) drawing. However, crystallization and orientation at a high temperature for a long time will cause a large decrease in molecular weight and optical purity due to thermal degradation and racemization of optically active LA units, respectively.

8.2
Pore Formation

Porous biodegradable materials are reported to enhance tissue regeneration, if they have appropriate pore size and porosity (Coombes and Meikle, 1994; Thomson et al., 1995a,b). The reported methods of preparing porous PLAs and their copolymers involve the removal of inorganic salts or organic low-molecular weight compounds from the melt-molded, solution-cast, gelled (Coombes and Heckman, 1992a,b; Coombes and Meikle, 1994), or frozen mixtures (Whang et al., 1995) of PLAs with additives or solvents and from the phase-separated mixtures of PLA/solvent/non-solvent systems (van de Witte et al., 1996a–f, 1997; Zoppi et al., 1999). Hence, a new method was proposed to prepare porous biodegradable polyesters by water-extraction of a water-soluble polymer such as PEO from their blends. It became evident that the pore size and porosity of the PLLA and poly(ε-caprolactone) (PCL) films are controllable by changing their mixing ratio and the molecular weight of the water-soluble polymer (Tsuji et al., 2000a; Tsuji and Ishizaka, 2001a). Another proposed novel method is to utilize selective enzymatic removal of one component from the aliphatic polyester blends (Tsuji and Ishizaka, 2001b).

9
Highly Ordered Structures

The highly ordered structures of PLAs include: (1) crystallinity (x_c); (2) crystalline thickness (L_c); (3) crystal structure; (4) spherulitic size and morphology; and (5) molecular orientation. As detailed in the literature, the highly ordered structures (1)–(5) of PLLA can be altered to some extent by varying the above-mentioned parameters for thermal treatment (Tsuji and Ikada, 1995b). Of these highly ordered structures, x_c and L_c are known to have crucial effects on the mechanical properties as well as the hydrolysis and biodegradation behavior of PLAs (Tsuji, 2000b).

9.1
Crystallization

PLAs and their copolymers crystallize into homo- and stereocomplex (racemic) crystallites in bulk as well as in solutions, in so far as their tacticities are sufficiently high. The homocrystallites consist of either LLA or DLA unit sequences, whereas the stereocomplex crystallites comprise both LLA and DLA unit sequences. PLAs and their copolymers can form eutectic and epitaxial crystallites with low-molecular weight organic compounds.

9.1.1
Homo-crystallization

The general crystallization behavior of PLLA in bulk and in solution has been studied by Fischer et al. (1973), Kalb and Pennings (1980), and Vasanthakumari and Pennings (1983). When crystallized in bulk directly from the melt, the PLLA spherulite growth rate becomes higher with decreasing PLLA molecular weight, and reaches a maximum at crystallization temperature around 130 °C (ca. 5 and 2.5 μm min^{-1} for PLLA with viscosity-average molecular weight (M_v) of 1.5 and 6.9×10^5, respectively) (Vasanthakumari and Pennings, 1983). However, the total crystallization rate monitored by total x_c increases with decreasing T_a, because the spherulite density increases with a decrease in T_a for a T_a range of 100–160 °C (Tsuji and Ikada, 1995b), and its maximum is observed at about 105 °C (Iannace and Nicolais, 1996). Vasanthakumari and Pennings (1983) found that the transition from regime II crystallization to regime I crystallization for PLLA occurs above 163 °C when monitored by spherulite growth rate, while Mazzullo et al. (1992) showed that the transition from regime III crystallization to regime II crystallization occurs above 140 °C when traced by total x_c and T_m of PLLA. Spherulite growth rate decreases and crystallization half-time becomes longer in the presence of comonomer unit in a polymer chain, as revealed for poly(L-lactide-*co*-meso-lactide) (Kolstad, 1996; Huang et al., 1998). The crystallization half-time and the time required for the completion of crystallization is reported to be dramatically decreased by addition of talc as a nucleating agent (Kolstad, 1996), and by quenching from the melt before crystallization (Tsuji and Ikada, 1995b), due to increased nuclear density of the spherulites.

Crystallization does not occur when the sequence length of L- or D-lactyl units becomes smaller than a critical value, which depends on the crystallization conditions and comonomer unit. The reported values are ca. 14 (Sarasua et al., 1998) and 15 (Tsuji and Ikada, 1996a) lactyl units for LA stereo-copolymers from LLA and DLA, assuming random addition of LA units during polymerization without transesterification, whereas poly(LLA-GA) or poly(DLA-GA) has a rather small critical sequence length of nine lactyl units when calculated under the same assumption (Tsuji and Ikada, 1994). When LLA polymer contains flexible CL units, the critical L-lactyl unit sequence length is lowered to 7.2, and the copolymers are crystallizable even at room temperature (25 °C) (Tsuji and Ikada, 2000a). Fischer et al. (1973) showed that DLA units of LLA-rich PLAs are partially included in the crystalline region during formation of their single crystals, while Zell et al. (1998) revealed that LLA content in the crystalline region is almost the same as that in the amorphous region for a crystallized DLA-rich PLA.

The crystallization of homopolymer PLLA or PDLA will proceed in the polymer blends as if it were in the nonblended specimens as far as the constituent polymers are phase-separated (Tsuji et al., 1998), while PLLA will not crystallize in the partially miscible polymer blends with amorphous PDLLA at low PLLA contents (Tsuji and Ikada, 1995a). The induction period until the start of crystallization, spherulite growth rate, and spherulite density of PLLA in the presence of PDLLA depends on the molecular weight of PDLLA (Tsuji and Ikada, 1996b) and the polymer mixing ratio (Tsuji and Ikada, 1995a).

9.1.2
Stereocomplexation (Racemic Crystallization)

Ikada et al. (1987) found that stereocomplexation or racemic crystallization takes place between LLA unit and DLA sequences. Numerous studies have been performed concerning this stereocomplex ever since, as shown in Table 2 (Tsuji, 2000a). Stereocomplex crystallites (racemic crystallites) of PDLA and PLLA have a melting temperature 50 °C higher than those of homocrystallites either of PLLA or PDLA. The stereocomplexation between PDLA and PLLA is known to occur in solution as well as in bulk from the melt. Stereocomplex crystallization from the melt completes in the blend specimen in a period as short as 2 min at 140 °C, probably due to the high-density nuclear formation of the stereocomplex spherulites, while it takes a much longer time (1 h) to complete homocrystallization in the nonblended PLLA or PDLA specimens at the same temperature (Tsuji and Ikada, 1993).

Because of the stoichiometric ratio of 1:1 (L-lactyl unit:D-lactyl unit) of the stereocomplex crystallites (see below), the excessive L- or D-lactyl unit sequences crystallize in homo-crystallites which consist of either L- or D-lactyl unit sequences, when the PDLA content $[X_D = PDLA/(PLLA + PDLA)]$ deviates from 0.5. Other important parameters affecting the stereocomplexation and homocrystallization are molecular weight and sequence lengths of L-lactyl and D-lactyl units of the polymers, and the solidification method and temperature (Tsuji, 2000a). Spinu and Gardner (1994) proposed a new approach of stereocomplexation during polymerization of monomer/polymer blends, while Radano et al. (2000) prepared 1:1 mixture containing stereocomplex phase directly by stereoselective polymerization of DLLA with a racemic catalyst.

9.1.3
Eutectic Crystallization

PLLA crystallizes into the eutectic crystals with pentaerythrityl tetrabromide (Vasanthakumari, 1981) and hexamethylbenzene (HMB) (Zwiers et al., 1983). Tonelli (1992) and Howe et al. (1994) showed that PLLA forms an inclusion compound with urea, where PLLA chains are trapped in the channels of the urea.

9.1.4
Epitaxial Crystallization

Cartier et al. (2000) revealed that PLA stereocomplex crystallizes epitaxially on HMB, while Brochu et al. (1995) reported that the stereocomplex crystallites are formed on those of homocrystallites composed either of L-lactyl or D-lactyl unit sequences in the blends from PLLA and D-lactide-rich PLA (D-lactide content = 80%) when the stereocomplex crystallization is slow and homocrystallization is rapid.

9.2
Crystal Structure

The cell parameters for non-blended PLLA and PLA stereocomplex are summarized in Table 3. Non-blended PLLA has been reported to crystallize in three forms: α (De Santis and Kovacs, 1968; Hoogsteen et al., 1990; Kobayashi et al., 1995), β (Hoogsteen et al., 1990; Puiggali et al., 2000), and γ (Cartier et al., 2000). The β-form of non-blended PLLA is reported to have a rather frustrated structure (Puiggali et al., 2000). The stereocomplex crystal has a triclinic unit cell, where the PLLA and PDLA chains taking a 3_1 helical conformation are packed side-by-side in parallel fashion.

Tab. 3 Unit cell parameters reported for nonblended PLLA and stereocomplex crystals

	Space group	Chain orientation	Number of helices per unit cell	Helical conformation	a [nm]	b [nm]	c [nm]	α [deg.]	β [deg.]	γ [deg.]
PLLAα-form (De Santis and Kovacs, 1968)	Pseudo-orthorhombic	–	2	10_3	1.07	0.645	2.78	90	90	90
PLLAα-form (Hoogsteen et al., 1990)	Pseudo-orthorhombic	–	2	10_3	1.06	0.61	2.88	90	90	90
PLLAα-form (Kobayashi et al., 1995)	Orthorhombic	–	2	10_3	1.05	0.61	2.88	90	90	90
PLLAβ-form (Hoogsteen et al., 1990)	Orthorhombic	–	6	3_1	1.031	1.821	0.90	90	90	90
PLLAβ-form (Puiggali et al., 2000)	Trigonal	Random up-down	3	3_1	1.052	1.052	0.88	90	90	120
PLLAγ-form (Cartier et al., 2000)	Orthorhombic	Antiparallel	2	3_1	0.995	0.625	0.88	90	90	90
Stereocomplex (Okihara et al., 199?)	Triclinic	Parallel	2	3_1	0.916	0.916	0.870	109.2	109.2	109.8

9.3

Morphology

The highly organized structures such as spherulites, single crystals, and gels are formed as result of the crystallization of PLAs and their copolymers.

9.3.1

Spherulites

The size and morphology of PLLA spherulites depend on the parameters such as crystallization temperature and time, copolymerization, and blending. PLLA spherulites become smaller with decreasing temperature and time (Marega et al., 1992; Tsuji and Ikada, 1995b). The spherulites of LA stereocopolymers retain well-defined structure, even when the monomer unit sequence length approaches the critical value for crystallization (Tsuji and Ikada, 1996a), whereas rather disturbed spherulites or assemblies of the crystallites were noted in the crystallizable PLLA/amorphous PDLLA blends having low PLLA contents (Tsuji and Ikada, 1995a). The morphology of the stereocomplex spherulites is similar to that of normal spherulites of the nonblended PDLA or PLLA when solely stereocomplexation takes place. However, the spherulites with complicated morphology are noted when stereocomplexation and homocrystallization occur simultaneously (Tsuji and Ikada, 1993, 1996a). The formation of isolated or aggregated stereocomplex spherulites occurs in suspended state in an acetonitrile mixed solution of PLLA and PDLA (Tsuji et al., 1992a).

9.3.2

Single Crystals

PLLA is reported to crystallize into lozenge-like (Fischer et al., 1973; Kalb and Pennings, 1980; Miyata and Masuko, 1997; Iwata and Doi, 1998) and hexagonal (Kalb and Pennings, 1980; Iwata and Doi, 1998) single crystals in dilute solutions, while single crystals of the stereocomplex of PLLA and PDLA having a peculiar triangular shape are formed in *p*-xylene when the solution concentration is as low as 0.04% (Okihara et al., 1991; Tsuji et al., 1992a). The mechanism of formation of the stereocomplex crystal has been proposed by Brizzolara et al. (1996a).

9.3.3

Gels

When the stereocomplexation of PDLA and PLLA occurs in concentrated chloroform and aqueous solution, the solution viscosity increases with time, and finally three-dimensional (3D) gelation or formation of microgels occurs as a result of the formation of stereocomplex microcrystallites which act as cross-links (Tsuji et al., 1991a; Tsuji, 2000c; de Jong et al., 2000).

9.3.4

Phase Structure

The microstructure analysis using high-resolution, solid-state ^{13}C-NMR spectroscopy was performed for nonblended PLLA and PLA stereocomplex specimens. Thakur et al. (1996) found five distinct resonance lines in the carbonyl region of the nonblended PLLA, which indicates the existence of five or more crystallographically inequivalent sites in the crystalline region of PLLA. These authors showed that solid-state NMR can be utilized to evaluate the x_c of PLLA and LA stereocopolymers. By contrast, we demonstrated from an evaluation of the spin lattice relaxation times (T_{1C}s) that the stereocomplexed PLA is composed of four regions: the racemic crystalline regions in both rigid and disordered states, the homocrystalline region, and the amorphous region (Tsuji et al., 1992b).

Tab. 4 Physical properties of some biodegradable aliphatic polyesters

	PLLA	PDLLA	Syn-PLA	PLA stereocomplex	PCL	R-PHB	PGA
T_m (°C)	170–190	–	151 Ovitt and Coates (1999)	220–230 Ikada et al. (1987)	60	5	225–230
T_m^0 (°C)	205 Tsuji and Ikada (1995b) 215 Kalb and Pennings (1980)	–	–	279 Tsuji and Ikada (1996a)	71, 79	180	–
T_g (°C)	50–65	50–60	34 Ovitt and Coates (1999)	65–72 Tsuji and Ikada (1999b)	–60	188, 197	40
$\Delta H_m(x_c = 100\%)$ (J g^{-1})	93 Fischer et al. (1973) 135 Miyata and Masuko (1998) 142 Locmis et al. (1990) 203 Jarrshidi et al. (1988)	–	–	142 Loomis et al. (1990)	142	146	180–207
Density (g cm^{-3})	1.25–1.29	1.27	–	–	1.06–1.13	1.177–1.260	1.50–1.69
Solubility parameter (δ_p) (25 °C) [J cm^{-3}]$^{0.5}$	19–20.5, 22.7	21.1	–	–	20.8	20.6	–
$[a]_{589}^{25}$ in chloroform (deg dm^{-1} g^{-1} cm^3)	–155 ± $^?$	0	–	–	0	+44[a]	–
WVTR[b] (g m^{-2} per day)	82–172	–	–	–	177	13[c]	–
σ_B[d] (kg mm^{-2})	12–230[f]	4–5[f]	–	90 Tsuji and Ikada (1999b)[e]	10–80[e]	18–20[e]	8–100[e]
E[g] (kg mm^{-2})	700–1000[e]	150–190[f]	–	880 Tsuji and Ikada (1999b)[e]	–	500–600[e]	400–1400[e]
ε_B[h] (%)	12–26[e]	5–10[f]	–	30 Tsuji and Ikada (1999b)[e]	20–120[e]	50–70[e]	30–40[e]

[a]300 nm, 23 °C; [b]Water vapor transmission rate at 25 °C; [c]PHB-HV 94/6; [d]Tensile strength; [e]Oriented fiber; [f]Nonoriented film; [g]Young's modulus; [h]Elongation-at-break.

10

Physical Properties

The physical properties of PLLA, PDLLA, and PLA stereocomplex are listed in Table 4, together with those of R-PHB, poly(ε-caprolactone) (PCL), and poly(glycolide) (PGA). As mentioned earlier, the physical properties of polymeric materials depend on their molecular characteristics, highly ordered structures, and material morphology. The changes in thermal and mechanical properties of PLAs upon hydrolysis are connected with the structural changes in the crystalline and amorphous regions.

10.1

Thermal Properties

The T_m, equilibrium melting temperature (T_m^0), glass transition temperature (T_g), and melting enthalpy of the crystal having infinite thickness $[\Delta H_m (x_c = 100\%)]$ of PLAs are listed in Table 4. The thermal properties such as T_m and melting enthalpy (ΔH_m) or x_c are very important parameters reflecting the highly ordered structures of L_c, fraction of crystalline region, polymer chain packing in the amorphous region of the materials, respectively, which influence the initial mechanical properties before hydrolysis, and their change during hydrolysis. The Thompson–Gibbs expression between T_m and L_c is given by the following equation (e.g. Gedde, 1995):

$$T_m = T_m^0 (1 - 2\sigma/\Delta h^0 \rho_c L_c) \qquad (1)$$

where σ, Δh^0, and ρ_c are the specific fold surface free energy, heat of fusion (per unit mass), and crystal density, respectively. Equation (1) infers that the T_m of the polymeric materials, which increases with L_c, can be an index of L_c. Using the ΔH_m ($x_c = 100\%$) values, x_cs of the stereocomplex

and homocrystals in specimens can be evaluated using the following equation:

$$x_c (\%) = 100 \times (\Delta H_c + \Delta H_m)/$$
$$[\Delta H_m (x_c = 100\%)] \qquad (2)$$

where ΔH_c is the enthalpy of cold crystallization. By definition, ΔH_c and ΔH_m are negative and positive, respectively.

T_m (L_c) and x_c become lower with decreasing crystallizable L- or D-lactyl unit sequence length in the LA homopolymers and copolymers including stereocopolymers, which can be caused by lowering the polymer molecular weight (Tsuji and Ikada, 1999b) and increasing the comonomer content (Tsuji et al., 1994). The T_g of LA stereocopolymers is almost constant at about 60 °C, irrespective of the LLA content (Tsuji and Ikada, 1996a), while the T_g of LA copolymers excluding stereocopolymers varies depending on the comonomer nature and its fraction, and the length of respective monomer sequences (Wada et al., 1991).

T_m and x_c increase with annealing (Migliaresi et al., 1991a,b; Tsuji and Ikada, 1995b) and orientation of PLLA (Jamshidi et al., 1988; Hyon et al., 1984; Fambri et al., 1997). The T_m and x_c of PLLA film can be altered in the range of 177–193 °C and 0–63%, respectively, by varying the pretreatment procedure, T_a, and t_a for thermal treatments (Migliaresi et al., 1991a,b; Tsuji and Ikada, 1995b). Similar to the effects of thermal treatment on T_m and x_c, Celli and Scandola (1992) and Cai et al. (1996) showed that T_g and the enthalpy of glass transition (ΔH_g) of PLLA and poly(L-lactide-co-D-lactide) [poly(LLA-DLA)] (96/4), respectively, increase with the increasing t_a at T_a below T_g. This means that PLLA and poly(LLA-DLA) (96/4) chains in the amorphous region become more densely packed by annealing at T_a below T_g.

In phase-separated PLA-based polymer blends, the thermal properties of PLA are insignificantly influenced by blending with other polymers in as much as the PLA content in the blends is high (Tsuji and Ikada, 1998a), while in partially miscible or miscible blends x_c and T_m of PLLA and T_g of the blends vary depending on the polymer mixing ratio (Koyama and Doi, 1997; Edlund and Albertsson, 2000; Ohkoshi et al., 2000). On the other hand, PLA stereocomplex crystallites have T_m and T_m^0 values much higher than those homocrystallites either of PDLA or PLLA (see Table 4). The T_g of the 1:1 blend films from PLLA and PDLA is ca. 5 °C higher than that of the nonblended PLLA and PDLA films in the M_w range from 5×10^4 to 1×10^5, where predominant stereocomplexation between PLLA and PDLA occurs in the blend film (Tsuji and Ikada, 1999b). The increased T_g is assignable to the strong interaction between L-lactyl and D-lactyl unit sequences in the amorphous region of the blend films, resulting in the dense chain packing that region.

The thermal stability of PLAs depends on the residual initiators and the molecular modification. Degée et al. (1997) demonstrated that the thermal stability is higher for PLLA synthesized using aluminum tri(iso-proxide) as initiator than for that synthesized using tin(II) bis(2-ethylhexanoate). Jamshidi et al. (1988) found that thermal stability of PLLA synthesized using tin(II) bis(2-ethylhexanoate) becomes higher when its terminal carboxyl group is acetylated, and that residual monomers enhance the PLLA thermal degradation.

10.2
Mechanical Properties

The mechanical properties are crucial when PLAs and their copolymers are utilized as bulk materials; however, such properties may be controlled by varying the material parameters such as molecular characteristics and highly ordered structures.

10.2.1
Effect of Molecular Characteristics

Molecular weight is an important parameter to determine mechanical properties as given by the following equation:

$$P = P_0 - K/M_n \tag{3}$$

where P is the physical property of a polymeric material, P_0 is P of the polymer with the infinite M_n, and K is a constant. We revealed that as-cast PLLA films have non-zero tensile strength (σ_B) below $1/M_n$ of 2.5×10^{-5} or above M_n of 4.0×10^4, and σ_B increases with $1/M_n$ (Tsuji and Ikada, 1999a) according to Eq. (3). Eling et al. (1982) showed the similar σ_B dependence of PLLA fibers on $1/M_v$, whilst Perego et al. (1996) reported that flexural strength becomes a plateau at M_v above 35,000 and 55,000 for PDLLA and amorphous-made PLLA, respectively. Ultimately, the σ_B and Young's modulus (E) of fibers from LA copolymers, poly(LLA-CL) and poly(LLA-DLA), are smaller than that from a homopolymer, PLLA, while elongation-at-break (ε_B) of poly(LLA-DLA) fiber becomes higher than that of PLLA fiber when they are melt-spun and thermally drawn (Penning et al., 1993). On the other hand, Grijpma et al. (1993) showed that the impact strength of PLLA can be increased by the cross-linking.

10.2.2
Effect of Highly Ordered Structures

The mechanical properties of PLLA vary depending on their highly ordered structure such as x_c and L_c (T_m) (Tsuji and Ikada, 1995b). An increase in x_c increases the σ_B and E of PLLA, but decreases the ε_B. The decrease in σ_B of PLLA films prepared at high T_a may be ascribed to the formation of large-sized spherulites and crystallites, in spite of their high x_c. These results indicate that the mechanical properties of PLLA can be controlled to some extent by altering the highly ordered structures. Similar to other polymers, the values of σ_B and E of PLLA fibers increase, but the ε_B value decreases with increasing degree of molecular orientation (Eling et al., 1982; Hyon et al., 1984; Horacek and Kalísek, 1994a,b; Fambri et al., 1997). Leenslag and Pennings (1987b) produced the PLLA fiber having $\sigma_B = 2.1$ GPa and E = 16 GPa by hot-drawing of a dry-spun fiber. Okuzaki et al. (1999) prepared the PLLA fiber with $\sigma_B = 275$ MPa and E = 9.1 GPa using a zone-drawing method from a LMW PLLA ($M_v = 13,100$). These results prove that molecular orientation, as well as molecular weight, has a major influence on the mechanical properties of PLLA.

10.2.3
Effects of Material Shapes
Pore formation is effective in lowering the Young's modulus of biodegradable polymers; reported examples relate to the compression modulus of poly(lactide-co-glycolide) [poly(LA-GA)] (50/50) (Thomson et al., 1995) and poly(HB-HV) (Chaput et al., 1995), and the tensile modulus of PCL (Tsuji and Ishizaka, 2001a).

10.2.4
Effect of Polymer Blending
In contrast to the mechanical properties of miscible polymer blends, those of the phase-separated polymer blends are discontinuous at a polymer mixing ratio where the inversion of the continuous and dispersed phases takes place. Even when phase separation occurs in the blends, σ_B, yield stress (σ_Y), and E of the blends of glassy PLLA (or PDLLA) with rubbery PEO (Sheth et al., 1997) or PCL (Tsuji and Ikada, 1996c, 1998a) can be widely varied by altering the polymer mixing ratio. The impact strength of PLLA is reported to become higher by the addition of rubbery biodegradable polymers such as PCL (Grijpma et al., 1994).

The stereocomplexation between PLLA and PDLA is reported to enhance the tensile properties of the blend films compared with those of the non-blended PLLA or PDLA film (Tsuji and Ikada, 1999b). Most likely, the microstructure formed by gelation and the inhibited growth of the spherulites may enhance the tensile properties of the blend film. The increased tensile properties of the blend films may be also caused by dense chain packing in the amorphous region due to a strong interaction between L- and D-unit sequences, as evidenced by the increased T_g of the blend films.

10.3
Electric Properties

The piezoelectric constants of PLLA, $-d_{14}$ and $-e_{14}$, increase with the draw ratio (DR) and become a maximum at a DR of about 4–5 (Ikada et al., 1996). Interestingly, the healing of fractured bone was found to be promoted under increased callus formation when drawn PLLA rods were implanted intramedullary in the cut tibiae of cats for its internal fixation (Ikada et al., 1996). This promotion is attributed to the piezoelectric current generated by the strains caused by leg movement of the cats. On the other hand, Pan et al. (1996) reported that the remnant polarization and coercive electric field of PLLA increased and decreased with temperature, respectively, their values at 130 °C being 96 mC m^{-2} and 20 MV m^{-1}, respectively. The pyroelectric constant of the PLLA specimen corona poled at an electric field of 50 MV m^{-1} and 130 °C for 5 min was 10 and 20 μC m^{-2} · K at temperatures lower and higher than T_g, respectively, but became almost zero at T_g (Pan et al., 1996).

10.4
Optical Properties

The specific optical rotations ($[a]^{25}_{589}s$) of PLLA and PDLA in chloroform are about -156 and 156 deg dm^{-1} g^{-1} cm^3, respectively (Tsuji and Ikada, 1992). The birefringence (Δn) of PLLA fibers increases from 0.015 to 0.038 with degree of molecular orientation (Hyon et al., 1984; Kobayashi et al., 1995). The intrinsic birefringence value estimated for highly drawn and annealed PLLA fibers containing solely α-form crystal using Stein's formula is 0.030–0.033 (Ohkoshi et al., 1999). Kobayashi et al. (1995) found that the gyration tensor component g_{33} of PLLA along the helical axis is extremely large $[(3.85 \pm 0.69) \times 10^{-2}]$, which corresponds to a rotary power of $(9.2 \pm 1.7) \times 10^3$ degree mm^{-1} and about two orders of magnitude larger than those of ordinary crystals. Similar rotary power of 7.2×10^3 degree mm^{-1} has also been reported by Tajitsu et al. (1999).

10.5
Surface Properties

The reported surface modification of PLAs to increase hydrophilicity are grafting hydrophilic chains such as acrylamide (AAm) (Tsuchiya et al., 1993) and peptide chains (Kimura and Yamaoka, 1996), and alkaline and enzymatic surface hydrolysis (Ishida and Tsuji, 2000). Alkaline and enzymatic treatment gives rise to scission of PLA chains through surface erosion, resulting in the increased densities of the hydrophilic terminal groups such as hydroxyl and carboxyl groups (Tsuji and Ikada, 1998b), resulting in the decrease in advancing and receding contact angles from 102 and 59° to 74 and 43°, respectively (Ishida and Tsuji, 2000). Another attempt to increase the hydrophilicity of PLLA can be made by coating PLA-b-

PEO containing a hydrophilic chain (Otsuka et al., 1998).

The hydrophilicity at the PLLA film surface is increased by simple physical blending with hydrophilic poly(vinyl alcohol). The advancing and receding contact angles can be controlled in the ranges of 61–95° and 28–59°, respectively, by altering the mixing ratio of the polymers (Tsuji and Muramatsu, 2001b).

10.6
Permeability

Shogren (1997) studied the effects of the x_c of PLLA on water vapor permeability, which is an important factor when PLLA is used as food packages, containers, and bottles. The water vapor transmission rate of an amorphous PLLA film (172 g mm^{-2} per day at 25 °C) is higher than that of a crystallized PLLA film (82 g mm^{-2} per day at 25 °C) due to the low permeability of the crystalline region. Pitt et al. (1992) studied the effects of polymer mixing ratio and molecular weight of solutes on the permeability coefficients of the solutes in blends from poly(DLLA-GA) and PVA.

By contrast, when PLAs are used as the matrices of DDSs, the permeability of the matrices in the water-swollen state is an important parameter to determine the rate of drug release. The overall rate will vary according to the changes in molecular characteristics, highly ordered structures, and morphology of the matrices caused by hydrolysis.

10.7
Swelling and Solubility

It was found that a PLLA film is durable to swelling solvents having solubility parameter (δ_s) values much lower or higher than the value range of 19–20.5 J$^{0.5}$ cm$^{-1.5}$ (e.g., cyclohexane, $\delta_s = 16.8 \, J^{0.5} \, cm^{-1.5}$; ethanol, $\delta_s = 26.0 \, J^{0.5} \, cm^{-1.5}$), and that the polymer solubility parameter (δ_p) for PLLA is in the range of $19-20.5 \, J^{0.5} \, cm^{-1.5}$ (Tsuji and Sumida, 2001). The solubility parameter value of PDLLA estimated by high-precision density measurements is $20.5 \, J^{0.5} \, cm^{-1.5}$ (Siemann, 1992). At room temperature, HMW PLLA, PDLA, and PDLLA are soluble in the solvents having δ_s values in the range of $19-20.5 \, J^{0.5} \, cm^{-1.5}$, such as chloroform, methylene dichloride, dioxane, and benzene, whereas these solvents become swelling-solvents when stereocomplexation occurs between PLLA and PDLA (Tsuji, 2000c). 1,1,1,3,3,3-Hexafluoro-2-propanol is a solvent for stereocomplexed PLAs, but these compounds become insoluble in it when the L_c increases. In addition to the above-mentioned solvents of PLLA, PDLLA is soluble in acetone. Well-known nonsolvents of PLAs are alcohols such as methanol and ethanol, and hence these are often used as precipitants.

10.8
Viscosity

The reported parameters affecting the intrinsic viscosity ($[\eta]$) of PLAs are M_w and M_n, L- and D-lactyl unit sequence of PLAs, and branching. The relationship between $[\eta]$ and M_n of PLLA and PDLLA is given by the following equations at 30 °C (Schindler and Harper, 1979):

For PLLA (4)
$$[\eta] = 5.45 \times 10^{-4} \, M_n^{0.73} \text{ (chloroform)}$$

$$[\eta] = 5.72 \times 10^{-4} \, M_n^{0.72} \text{ (benzene)} \quad (5)$$

For PDLLA (6)
$$[\eta] = 2.21 \times 10^{-4} \, M_n^{0.77} \text{ (chloroform)}$$

$$[\eta] = 2.27 \times 10^{-4} \, M_n^{0.75} \text{ (benzene)} \quad (7)$$

For branched PLLA (five arms) synthesized using pentaerythritol as initiator, the relationship between $[\eta]$ and M_n is given by the following equation (Kim and Kim, 1994):

$$[\eta] = 2.04 \times 10^{-4} \, M_n^{0.77} \text{ (chloroform)} \quad (8)$$

Cooper-White and Mackay (1999) reported that PLLA melts have a critical molecular weight for entanglement of 16,000 g mol^{-1} and an entanglement density of 0.16 mmol cm^{-3} at 25 °C, and showed a dependence of terminal viscosity (η_0) on chain length to the fourth power, while equilibrium compliance (J^0_e) is independent of the molecular weight in the terminal region. By contrast, for branched PLLAs it was found that a critical molecular weight for entanglement is about four-fold that of linear PLLA, and a dependence of η_0 increase to the power of 4.6 for molecular weight (Dorgan et al., 1999).

11
Hydrolysis

PLAs, which are water-insoluble when their molecular weights are sufficiently high, are hydrolyzed at the ester groups to form LMW, water-soluble oligomers and monomers of lactic acids. The hydrolysis mechanisms and behaviors of PLAs are affected by numerous factors, including the materials and the hydrolysis media.

11.1
Hydrolysis Mechanisms

The hydrolysis mechanisms of PLAs may be separated into those of molecular chains and bulk materials.

11.1.1
Hydrolysis Mechanisms of Molecular Chains
The hydrolysis mechanisms of PLA chains are classified into two groups: catalytic and noncatalytic hydrolysis, or enzymatic and nonenzymatic hydrolysis (Figure 3) (Tsuji,

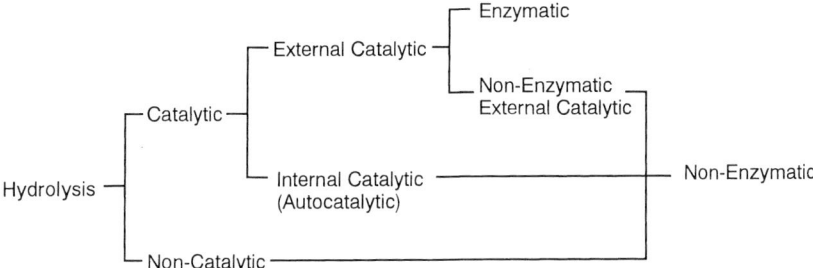

Fig. 3 Hydrolysis mechanisms of PLA chains.

2000b). Catalytic hydrolysis involves catalysis by external and internal substances (abbreviated as the external and internal catalytic hydrolysis, respectively). The representative internal catalytic hydrolysis mediated by the terminal carboxy groups is generally called autocatalytic or autocatalyzed hydrolysis.

External Catalytic Hydrolysis

The representative external catalysts for PLA hydrolysis are enzymes and alkalis. The typical hydrolysis enzymes of ester group are esterases, such as lipases. Endo- and exo-esterases catalyze the hydrolytic scission of an ester group randomly, irrespective of its position in a polymer chain (endo-chain scission) and that of ester group around a polymer chain end (exo-chain scission), respectively. However, the lipases have no significant catalytic effect on PLA hydrolysis. Williams (1981) found that proteinase K, which is a well-known hydrolytic enzyme that catalyzes endo-chain scission of poly(amino acids), can catalyze the hydrolysis of PLLA, while Makino et al. (1985) and Ivanova et al. (1997) studied carboxylic esterase (EC. 3.1.1.1) and cutinase-accelerated hydrolysis of PDLLA, respectively.

In contrast, alkali- (or base-) catalyzed hydrolysis of PDLLA (Shih, 1995) and PLLA (Tsuji and Ikada, 1998b) proceeds via an endo-chain scission mechanism, whereas acid-catalyzed hydrolysis of PDLLA demonstrated exo-chain scission which was more rapid than endo-chain scission (Shih, 1995).

Autocatalytic Hydrolysis

The hydrolytic scission of aliphatic polyester chains without any external catalysts may proceed via the combination of autocatalyzed and noncatalyzed mechanisms. For PLAs, it has been reported that the former mechanism prevails the latter mechanism. In the former mechanism, the hydrolysis is catalyzed by the terminal carboxy groups of PLAs, and its rate is proportional to the concentrations of carboxy and ester groups and water. The kinetic equation expressing the scission of molecular chains by autocatalytic hydrolysis can be derived under the above-mentioned assumption (Pitt et al., 1979):

$$d\,[COOH]/dt = k'\,[COOH]\,[H_2O]\,[ester] \quad (9)$$

where $[COOH]$, $[H_2O]$, and $[ester]$ are the concentrations of terminal carboxy group, water, and whole ester group in PLAs or their copolymers, respectively. If $k'\,[H_2O]\,[ester]$ is assumed to be constant, integration of Eq. (9) will give Eq. (10), coupled with the relationship of $[COOH] \propto M_n^{-1}$:

$$\ln M_{n,t} = \ln M_{n,0} - k_1\,t \quad (10)$$

where $M_{n,t}$ and $M_{n,0}$ are M_n of the polymer at hydrolysis times $= t$ and 0, respectively, and the hydrolysis rate constant (k_1) is equal to $k'\,[H_2O]\,[ester]$. The k_1 value estimated for

PLLA is approximately $2-7 \times 10^{-3}$ per day, depending on the initial highly ordered structure (Tsuji et al., 2000b).

Noncatalytic Hydrolysis

When there is no catalytic effect, the kinetic equation for scission of molecular chains during hydrolysis can be expressed by the following equation:

$$d\,[COOH]/dt = k'\,[H_2O]\,[ester] \quad (11)$$

Since the concentration terms in Eq. (11) can be assumed to be constant during initial stage of hydrolysis when the polymer molecular weight is sufficiently high, integration of Eq. (11) will give Eq. (12), coupled with the relationship of $[COOH] \propto M_n^{-1}$:

$$M_{n,t}^{-1} = M_{n,0}^{-1} + k_2\,t \quad (12)$$

where the hydrolysis rate constant (k_2) is equal to $k'\,[H_2O]\,[ester]$.

11.1.2
Hydrolysis Mechanisms of Bulk Materials

The hydrolysis of water-insoluble bulky PLAs proceeds through surface and/or bulk erosion mechanisms, which are illustrated schematically in Figure 4 (Tsuji, 2000b). More detailed erosion mechanisms of hydrolyzable polymers are described in other monographs (e.g., Göpferich, 1997).

Surface Erosion

Surface erosion is the main route of hydrolysis when the hydrolysis media contain external catalysts, or the hydrolysis rate of the materials is extremely high compared with the diffusion rate of the hydrolysis medium. On the other hand, bulk erosion occurs when no such external catalysts are present in the hydrolysis media, or the hydrolysis rate is low. The relative contribution of these two mechanisms depends on the nature of polymer and the hydrolysis

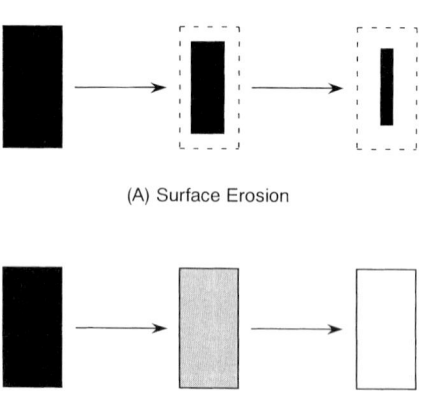

(A) Surface Erosion

(B) Bulk Erosion

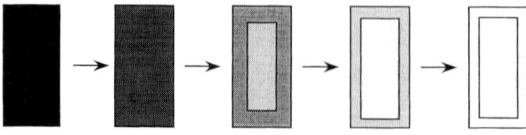

(C) Core-Accelerated Bulk Erosion

Fig. 4 Hydrolysis mechanisms of PLA bulk materials.

medium and conditions. In the case of surface erosion, hydrolytic chain scission occurs solely at the material surface and forms LMW, water-soluble oligomers, which will be removed from their surface or diffuse into the medium, while the core of the materials will remain unhydrolyzed (Figure 4A).

Bulk Erosion

The autocatalytic hydrolysis of PLLA as in phosphate-buffered solution occurs homogeneously along the material cross-section via a bulk erosion mechanism, as long as the material thickness is <2 mm (Figure 4B) (Tsuji and Ikada, 2000b; Tsuji et al., 2000b), even when the temperature is raised to 97 °C, which is higher than the T_g (60 °C) (Tsuji and Nakahara, 2001). Li et al. (1990) and Grizzi et al. (1995) demonstrated that catalytic LMW oligomers and monomers formed by hydrolysis remain at the core of PLA materials, and consequently they accelerate the hydrolysis in the core region of the PLA materials when they have a thickness >2 mm (Figure 4C). Due to this effect, the hydrolysis proceeds via two modes – slow and rapid hydrolysis at the outer layer and at the core, respectively.

Hydrolysis of Crystallized Specimens

The PLA chains in the crystalline region are more hydrolysis-resistant than those in the amorphous region. This causes predominant hydrolysis and removal of the chains in the amorphous region of the crystallized PLA, leaving the chains in the crystalline region. In addition, crystallization of PLA chains will occur during hydrolysis under increased molecular mobility in the presence of water molecules at a raised temperature, resulting in the increased x_c during hydrolysis (Li et al., 1990; Migliaresi et al., 1994; Pistner et al., 1994; Duek et al., 1999; Tsuji and Ikada, 2000b). Spherulitic structure was noticed in scanning electron

microscopy (SEM) observation of the PLLA films crystallized and hydrolyzed via surface erosion in alkaline and enzyme solutions (Tsuji and Ikada, 1998b; Tsuji and Miyauchi, 2001a), which resulted from preferred hydrolysis and removal of the chains in the amorphous region inside and outside the PLLA spherulites.

The hydrolysis of the chains in the amorphous region of crystallized PLLA is accelerated in phosphate-buffered solution compared with that of the chains in the free amorphous region, as in completely amorphous specimens (Tsuji et al., 2000b). This is attributed to the increased density of catalytic terminal carboxy group in the amorphous region of the crystallized PLLA specimens during crystallization compared with that of the completely amorphous specimens.

Figure 5 shows the gel permeation chromatography (GPC) curves of crystallized PLLA films hydrolyzed enzymatically in proteinase K/Tris-buffered solution (Figure 5A) (Tsuji and Miyauchi, 2001a,b), together with those hydrolyzed in alkaline solution (Figure 5B) (Tsuji and Ikada, 1998b) and phosphate-buffered solution (Figure 5C) (Tsuji et al., 2000b; Tsuji and Ikada, 2000b). For enzymatic hydrolysis (Figure 5A), the peak ascribed to the unhydrolyzed core part remains at the same position, while LMW specific peaks appear at the molecular weights of 1, 2, and 3×10^4, which are attributed to the one, two, and three folds of PLLA chain in the crystalline region, respectively. The fact that peak ascribed to the two or three folds of PLLA chain in the crystalline region remains for as long as 70 h, even when the height of initial main peak decreased to 20% of its initial value, strongly suggests that the folding chains in the restricted amorphous region between the crystalline regions are hydrolysis-resistant in the presence of proteinase K

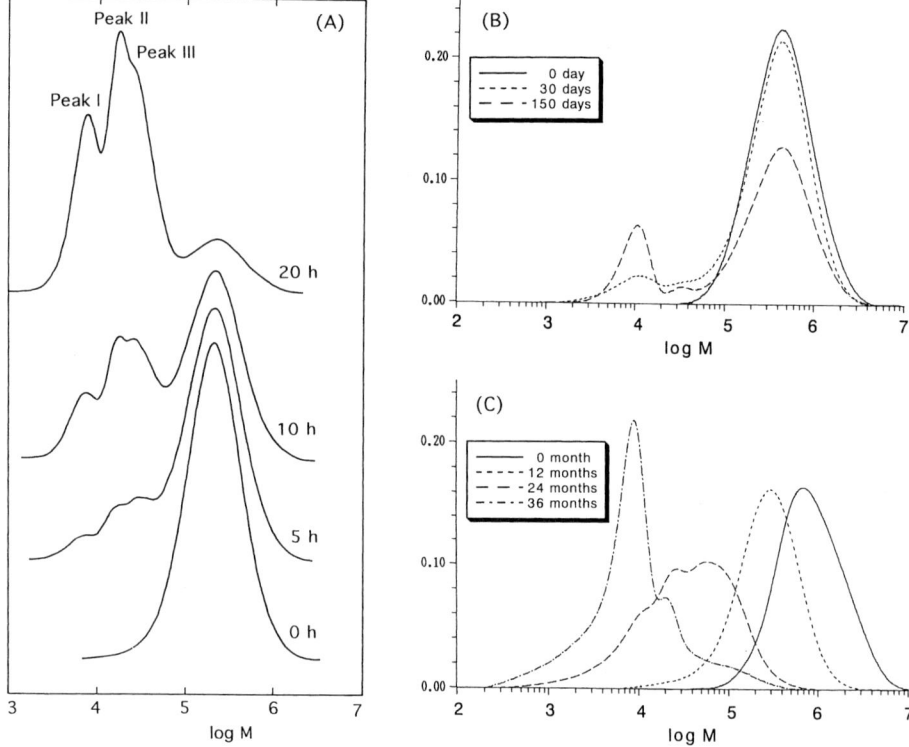

Fig. 5 Gel permeation chromatography curves of crystallized PLLA hydrolyzed in enzymatic (A), alkaline (B), and phosphate-buffered (C) solutions.

compared with tie chains and the chains with free ends. On the other hand, only the lowest molecular weight specific peak remained after long-term hydrolysis in alkaline and phosphate-buffered solutions (Figure 5B and C), representing non-selective or random scission of the chains (endo-chain scission) in the amorphous region. The dependence of x_c on the proteinase K-catalyzed hydrolysis rate of PLLA revealed that the PLLA chains in the restricted amorphous region between the crystalline regions in crystallized specimens are much more hydrolysis-resistant than those in the free amorphous region, as in a completely amorphous specimen (Tsuji and Miyauchi, 2001a,b). Iwata and Doi (1998) demonstrated that the enzymatic hydrolysis of PLLA

single crystals in the presence of proteinase K occurs at their crystal edges rather than the chain fold lamellar surfaces.

Hydrolysis of amorphous specimens
Similar to the crystallized PLLA, crystallization during hydrolysis is also reported for initially amorphous PLLA specimens, resulting in the increased x_c with hydrolysis time (Li et al., 1990; Migliaresi et al., 1994; Pistner et al., 1994; Duek et al., 1999; Gonzalez et al., 1999; Tsuji et al., 2000b). Li and McCarthy (1999b) found that stereocomplexation or racemic crystallization occurs when amorphous PDLLA is hydrolyzed for a long period, or at a high temperature, to have a LMW. In contrast, due to the stabilized chain packing in the amorphous region, the

increase in T_g occurs at the initial stage of hydrolysis, followed by the rapid decrease in T_g due to decreased molecular weight or enhanced molecular mobility at the late stage of hydrolysis (Gonzalez et al., 1999; Tsuji et al., 2000b).

11.2
Effects of Surrounding Media

Mason et al. (1981) studied a wide variety effects of the surrounding media on the hydrolysis of PDLLA. It is well known that the hydrolysis of PLAs is accelerated in aqueous media at high temperature (Jamshidi et al., 1986; Tsuji and Nakahara, 2001) and deviation from pH 7 (Makino et al., 1985; Tsuji and Ikada, 1998b) and in a temperature/humidity-controlled chamber at high temperature and humidity (Ho et al., 1999). Pitt and Gu (1987) studied the degradation behavior of PLLA, poly(LLA-GA), and PCL films at 37 °C in water, alcohols, and acidic and basic reagents. These authors found significant effects of the nature of degradation mediums on the degradation rates of poly(LLA-GA) and PCL, but no such effects for the PLLA degradation when their degradation was estimated by molecular weight. However, crystallization of an initially amorphous PLLA was observed during degradation in ethanol. The rapid crystallization at high temperature reduces the difference in the hydrolytic behavior between the PLLA specimens having different initial x_cs and Lc prepared by different annealing or crystallization conditions when PLLA is hydrolyzed autocatalytically in a phosphate-buffered solution (Tsuji and Nakahara, 2001).

Water-soluble oligomers and monomers of PLAs formed by hydrolysis will diffuse into surrounding media, resulting in weight loss of the materials. The water solubility of the lactic acid oligomers depend on the pH and temperature of surrounding media, which will alter the weight loss behavior of PLAs during hydrolysis. Kamei et al. (1992), using high-performance liquid chromatography, showed that 1- to 7-mers of L-lactic acid are soluble in distilled water, while Braud et al. (1996) used capillary electrophoresis to show that 1- to 9-mers of DL-lactic acid are soluble in phosphate-buffered solution at pH 6.8; however, lactic acid was the only component released from the mother PDLLA materials into aqueous medium during hydrolysis. The latter finding means that the oligomers having a degree of polymerization >2 are trapped in the materials during hydrolysis. Karlsson and Albertsson (1998) reported that high pH and high temperature are favorable for the water-soluble oligomers and monomers to diffuse out from the materials.

11.3
Effects of Material Factors

The hydrolysis behaviors of PLA materials can be controlled by altering their material parameters such as molecular characteristics and highly ordered structures.

11.3.1
Molecular Characteristics
Reducing the molecular weight of PLAs increases the density of hydrophilic terminal carboxy and hydroxy groups and molecular mobility; this will increase the diffusion rate and concentration of water in the materials and the probability of formation of water-soluble oligomers and monomers upon hydrolysis, resulting in their accelerated hydrolysis. The autocatalytic (Figure 6) and proteinase-K catalyzed (Figure 7) hydrolysis rates of initially amorphous PLLA become lower with the initial M_n (Tsuji et al., 2000b; Tsuji and Muramatsu, 2001a; Tsuji and Miyauchi, 2001c). On the other hand,

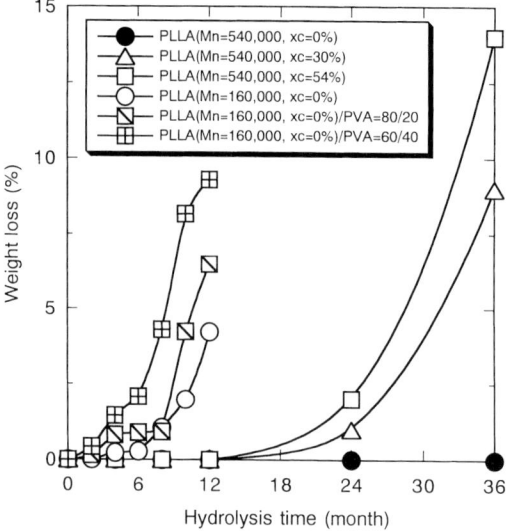

Fig. 6 Weight loss of PLAs by autocatalytic hydrolysis in phosphate-buffered solution.

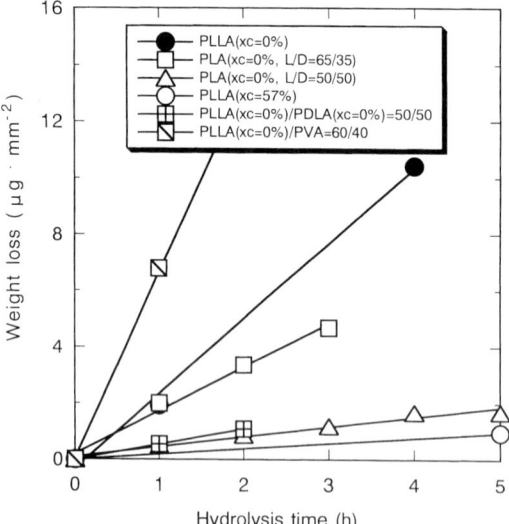

Fig. 7 Weight loss of PLAs by enzymatic hydrolysis in the presence of proteinase K.

LAs remaining unreacted during polymerization and/or formed during thermal processes accelerate the autocatalytic hydrolysis of PLA materials (e.g., Tsuji and Suzuki, 2001). Nakamura et al. (1989) and Zhang et al. (1994) confirmed the accelerated autocatalytic hydrolysis of PLAs upon addition of LAs.

The tacticity and the ratio and sequence lengths of chiral monomer units of PLAs have no significant effect on their autocatalytic hydrolysis, while proteinase K-catalyzed hydrolysis rate decreases significantly with decreasing L-lactyl unit sequence length (Reeve et al., 1994; MacDonald et al., 1996; Tsuji and Miyauchi, 2001c). Li et al. (2000)

found that the secondary effect of water-absorption of amorphous PLAs originated from the tacticity difference influences the enzymatic hydrolysis. In contrast, the auto-catalytic hydrolysis rate of amorphous poly-(DLLA-GA)s becomes higher with a rise in hydrophilic GA content (Hyon et al., 1998).

11.3.2
Highly Ordered Structures

Among the highly ordered structures, crystallinity is the most important parameter to determine the hydrolysis rate of PLA materials at the initial stage, whereas crystalline thickness is the dominant parameter to affect their hydrolysis rate at the late stage.

Crystallinity

When PLLA is hydrolyzed autocatalytically *in vivo* (Nakamura et al., 1989; Pistner et al., 1993, 1994) and in phosphate-buffered solution (Li et al., 1990; Duek et al., 1999; Tsuji and Ikada, 2000b; Tsuji et al., 2000b), the induction period until the start of decrease in mechanical properties and remaining weight decreases with the initial x_c. The decreases in mechanical properties, molecular weight, and remaining weight of PLLA after hydrolysis become higher with the initial x_c, when compared on the basis of the same hydrolysis time longer than 2 months (Duek et al., 1999) and 24 months (Tsuji and Ikada, 2000b; Tsuji et al., 2000b). On the other hand, when PLLA films are hydrolyzed in enzyme solution (Cai et al., 1996, Li and McCarthy, 1999a; Tsuji and Miyauchi, 2001a,b) or alkaline solution (pH 12) (Cam et al., 1995; Tsuji and Ikada, 1998b), the weight loss rate is lower for PLLA films having a higher initial x_c. Due to the high hydrolysis-resistance of the chains in the restricted amorphous region compared with that in the free amorphous region, the effect of x_c on enzymatic hydrolysis is strong and weak for the PLLA specimens having low

and high x_cs, respectively (Tsuji and Miyauchi, 2001b). The hydrolysis rates of crystallized PLAs and their copolymers are determined by the combination of the relatively strong x_c effect and the weak effects of molecular characteristics such as molecular weight (PLLA) (Cam et al., 1995), L-lactyl unit sequence length (LA stereocopolymers) (Reeve et al., 1994), and copolymer composition [poly(LLA-GA)s] (Reed and Gilding, 1981).

Crystalline Thickness

Fischer et al. (1973) prepared the single crystals of L-lactide-rich PLAs in a xylene dilute solution at different temperatures to have different L_cs, and their external catalytic hydrolysis was studied in a dilute alkaline solution. In contrast, we prepared PLLA films having different L_cs by annealing at different T_as from the melt (Tsuji and Ikada, 1995b) and their external catalytic and autocatalytic hydrolysis was investigated in alkaline (Tsuji and Ikada, 1998b) and enzyme (Tsuji and Miyauchi, 2001a) solutions and a phosphate-buffered solution (Tsuji and Ikada, 2000b), respectively. The size of crystalline residue at the late stage of hydrolysis, when a large weight loss occurs, depends on the initial L_c before hydrolysis, as evidenced by the fact that the lowest molecular weight specific peak ascribed to one fold of the PLLA chain in the crystalline residue increases with the initial T_m (L_c) before hydrolysis, irrespective of the hydrolysis media. The peak molecular weight of the lowest molecular weight specific peak ($M_{n,s}$) can be converted to the L_c of PLLA lamella after hydrolysis using the following equation, assuming that the PLLA chains take 10_3 helix in the α-form unit cell having a dimension $c = 2.78$ (or 2.88) nm (fiber axis) (Hoogsteen et al., 1990; Okihara et al., 1991):

$$L_c \ (nm) = 0.278 \ (or \ 0.288) \times M_{n,s}/72.1 \qquad (13)$$

where 72.1 is the mass per mole of lactyl unit (a half of lactide unit). The L_c values calculated from Eq. (13) are 29 (30), 36 (37), and 54 (56) nm for the PLLA specimens having T_m values of 172, 177, and 184 °C, respectively, after hydrolysis in phosphate-buffered solution, and 23 (24), 32 (33), and 36 (37) nm for those having T_m values of 170, 177, and 182 °C, respectively, after enzymatic hydrolysis (Tsuji and Ikada, 2000b; Tsuji and Miyauchi, 2001a). Here, the values before and in parentheses were calculated using the c values of 2.78 and 2.88 nm, respectively. The obtained L_c values are comparable with 10 and 11 nm for poly-(LLA-GA) (95/5) crystallized at 120 and 140 °C, respectively, from the melt (Wang et al., 2000), and also with 11 and 16 nm for PLLA single crystals formed at 82 and 102 °C, respectively, in a dilute solution (Fischer et al., 1973). The L_c values for c = 2.78 nm give the following relationships:

$$T_m \ (K) = 471 \ [1 - 1.59/L_c \ (nm) \ ] \qquad (14)$$
(hydrolyzed in phosphate-buffered solution)

$$T_m \ (K) = 472 \ [1 - 1.46/L_c \ (nm) \ ] \qquad (15)$$
(hydrolyzed in the presence of proteinase K)

Comparisons between Eqs. (1) and (14) or (15) give 198 and 199 °C as T_m^0 values, which are slightly lower but almost in agreement with those obtained from the Hoffman–Weeks procedure for the melt-crystallized PLLA specimens by Kalb and Pennings (1980) (215 °C) and by ourselves (Tsuji and Ikada, 1995b) (205 °C). This agreement on T_m^0 estimated with different methods also confirms our assumption that $M_{n,s}$ is the molecular weight of one fold of the PLLA chain in the crystalline region.

Spherulite Size
It has been concluded that there is no significant effect of the spherulite size on the external catalytic and autocatalytic hydrolysis of PLLA in alkaline and phosphate-buffered solutions, respectively (Tsuji and Ikada, 1998b, 2000b), which is comparable with the result reported for the enzymatic hydrolysis of R-PHB (Kumagai et al., 1992).

Orientation
Jamshidi et al. (1986) reported that the decrease rate of residual tensile strength and remaining weight of PLLA fibers during hydrolysis in phosphate-buffered solution decreases with DR or molecular orientation, and that the disk- and column-shaped crystalline residues were formed by the hydrolysis. The formation of the disk- and column-shaped crystalline residues is attributed to the selective hydrolysis and removal of the chains in the amorphous region between the crystalline regions which are located at a common cross-sectional plane normal to the fiber axis, as proposed for PGA (Chu and Campbell, 1982).

11.3.3
Material Shapes
The porous structure of PLAs has been reported to retard their bulk autocatalytic hydrolysis in a phosphate-buffered solution, which was ascribed to the enhanced diffusion of catalytic oligomers into a surrounding medium resulting from a reduced average distance of the porous material for the elution of catalytic oligomers from the material surface compared with that of the non-porous one (Lam et al., 1994). On the other hand, in the case of the surface hydrolysis in enzymatic and alkaline solutions, the hydrolysis rate is expected to increase with increasing porosity and decreasing pore size due to the increased surface area per unit weight on the basis of the finding for porous PCL specimens (Tsuji and Ishizaka, 2001a).

11.3.4

Polymer Blending

With regard to PLAs, no polyesters are reported to be highly miscible with them in the blends, excluding the self blends from HMW and LMW PLAs and the enantiomeric and diastereoisomeric PLA blends, as long as the molecular weights of PLAs and the other polyesters to be blended are sufficiently high ($M_n > 10^5$). Therefore, most of the PLA-based polymer blends are at least partially phase-separated. The effects of polymer blending depend on the hydrolysis medium, and the effects on the hydrolysis rate of PLAs are detailed below for two different media, namely phosphate-buffered and enzyme solutions.

The hydrolysis of PLAs in a phosphate-buffered solution will be accelerated by the presence of the polymer having either a high hydrophilicity, or/and a high density of carboxyl groups. Polymers such as HMW water-insoluble PVA and PEO, which satisfy the first requirement, will increase the diffusion rate and concentration of water in the materials, resulting in accelerated hydrolysis of PLAs. This acceleration was recognized for both the partially miscible (Pitt et al., 1992; Nijenhuis et al., 1996b) and phase-separated (Tsuji and Muramatsu, 2001a) blends. The polymers such as LMW aliphatic polyesters having a high density of catalytic carboxyl groups, which satisfy the second as well as the first requirements, will enhance the autocatalytic hydrolysis. The high density of hydrophilic terminal groups will increase the water concentration and diffusion rate into the materials, resulting in accelerated hydrolysis of PLAs. Examples include the blends of HMW PLLA with LMW PLLA (von Recum et al., 1995) or LMW PCL (Tsuji and Ikada, 1998a), and those from HMW and LMW PDLLAs (Bodmeier et al., 1989; Asano et al., 1991; Mauduit et al., 1996).

By contrast, the hydrolysis of PLAs in a phosphate-buffered solution will be retarded when a specimen contains the PLA stereocomplex (Tsuji, 2000c). The retarded hydrolysis of the well-stereocomplexed PLA blend compared with that of the nonblended PLLA and PDLA is ascribed mainly to the peculiarly strong interaction between L-lactyl and D-lactyl unit sequences in the amorphous region and/or the three-dimensional micronetwork structure in the blend formed by stereocomplexation in the course of solvent evaporation. They will reduce the diffusion rate of water molecules into the amorphous region and/or the concentration of catalytic oligomers formed by hydrolysis, respectively.

The proteinase K-catalyzed hydrolysis of PLLA is reported to be accelerated in the phase-separated blends by the presence of hydrophilic PEO (Sheth et al., 1997), cellulose (Nagata et al., 1998), PVA (Tsuji and Muramatsu, 2001a) and hydrophobic PCL (Tsuji and Ishizaka, 2001b), whereas decelerated in the miscible blends from L-rich PLA and hydrophobic poly(vinyl acetate) (PVAc) (Gajria et al., 1996). This means that the phase-structure of the blend rather than the hydrophilicity of the second polymers has a crucial effect on the enzymatic hydrolysis rate of PLLA, and strongly suggests that the enhanced enzymatic hydrolysis in the phase-separated blends is due to the occurrence of enzymatic hydrolysis at the interfaces of the two polymer phases as well as at the specimen surface.

12

Biodegradation

In natural environments, nonenzymatic hydrolysis is recognized as the main route of PLA degradation. However, it was confirmed that biodegradation also occurs in the presence of some microbes in natural environments (Torres et al., 1996a–c, 1999; Pranamunda et al., 1997; Karjomaa et al., 1998; Tsuji et al., 1998; Pranamunda and Tokiwa, 1999; Tokiwa et al., 1999), and that the mineralization of PLLA to carbon dioxide, rather than to lactic acid, occurs in soils (Ho and Pometto, 1999) and composts (Meinander et al., 1997).

and the companies producing lactic acids, lactides, PLAs, and PLA end-products are described in Chapters 8 and 9 of this volume, and also in review articles (e.g., Datta et al., 1995; Bogaert and Coszach, 2000; Middleton and Tipton, 2000).

13
Applications

PLAs have been regarded as biomedical materials and matrices in the forms of rods, plates, films, meshes, and microspheres for tissue engineering and drug delivery systems. Due to the lowered prices of PLAs, they are recently recognized as substitutes for the general-purpose "petro" polymers such as polyethylene, polypropylene, and polystyrene. The medical and ecological applications of PLAs are summarized in Tables 5 and 6 (Ikada and Tsuji, 2000), respectively. More detailed applications in medical and pharmaceutical areas are described in several monographs (e.g., Kharas et al., 1994; Hartmann, 1998; Coombes and Meikle, 1994), and the end-products of PLAs

14
Outlook and Perspectives

The physical properties and hydrolysis and biodegradation behavior of PLAs are controllable by varying molecular characteristics, highly ordered structures, and material shapes, and also by polymer blending. It seems that PLA-based polymeric materials having excellent physical properties will be widely used as commodity plastics in the near future when PLLA is produced probably under a $1.00/lb cost by major US companies such as Cargill-Dow Polymers. The most appropriate biodegradable polymer for the respective end uses can be selected from the PLA based polymeric materials having different physical properties, enzymatic and non-enzymatic hydrolytic behavior, and cost/performance.

Tab. 5 Hitherto medical applications of PLAs and their copolymers (Ikada and Tsuji, 2000)

Function	Purpose	Examples
Bonding	Suturing	Vascular and intestinal anastomosis
	Fixation	Fractured bone fixation
	Adhesion	Surgical adhesion
Closure	Covering	Wound cover, local hemostasis
	Occlusion	Vascular embolization
Separation	Isolation	Organ protection
	Contact inhibition	Adhesion prevention
Scaffold	Cellular proliferation	Skin reconstruction, blood vessel reconstruction
	Tissue guide	Nerve reunion
Capsulation	Controlled drug delivery	Sustained drug release

Tab. 6 Hitherto ecological applications of PLAs and their copolymers (Ikada and Tsuji, 2000)

Application	Fields	Examples
Industrial applications	Agriculture, forestry	Mulch films, temporary replanting pots, Delivery system for fertilizers and pesticides,
	Fisheries	Fishing lines and nets, fishhooks, fishing Gears
	Civil engineering and construction industry	Forms, vegetation nets and sheets, water-retention sheets
	Outdoor sports	Golf tees, disposable plates, cups, bags, and Cutlery
Composting	Food package	Package, containers, wrappings, bottles, bags, films, retail bags, six-pack rings
	Toiletry	Diapers, feminine hygiene products
	Daily necessities	Refuge bags, cups

15

Patents

The important patent articles concerning the synthesis and applications of PLAs and their copolymers are summarized in Table 7. A large number of patent articles have been issued for lactic acids, LAs formation, and synthesis of PLAs. Lowe (1954) disclosed the improved LAs purification techniques, which led to synthesis of HMW PLAs, while Selman (1967) later proposed the distillation process to obtain optically pure LAs from the mixture of LMW PLAs and LAs. Bellis (1988) claimed the recrystallization method using toluene or ethyl acetate for the purification of LAs. Datta (1989) proposed an efficient and economical process for lactic acid production and purification. Gruber and Hall (1992) claimed the continuous closed-loop system to synthesize LAs from lactic acids without waste products other than water, which leads to low-cost production of PLAs. Ohara and Okamoto (1997) proposed the recovery method of LA from high-molecular weight PLA, which may be intended to obtain LA from PLA wastes.

Enomoto et al. (1994) proposed a method of synthesizing HMW PLAs and their copolymers by direct condensation of lactic acids and other hydroxycarboxylic acids using azeotropic solvents such as diphenylether. Seppälä and Selin (1995) claimed the method to produce lactic acid-based polyurethane, which enables the chain length of the LMW PLAs and their copolymers to be extended. Vitalis (1959) and Casey and Huffman (1984) suggested the basic method to synthesize the block copolymers of poly(α-hydroxylic acid)s with flexible hydrophilic polyethers such as PEO. These block copolymers are utilized as DDS matrices containing hydrophilic drugs.

In terms of processes and applications of the materials from PLAs and their copolymers, numerous patents have been issued. Shmitt and Polistina (1967) and Shneider (1972) proposed the applications of aliphatic polyesters and poly(LA-GA), respectively, to surgical sutures. Michaels (1976) suggested the applications of PLAs to DDS matrices, whereas Pitt and Schindler (1979) proposed the application of poly(LA-CL) to the matrices for sustained subdermal delivery system. Törmälä et al. (1988) claimed the preparation method of self-reinforced high-strength PLAs, which are useful as osteosynthesis devices. Murdoch and Loomis (1988)

Tab. 7 Patent articles of PLAs and copolymers

Number of patent	Patent holder	Inventor	Title of Patent	Date of publication
USP 2 668 162	Du Pont	Lowe, C.E.	Preparation of high molecular weight polyhydroxyacetic ester	Feb. 3, 1954
USP 4 438 253	American Cyanamid	Vitalis, E.A.	Polyglycol-polyacid ester treatment of textiles	Dec. 15, 1959
USP 3 297 033	American Cyanamid	Shmitt, E.E., Polistina, R.A.	Surgical sutures	Jan. 10, 1967
USP 3 322 791	Ethicon	Selman, S.	Preparation of optically active lactide	May 30, 1967
USP 3 636 956	American Cyanamid	Shneider, A.K.	Polylactide sutures	Jan. 10, 1972
USP 3 962 414	Alza	Michaels, A.S.	Structured bioerodible drug delivery device	June 8, 1976
USP 4 148 871	–	Pitt, C.G., Schindler, A.E.	Sustained subdermal delivery of drugs using poly(ε-caprolactone) and its copolymers	Apr. 10, 1979
USP 4 438 253	American Cyanamid	Casey, D.J., Huffman, K.R.	Polyglycolic acid) / polyalkylene glycol) block copolymers and method of manufacturing the same	Mar. 20, 1984
USP 4 719 246	Du Pont	Murdoch, J.R., Loomis, G.L.	Polylactide compositions	Jan. 12, 1988
USP 4 727 163	Du Pont	Bellis, H.E.	Process for preparing highly pure cyclic esters	Feb. 23, 1988
USP 4 743 257	Materials Consultant	Törmälä, P., Rokkanen, P., Laiho, J., Tammimmäki, M., Vainionpää, S.	Materials for osteosynthesis devices	May 10, 1988
USP 4 863 506	Union Oil Company of California	Young, D.C.	Methods for regulating of the growth of plants and growth regulant composition	Sep. 5, 1989
USP 4 885 247	Michigan Biotechnology Institute	Datta R.	Recovery and purification of lactate salts from whole fermentation broth by electrodialysis	Dec. 5, 1989
Jpn. Kokai Tokkyo Koho, JP 03 183 428	Gunze	Saito, Y., Ikada, Y., Gen, J., Suzuki, M.	Hydrolyzable polylactide fishing lines and their manufacture	Aug. 9, 1991
Fr. Demande, FR 2 657 255 A1	Sederma, S.A.	Greff, D.	Description de préparations cosmétiques originales dont les principes actifs sont piégés dans un réseau polymérique greffé à la surface de particules de silice	Dec. 27, 1991
PCT Int. Appl. WO 9 201 737	Du Pont	Hammel, H.S., York, R.O.	Degradable foam materials	Feb. 6, 1992
PCT Int. Appl. WO 9 204 412 A1	Du Pont	Ostapchenko, G.J.	Films containing polyhydroxy acid	Mar. 19, 1992
EP 467 478 A2	DSM	Van den Berg, H.J.	Method for the manufacture of polymer products from cyclic esters	Jan. 22, 1992
USP 5 142 023	Cargil	Gruber, P.R., Hall, E.S.	Continuous Process for manufacture of lactide polymers with controlled optical purity	Aug. 25, 1992
USP 5 298 602	Takiron	Shikinami, Y., Hata, K.	Polymeric piezoelectric material	Mar. 29, 1994
USP 5 310 865	Mitsui Toatsu Chemicals	Enomoto, K. Ajioka, M., Yamaguchi, A.	Polyhydroxycarboxylic acid and preparation process thereof	May 10, 1994
USP 5 380 813	.	Seppälä, J., Selin, J.F.	Method for producing lactic acid based polyurethane	Jan. 10, 1995
DE 19 637 404	Shimadzu	Ohara, H., Okamoto, T.	Recovery of lactide from high-molecular-weight polylactic acid	Mar. 20, 1997
Jpn. Kokai Tokkyo Koho, JP 09 286 909 A2	Mitsui Toatsu Chemicals	Kobayashi, N., Imon, S., Kono, A., Kuroki, T., Wanibe, H.,	Lactic acid-based polymer films for agriculture	Nov. 4, 1997
Jpn. Kokai Tokkyo Koho, JP 09 110 968 A2	Shimadzu	Tasaka, S., Kosei, E.	Polymeric electret material and its manufacture	Apr. 28, 1998

proposed the utilization of stereocomplexation to enhance the mechanical properties of PLA materials. Young (1989) suggested the utilization of L-lactic acid solution containing its dimer and oligomer to enhance growth of plants. Hammel and York (1992) disclosed the preparation method of food-grade cellular materials. Van den Berg (1992) claimed the *in-situ* polymerization method of LAs to synthesize PLA materials for reconstructive orthopedics. Shikinami and Hata (1994) proposed that piezoelectric materials could be prepared by uniaxial drawing, which is also useful for treating bone fractures, in ultrasonographic devices, and in ultrasonic flaw detectors. The other patent articles issued relating to applications of PLAs have included those for fishing lines (Saito et al., 1991), cosmetic preparations (Greff, 1991), films useful in disposable shrink-wrap packing (Ostapchenko, 1992), films for agriculture (Kobayashi et al., 1997), and electret materials (Tasaka and Kosei, 1998).

Acknowledgments

The author wishes to thank Dr. Carlos Adriel Del Carpio, Department of Ecological Engineering, Faculty of Engineering, Toyohashi University of Technology, for his helpful and kind comments on the manuscript.

16
References

Agarwal, S., Brandukova-Szmikowski, N. E., Greiner, A. (1999) Samarium (III)-mediated graft polymerization of ε-caprolactone and L-lactide in functionalized poly(p-xylylene)s: model studies and polymerization, *Polym. Adv. Technol.* **10**, 528–534.

Agrawal, C. M., Athanasiou, K. A., Heckman, J. D. (1997) Biodegradable PLA-PGA polymers for tissue engineering in orthopedics, *Mater. Sci. Forum* **250**, 115–128.

Ajioka, M., Enomoto, E., Suzuki, K., Yamaguchi, A. (1995) Basic properties of polylactic acid produced by the direct condensation polymerization of lactic acid, *Bull. Chem. Soc. Jpn.* **68**, 2125–2131.

Arvanitoyannis, I., Nakayama, A., Kawasaki, N., Yamamoto, N. (1995) Novel star-shaped polylactide with glycerol using stannous octoate or tetraphenyl tin as catalyst: 1. Synthesis, characterization and study of their biodegradability, *Polymer* **36**, 2947–2956.

Arvanitoyannis, I., Nakayama, A., Psomiadou, E., Kawasaki, N., Yamamoto, N. (1996) Synthesis and degradability of a novel aliphatic polyester based on L-lactide and sorbitol:3, *Polymer* **37**, 651–660.

Asano, M., Fukuzaki, H., Yoshida, M., Kumakura, M., Mashimo, T., Yuasa, H., Imai, K., Yamanaka, H., Kawahara, U., Suzuki, K. (1991) *In vivo* controlled release of a luteinizing hormone-releasing hormone agonist from poly(DL-lactic acid) formulations of varying degradation pattern, *Int. J. Pharm.* **67**, 67–77.

Aslan, S., Calandrelli, L., Lauriezo, P., Malinconico, M., Migliaresi, C. (2000) Poly(D,L-lactide)/poly-(ε-caprolactone) blend membranes: preparation and morphological characterisation, *J. Mater. Sci.* **35**, 1615–1622.

Atthoff, B., Trollsås, M., Claesson, H., Hedrick, J. L. (1999) Poly(lactides) with controlled molecular architecture initiated from hydroxy functional dendrimers and the effects on the hydrodynamic volume, *Macromol. Chem. Phys.* **200**, 1333–1339.

Blümm, E., Owen, A. J. (1995) Miscibility, crystallization and melting of poly(3-hydroxybutyrate)/poly(L-lactide) blends, *Polymer* **36**, 4077–4082.

Bodmeier, R., Oh, K. H., Chen, H. (1989) The effect of the addition of low molecular weight poly(DL-lactide) on drug release from biodegradable poly(DL-lactide) drug delivery systems, *Int. J. Pharm.* **51**, 1–8.

Bogaert, J.-C., Coszach, P. (2000) Poly(lactic acids): a potential solution to plastic waste dilemma, *Macromol. Symp.* **153**, 287–303.

Braud, C., Devarieux, R., Garreau, H., Vert, M. (1996) Capillary electrophoresis to analyze water-soluble oligo(hydroxyacids) issued from degraded or biodegraded aliphatic polyesters, *J. Environ. Polym. Degrad.* **4**, 135–148.

Breitenbach, A., Kissel, T. (1998) Biodegradable comb polyesters: Part 1 Synthesis, characterization and structural analysis of poly(lactide) and poly(lactide-co-glycolide) grafted onto water-soluble poly(vinyl alcohol) as backbone, *Polymer* **39**, 3261–3271.

Brizzolara, D., Cantow, H.-J., Diederichs, K., Keller, E., Domb, A. J. (1996a) Mechanism of stereocomplex formation between enantiomeric poly(lactide)s, *Macromolecules*, **29** 191–197.

Brizzolara, D., Cantow, H.-J., Mülhaupt, R., Domb, A. J. (1996b) Novel materials through stereocomplexation, *J. Computer-Aided Mater. Design* **3**, 341–350.

Brochu, S. Prud'homme, R. E., Barakat, I., Jérome, R. (1995) Stereocomplexation and morphology of polylactides, *Macromolecules* **28**, 5230–5239.

Cai, H. Dave, V. Gross, R. A., McCarthy, S. P. (1996) Effects of physical aging, crystallinity, and orientation on the enzymatic degradation of poly-

(lactic acid), *J. Polym. Sci.: Part B: Polym. Phys.* **34**, 2701–2708.

Cam, D., Hyon, S.-H., Ikada, Y. (1995) Degradation of high molecular weight poly(L-lactide) in alkaline medium, *Biomaterials* **16**, 833–843.

Carothers, W. H., Dorough, G. L., Van Natta, F. J. (1932) Studies of polymerization and ring formation. X. The reversible polymerization of six-membered cyclic esters, *J. Am. Chem. Soc.* **54**, 761–772.

Cartier, L. Okihara, T., Lotz, B. (1997) Triangular polymer single crystals: stereocomplexes, twins, and frustrated structures, *Macromolecules* **30**, 6313–6322.

Cartier, L., Okihara, T., Ikada, Y., Tsuji, H., Puiggali, J., Lotz, B. (2000) Epitaxial crystallization and crystalline polymorphism of polylactides, *Polymer* **41**, 8909–8919.

Celli, A., Scandola, M. (1992) Thermal properties and physical aging of poly(L-lactic acid), *Polymer* **33**, 2699–2703.

Cha, Y. Pitt, C. G. (1990) The biodegradability of polyester blend, *Biomaterials* **11**, 108–112.

Chaput, C., DesRosiers, E. A., Assad, M., Brochu, M., Yahia, L'H., Selmani, A., Rivard, C.-H. (1995) Processing biodegradable natural polyesters for porous soft-materials, in: *Advances in Materials Science and Implant Orthopedic Surgery* (Kossowski, R., Kossovsky, N., Eds.), Dordrecht: Kluwer Academic Publishers, 229–245.

Chen, X., McGurk, S. L., Davies, M. C., Roberts, C. J., Shakesheff, K. M., Tendler, S. J. B., Williams, P. M., Davies, J., Dawkes, A. C., Domb, A. (1998) Chemical and morphological analysis of surface enrichment in a biodegradable polymer blend by phase-detection imaging atomic force microscopy, *Macromolecules* **31**, 2278–2283.

Chisholm, M. H., Iyer, S. S., Matison, M. E., McCollum, D. G., Pagel, M. (1997) Concerning the stereochemistry of poly(lactide), PLA. Previous assignments are shown to be incorrect and a new assignment is proposed, *Chem. Commun.* 1999 2000.

Cho, K. Y., Kim. C.-H., Lee, J.-W., Park, J.-K. (1999) Synthesis and characterization of poly(ethyleneglycol) grafted poly(L-lactide), *Macromol. Rapid Commun.* **20**, 598–601.

Choi, E.-J, Park, J.-K., Chang, H.-N. (1994) Effect of polymerization catalysts on the microstructure of P(LLA-co-εCL), *J. Polym. Sci.: Part B: Polym. Phys.* **32**, 2481–2489.

Chu, C. C., Campbell, N. D. (1982) Scanning electron microscopic study of the hydrolytic degradation of poly(glycolic acid), *J. Biomed. Mater. Res.* **16**, 417–430.

Cook, A. D., Pajvani, U. B., Hrkach, J. S., Cannizzaro, S. M., Langer, R. (1997) Calorimetric analysis of surface reactive amino group on poly(lactic acid-co-lysin): poly(lactic acid) blends, *Biomaterials* **18**, 1417–1424.

Coombes, A. G. A, Heckman, J. D. (1992a) Gel casting of resorbable polymers. 1. Processing and applications, *Biomaterials* **13**, 217–224.

Coombes, A. G. A, Heckman, J. D. (1992b) Gel casting of resorbable polymers. 2. In-vitro degradation of bone graft substitutes, *Biomaterials* **13**, 297–307.

Coombes, A. G. A., Meikle, M. C. (1994) Bioabsorbable synthetic polymers as replacements for bone graft, *Clin. Mater.* **17**, 35–67.

Cooper-White, J. J., Mackay, M. E. (1999) Rheological properties of poly(lactides). Effect of molecular weight and temperature on the viscoelasticity of poly(L-lactic acid), *J. Polym. Sci.: Part B: Polym. Phys.* **37**, 1803–1814.

Datta, R., Tsai, S.-P., Bonsignore, P., Moon, S.-H., Frank, J. R. (1995) Technological and economic potential of poly(lactic acid) and lactic acid derivatives, *FEMS Microbiol. Rev.* **16**, 221–231.

Davies, M. C., Shakesheff, K. M., Shard, A. G., Domb, A., Roberts, C. J., Tendler, S. J. B., Williams, P. M. (1996) Surface analysis of biodegradable polymer blends of poly(sebacic anhydride) and poly(DL-lactic acid), *Macromolecules* **29**, 2205–2212.

Degée, P., Dubois, P., Jérome, R. (1997), Bulk polymerization of lactides initiated by aluminium isopropoxide, 3. Thermal stability and viscoelastic properties, *Macromol. Chem. Phys.* **198**, 1985–1995.

de Jong, S. J., van Dijk-Wolthuis, W. N. E., Kettenes-van den Bosch, J. J., Schuyl, P. J. W., Hennink, W. E. (1998) Monodisperse enantiomeric lactic acid oligomers: preparation, characterization, and stereocomplex formation, *Macromolecules* **31**, 6397–6402.

de Jong, S. J., Smedt, S. C. De, Wahls, M. W. C., Demeester, J., Kettenes-van den Bosch, J. J., Hennink, W. E. (2000) Novel self-assembled hydrogels by stereocomplex formation in aqueous solution of enantiomeric lactic acid oligomers grafted to dextran, *Macromolecules* **33**, 3680–3686.

De Santis, P., Kovacs, A. J. (1968) Molecular conformation of poly(S-lactic acid), *Biopolymer* **6**, 299–306.

Dijkstra, P. J. Bulte, A., Feijen, J. (1991) Block copolymers of L-lactide, D-lactide, and ε-caprolactone, *The 17th Annual Meeting of the Society for*

Biomaterials, Scottsdale, Arizona. Minneapolis, Minnesota: Society for Biomaterials.

Domb, A. J. (1993) Degradable polymer blends. I. Screening of miscible polymers, *J. Polym. Sci.: Part A: Polym. Chem. Ed.* **31**, 1973–1981.

Dorgan, J. R., Williams, J. S., Lewise, D. N. (1999) Melt rheology of poly(lactic acid): entanglement and chain architecture effects, *J. Rheol.* **43**, 1141–1155.

Dubois, P., Degée, P., Ropson, N., Jérome, R. (1997) Macromolecular engineering of polylactones and polylactides by ring-opening polymerization, in: *Macromolecular Design of Polymeric Materials* (Hatada, K., Kitayama, T., Vogl, O., Eds.), New York: Marcel Dekker, Inc., 247–272.

Duek, E. A. R., Zavaglia, C. A. C, Belangero, W. D. (1999) In vitro study of poly(lactic acid) pin degradation, *Polymer* **40**, 6465–6473.

Edlund, U., Albertsson, A.-C. (2000) Microspheres from poly(D,L-lactide)/poly(1,5-dioxepan-2-one) miscible blends for controlled drug delivery, *J. Bioact. Compatible Polym.* **15**, 214–229.

Eguiburu, J. L., Fernandez-Berridi, M. J., Roman, J. S. (1996) Graft copolymers for biomedical applications prepared by free radical polymerization of poly(L-lactide) macromonomers with vinyl and acrylic monomers, *Polymer* **37**, 3615–3622.

Eling, B., Gogolewski, S., Pennings, A.J. (1982) Biodegradable materials of poly(L-lactic acid): 1. Melt-spun and solution spun fibres, *Polymer* **23**, 1587–1593.

Fambri, L., Pegoretti, A., Fenner, R., Incardona, S. D., Migliaresi, C. (1997) Biodegradable fibres of poly(L-lactic acid) produced by melt-spinning, *Polymer* **38**, 79–85.

Fischer, E. W., Sterzel, H. J., Wegner, G. (1973) Investigation of the structure of solution grown crystals of lactide copolymers by means of chemical reaction, *Kolloid-Z. u. Z. Polym.* **251**, 980–990.

Fukuzaki, H., Yoshida, M., Asano, M., Kumakura, M. (1989) Synthesis of copoly(D,L-lactic acid) with relatively low molecular weight and *in vitro* degradation, *Eur. Polym. J.* **25**, 1019–1026.

Gajria, A. M., Davé, V., Gross, R. A., McCarthy, S. P. (1996) Miscibility and biodegradability of blends of poly(lactic acid) and poly(vinyl acetate), *Polymer* **37**, 437–444.

Gan, Z., Yu, D., Zhong, Z., Liang, Q., Jing, X. (1999) Enzymatic degradation of poly(ε-caprolactone)/poly(DL-lactide) blends in phosphate-buffer solution, *Polymer* **40**, 2859–2862.

Gedde, U. W. (1995) *Polymer Physics*, London: Chapman & Hall, Chapters 7 and 8, 131–198.

Gilding, D. K., Reed, A. M. (1979) Biodegradable polymers for use in surgery-polyglycolic/poly(lactic acid) homo- and copolymers: 1, *Polymer* **20**, 1459–1464.

Göpferich, A (1997) Mechanisms of polymer degradation and elimination, in: *Handbook of Biodegradable Polymers* (Domb, A. J., Kost, A., Wiseman, D. M., Eds.), Amsterdam: Harwood Academic Publishers, 451–471.

Gonzalez, M. F., Ruseckaite, R. A., Cuadrado, T. R. (1999) Structural changes of polylactic-acid (PLA) microspheres under hydrolytic degradation, *J. Appl. Polym. Sci.* **71**, 1223–1230.

Grijpma, D. W., Kroeze, E., Nijenhuis, A. J., Pennings, A. J. (1993) Poly(L-lactide) crosslinked with spiro-bis-dimethylene-carbonate, *Polymer* **34**, 1496–1503.

Grijpma, D. W., van Hofslot, R. D. A., Supèr, H., Nijenhuis, A. J., Pennings, A. J. (1994) Rubber toughening of poly(lactide) by blending and block copolymerization, *Polym. Eng. Sci.* **34**, 1674–1684.

Grizzi, I., Garreau, H., Li, S., Vert, M. (1995) Hydrolytic degradation of devices based on poly(DL-lactic acid) size-dependence, *Biomaterials* **16**, 305–311.

Hartmann, M. H. (1998) High molecular weight polylactic acid polymers, in: *Biopolymers from Renewable Resources* (Kaplan, D.L., Ed.), Berlin: Springer-Verlag, 367–411.

Hartmann, M. (1999) Advances in the commercialization of poly(lactic acid), *Polym. Prepr. (Am. Chem. Soc., Div. Polym. Chem.)* **40**, 570–571.

Higashi, S., Yamamuro, T., Nakamura, T., Ikada, Y., Hyon, S.-H., Jamshidi, K. (1986) Polymer-hydroxyapatite composites for biodegradable bone fillers, *Biomaterials* **7**, 183–187.

Hiki, S., Miyamoto, M., Kimura, Y. (2000) Synthesis and characterization of hydroxy-terminated (RS)-poly(3-hydroxybutyrate) and its utilization to block copolymerization with l-lactide to obtain a biodegradable thermoplastic elastomer, *Polymer* **41**, 7369–7379.

Hiljanen-Vainio, M., Varpomaa, P., Seppälä, J., Törmälä, P. (1996) Modification of poly(L-lactides) by blending: mechanical and hydrolytic behavior, *Macromol. Chem. Phys.* **197**, 1503–1523.

Ho, K.-L. G., Pometto, A. L., III (1999) Temperature effects on soil mineralization of polylactic acid plastic in laboratory respirometers, *J. Environ. Polym. Degrad.* **7**, 101–108.

Ho, K.-L. G., Pometto, A.L., III, Hinz, P. N. (1999) Effects of temperature and relative humidity on polylactic acid plastic degradation, *J. Environ. Polym. Degrad.* **7**, 83–92.

Hoogsteen, W., Postema, A. R., Pennings, A. J., ten Brinke, G., Zugenmaier, P. (1990) Crystal structure, conformation, and morphology of solution-spun poly(L-lactide) fibers, *Macromolecules* **23**, 634–642.

Horacek, I., Kalísek, V. (1994a) Polylactide. I. Continuous dry spinning-hot drawing preparation of fibers, *J. Appl. Polym. Sci.* **54**, 1751–1757.

Horacek, I., Kalísek, V. (1994b) Polylactide. II. Discontinuous dry spinning-hot drawing preparation of fibers, *J. Appl. Polym. Sci.* **54**, 1759–1765.

Horacek, I., Kalísek, V. (1994c) Polylactide. III. Fiber preparation by spinning in precipitant vapor, *J. Appl. Polym. Sci.* **54**, 1767–1771.

Hori, Y., Takahashi, Y. Yamaguchi, A., Nishishita, T. (1993) Ring-opening copolymerization of optically active β-butyrolactone with several lactones catalyzed by distannoxane complexes: synthesis of new biodegradable polyesters, *Macromolecules* **26**, 4388–4390.

Howe, C., Vasanthan, N., MacClamrock, C., Sankar, S., Shin, I. D., Simonsen, I. K., Tonelli, A. E. (1994) Inclusion compound formed between poly(L-lactic acid) and urea, *Macromolecules* **27**, 7433–7436.

Hsu, S.-H., Chen, W.-C. (2000) Improved adhesion by plasma-induced grafting of l-lactide onto polyurethane surface, *Biomaterials* **21**, 359–368.

Hu, L.-C., Nakata, H., Yamane, H., Kitada, T. (1994) The role of poly(L-lactide) in the degradation of poly(ε-amino caproic acid), *Kobunshi Ronbunshu* **51**, 486–492.

Hu, L.-C., Shinoda, H., Yoshida, E., Kitada, T. (1995) Hydrolysis of polyblends comprising poly(L-lactide) and poly(ε-amino-caproic acid), *Kobunshi Ronbunshu* **52**, 114–120.

Huang, J., Lisowski, M. S., Runt, J., Hall, E. S., Kean, R. T., Buehler, N., Lin, J. S. (1998) Crystallization and microstructure of poly(L-lactide-co-meso-lactide) copolymers, *Macromolecules* **31**, 2593–2599.

Hyon, S.-H., Jamshidi, K., Ikada, Y. (1984) Melt spinning of poly-L-lactide and hydrolysis of the fiber *in vitro*, in: *Polymers as Biomaterials* (Shalaby, S. W., Hoffman, A. S., Ratner, B. D., Horbett, T A., Eds.), New York: Plenum Press, 51–65.

Hyon, S.-H., Jamshidi, K., Ikada, Y., Higashi, S., Kakutani, Y., Yamamuro, T. (1985) Bone filler from poly(lactic acid)-hydroxyapatite composites, *Kobunshi Ronbunshu* **42**, 771–776.

Hyon, S.-H., Jamshidi, K., Ikada, Y. (1997) Synthesis of polylactides with different molecular weights, *Biomaterials* **18**, 1503–1508.

Hyon, S.-H., Jamshidi, K., Ikada, Y. (1998) Effects of the residual monomer on the degradation of DL-lactide copolymer, *Polym. Int.* **46**, 196–202.

Iannace, S., Nicolais, L. (1996) Isothermal crystallization and chain mobility of poly(L-lactide), *J. Appl. Polym. Sci.* **64**, 911–919.

Iannace, S., Ambrosio, L., Huang, S. J., Nicolais, L. (1994) Poly(3-hydroxybutyrate)-co-(3-hydroxyvalerate)/poly-L-lactide blends: thermal and mechanical properties, *J. Appl. Polym. Sci.* **54**, 1525–1536.

Iannace, S., Ambrosio, L., Huang, S. J., Nicolais, L. (1995) Effect of degradation on the mechanical properties of multiphase polymer blends: PHBV/PLLA, *J. Macromol. Sci.-Pure Appl. Chem.* **A32**, 881–888.

Ibim, S. E. M., Ambrosio, A. M. A., Kwon, M. S., El-Amin, S. F., Allcock, H. R., Laurencin, C. T. (1997) Novel polyphosphazene/poly(lactide-co-glycolide) blends: miscibility and degradation studies, *Biomaterials* **18**, 1565–1569.

Ikada, Y. (1998) Tissue Engineering research trends at Kyoto University, in: *Tissue Engineering for Therapeutic Use 1* (Ikada, Y., Yamaoka, Y., Eds.), Washington, DC, Am. Chem. Soc., 1–14.

Ikada, Y., Tsuji. H. (2000) Biodegradable polyesters form medical and ecological applications, *Macromol. Rapid Commun.* **21**, 117–132.

Ikada, Y., Jamshidi, K., Tsuji, H., Hyon, S.-H. (1987) Stereocomplex formation between enantiomeric poly(lactides), *Macromolecules* **20**, 904–906.

Ikada, Y., Shikinami, Y., Hara, Y., Tagawa, M., Fukada, E. (1996) Enhancement of bone formation by drawn poly(L-lactide), *J. Biomed. Mater. Res.* **30**, 553–558.

Inoue, K., O-oya, s., Mun-soo, L., Akasaka, S., Asai, S., Sumita, M. (1998) Fractal and degradation process of biodegradable polyester blends, *Sen'i Gakkaishi* **54**, 277–284.

Ishida, T., Tsuji, H. (2000) Hydrolysis mechanisms of PLLA in enzyme and alkaline solutions, *Polym. Prepr. Jpn.* **49**, 1008.

Ivanova, T. Z., Panaiotov, I., Boury, F., Proust, J. E., Verger, R. (1997) Enzymatic hydrolysis of poly(D,L lactide) spread monolayers by cutinase, *Colloid Polym. Sci.* **275**, 449–457.

Iwata, T., Doi, Y. (1998) Morphology and enzymatic degradation of poly(L-lactic acid) single crystals, *Macromolecules* **31**, 2461–2467.

Jamshidi, K., Hyon, S.-H., Nakamura, T., Ikada, Y., Shimidu, Y., Teramatsu, T. (1986) In vitro and in vivo degradation of poly-L-lactide fibers, in: *Biological and Biomechanical Performance of Biomaterials* (Christel, P., Meunier, A., Lee, A. J. C., Eds.), Amsterdam, The Netherlands: Elsevier Science Publisher B.V., 227–232.

Jamshidi, K., Hyon, S.-H., Ikada, Y. (1988) Thermal characterization of polylactides, *Polymer* **29**, 2229–2234.

Jorda, R., Wilkes, G. L. (1988) A novel use of physical aging to distinguish immiscibility in polymer blends, *Polym. Bull.* **20**, 479–485.

Kalb, B., Pennings, A. J. (1980) General crystallization behaviour of poly(L-lactic acid), *Polymer* **21**, 607–612.

Kamei, S., Inoue, Y., Okada, H., Yamada, M., Ogawa, Y., Toguchi, H. (1992) New method for analysis of biodegradable polyesters by high-performance liquid chromatography after alkaline hydrolysis, *Biomaterials* **13**, 953–958.

Karjomaa, S., Suortti, T., Lempiäinen, R., Selin, J.-F., Itävaara, M. (1998) Microbial degradation of poly-(L-lactic acid) oligomers, *Polym. Degrad. Stabil.* **59**, 333–336.

Karlsson, S., Albertsson, A.-C. (1998) Abiotic and biotic degradation of aliphatic polyesters from "petro" versus "green" resources, *Macromol. Symp.* **127**, 219–225.

Kasperczyk, J. E. (1995) Microstructure analysis of poly(lactic acid) obtained by lithium tert-butoxide as initiator, *Macromolecules* **28**, 3937–3939.

Kasperczyk, J. (1996) Microstructural analysis of poly[(L,L-lactide)-co-(glycolide)] by ¹H and ¹³C n.m.r. spectroscopy, *Polymer* **37**, 201–203.

Kasperczyk, J., Bero, M. (1991) Coordination polymerization of lactides, 2. Micro structure determination of poly[(L,L-lactide)-co-(ε-caprolactone)] with ¹³C nuclear magnetic resonance spectroscopy, *Makromol. Chem.* **192**, 1777–1787.

Kharas, G. B., Sanchez-Riera, F., Severson, D. K. (1994) Polymer of lactic acids, in: *Plastics from Microbes* (Mobley, D. P., Ed.), New York: Hanser Publishers, 93–137.

Kim, C.-H., Cho, K. Y., Choi, E.-J., Park, J.-K. (2000) Effect of P(lLA-co-εCL) on the compatibility and crystallization behavior of PCL/PLLA blends, *J. Appl. Polym. Sci.* **77**, 226–231.

Kim, S. H., Kim, Y. H. (1994) Biodegradable star-shaped poly-L-lactide, in: *Biodegradable Plastics and Polymers* (Doi, Y, Fukuda, K., Eds.), Amsterdam: Elsevier Science B.V., 464–469.

Kimura, Y., Yamaoka, T. (1996) Surface modification and properties of biodegradable polymers based on poly-L-lactides, in: *Surface Science of Crystalline Polymers* (Yui, N., Terano, M., Eds.), Japan: Kodansha Scientific Ltd., 163–172

Kimura, Y., Matsuzaki, Y., Yamane, H., Kitao, T. (1989) Preparation of block copoly(ester-ether) comprising poly(L-lactide) and poly(oxypropy-

lene) and degradation of its fibre *in vitro* and in vivo, *Polymer* **30**, 1342–1349.

Kobayashi, J., Asahi, T., Ichiki, M., Okikawa, A., Suzuki, H., Watanabe, T. Fukada, E., Shikinami, Y. (1995) Structural and optical properties of poly lactic acids, *J. Appl. Phys.* **77**, 2957–2973.

Kohn, F. E., Van Den Berg, J. W. A., Van De Ridder, G., Feijen, J. (1984) The ring-opening polymerization of D,L-lactide in the melt initiated with tetraphenyltin, *J. Appl. Polym. Sci.* **29**, 4265–4277.

Kolstad, J. J. (1996) Crystallization kinetics of poly(L-lactide-co-meso-lactide), *J. Appl. Polym. Sci.* **62**, 1079–1091.

Koyama, N., Doi, Y. (1995) Morphology and biodegradability of a binary blend of poly((R)-3-hydroxybutyric acid) and poly((R,S)-lactic acid), *Can. J. Microbiol.* **41**(Suppl.1), 316–322.

Koyama, N., Doi, Y. (1996) Miscibility, thermal properties, and enzymatic degradability of binary blends of poly((R)-3-hydroxybutyric acid) with poly(ε-caprolactone-co-lactide), *Macromolecules* **29**, 5843–5851.

Koyama, N., Doi, Y. (1997) Miscibility of binary blends of poly((R)-3-hydroxybutyric acid) and poly((S)-lactic acid), *Polymer* **38**, 1589–1594.

Kricheldorf, H. R., Kreiser, I. (1987a) Polylactones. 11. Cationic copolymerization of glycolide with L,L-dilactide, *Makromol. Chem.* **188**, 1861–1873.

Kricheldorf, H. R., Boettcher, C. (1993) Polylactones. XXV. Polymerizations of racemic- and meso-D,L-lactide with Zn, Pb, Sb, and Bi salts – stereochemical aspects, *J. Macromol. Sci., Pure Appl. Chem.* **A30**, 441–448.

Kricheldorf, H. R., Kreiser, I. (1987b) Polylactones. 13. Transesterification of poly(L-lactide) with poly(glycolide), poly(β-propiolactone), and poly(ε-caprolactone), *J. Macromol. Sci.-Chem.* **A24**, 1345–1356.

Kricheldorf, H. R., Kreiser-Saunders, I., Scharnagl, N. (1990), Anionic and pseudoanionic polymerization of lactones – A comparison, *Makromol. Chem. Macromol. Symp.* **32**, 285–298.

Kricheldorf, H. R., Kreiser-Saunders, I. (1996) Polylactides – synthesis, characterization and medical application, *Macromol. Symp.* **103**, 85–102.

Kricheldorf, H. R., Sumbél, M. (1989) Polylactones-18. Polymerization of L,L-lactide with Sn(II) and Sn(IV) halogenides, *Eur. Polym. J.* **25**, 585–591.

Kumagai, Y., Kanesawa, Y., Doi, Y. (1992) Enzymatic degradation of microbial poly(3-hydroxy-butyrate) films, *Makromol. Chem.* **193**, 53–57.

Lam, K. H., Nieuwenhuis, P., Molenaar, I., Essel-brugge, H., Feijen, J., Dijkstra, P. J., Shakenraad, J. M. (1994) Biodegradation of porous versus non-porous poly(ʟ-lactic acid) films, *J. Mater. Sci. Mater. Med.* **5**, 181–189.

Leenslag., J. W., Pennings, A. J. (1987a) Synthesis of high-molecular-weight poly(ʟ-lactide) initiated with tin 2-ethylhexanoate, *Makromol. Chem.* **188**, 1809–1814.

Leenslag., J. W., Pennings, A. J. (1987b) High-strength poly(ʟ-lactide) fibres by a dry-spinning/hot-drawing process, *Polymer* **28**, 1695–1702.

Lewis, D. H. (1990) Controlled release of bioactive agents from lactide/glycolide polymers, in: *Biodegradable Polymers as Drug Delivery Systems* (Chasin, M., Langer, R., Eds.), New York: Marcel Dekker, 1–42.

Li, S., Vert, M. (1995) Biodegradation of aliphatic polyesters, in: *Degradable Polymers – Principles and Applications* (Scott, G., Gilead, D., Eds.), London: Chapman & Hall, 43–87.

Li, S., McCarthy, S. (1999a) Influence of crystallinity and stereochemistry on the enzymatic degradation of poly(lactide)s, *Macromolecules* **32**, 4454–4456.

Li, S., McCarthy, S. (1999b) Further investigations on the hydrolytic degradation of poly(ᴅʟ-lactide), *Biomaterials* **20**, 35–44.

Li, S., Garreau, H., Vert, M. (1990) Structure-property relationships in the case of the degradation of massive poly(α-hydroxy acids) in aqueous media, *J. Mater. Sci., Mater. Med.* **1**, 198–206.

Li, S., Tenon, M., Garreau, H., Braud, C., Vert, M. (2000) Enzymatic degradation of stereocopolymers derived from ʟ-, ᴅʟ- and meso-lactides, *Polym. Degrad. Stabil.* **67**, 85–90.

Li, Y., Nothnagel, N., Kissel, T. (1997) Biodegradable brush-like graft polymers from poly(ᴅ,ʟ-lactide) or poly(ᴅ,ʟ-lactide-co-glycolide) and charge-modified, hydrophilic dextrans as backbone – synthesis, characterization and *in vitro* degradation properties, *Polymer* **38**, 6197–6206.

Lillie, E. Schulz, R. C. (1975) ¹H- and ¹³C-{¹H}-NMR spectra of stereocopolymers of lactide, *Makromol. Chem.* **176**, 1901–1906.

Liu, X., Dever, M., Fair, N., Benson, R. S. (1997) Thermal and mechanical properties of poly(lactic acid) and poly(ethylene/butylene succinate) blends, *J. Environ. Polym. Degrad.* **5**, 225–235.

Loomis, G. L. Murdoch, J. R. (1990) Polylactide compositions, US Patent 4 902 515.

Loomis, G. L., Murdoch, J. R., Gardner, K. H. (1990) Polylactide stereocomplexes, *Polym. Prepr.* **31**, 55.

Lostocco, M. R., Huang, S. J. (1997) Aliphatic polyester blends based upon poly(lactic acid) and oligomeric poly(hexamethylene succinate), *J. Macromol. Sci. -Pure Appl. Chem.* **A34**, 2165–2175.

Lostocco, M. R., Borzacchiello, A., Huang, S. J. (1998) Binary and ternary poly(lactic acid)/poly(ε-caprolactone) blends: the effects of oligo-ε-caprolactones upon mechanical properties, *Macromol. Symp.* **130**, 151–160.

Lowe, C. E. (1954) Preparation of high molecular weight polyhydroxyacetic ester, US Patent 2 668 162.

MacDonald, R. T., McCarthy, S. P., Gross, R. A. (1996) Enzymatic degradability of poly(lactide): effects of chain stereochemistry and material crystallinity, *Macromolecules* **29**, 7356–7361.

Majola, A., Vainionpää, S., Rokkanen, P., Mikkola, H.-M., Törmälä, P. (1992) Absorbable self-reinforced polylactide (SR-PLA) composite rods for fracture fixation: strength and strength retention in the bone and subcutaneous tissue of rabbits, *J. Mater. Sci. Mater. Med.* **3**, 43–47.

Makino, K., Arakawa, M., Kondo, T. (1985) Preparation and *in vitro* degradation properties of polylactide microcapsules, *Chem. Pharm. Bull.* **33**, 1195–1201.

Mallardé, D., Valière, M., David, C., Menet, M., Guérin, Ph. (1998) Hydrolytic degradability of poly(3-hydroxyoctanoate) and of a poly(3-hydroxyoctanoate)/poly(R,S-lactic acid) blend, *Polymer* **39**, 3387–3392.

Marega, C., Marigo, A., Di Noto, V., Zannetti, R., Martorana, A., Paganetto, G. (1992) Structure and crystallization kinetics of poly(ʟ-lactic acid), *Makromol. Chem.* **193**, 1599–1606.

Mason, N. S., Miles, C. S., Sparks, R. E. (1981) Hydrolytic degradation of poly ᴅʟ-(lactide), *Polym. Sci. Technol.* **14**, 279–291.

Matsumura, S., Mabuchi, K., Toshima, K. (1998) Novel ring-opening polymerization of lactide by lipase, *Macromol. Symp.* **130**, 285–304.

Matsumoto, A., Matsukawa, Y., Suzuki, T., Yoshino, H., Kobayashi, M. (1997) The polymer-alloys method as a new preparation method of biodegradable microspheres: principle and application to cisplatin loaded microspheres, *J. Controlled Release* **48**, 19–27.

Mauduit, J., Pérouse, E., Vert, M. (1996) Hydrolytic degradation of films prepared from blends of high and low molecular weight poly(ᴅʟ-lactic acid)s, *J. Biomed. Mater. Res.* **30**, 201–207.

Mazzullo, S., Paganetto, G., Celli, A. (1992) Regime III crystallization in poly-(ʟ-lactic) acid, *Progr. Colloid Polym. Sci.* **87**, 32–34.

Meredith, J. C., Amis, E. J. (2000) LCST phase separation in biodegradable polymer blends: poly(ᴅ,ʟ-lactide) and poly(ε-caprolactone), *Macromol. Chem. Phys.* **201**, 733–739.

Meinander, K., Niemi, M., Hakola, J. S., Selin, J.-F. (1997), Polylactides – degradable polymers for fibers and films, *Macromol. Symp.* **123**, 147–153.

Middleton, J. C., Tipton, A. J. (2000) Synthetic biodegradable polymers as orthopedic devices, *Biomaterials* **21**, 2335–2346.

Migliaresi, C., Cohn, D., De Lollis, A., Fambri, L. (1991a) Dynamic mechanical and calorimetric analysis of compression-molded PLLA of different molecular weights: effect of thermal treatments, *J. Appl. Polym. Sci.* **43**, 83–95.

Migliaresi, C., De Lollis, A., Fambri, L., Cohn, D. (1991b) The effect of thermal history on the crystallinity of different molecular weight PLLA biodegradable polymers, *Clin. Mater.* **8**, 111–118.

Migliaresi, C., Fambri, L., Cohn, D. (1994) A study on the *in vitro* degradation of poly(lactic acid), *J. Biomater. Sci., Polym. Ed.* **5**, 591–606.

Miyata, T., Masuko, T. (1997) Morphology of poly(L-lactide) solution-grown crystals, *Polymer* **38**, 4003–4009.

Miyata, T., Masuko, T. (1998) Crystallization behavior of poly(L-lactide), *Polymer* **39**, 5515–5521.

Moon, S. I., Lee, C. W., Miyamoto, M., Kimura, Y. (2000) Melt polycondensation of L-lactic acid with Sn(II) catalysts activated by various proton acids: a direct manufacturing route to high molecular weight poly(L-lactic acid), *J. Polym. Sci., Part A: Polym. Chem.* **38**, 1673–1679.

Murdoch, J. R., Loomis, G. L. (1988a) Polylactide compositions, US Patent 4 719 246.

Murdoch, J. R., Loomis, G. L. (1988b) Polylactide compositions, US Patent 4 766 182.

Murdoch, J. R., Loomis, G. L. (1989) Polylactide compositions, US Patent 4 800 219.

Nagata, M., Okano, F., Sakai, W., Tsutsumi, N. (1998) Separation and enzymatic degradation of blend films of poly(L-lactic acid) and cellulose, *J. Polym. Sci.: Part A: Polym. Chem.* **36**, 1861–1864.

Nakafuku, C. (1994) High pressure crystallization of poly(L-lactic acid) in a binary mixture with poly(ethylene oxide), *Polym. J.* **26**, 680–687.

Nakafuku, C. (1996) Effects of molecular weight on the melting and crystallization of poly(L-lactic acid) in a mixture with poly(ethylene oxide), *Polym. J.* **28**, 568–575.

Nakafuku, C., Sakoda, M. (1993) Melting and crystallization of poly(L-lactic acid) and poly(ethylene oxide) binary mixture, *Polym. J.* **25**, 909–917.

Nakamura, T., Hitomi, S., Watanabe, S., Shimizu, Y., Jamshidi, K., Hyon, S.-H., Ikada, Y. (1989) Bioabsorption of polylactides with different molecular properties, *J. Biomed. Mater. Res.* **23**, 1115–1130.

Nef, J. U. (1914) Dissoziationsvorgänge in der zuckergruppe, *Liebigs. Ann. Chem.* **403**, 204–383.

Nijenhuis, A. J. Grijpma, D. W., Pennings, A. J. (1996a) Crosslinked poly(L-lactide) and poly(ε-caprolactone), *Polymer* **37**, 2783–2791.

Nijenhuis, A. J., Colstee, E., Grijpma, D. W., Pennings, A. J. (1996b) High molecular weight poly(L-lactide) and poly(ethylene oxide) blends: thermal characterization and physical properties, *Polymer* **37**, 5849–5857.

Ohkoshi, Y., Shirai, H., Gotoh, Y., Nagura, M. (1999) Instrinsic birefringence of poly(L-lactide), *Sen'i Gakkaishi* **55**, 62–68.

Ohkoshi, I., Abe, H., Doi, Y. (2000) Miscibility and solid-state structures for blends of poly[(S)-lactide] with atactic poly[(R,S)-3-hydroxybutyrate], *Polymer* **41**, 5985–5992.

Ohya, Y., Sugitou, T., Ouchi, T. (1995) Polycondensation of α-hydroxy acids by enzymes or PEG-modified enzymes in organic media, *J. Macromol. Sci.-Pure Appl. Chem.* **A32**, 179–190.

Ohya, Y., Maruhashi, S., Ouchi, T. (1998a) Graft polymerization of L-lactide on pullulan through the trimethylsilyl protection method and degradation of the graft copolymers, *Macromolecules* **31**, 4662–4665.

Ohya, Y., Maruhashi, S., Ouchi, T. (1998b) Preparation of poly(L-lactide)-grafted amylose through the trimethylsilyl protection method and degradation of the graft copolymers, *Macromol. Chem. Phys.* **199**, 2017–2022.

Okihara, T., Tsuji, M., Kawaguchi, A., Katayama, K., Tsuji, H., Hyon, S.-H., Ikada, Y. (1991) Crystal structure of stereocomplex of poly(L-lactide) and poly-(D-lactide), *J. Macromol. Sci.-Phys.* **B30**, 119–140.

Okuzaki, H., Kubota, I., Kunugi, T. (1999) Mechanical properties and structure of the zone-drawn poly(L-lactic acid) fibers, *J. Polym. Sci., Part B: Polym. Phys.* **37**, 991–996.

Onyari, J. M., Huang, S. J. (1996) Graft copolymers of poly(vinyl alcohol) and poly(lactic acid), *Polym. Prepr. (Am. Chem. Soc., Div. Polym. Chem.)* **37**, 145–146.

Otsuka, H., Nagasaki, Y., Kataoka, K., Okano, T., Sakurai, Y. (1998) Reactive-PEG-polylactide block copolymer for tissue engineering, *Polym. Prepr. (Am. Chem. Soc., Div. Polym. Chem.)* **39**, 128–129.

Ovitt, T. M., Coates, G. W. (1999) Stereoselective ring-opening polymerization of *meso*-lactide: synthesis of syndiotactic poly(lactic acid), *J. Am. Chem. Soc.* **121**, 4072–4073.

Pan, Q. Y., Tasaka, S., Inagaki, N. (1996) Ferroelectric behavior in poly-L-lactic acid. *Jpn. J. Appl. Phys.* **35**, L1442–L1445.

Park, T. G., Cohen, S., Langer, R. (1992) Poly(L-lactide)/Pluronic blends: characterization of phase separation behavior, degradation, and morphology and use as protein-releasing matrices, *Macromolecules* **25**, 116–122.

Pelouze, P. M. J. (1845) Mémoire sur l'acide lactique, *Ann. Chim. Phys., Ser.3* **13**, 257–268.

Penning, J. P., Dijkstra, H., Pennings, A. J. (1993) Preparation and properties of absorbable fibers from L-lactide copolymers, *Polymer* **34**, 942–951.

Perego, G., Cella, G. D., Bastioli, C. (1996) Effects of molecular weight and crystallinity on poly(lactic acid) mechanical properties, *J. Appl. Polym. Sci.* **59**, 37–43.

Pistner, H., Bendix, D. R., Mühling, J., Reuther, J. F. (1993) Poly(L-lactide): a long-term degradation study in vivo. Part III: Analytical characterization, *Biomaterials* **14**, 291–298.

Pistner, H., Stallforth, H., Gutwald, R., Mühling, J., Reuther, J., Michel, C. (1994) Poly(L-lactide): a long-term degradation study in vivo, Part II: Physico-mechanical behaviour of implants, *Biomaterials* **15**, 439–450.

Pitt, C. G. (1990) Poly(ε-caprolactone) and its copolymers, in: *Biodegradable Polymers as Drug Delivery Systems* (Chasin, M., Langer, R., Eds.), New York: Marcel Dekker, 71–120.

Pitt, C. G., Gu, Z.-W. (1987) Modification of the rates of chain cleavage of poly(ε-caprolactone) and related polyesters in the solid state, *J. Controlled Release* **4**, 283–292.

Pitt, C. G., Jeffcoat, A. R., Zweidinger, R. A., Schindler, A. (1979) Sustained drug delivery system. I. The permeability of poly(ε-caprolactone), poly(DL-lactic acid), and their copolymers, *J. Biomed. Mater. Res.* **13**, 497–507.

Pitt, C. G., Cha, Y., Shah, S. S., Zhu, K. J. (1992) Blends of PVA and PGLA: control of the permeability and degradability of hydrogels by blending, *J. Controlled Release* **19**, 189–200.

Pitt, C. G., Wang, J., Shah, S. S., Sik, R., Chignell, C. F. (1993) ESR spectroscopy as a probe of the morphology of hydrogels and polymer-polymer blends, *Macromolecules* **26**, 2159–2164.

Pranamuda, H., Tokiwa, Y. (1999) Degradation of poly(L-lactide) by strains belonging to genus *Amycolatopsis*, *Biotechnol. Lett.* **21**, 901–905.

Pranamuda, H., Tokiwa, Y., Tanaka, H. (1997) Polylactide degradation by an *Amycolatopsis* sp., *Appl. Environ. Microbiol.* **63**, 1637–1640.

Puiggali, J., Ikada, Y., Tsuji, H., Cartier, L., Okihara, T., Lotz, B. (2000) The frustrated structure of poly(L-lactide), *Polymer* **41**, 8921–8930.

Radano, C. P., Baker, G. L., Smith, M. R., III (2000) Stereoselective polymerization of a racemic monomer with a racemic catalyst: direct preparation of the polylactide stereocomplex from racemic lactide, *J. Am. Chem. Soc.* **122**, 1552–1553.

Rafler, G., Jobmann, M. (1994) Controlled release systems of biodegradable polymers. 3rd communication: Degradation and release properties of poly(glycolide(50)-co-lactide(50))s with random and non-random monomer distribution, *Drug Made in Germany* **38**, 20–22.

Reed, A. M., Gilding, D. K. (1981) Biodegradable polymers for use in surgery – Poly(glycolic)/poly(lactic acid) homo and copolymers: 2. *In vitro* degradation, *Polymer* **22**, 494–498.

Reeve, M. S., McCarthy, S. P., Downey, M. J., Gross, R. A. (1994) Polylactide stereochemistry: effect on enzymatic degradability, *Macromolecules* **27**, 825–831.

Renstad, R., Karlsson, S., Sandgren, Å., Albertsson, A.-C. (1998) Influence of processing additives on the degradation of melt-pressed films of poly(ε-caprolactone) and poly(lactic acid), *J. Environ. Polym. Degrad.* **6**, 209–221.

Sarasua, J.-R., Prud'homme, R. E., Wisniewski, M., Le Borgne, A., Spassky, N. (1998) Crystallization and melting behavior of polylactides, *Macromolecules* **31**, 3895–3905.

Schindler, A., Gaetano, K. D. (1988) Poly(lactate) III. Stereoselective polymerization of meso dilactide, *J. Polym. Sci.: Part C: Polym. Lett.* **26**, 47–48.

Schindler, A., Harper, D. (1979) Polylactide. II. Viscosity-molecular weight relationships and unperturbed chain dimensions, *J. Polym. Sci., Polym. Chem. Ed.* **17**, 2593–2599.

Schindler, A., Hibionada, Y. M., Pitt, C. G. (1982) Aliphatic polyesters. III. Molecular weight and molecular weight distribution in alcohol-initiated polymerizations of ε-caprolactone, *J. Polym. Sci. Polym. Chem.* **20**, 319–326.

Serizawa, T., Yamashita, H., Fujiwara, T., Kimura, Y., Akashi, M. (2001), Stepwise assembly of enantiomeric poly(lactide)s on surfaces, *Macromolecules* **34**, 1996–2001.

Shneider, A. K. (1972) Polylactide sutures, US Patent 3 636 956.

Shakesheff, K. M., Davies, M. C., Roberts, C. J., Tendler, S. J. B., Shard, A. G., Domb, A. (1994) In situ atomic force microscopy imaging of polymer degradation in an aqueous environment, *Langmuir* **10**, 4417–4419.

Shen, Y., Zhu, K. J., Shen, Z., Yao, K.-M. (1996) Synthesis and characterization of highly random copolymer of ε-caprolactone and D,L-lactide using rare earth catalyst, *J. Polym. Sci.: Part A: Polym. Chem.* **34**, 1799–1805.

Sheth, M., Kumar, R. A., Davé, V., Gross, R. A., McCarthy, S. P. (1997) Biodegradable polymer blends from poly(lactic acid) and poly(ethylene glycol), *J. Appl. Polym. Sci.* **66**, 1495–1505.

Shih, C. (1995) A graphical method for the determination of the mode of hydrolysis of biodegradable polymers, *Pharm. Res.* **12**, 2036–2040.

Shogren, R. (1997) Water vapor permeability of biodegradable polymers, *J. Environ. Polym. Degrad.* **5**, 91–95.

Siemann, U. (1992) The solubility parameter of poly(DL-lactic acid), *Eur. Polym. J.* **28**, 293–297.

Södergård, A. (1998) Preparation of poly(L-lactide-graft-acrylic acid), *Polym. Prepr. (Am. Chem. Soc. Div., Polym. Chem.)* **39**, 214–215.

Song, C. X., Feng, X. D. (1984) Synthesis of ABA triblock copolymers of ε-caprolactone and DL-lactide, *Macromolecules* **17**, 2764–2767.

Spassky, N., Pluta, C., Simic, V., Thiam, M., Wisniewski, M. (1998) Stereochemical aspects of the controlled ring-opening polymerization of chiral cyclic esters, *Macromol. Symp.* **128**, 39–51.

Spinu, M., Gardner, K. H. (1994) A new approach to stereocomplex formation between L- and D-poly(lactic acid)s, *Abstracts of Papers of the American Chemical Society*, **208**, No. Pt. 2, pp. 12 (PMSE).

Spinu, M., Jackson, C., Keating, M. Y., Gardner, K. H. (1996) Material design in poly(lactic acid) system: block copolymers, star homo- and co-polymers, and stereocomplexes, *J. Macromol. Sci.-Pure Appl. Chem.* **A33**, 1497–1530.

Stevels, W. M., Ankoné, M. J. K., Dijkstra, P. J., Feijen, J. (1995) Stereocomplex formation in ABA triblock copolymers of poly(lactide) (A) and poly(ethylene glycol) (B), *Macromol. Chem. Phys.* **196**, 3687–3694.

Storey, R. F., Hickey, T. P. (1994) Degradable polyurethane networks based on D,L-lactide, glycolide, ε-caprolactone, and trimethylene carbonate homopolyester and copolyester triols, *Polymer* **35**, 830–838.

Storey, R. F., Warren, S. C., Allison, C. J., Wiggins, J. S., Puckett, A. D. (1993) Synthesis of bioabsorbable networks from methacrylate-endcapped polyesters, *Polymer* **34**, 4365–4372.

Stridsberg, K, Albertsson, A.-C. (2000) Changes in chemical and thermal properties of the tri-block copolymer poly(L-lactide-b-1,5-dioxepan-2-one-b-L-lactide) during hydrolytic degradation, *Polymer* **41**, 7321–7330.

Tajitsu, Y., Hosoya, R., Murayama, T., Aoki, M., Shikinami, Y., Date, M., Fukada, E. (1999) Huge optical rotatory power of uniaxially oriented film of poly-L-lactic acid, *J. Mater. Sci. Lett.* **18**, 1785–1787.

Thakur, K. A. M., Kean, R. T., Zupfer, J. M., Buehler, N. U., Doscotch, M. A., Munson, E. J. (1996) Solid state ^{13}C NMR studies of the crystallinity and morphology of poly(L-lactide), *Macromolecules* **29**, 8844–8851.

Thomson, R. C., Wake, M. C., Yaszemski, M. J., Mikos, A. G. (1995a) Biodegradable polymer scaffolds to regenerate organs, in: *Advances in Polymer Science: Biopolymer II* (Peppas, N. A., Langer, R. S., Eds.), Berlin: Springer-Verlag, 245–274, vol. 122.

Thomson, R. C, Yaszemski, M. J., Powers, J. M., Mikos, A. G. (1995b) Fabrication of biodegradable polymer scaffolds to engineer trabecular bone, *J. Biomater. Sci.: Polym. Ed.* **7**, 23–38.

Tokiwa, Y., Konno, M., Nishida, H. (1999) Isolation of silk degrading microorganisms and its poly(L-lactide) degradability, *Chem. Lett.* 355–356.

Tonelli, A. E. (1992) Polylactides in channels, *Macromolecules* **25**, 3581–3584.

Törmälä, P. (1992) Biodegradable self-reinforced composite materials; manufacturing structure and mechanical properties, *Clin. Mater.* **10**, 29–34.

Torres, A., Li, S., Roussos, S., Vert, M. (1996a) Poly(lactic acid) degradation in soil or under controlled conditions, *J. Appl. Polym. Sci.* **62**, 2295–2302.

Torres, A., Li, S., Roussos, S., Vert, M. (1996b) Degradation of L- and DL-lactic aid oligomers in the presence of *Fusarium moniliforme* and *Pseudomonas putida*, *J. Environ. Polym. Degrad.* **4**, 213–223.

Torres, A., Li, S., Roussos, S., Vert, M. (1996c) Screening of microorganisms of biodegradation of poly(lactic acid) and lactic acid-containing polymers, *Appl. Environ. Microbiol.* **62**, 2393–2397.

Torres, A., Li, S., Roussos, S., Vert, M. (1999) Microbial degradation of a poly(lactic acid) as a model of synthetic polymer degradation mechanisms in outdoor conditions, *ACS Symp. Ser.*, **723**, 218–226.

Trofimoff, L., Aida, T., Inoue, S. (1987) Formation of poly(lactide) with controlled molecular weight. Polymerization of lactide by aluminium porphyrin, *Chem. Lett.* 991–994.

Tsuchiya, F., Tomida, Y., Fujimoto, K., Kawaguchi, H. (1993) Surface modification of poly(L-lactic acid) by plasma discharge, *Polym. Prepr. Jpn.* **42**, 4916–4918.

Tsuji, H. (2000a) Stereocomplex from enantiomeric polylactides, in: *Research Advances in Macromolecules* (Mohan, R. M., Ed.), Trivandrum, India: Grobal Research Network, 25–48, Vol. 1.

Tsuji, H. (2000b) Hydrolysis of biodegradable aliphatic polyesters, in: *Recent Research Developments in Polymer Science* (Salamone, A. B., Brandrup, J., Ottenbrite, R.M., Editorial advisors), Trivandrum, India: Transworld Research Network, 13–37, Vol. 4.

Tsuji, H. (2000c) *In vitro* hydrolysis of blends from enantiomeric poly(lactide)s. Part 1. Well-stereocomplexed blend and non-blended films, *Polymer* **41**, 3621–3630.

Tsuji, H., Ikada, Y. (1992) Stereocomplex formation between enantiomeric poly(lactic acid)s. 6. Binary blends from copolymers, *Macromolecules* **25**, 5719–5723.

Tsuji, H., Ikada, Y. (1993) Stereocomplex formation between enantiomeric poly(lactic acid)s. 9. Stereocomplexation from the melt, *Macromolecules* **26**, 6918–6926.

Tsuji, H., Ikada, Y. (1994) Stereocomplex formation between enantiomeric poly(lactic acid)s. X. Binary blends from poly(D-lactide-co-glycolide) and poly(L-lactide-co-glycolide), *J. Appl. Polym. Sci.* **53**, 1061–1071.

Tsuji, H., Ikada, Y. (1995a) Blends of isotactic and atactic poly(lactide). I. Effects of mixing ratio of isomers on crystallization of blends from melt, *J. Appl. Polym. Sci.* **58**, 1793–1802.

Tsuji, H., Ikada, Y. (1995b) Properties and morphologies of poly(L-lactide): 1. Annealing condition effects on properties and morphologies of poly(L-lactide), *Polymer* **36**, 2709–2716.

Tsuji, H., Ikada, Y. (1996a) Crystallization from the melt of poly(lactide)s with different optical purities and their blends, *Macromol. Chem. Phys.* **197**, 3483–3499.

Tsuji, H., Ikada, Y. (1996b) Blends of isotactic and atactic poly(lactide): 2. Molecular-weight effects of atactic component on crystallization and morphology of equimolar blends from the melt, *Polymer* **37**, 595–602.

Tsuji, H., Ikada, Y. (1996c) Blends of aliphatic polyesters. I. Physical properties and morphologies of solution-cast blends from poly(DL-lactide) and poly(ε-caprolactone), *J. Appl. Polym. Sci.* **60**, 2367–2375.

Tsuji, H., Ikada, Y. (1997) Blends of crystallizable and amorphous poly(lactide). III. Hydrolysis of

solution-cast blend films, *J. Appl. Polym, Sci.* **63**, 855–863.

Tsuji, H., Ikada, Y. (1998a) Blends of aliphatic polyesters. II. Hydrolysis of solution-cast blends from poly(L-lactide) and poly(ε-caprolactone) in phosphate-buffered solution, *J. Appl. Polym. Sci.* **67**, 405–415.

Tsuji, H., Ikada, Y. (1998b) Properties and morphology of poly(L-lactide). II. Hydrolysis in alkaline solution, *J. Polym. Sci.: Part A: Polym. Chem.* **36**, 59–66.

Tsuji, H., Ikada, Y. (1999a) Physical properties of polylactides, in: *Current Trends in Polymer Science* (DeVries, K. L. et al., Editorial Advisory Board), Trivandrum, India: Research Trends, 27–46, Vol. 4.

Tsuji, H., Ikada, H. (1999b) Stereocomplex formation between enantiomeric poly(lactic acid)s. XI. Mechanical properties and morphology, *Polymer* **40**, 6699–6708.

Tsuji, H., Ikada, Y. (2000a) Enhanced crystallization of poly(L-lactide-co-ε-caprolactone) during storage at room temperature, *J. Appl. Polym. Sci.* **76**, 947–953.

Tsuji, H., Ikada, Y. (2000b) Properties and morphology of poly(L-lactide). 4. Effects of structural parameters on long-term hydrolysis of poly(L-lactide) in phosphate-buffered solution, *Polym. Degrad. Stabil.* **67**, 179–189.

Tsuji, H., Ishizaka, T. (2001a) Porous biodegradable polyesters. 2. Physical properties, morphology, and enzymatic and alkaline hydrolysis of porous poly(ε-caprolactone) films, *J. Appl. Polym. Sci.* **80**, 2281–2291.

Tsuji, H., Ishizaka, T. (2001b) Porous biodegradable polyesters, 3. Preparation of porous poly(ε-caprolactone) films from blends by selective enzymatic removal of poly(L-lactide), *Macromol. Biosci.* **1**, 59–65.

Tsuji, H., Miyauchi, S. (2001a) Poly(L-lactide). 6. Effects of crystallinity on enzymatic hydrolysis of poly(L-lactide) without free amorphous region, *Polym. Degrad. Stabil.* **71**, 415–424.

Tsuji, H., Miyauchi, S. (2001b) Poly(L-lactide). 7. Enzymatic hydrolysis of free and restricted amorphous regions in poly(L-lactide) films with different crystallinities and a fixed crystalline thickness, *Polymer* **42**, 4465–4469.

Tsuji, H., Miyauchi, S. (2001c) Enzymatic hydrolysis of polylactides. Effects of molecular weight, L-lactide content, and enantiomeric and diastereoisomeric polymer blending, *Biomacromolecules* **2**, 597–604.

Tsuji, H., Muramatsu, H. (2001a) Blends of aliphatic polyesters. 5. Non-enzymatic and enzymatic hydrolysis of blends from hydrophobic poly(L-lactide) and hydrophilic poly(vinyl alcohol), *Polym. Degrad. Stabil.* **71**, 403–413.

Tsuji, H., Muramatsu, H. (2001b) Blends of aliphatic polyesters. 4. Morphology, swelling behavior, and surface and bulk properties of blends from hydrophobic poly(L-lactide) and hydrophilic poly(vinyl alcohol), *J. Appl. Polym. Sci.* **81**, 2151–2160.

Tsuji, H., Nakahara, K. (2001) Poly(L-lactide). 8. High-temperature hydrolysis of poly(L-lactide) films with different crystallinities and crystalline thicknesses in phosphate-buffered solution, *Macromol. Mater. Eng.* **286**, 398–406.

Tsuji, H., Sumida, K. (2001) Poly(L-lactide). V. Effect of storage in swelling solvents on physical properties and structure of poly(L-lactide), *J. Appl. Polym. Sci.* **79**, 1582–1589.

Tsuji, H., Suzuki, M. (2001) In vitro hydrolysis of blends from enantiomeric poly(lactide)s. 2. Well-stereocomplexed fibers and films, *Sen'i Gakkaishi* **57**, 198–202.

Tsuji, H., Hyon, S.-H., Ikada, Y. (1991a) Stereocomplex formation between enantiomeric poly(lactic acid)s. 2. Stereocomplex formation in concentrated solutions, *Macromolecules* **24**, 2719–2724.

Tsuji, H., Hyon, S.-H., Ikada, Y. (1991b) Stereocomplex formation between enantiomeric poly(lactic acid)s. 3. Calorimetric studies on blend films cast from dilute solution, *Macromolecules* **24**, 5651–5656.

Tsuji, H., Hyon, S.-H., Ikada, Y. (1991c) Stereocomplex formation between enantiomeric poly(lactic acid)s. 4. Differential scanning calorimetric studies on precipitates from mixed solutions of poly(D-lactic acid) and poly(L-lactic acid), *Macromolecules* **24**, 5657–5662.

Tsuji, H., Hyon, S.-H., Ikada, Y. (1992a) Stereocomplex formation between enantiomeric poly(lactic acid)s. 5. Calorimetric and morphological studies on the stereocomplex formed in acetonitrile solution, *Macromolecules* **25**, 2940–2946.

Tsuji, H., Horii, F., Nakagawa, M., Ikada, Y., Odani, H., Kitamaru, R. (1992b) Stereocomplex formation between enantiomeric poly(lactic acid)s. 7. Phase structure of the stereocomplex crystallized from a dilute acetonitrile solution as studied by high-resolution solid-state ^{13}C NMR spectroscopy, *Macromolecules* **25**, 4114–4118.

Tsuji, H., Ikada, Y., Hyon, S.-H., Kimura, Y., Kitada, T. (1994) Stereocomplex formation between enantiomeric poly(lactic acid)s. VIII. Complex fiber spun from mixed solution of poly(D-lactic acid) and poly(L-lactic acid), *J. Appl. Polym. Sci.* **51**, 337–344.

Tsuji, H., Mizuno, A., Ikada, Y. (1998) Blends of aliphatic polyesters. III. Biodegradation of solution-cast blends from poly(L-lactide) and poly(ε-caprolactone), *J. Appl, Polym, Sci.* **70**, 2259–2268.

Tsuji, H., Smith, R., Bonfield, W., Ikada, Y. (2000a) Porous biodegradable polyesters. I. Preparation of porous poly(L-lactide) by extraction of poly(ethylene oxide) from their blends, *J. Appl. Polym. Sci.* **75**, 629–637.

Tsuji, H., Mizuno, A., Ikada, Y. (2000b) Properties and morphologies of poly(L-lactide). III. Effects of initial crystallinity on long-term *in vitro* hydrolysis of high molecular weight poly(L-lactide), *J. Appl. Polym. Sci.* **77**, 1452–1464.

Vainionpää, S., Rokkanen, P., Törmälä, P. (1989) Surgical applications of biodegradable polymers in human tissues, *Prog. Polym. Sci.* **14**, 679–716.

van de Witte, P., Esselbrugge, H., Dijkstra, P.J., van den Berg, J. W. A., Feijen, J. (1996a) Phase transitions during membrane formation of polylactides, I. A morphological study of membranes obtained from the system polylactide-chloroform-methanol, *J. Membrane Sci.* **113**, 223–236.

van de Witte, P., Boorsma, A., Esselbrugge, H., Dijkstra, P. J., van den Berg, J. W. A., Feijen, J. (1996b) Differential scanning calorimetry study of phase transitions in poly(lactide)-chloroform-methanol system, *Macromolecules* **29**, 212–219.

van de Witte, P., Dijkstra, P. J., van den Berg, J. W. A., Feijen, J. (1996c) Phase separation processes in polymer solutions in relation to membrane formation, *J. Membrane Sci.* **117**, 1–31.

van de Witte, Dijkstra, P. J., van den Berg, J. W. A., Feijen, J. (1996d) Phase behavior of polylactides in solvent-nonsolvent mixtures, *J. Polym. Sci., Part B: Polym. Phys.* **34**, 2553–2568.

van de Witte, P., Esselbrugge, H., Dijkstra, P. J., van den Berg, J. W. A., Feijen, J. (1996e) A morphological study of membranes obtained from the systems polylactide-dioxane-methanol, polylactide-dioxane-water, polylactide-N-methyl pyrrolidone-water, *J. Polym. Sci., Part B: Polym. Phys.* **34**, 2569–2578.

van de Witte, P., van den Berg, J. W. A., Feijen, J., Reeve, J. L., Mchugh, A. J. (1996f) In situ analysis of solvent/nonsolvent exchange and phase separation processes during membrane formation of polylactides, *J. Appl. Polym. Sci.* **61**, 685–695.

van de Witte, P., Dijkstra, P. J., van den Berg, J. W. A., Feijen, J. (1997) Metastable liquid-liquid and solid-liquid phase boundaries in polymer-solvent-nonsolvent system, *J. Polym. Sci., Part B: Polym. Phys.* **35**, 763–770.

Vasanthakumari, R. (1981) Eutectic crystallization of poly(L-lactic acid) and pentaerythrityl tetra-bromide, *Polymer* **22**, 862–865.

Vasanthakumari, R., Pennings, A. J. (1983) Crystallization kinetics of poly(L-lactic acid), *Polymer* **24**, 175–178.

von Recum, H. A., Cleek, R. L. Eskin, S. G., Mikos, A. G. (1995) Degradation of polydispersed poly(L-lactic acid) to modulate lactic acid release, *Biomaterials* **16**, 441–447.

Wada, R., Hyon, S.-H., Nakamura, T., Ikada, Y. (1991) In vitro evaluation of sustained drug release from biodegradable elastomer, *Pharm. Res.* **8**, 1292–1296.

Wang, L., Ma, W. Gross, R. A., McCarthy, S. P. (1998) Reactive compatibilization of biodegradable blends of poly(lactic acid) and poly(ε-caprolactone), *Polym. Degrad. Stabil.* **59**, 161–168.

Wang, Z.-G., Hsiao, B. S., Zong, X.-H., Yeh, F., Zouh, J. J., Dormier, E., Jamiolkowski, D. D. (2000) Morphological development in absorbable poly(glycolide), poly(glycolide-co-lactide) and poly(glycolide-*co*-caprolactone) copolymers during isothermal crystallization, *Polymer* **41**, 621–628.

Whang, K., Thomas, C. H., Healy, K. E., Nuber, G. (1995) A novel method to fabricate bioabsorbable scaffolds, *Polymer* **36**, 837–842.

Williams, D. F. (1981) Enzymatic hydrolysis of polylactic acid, *Eng. Med.* **10**, 5–7.

Wise, D. L., Fellmann, T. D., Sanderson, J. E., Wentworth, R. L. (1979) Lactic/glycolic acid polymers, in: *Drug Carriers in Biology and Medicine* (Gregoriadis, G., Ed.), London, New York: Academic Press, 237–270.

Wisniewski, M., Le Borgne, A., Spassky, N. (1997) Synthesis and properties of (D)- and (L)-lactide stereocopolymers using the system achiral Shiff's base/aluminium methoxide as initiator, *Macromol. Chem. Phys.* **198**, 1227–1238.

Yamaoka, T., Takahashi, Y., Ohta, T., Miyamoto, M., Murakami, A., Kimura, Y. (1999) Synthesis and properties of multiblock copolymers consisting of poly(L-lactic acid) and poly(oxypropylene-*co*-oxyethylene) prepared by direct polycondensation, *J. Polym. Sci.: Part A: Polym. Chem.* **37**, 1513–1522.

Yang, J.-M., Chen, H.-L., You, J.-W., Hwang, J. C. (1997) Miscibility and crystallization of poly(L-lactide)/poly(ethylene glycol) and poly(L-lactide)/poly(ε-caprolactone), *Polym. J.* **29**, 657–662.

Yoon, J.-S., Lee, W.-S., Kim, K.-S., Chin, I.-J., Kim, M.-N., Kim, C. (2000) Effect of poly(ethylene glycol)-block-poly(L-lactide) on the poly((R)-3-hydroxybutyrate)/poly(L-lactide) blends, *Eur. Polym. J.* **36**, 435–442.

Younes, H., Cohn, D. (1987) Morphological study of biodegradable PEO/PLA block copolymers, *J. Biomed. Mater. Res.* **21**, 1301–1316.

Younes, H., Cohn, D. (1988) Phase separation in poly(ethylene glycol)/poly(lactic acid), *Eur. Polym. J.* **24**, 765–773.

Youxin, L., Volland, C., Kissel, T. (1998) Biodegradable brush-like graft polymers from poly(D,L-lactide) or poly(D,L-lactide-co-glycolide) and charge-modified, hydrophilic dextrans as backbone – *in vitro* degradation and controlled release of hydrophilic macromolecules, *Polymer* **39**, 3087–3097.

Yui, N., Dijkstra, P. J., Feijen, J. (1990) Stereo block copolymers of L- and D-lactides, *Makromol. Chem.* **191**, 481–488.

Zell, M. T., Padden, B. E., Paterick, A. J., Hillmyer, M. A., Kean, R. T., Thakur, K. A. M., Munson, E. J. (1998) Direct observation of stereodefect sites in semicrystalline poly(lactide) using ^{13}C solid-state NMR, *J. Am. Chem. Soc.* **120**, 12672–12673.

Zhang, L., Xiong, C., Deng, X. (1995) Biodegradable polyester blends for biomedical application, *J. Appl. Polym. Sci.* **56**, 103–112.

Zhang, L., Xiong, C., Deng, X. (1996) Miscibility, crystallization and morphology of poly(β-hydroxybutyrate)/poly(d,l-lactide) blends, *Polymer* **37**, 235–241.

Zhang, L., Deng, X., Zhao, S. Huang, Z. (1997) Biodegradable polymer blends of poly(3-hydroxybutyrate) and poly(DL-lactide)-co-poly(ethylene glycol), *J. Appl. Polym. Sci.* **65**, 1849–1856.

Zhang, X., Wyss, U. P., Pichora, D., Goosen, M. F. A. (1994) An investigation of poly(lactic acid) degradation, *J. Bioact. Compat. Polym.* **9**, 80–100.

Zoppi, R. A., Contant, S., Duek, E. A. R., Marques, F. R., Wada, M. L. F., Nunes, S. P. (1999) Porous poly(L-lactide) films obtained by immersion precipitation process: morphology, phase separation and culture of VERO cells, *Polymer* **40**, 3275–3289.

Zwiers, R. J. M., Gogolewski, S., Pennings, A. J. (1983) General crystallization behaviour of poly(L-lactic acid) PLLA: 2 Eutectic crystallization of PLLA, *Polymer* **24**, 167–174.

6

Polyglycolide and Copolyesters with Lactide

Prof. Dr. Michel Vert

CRBA-UMR CNRS 5473, Faculty of Pharmacy, University Montpellier 1, 15 Avenue Charles Flahault, 34060 Montpellier Cedex 2, France; Tel: + 33-467-418260; Fax: + 33-467-520998; E-mail: vertm@pharma.univ.montpl.fr

NMR   nuclear magnetic resonance
PGA   poly(glycolic acid) (= polyglycolide)
PLGA or PLAGA or PLA$_x$GA$_y$ poly(glycolic acid-*co*-lactic acid) copolymers (= poly(glyco-
                           lide-*co*-lactide) copolymers)

1
Introduction

Polyglycolide, also named poly(glycolic acid), and poly(glycolide-*co*-lactide) copolymers are polymers of the poly(α-hydroxy acid) type that are composed of building block or repeating units derived from glycolic and lactic acids. These polymers and copolymers that are sensitive to moisture became of interest as polymeric materials rather late as compared with other polymers like polyethylene or poly(vinyl chloride). When people realized that advantage could be taken of the degrading activity of water, devices were fabricated that can be degraded in an animal's body. Since then lactic acid and glycolic acid-based copolymers have attracted the attention of both scientists and industrialists in the biomedical and the pharmaceutical fields. In contrast to their homologs based only on lactic acid enantiomers that are being commercialized as therapeutic materials and environment-friendly materials as well, glycolic acid-containing copolymers have not found other fields of applications other surgery and pharmacology. In polymer science, trying to enlarge the range of properties of homopolymers by copolymerization is a basic strategy. In the case of glycolide, a number of copolymers with lactide enantiomers have been investigated.

2
Historical Outline

Glycolic acid-containing polymer chains were first synthesized by step-growth polymerization (polycondensation) of glycolic and lactic acid at the time other polycondensates like Nylon®-type polyamide were investigated (Higgins, 1950). The resulting poor polymers were rather sensitive to humidity and heat, and they were considered as useless. Fortunately, advances in polymer chemistry led chemists to discover another synthetic route, i.e. the ring-opening polymerization of the cyclic dimers of glycolic and lactic acids (Lowe, 1954). Soon after, poly(glycolic acid) was proposed by American Cyanamid as suitable polymers to make sutures (Schmitt and Polistina, 1967). Commercial sutures were developed by Davis and Geek under the registered name Dexon® in the USA and Ercedex® in Europe (Frazza and Schmitt, 1971). Soon after, Ethicon came up with a concurrent product that was a copolymer of glycolide with 8% lactide known as Glactine910® or Vicryl®, and processed to sutures and other fiber-based devices. In parallel, lactic acid-based polymers were studied from a scientific viewpoint (Kulkarni et al., 1966) until they became of industrial interest for bone surgery in the 1990s (Barber, 1998) after their known properties had been considerably improved (Kulkarni et al., 1971; Vert et al., 1984). In the meantime pharmacists found poly(glycolic acid-*co*-lactic acid) copolymers of interest to deliver drugs in a more or less controlled manner (Holland et al., 1986; Brannon-Peppas and Vert, 2000). Among the poly(glycolic acid-*co*-lactic acid) copolymer family, lactoyl-rich semi-crystalline polymers were preferred for the temporary internal fixation of bone fractures, whereas amorphous polymers appeared more convenient for tissue reconstruction and drug delivery.

3
Chemical Structures

Poly(glycolic acid-*co*-lactic acid) copolymers belong to the poly(α-hydroxy acid) family that is characterized by the following general formula:

where R is either H (glycoloyl) or CH_3 (lactoyl).

Formula A obeys the IUPAC rules, whereas formula B is the commonly used one, mostly because it is more symmetrical and emphasizes the ester links.

In literature, different acronyms are currently used to name these different basic chains, the most common being PLGA and PLAGA. Our preference for PLA_xGA_y was mentioned as early as 1981 (Vert et al., 1981) when we realized that the properties of a given member of the poly(α-hydroxy acid) family depend very much on many factors and that a convenient acronym has to reflect at least the gross composition in repeating units. Since then, we have identified more than 20 factors that can affect the behavior of

a PLA_xGA_y polymer, i.e. much more than in the case of a regular polymer primarily because of the sensitivity to water and heat, the particular pair-addition mechanism of the ring-opening polymerization due to the dimeric structure of the cyclic monomers, and the presence of chiral centers in lactic acid-based polymers and stereocopolymers and glycolic acid-copolymers (Li and Vert, 1995; Vert et al., 1995). Of course, it is not possible to reflect the contribution of all these factors within a simple acronym, but we strongly believe that everyone ought to use the PLA_xGA_y acronyms because they reflect both the chemical composition and the enantiomeric composition.

The basic chain structures of PLA_xGA_y polymers are shown in Table 1 (Vert et al., 1984).

Lactic acid-containing stereocopolymers and copolymers are outstanding compounds because of the presence of chiral centers along the chain that generate another source of differentiation between PLA_xGA_y polymers of similar gross compositions, i.e. the distribution of chiral centers. This distribution depends on many more or less interrelated factors, including the gross composition, the nature of the monomer, the mechanism of polymerization, the differences between the reactivities of glycolic acid

Tab. 1 The PLA_xGA_y family composed of homopolymers, stereocopolymers and copolymers

and lactic acid-type monomers, and the occurrence of configurational rearrangements caused by intrachain and interchain transesterification reactions (Lillie and Schulz, 1975; Chabot et al., 1983).

There are two main routes to make PLA_xGA_y polymer chains, i.e. step-growth polymerization or copolycondensation of glycolic and lactic acids, and the chain-growth polymerization or ring-opening polymerization of the glycolide and lactide cyclic dimers.

3.1
Step-growth Polymerization of Glycolic Acid

The step-growth polymerization of glycolic acid (GlcAc) corresponds to the following equilibrium reaction (see below)

The resulting PGA homopolymers have low molecular weight although they already present the characteristics of high molecular weight PGA because of their high crystallinity and insolubility in common solvents. Although nobody seems to have investigated the relative reactivities of glycolic acid and of lactic acid enantiomers, it is likely that the distribution of the glycoloyl and lactoyl repeating units is not random. Anyhow, the distribution resulting from the polycondensation of the hydroxy acids will be always different from that resulting from the ring-opening polymerization of glycolide with L- and/or D-lactide for reason underlined below. Under normal conditions PLA_xGA_y polymer chains obtained by polycondensation of the hydroxy acids are terminated by OH and COOH end groups in equivalent amounts.

3.2
Chain-growth Polymerization of Glycolide, the Cyclic Dimer of Glycolic Acid

The structure of the cyclic dimer of glycolic acid is symmetrical and includes two ester bonds that are equivalent provided the ring is closed. As soon as one of the ester bond is broken due to the action of a reagent, the second ester bond is stabilized and can no longer be involved in the polymerization process.

Spatial   Planar

diglycolide (glycolide or GA)

The official IUPAC nomenclature is 1,4-dioxane-2,5-dione or diglycolide. In literature people generally use glycolide for the sake of simplicity.

The polymer resulting from ring-opening polymerization is actually a polydimer according to chain growth by the addition of pairs of repeating units (Lillie and Schulz, 1975; Vert et al., 1984).

GA   HMW polydimeric PGA

The ring-opening polymerization of glycolide can be initiated by many compounds and leads to rather high molecular weight

GlcAc   LMW PGA

PGA polymers whose molecular weights are difficult to measure because of the lack of practical solvents. In the case of PGA, there is no difference in the distribution of repeating units resulting from step-growth and chain-growth polymerizations. Only differences at chain ends can occur, the polycondensation leaving OH and COOH chain ends, whereas ring-opening polymerization leads to substitution by an initiator or a catalyst residue at one terminal unit, a behavior usually deduced from the features collected from lactide polymerizations (Dubois et al., 1991; Kowalski et al., 2000; Kricheldorf et al., 2000). Chain end differences, that are generally neglected in common polymers, turn out to be critical in the case of PLA$_x$GA$_y$ polymers, as it will be discussed in the section on degradation. Insofar as the chain structure is concerned, the situation is much more complex when one deals with lactic acid-containing glycolic acid copolymers.

3.3
Step-growth Copolymerization of Glycolic Acid with L- and D-Lactic Acid

Low molecular weight PLA$_x$GA$_y$ copolymers can be synthesized by polycondensation starting from mixtures of glycolic acid and D- and L-lactic acids. Except for the glycolic acid-rich PLA$_x$GA$_y$ copolymers that are semicrystalline, most of the PLA$_x$GA$_y$ copolymers are amorphous (Miller 1977). The reaction proceeds as in the case of glycolic acid alone. However, whereas L- and D-lactic acids react at similar rates basically, glycolic acid is likely to react differently thus leading to non random unit distributions. The unit distributions in low molecular weight PLA$_x$GA$_y$ chains obtained by polycondensation has not been investigated so far. The chain structures of PLA$_x$GA$_y$ copolymers obtained through chain-growth ring-opening poly-

merization are dramatically more complex than those of the chains obtained by polycondensation.

3.4
Chain-growth Copolymerization of Glycolide with L- and D-Lactides

As the lactide molecule bears two asymmetric carbon atoms, lactide exists under the form of three diastereoisomers and a racemate whose structures are shown below, see p. 184 (Holten, 1971).

As in the case of the cyclic dimer of glycolic acid, the lactic acid cyclic dimers should be named after 2,6-dimethyl-1,4-dioxane-2,5-dione or referred to as dilactides. Instead people use lactides as indicated in the formula above. Accordingly, D-lactide, L-lactide and *meso*-lactide are composed of two D-lactoyl, two L-lactoyl, and one D- and one L-lactoyl units, respectively, whereas *rac*-lactide is a 50/50 mixture of L- and D-lactides.

The diastereoisisomeric structure of dilactides and the pair addition mechanism, and the difference of reactivity between glycolic acid and lactic acid cyclic monomers lead to irregular unit distributions in a given chain. The use of high-field nuclear magnetic resonance (NMR) is a good means to distinguish between PLA$_x$GA$_y$ copolymers having different chain compositions and chiral and achiral distributions, although no one has tried to fully assign the fine structures so far. Nevertheless, Figure 1 shows that NMR is very sensitive to the gross composition and to the chirality, as shown by the spectra of various PLA$_x$GA$_y$ copolymers (left hand side), the comparison of copolymers of glycolide with D- and L-lactide (first top line), and the spectrum of a PLA$_{50}$GA$_{50}$ obtained by ring-opening polymerization of 3-methyl-1,4-dioxane-2,5-dione, a cyclic monomer composed of one lactic acid and one glycolic acid residues. The

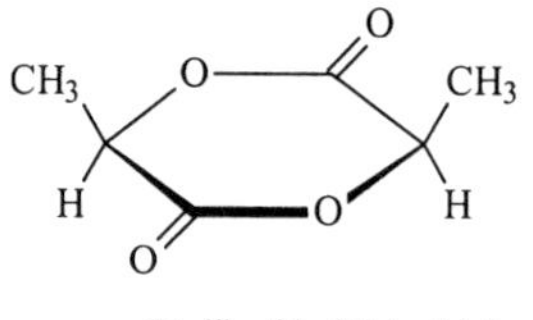

D-dilactide (D-lactide)

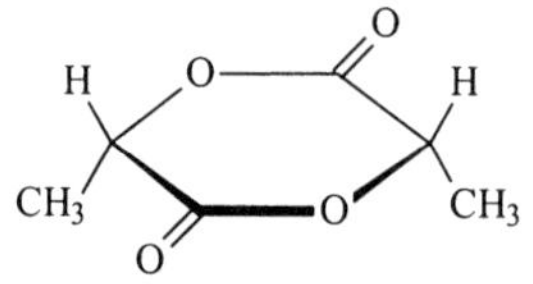

L-dilactide (L-lactide)

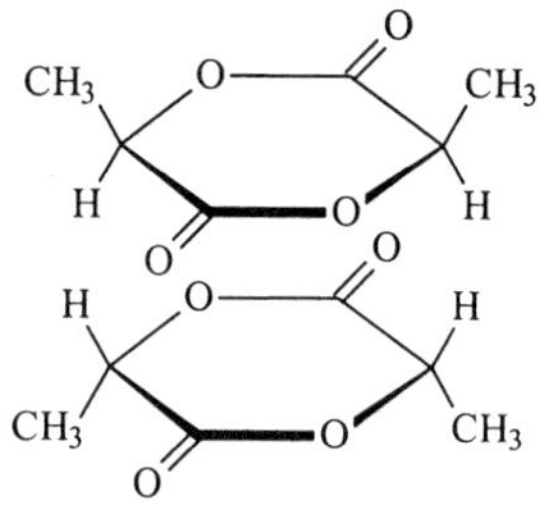

rac-dilactide (*rac*-lactide)

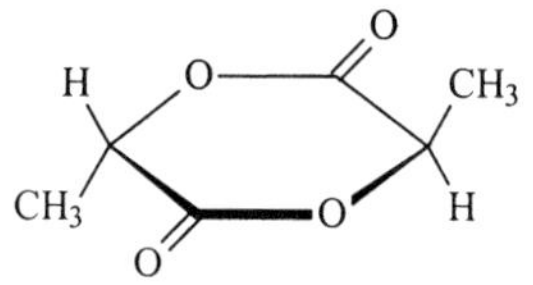

meso-dilactide (*meso*-lactide)

combination of mass spectrometry and molecule fragmentation might be good complementary tools to determine the chain fine structures (Chen et al., 2000; Adamus et al., 2000).

The characterization of PLA$_x$GA$_y$ copolymer chain structures in relation to the gross composition and repeating unit distribution is very important because these factors can have dramatic consequences on the hydrolytic degradation of corresponding polyester chains as we will see later on. The ring-opening polymerization can also affect the hydrolytic degradation because of the chain end modification by initiator or catalyst residues. This was shown during the past years in the case of lactides initiated by zinc metal, zinc lactate, and stannous octoate comparatively under conditions allowing chain end detection by NMR (Schwach et al., 1997, 1998a). Last, but not least, the characterization of PLA$_x$GA$_y$ copolymer fine structure prior to any other characterization is also important because of the possible occurrence of transesterification affecting repeating unit distributions or racemization at lactoyl sites affecting the chiral unit distribution, two factors of dramatic significance with respect to hydrolytic degradation.

3.5
Transesterification Rearrangements

As we have seen before, the fine chain structure of PLA$_x$GA$_y$ copolymers is rather complex. In contrast to the cases of vinylics or acrylics copolymerization, where one can basically calculate unit distributions using simple kinetics equations, the copolymerization of glycolic acid with L- and D-lactic acid does not allow simple calculation, partly because the achiral and chiral unit distributions can be rearranged during the polymerization reaction, or even later on during any processing, especially in the presence of compounds that can catalyze transesterification rearrangements. The risk of transesterification reactions is very high when polymerization is conducted at high temperature and in the bulk or in solution in the

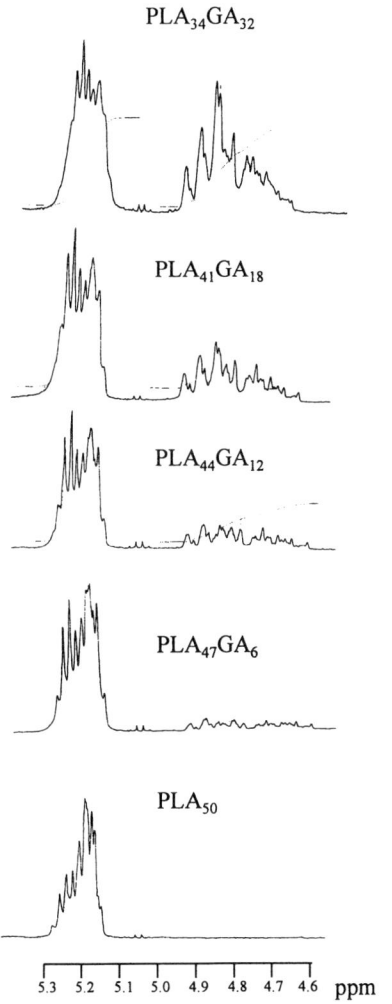

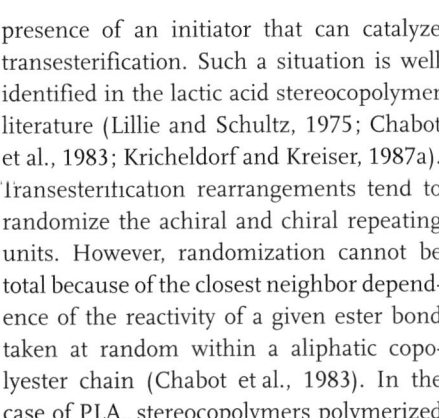

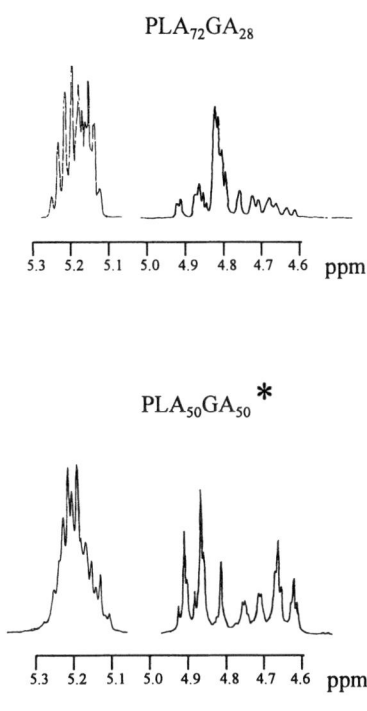

Fig. 1 The 400 MHz ^{1}H-spectra of different PLA$_{x}$-GA$_{y}$ copolymers in CDCl$_{3}$ (Granger et al., unpublished data). The PLA$_{50}$GA$_{50}$* copolymer was kindly supplied by Y. Kimura from the Tokyo Institute of Technology. (Lactic acid fine structures between 5.1 and 5.3 ppm; glycolic acid fine structures between 4.6 and 4.9 ppm, resonances of the residual lactide serving as internal standard.)

presence of an initiator that can catalyze transesterification. Such a situation is well identified in the lactic acid stereocopolymer literature (Lillie and Schultz, 1975; Chabot et al., 1983; Kricheldorf and Kreiser, 1987a). Transesterification rearrangements tend to randomize the achiral and chiral repeating units. However, randomization cannot be total because of the closest neighbor dependence of the reactivity of a given ester bond taken at random within a aliphatic copolyester chain (Chabot et al., 1983). In the case of PLA$_{x}$ stereocopolymers polymerized in the bulk in the presence of zinc metal, transesterification always leads to the same NMR spectrum resulting from the particularities of both polymerization and transesterification processes occurring almost simultaneously (Chabot et al., 1983). Transesterification rearrangements appeared different in the cases of stannous octoate and zinc metal initiated poly(DL-lactides) (Schwach et al., 1984). Polymer chains are also sensitive to the radiation generally used for sterilization (Chu and Williams, 1983; Vert et al., 1992)

4

Occurrence

PGA and PLA$_x$GA$_y$ copolymers are synthetic polymers that do not have any equivalent in nature. As previously shown they can be produced from glycolic acid, and from mixtures of glycolic acid and D- and or L-lactic acids by step-growth polymerization or from mixtures of glycolide and lactides by chain-growth polymerization. The literature is sometimes misleading because authors do not say whether the composition of the monomer feed is defined in mol/mol or in w/w. Another source of confusion exists at the level of the gross chain composition that is defined either from the composition of the monomer feed or from the analysis of the chain composition by NMR. Any scientific contribution dealing with PLA$_x$GA$_y$ copolymers should mentioned as accurately as possible the identity of the investigated compounds – a requirement underlined 20 years ago and still not respected (Vert et al., 1981)

necessary to obtain the highly pure monomers required to polymerize glycolide up to high molecular weight PGA (Neuwenjhuis, 1992).

Similarly, L- and D-lactide are generated during the thermal degradation of lactic acid oligomers obtained by polycondensation of L- and D-lactic acid, respectively. They melt at 97–98 °C and are soluble in many organic solvents (e.g. acetone, ethyl acetate, and benzene) that can be used for purification by crystallization from hot concentrated solutions.

rac-lactide ($M_p = 126$ °C) is obtained by the same method starting from *rac*-lactic acid. This lactide can be readily purified by recrystallization from hot saturated solutions in organic solvents.

meso-lactide ($M_p = 46$ °C) is usually extracted from the cold filtrate solution remaining after recrystallization of *rac*-lactide after evaporation of most of the solvent to end up with a hot concentrated solution. It is rather difficult to obtain pure *meso*-lactide (Chabot et al., 1983).

4.1

Monomers

Glycolic, and L-, D- and *rac*-lactic acids are commercially available either as solids or as concentrated aqueous solutions. The acid aqueous solutions usually contain the dimer and trimer oligomers at equilibrium with the hydroxy acid itself (Braud et al., 1996).

Glycolide (melting temperature $M_p = 86$ °C) is synthesized from synthetic glycolic acid. There are different routes but the most common consists in thermal degradation of low molecular weight PGA or PLA$_x$GA$_y$ copolymers obtained by polycondensation. These oligomers are heated under vacuum in the presence of a catalyst to yield a crude monomer. This process is usually referred to as ring closure. Several purification steps are

4.2

Polymers

PGA and PLA$_x$GA$_y$ copolymers can be obtained by atmospheric or vacuum distillation of water starting from a defined monomer feed composed of glycolic and lactic acids. This route leads to compounds having very low molecular weights that are brittle and glassy or waxy and sticky depending on the composition and on the molecular weight of the copolymers.

High molecular weight PGA and PLA$_x$GA$_y$ copolymers are synthesized by ring-opening polymerization of glycolide and lactides in convenient proportions either under an inert atmosphere or under vacuum. Under vacuum people ought to be careful not to distillate glycolide through the vacuum line otherwise the composition of the resulting polymer can be largely different from that of the feed. The polymerization reaction can be conducted either in the melt, i.e. at temperatures higher than the melting temperature of the resulting polymer (melt polymerization), below the polymer melting temperature (bulk polymerization), in suspension or in solution (Nieuwenhuis, 1992).

Only low molecular weight poly(α-hydroxy acid)s had been synthesized by melt polymerization at $250-270\,^\circ$C when, in 1954, a route to high molecular weight PGA using $ZnCl_2$ as initiator was patented. The procedure included purification of the monomer by recrystallization and degassing of the reaction mixture prior to bulk polymerization under vacuum (Lowe, 1954).

Since then many initiator systems have been reported in literature, primarily for lactide polymerization (Kleine and Kleine, 1959). In comparison, little attention has been paid to glycolide polymerization (Kohn et al., 1983). Except for our group which used zinc metal and more recently zinc lactate to copolymerize glycolic acid and lactic acid monomers, most of the polymers reported in the literature were made using stannous octoate ($Sn(Oct)_2$) either alone or with a co-initiator of the alcohol type. Both zinc lactate and stannous octoate are approved by the American Food and Drug Administration for surgical and pharmacological applications. We selected zinc many years ago because stannous octoate is unstable and usually contains impurities, especially when the industrial product is used as received (Schwach et al., 1997). Zinc metal was also retained because it is an oligoelement with daily allowance for the metabolism of mammalian bodies (Leray et al., 1976). Stannous octoate is generally preferred because it gives faster polymerization, higher conversion ratios and higher molecular weights than zinc metal or zinc

lactate (Schwach et al., 1994, 1996a). It has been also shown that stannous octoate is slightly cytotoxic (Tanzi et al., 1994). In solution polymerization and in melt and bulk polymerizations, analyses of the reaction media showed that polymers are at equilibrium with the monomers because of a ceiling temperature evaluated at $270\,^\circ$C (Nieuwenhuis, 1992; Leenslag and Pennings, 1987). Thermal degradation showed that PGA- and PLA-type polymers degrade to several degradation products including carbon dioxide, monomers, and aldehydes, i.e. acetaldehyde in the case of PLA and formaldehyde in the case of PGA (McNeill and Leiper, 1985). Last, but not least, the presence of residues issued from stannous octoate was recently shown for the bulk polymerization of *rac*-lactide (Schwach et al., 1998b) and L-lactic acid-rich PLA stereocopolymer. Among these residues hydroxy tin octanoate that can initiate lactic acid polymerization and ethyl-2 hexanoic acid that hydrophobizes the polymer mass have been identified in the case of polymerizations conducted at high initiator/monomer ratios. It seems that hydrophobic residues issued from stannous octoate are difficult to remove by precipitation from organic solution either by methanol, ethanol, or water (Schwach et al., 1998b).

According to literature, the polymerization of glycolide and lactides can proceed through anionic, cationic, or coordination mechanisms. A great deal of work has been devoted to the understanding of polymerization mechanisms because they determine the chain structures and to some extent, the degradability. However, many unknowns still remain in the literature due to the complexity of lactone polymerization chemistry (Li and Vert, 1995; Li, 1999). Recently, the polymerization of lactides in the presence of stannous octoate was revisited (Kowalski et al., 2000; Kricheldorf et al.,

2000). However, the authors did not pay attention to the difference between polymerizations carried out in solution and in the bulk, a point that would deserve deeper investigation in the future because industrial polymers are synthesized through bulk polymerization. It is of interest to note that most of the kinetics investigations reported in literature dealt with lactide polymerization.

5
Biochemistry

PLA$_x$GA$_y$ polymers are not biopolymers. So far no one has found any poly(α-hydroxy acid)-type polymers in living systems. Therefore, these compounds have no recognized biological or physiological function. However, because they degrade in aqueous media and because living systems function in aqueous environments, PLA$_x$GA$_y$ polymers undergo hydrolytic degradation in the presence of living organisms or microorganisms. At the minimum they then generate glycolic and lactic acids as degradation by-products. These metabolites are finally bioassimilated. Whether enzymes are involved in the degradation of PLA$_x$GA$_y$ polymers is still a matter of debates in literature.

6
Degradation in Abiotic and Biotic Aqueous Media

Basically the degradation of polymeric materials in a living environment can result from enzymatically mediated and/or chemically mediated cleavage. Although animal or human *in vivo* environments are different from environmental outdoor ones, there is no fundamental difference between the biodegradation of a polymer by animal cells or by microorganisms. Both involve water, enzymes, metabolites, ions, etc., which interact with the material (Vert et al., 1994a). From a general viewpoint, degradation in the presence of enzymes with or without the cells that produce these enzymes is regarded as biodegradation and can proceed through unzipping from chain ends (exoenzymes) or cleavages within main chains (endoenzymes). Anyhow, the term "degradation" should be reserved to degradations in the absence of any living cells or isolated enzymes or when the mechanisms of degradation is unknown (Li and Vert, 1995).

6.1
Degradation in Abiotic Media

For almost 20 years after the early work on high molecular weight PLA$_x$GA$_y$ aliphatic polyesters, hydrolytic degradation in the absence of enzyme was regarded as homogeneous (bulk erosion) (Pitt et al., 1981), although surface erosion was claimed in a few cases (Kronenthal, 1974). This was primarily due to the fact that molecular weight changes were monitored by viscometry. During the last 12 years, the understanding of the hydrolytic degradation of PLA$_x$GA$_y$ polymers advanced significantly due to the use of size exclusion chromatography that revealed the presence of two populations of partially degraded PLA$_x$GA$_y$ devices of rather large sizes. This discovery led us to introducing the concept of heterogeneous degradation related to diffusion–reaction–dissolution phenomena (Li et al., 1990a; Vert, 1990)

To summarize, once a device made of a PLA$_x$GA$_y$ polymer is placed in contact with an aqueous medium, water penetrates into the specimen and the hydrolytic cleavage of ester bonds starts (Figure 2).

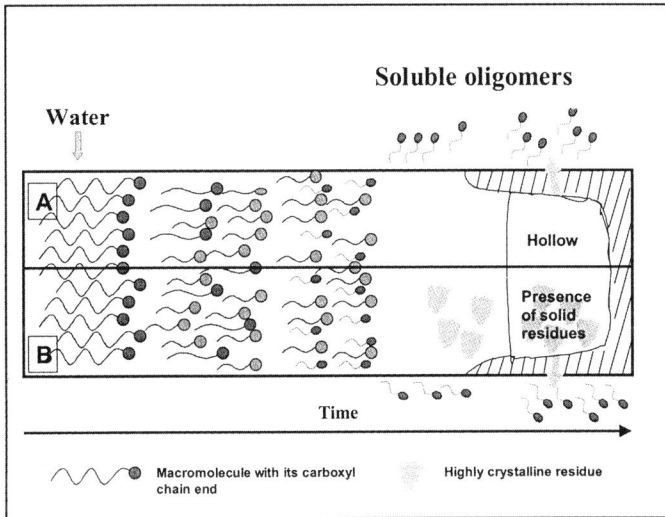

Fig. 2 Schematic representation of the fate of macromolecules forming a parallel-sided device made of a PLA$_x$GA$_y$ polymer as a function of time (case A: initially amorphous device composed of a stereo-irregular PLA$_x$GA$_y$ polymer that does not lead to crystalline residues; case B: either an initially semi-crystalline device or a device that is initially amorphous but composed of PLA$_x$GA$_y$ polymer stereoregular enough to generate semi-crystalline low molecular weight residues because of faster degradation in the most disordered or glycolic acid-rich part of copolymer chains

Water absorption is thus a critical factor (Schmitt et al., 1994). Each cleaved ester bond generates a new carboxyl end group that, in principle, can catalyze the hydrolytic reaction of other ester bonds as proposed in the case of the homogeneous degradation mechanism (Pitt et al., 1981). For a time the partially degraded macromolecules remain insoluble in the surrounding aqueous medium, regardless of its nature and the degradation proceeds homogeneously. However, as soon as the molecular weight of some of the partially degraded macromolecules becomes low enough to allow dissolution in the aqueous medium, diffusion starts within the whole bulk, with the soluble compounds moving slowly to and off the surface while they continue to degrade. This process combining diffusion, chemical reaction and dissolution phenomena results in a differentiation between the rates of the degradations at surface and interior of the

matrix. Indeed, oligomers can escape from the surface before total degradation, in contrast to those oligomers that are far inside. The result is a smaller autocatalytic effect at the carboxyl-depressed surface with respect to the bulk. In buffered media, neutralization of terminal carboxyl groups might also contribute to discriminate the surface degradation rate. This mechanism well accounted for the formation of two populations of partially degraded macromolecules with different molecular weight and thus the observation of bimodal size exclusion chromatography traces. At the end, hollow structures were observed occasionally, especially in the case A in Figure 2 representing the fate of amorphous matrices that cannot generate semi-crystalline residues upon degradation (Li et al., 1990b,c). Indeed it has been shown that degradation of PLA$_x$GA$_y$ is faster within glycolic acid-rich intrachain segments (Li et al., 1990b) and in highly disordered parts of the copolymer chains. Therefore, PLA_xGA_y polymer chains lacking intrachain order cannot generate crystallizable residues. In contrast, in the case of initially crystalline or of initially amorphous but rather stereoregular PLA_xGA_y polymer chains, no hollow structures is observable despite the faster inner degradation, the crystallization of inside degradation products leading to a more degraded but solid interior (case B in Figure 2) (Li et al., 1990b,c). The formation or not of crystalline residues was mapped and discussed in details, at least for zinc-derived PLA_xGA_y polymer (Vert et al., 1994b; Li and Vert, 1999).

Basically there are four main factors which condition the diffusion-reaction-dissolution phenomena: (1) the hydrolysis rate constant of the ester bond, (2) the diffusion coefficient of water within the matrix, (3) the diffusion coefficient of chain fragments within the polymeric matrix, and (4) the solubility of degradation products, generally oligomers, within the surrounding liquid medium from which penetrating water is issued. Any additional factors such as temperature, additives in the polymeric matrix, additives in the surrounding medium, pH, buffering capacity, size and processing history, quenching or annealing, steric hindrance, porosity, and other variables affect the general balance through their effects on the main factors listed above (Vert, 1998).

The discovery of the heterogeneous degradation mechanism helped understanding the effect of many factors, i.e. matrix morphology, chemical composition and configurational structure, molar mass, size and shape, distribution of chemically reactive compounds within the matrix, and nature of the degradation media (Vert, 1990).

The effects of chemical composition, molecular weight, molecular weight and repeating units distributions, and crystallinity are well documented in the literature (Holland et al., 1986). Recently, the differences between the degradation rates observed when comparing the degradation characteristics of films, powder, and microspheres issued from the same batch of PLA_{50} polymer were accounted for (Grizzi et al., 1995). This work shows conclusively that the smaller the size of a polymer device, the slower the degradation rate. From the same logic, one can conclude that porous systems degrade at a slower rate than plain ones, especially if dimensions of the considered device are millimetric. Glycolic acid-rich PLA_xGA_y copolymers and PGA being more hydrophilic and degrading faster in aqueous media than PLA and lactic acid-rich PLA_xGA_y copolymers, the leaching of the water-soluble oligomers from the surface of large-size devices made of a glycolic acid-rich PLA_xGA_y occurs relatively earlier than for lactic acid-rich analogs and thus the skin is thinner.

The heterogeneous degradation mechanism provides also fruitful information on the behavior one can expect from a PLA_xGA_y polymer chain in which additives or foreign molecules have been introduced. Indeed any electrostatic interaction between a PLAGA matrix and acidic, basic or amphoteric molecules can drastically affect the degradation characteristics by changing the natural acid–base equilibrium of the matrix due to the presence of chain end carboxylic groups. For acidic compounds, one can expect faster hydrolysis of ester bonds. In contrast, in the case of basic compounds, two effects can be observed: base catalysis of ester bond cleavage when the basic compound is in excess with respect to acid chain ends and decrease of the degradation rate in the opposite situation (Li et al., 1996). Examples were reported that deal with oligomers (Mauduit et al., 1996), residual monomers (Cordewener et al., 1995), coral (Li and Vert, 1996), caffeine (Li et al., 1996), and gentamycin (Mauduit et al., 1993a,b). Reviews are available (Li, 1999; Li and Vert, 1995, 1999; Brannon-Peppas, 2000)

Recently, attention was paid to the effects of the initiator on the degradation characteristics of some PLA_xGA_y polymers. It has been shown that dramatic differences in repeating unit distribution can be observed in the case of stereocopolymers such as PLA_{50} polymerized in the presence of stannous octoate or zinc metal (Schwach et al., 1994). Later on it was suggested that stannous octoate causes more or less chain end modification by esterification of alcoholic end groups by octanoic acid and generates hydrophobic residues (Zhang et al., 1994; Schwach et al., 1997) which resist purification by precipitation from an organic solution with ethanol or water (Schwach et al., 1997). Such modifications do not occur in the case of zinc metal or zinc lactate initiations. Whether similar effects can be observed in the case of polymers obtained using other initiating systems, or through solution polymerization, or in the presence of alcohols is still unknown. Indeed, scientists rarely investigated the effect of resulting structures on degradation characteristics and comparison of literature data is seldom feasible significantly. In the absence of accurate information on structure and history, it is rather difficult to imagine how much the various factors mentioned above contribute to the degradation and the release characteristics. Nevertheless, the influence of chain end hydrophobization on degradation characteristics of PLA_xGA_y polymers is well exemplified by the difference between the so-called "H" and "non-H" polymers appearing in the catalog of one of the greatest supplier of PLA_xGA_y polymers, i.e. Boeringher Ingelheim (Germany). At this point people ought to keep in mind that the degradation and release characteristics of PLA_xGA_y depend on experimental conditions and on the device's history much more

than for any other type of polymeric matrix. It is then strongly recommended again that careful identification of the investigated compounds be included in any future paper in the field.

Another critical recommendation is to be careful in modeling the conditions found in real applications. To model body fluids, it is recommended to take into account physiological or outdoor environmental pH, ionic strength, and temperature as much as possible. Using 0.13 M pH 7.4 isosmolar phosphate buffer at 37 °C is a convenient choice to model human body fluids. If phosphate ions can perturb the system, the use of any other pH 7.4 isosmolar buffered medium is possible. However, one must keep in mind that, in such media, species can be present that can diffuse and modify the matrix physical characteristics and the chain end chemistry, thus affecting the general behavior through the various interrelated factors mentioned above. As the glass temperature of PGA and of PLA_xGA_y polymers is a critical factor for water uptake and for diffusion of soluble oligomers, raising the temperature to shorten degradation time can be the source of unexpected phenomena that change the regime of degradation in a glassy state to a regime of degradation in a rubbery state.

The hydrolytic degradation of PGA fibers was extensively investigated by several authors, primarily by Chu and co-workers. It was shown that degradation occurs faster in amorphous domains, thus leading to crystallinity increase. The effects of other parameters like pH, temperature, etc., were also studied (Chu et al., 1981; Chu and Browning, 1988). It is worth noting that the main characteristics of the above mechanism of heterogeneous degradation well agree with the features reported for small-size PGA sutures. Indeed, the small size and the high crystallinity of fibers and suture threads prevent the occurrence of any surface-center differentiation and any formation of hollow structures. Attempts have been made to predict molecular weight changes on random hydrolysis (Nishida et al., 2000). Successful predictions are reported for fims, i.e. for small-size devices. In contrast no theoretical derivation has been found to reflect the heterogeneous degradation mechanism, mostly because all parameters change with time and swelling (Vert, 1998).

6.2
Degradation in Biotic Media

One can define different cases where degradation occurs in the presence of living systems. The problem is then showing the involvement of enzymes in the presence of the complex abiotic hydrolysis described previously.

6.2.1
Enzyme-containing Media

Whether enzymes in aqueous solutions can degrade PLA$_x$GA$_y$ polymers *in vitro* or *in vivo* has been a matter of debates in literature. Some authors claimed occurrence of enzymatic degradation (Herrman et al., 1970; Williams, 1977; Reed, 1978), while most people excluded biodegradation or relegated it to a second role (Salthouse and Matlaga, 1976; Gilbert et al., 1982; Zaikov, 1985; Schakenraad et al., 1990).

In order to make the situation clearer, one must consider different cases. Some enzymes have been found that can attack PLA$_x$GA$_y$ polymer chains *in vitro* and in the absence of the cells that normally produce the enzymes. Among these enzymes, bacterial proteinase K was shown to be able to strongly accelerate the degradation rate of PLA stereocopolymers (Williams, 1981; Fukuzaki et al., 1989; Ashley and McGinity, 1989; Reeve et al., 1994; MacDonald et al.,

1996; Li, 1999). L-Lactic acid units are preferentially degraded as compared to D-lactic acid ones (Reeve et al., 1994; MacDonald et al., 1996). Enzymes such as tissue esterases, pronase, and bromelain are also able to affect PLA degradation (Herrman et al., 1970; Williams, 1981). It is important to note that there is usually competition between hydrolytic and enzymatic degradations. On the other hand, enzymes can be inactive on high molecular weight compounds and become active at the later stages of degradation when the chain fragments become small and soluble in surrounding fluids.

Because glycolic acid-containing, and especially glycolic acid-rich PLA$_x$GA$_y$ polymers, are very sensitive to water and heat, showing a contribution specific to enzymes is usually difficult. Enzymatic degradation should not be claimed unless mass loss and dimensional change without molar mass decrease are conclusively shown since enzymes can hardly penetrate deep in a dense polymer matrix. However, one must keep in mind that if, for any reason, the rate of nonenzymatic hydrolysis at the surface of a hydrolytically degradable matrix is greater than the rate of diffusion of water into the polymeric mass, surface erosion can be observed, and thus confusion is possible between enzymatic surface degradation and surface degradation by water in the absence of enzymes. This is unlikely in the case of PLA$_x$GA$_y$ polymers for which water uptake is always rather fast as compared with degradation rates.

6.2.2
Degradation by Microorganisms

In the past decades increasing attention has been paid to the degradation of aliphatic polyesters of lactic acid-rich PLA$_x$GA$_y$ polymers in relation with the problems of the accumulation of solid plastic waste in the environment. Whether these polymers can biodegrade in a wild environment or whether it is the by-products generated by abiotic hydrolysis that can be bioassimilated by fungi or bacteria are two questions currently being investigated, in particular with regard to composting.

The ability of some microorganisms to use lactic acids, PLA oligomers and polymers, and PLA_xGA_y copolymers as sole carbon and energy sources under controlled or natural conditions was investigated in details (Vert et al., 1994c; Torres et al., 1996a–c). Among 14 filamentous fungal strains tested, two strains of *Fusarium moniliforme* and one strain of *Penicillium roqueforti* were identified as able to totally assimilate L- and DL-lactic acids as well as PLA_{50} oligomers (Torres et al., 1996a,b). In contrast, PLA_{100} oligomers appeared biostable because of crystallinity. A synergistic effect was observed when both microorganisms were present in the same culture medium containing PLA_{50} oligomers (Torres et al., 1996b).

High molar mass polymers were also considered via degradation tests in soil and in selected culture media. *F. moniliforme* filaments were found to grow on the surface of a $PLA_{37.5}GA_{25}$ copolymer and through the bulk. On the other hand, five strains of different filamentous fungi were isolated from the soil as capable of bioassimilating PLA_{50} soluble oligomers in mixed cultures. Based on the results obtained by scanning electron microscopy, size exclusion chromatography, pH, and water absorption measurements, the authors proposed a mechanism including abiotic hydrolysis of macromolecules followed by bioassimilation of soluble oligomers (Torres et al., 1996c). To summarize, when a PLA_xGA_y material is placed in a degradation medium in the presence of microorganisms, only abiotic hydrolytic degradation occurs unless the few enzymes mentioned above as capable of degrading high molar mass PLA are present. Similar investigations were reported recently that came to similar conclusions (Hakkarainen et al., 2000).

Whether solid fermentation of PLA_xGA_y polymers in compost or in soil proceeds through homogeneous or heterogeneous degradation is of interest. It should depend on the size of the material as discussed before (Grizzi et al., 1985). However, one must consider also the physical state of the medium (liquid or moisture). In a liquid medium, degradation of large-size devices will be heterogeneous due to greater autocatalysis inside, whereas, in the case of a gaseous water environment, degradation of similar devices will be homogeneous since the soluble oligomers cannot diffuse off the matrix. Anyhow, sooner or later, the whole compound will be degraded to low molecular weight assimilatable oligomers. The bioassimilation process will then take place. At this time, fungal filaments will penetrate the partially degraded mass to take advantage of the internal oligomers, thus turning the abiotic degradation to a biotic one. Under wild conditions, this should happen after a long lag period. However, PLA_xGA_y polymers will be definitely degraded and assimilated since the ultimate degradation products (L- and D-lactic acids) have been shown to be metabolized by some common microorganisms (Torres et al., 1996b; Hakkarainen, 2000). In particular, it is now well known that industrial lactic acid-based polymers like Dow-Cargill ECOPLA® degrade completely and rather rapidly in a compost provided the temperature of the compost raises up to 70 °C. At lower temperature composting is much slower. Information on the biodegradation of glycolic acid-containing polymers under composting conditions is not yet available, probably because such polymers have found no potential use in the environment so far.

6.2.3

Degradation in an Animal Body or in Human
The most efficient enzyme that is capable of degrading PLA_xGA_y polymers *in vitro*, i.e. proteinase K, is a protease of bacterial origin, and thus is not present in animal and human bodies. During the pre-industrial development of PGA sutures, information on the *in vivo* behavior and on the biocompatibility of these sutures was proprietary. It is in the late 1960s that people started releasing information. In 1966 appeared a remarkable contribution dealing with poly(L-lactide) and PLA_{50} stereocopolymers (Kulkarni et al., 1966). Most of the basic information regarding biocompatibility, configuration-dependence of degradability and crystallinity, and routes of elimination after ^{14}C labeling were identified. Until 1981, when it was shown that PLA_{100} and lactic acid-rich stereocopolymers prepared in the presence of zinc metal and carefully purified can last for a year *in vivo* with less than 10% molecular weight decrease (Chabot et al., 1981), the literature had reported much shorter half-lifetime systematically (Cutright et al., 1974; Miller et al., 1977; Guilding and Reed, 1979, 1981). Despite this difference of extreme lifetimes, it had been shown that the *in vivo* degradability of PLA_xGA_y polymers, copolymers, and stereocopolymers depends on the gross composition, and on the content in chiral units (Sedel et al., 1978; Reed, 1979). Since then, most of the contributions concluded to excellent biocompatibility of PLA_xGA_y polymers (Ignatius and Claes, 1996). Year after year the literature was enriched in information on devices made of PLA_xGA_y, such as sutures (Chu and Browning, 1988), implants for bone surgery (Hollinger, 1989) or microspheres and nanospheres for drug delivery (Brannon-Peppas, 1995). The literature is now rich of thousands of articles concerning PLA_xGA_y polymers. From a general viewpoint a majority of

authors agree to say that PLA_xGA_y polymers do not degrade enzymatically in an animal body. In contrast, a minority claim enzymatic contribution (Li and Vert, 1995, 1999). The demonstration is usually based on the comparison of *in vivo* and *in vitro* degradation characteristics – differences being assigned to effects due to enzymes. However, the correlation cannot be straightforward. Indeed the heterogeneous degradation mechanism tells us that one of the driving phenomena is the dissolution of oligomers in the surrounding liquid medium. One can find many reasons to see different oligomer diffusion and solubilization between a device implanted *in vivo* and the same device allowed to age in a phosphate buffer medium modeling body fluids. Phenomena like adsorption of proteins, absorption of lipids, and greater solubility of PLA_xGA_y oligomers in blood are examples of sources of differences (Sharma and Williams, 1981). Superoxide ions also contribute to *in vivo* degradation of synthetic sutures (Lee and Chu, 2000). Anyhow, it has been found that *in vitro* and *in vivo* degradation characteristics are comparable in the case of small PLA_xGA_y cylinders implanted subcutaneously in rats (Kenley et al., 1987). In agreement with these findings, PLA_xGA_y parallel-sided plates implanted intramuscularly in rabbits and similar plates allowed to age in isoosmolar pH 7.4 phosphate buffer showed similar degradation characteristics, except shape modifications that were greater *in vivo* than *in vitro* according to the effect of the mechanical stresses generated by the muscles (Therin et al., 1993). Several authors found differences and the subject will be the source of endless debate (Gogolewski et al., 1993; Mainil-Varlet et al., 1997) unless people characterize carefully the investigated compounds including processing and sterilization protocols and consequences. In the previous sections it was mentioned for PLA_{50}

matrices that stannous octoate lead to more hydrophobic polymers than zinc metal and zinc lactate (Schwach et al., 1997). Similar investigations performed on PLA_{100} tin- and PLA_{98} zinc-derived interference screws currently commercialized to fix anterior cruciate ligament autografts in human and implanted in sheep knees showed dramatic differences of behavior insofar as water absorption, molecular weight decreases, decrease of mechanical properties, and rate of degradation were concerned (Schwach and Vert, 2000). Similar features should be found for glycolic acid-containing polymers.

Nowadays everybody agrees that PLA_xGA_y polymers are remarkably biocompatible. However, it appears that glycolic acid-containing copolymers are more inflammatory than PLA_x stereocopolymers, the most critical factor being the presence of crystalline tiny particles that cause well-known foreign body reactions to particles. Accordingly, taking advantage of polymer science to keep the polymer mass amorphous during the *in vivo* degradation process is always profitable in terms of biocompatibility since it keeps the inflammatory response low. Anyhow, once formed during degradation, any tiny PLA_xGA_y crystalline particles can be phagocytosed to undergo intracellular degradation (Woodward et al., 1985; Tabata and Ikada, 1988).

In the future, one of the key points in degradable and biodegradable polymer science is going to be the demonstration of the fate of degradation by-products in complex living media, especially mineralization to carbon dioxide and water, bioassimilation, and/or subsistence as biostable residues. The best method to monitor these fates within the complex living media (animal body, microorganism, outdoor environment) is radiolabeling. Recently, a method to tritiate lactide and glycolide by catalytic isotopic exchange in the presence of palladium on calcium carbonate was reported

that paves the route to the synthesis of radioactive PLA_xGA_y polymers (Dos Santos et al., 1998; Dos Santos et al., 1999).

In conclusion, it is recommended that the demonstration of the mechanism of degradation of PLA_xGA_y polymers under any condition be based on the concordance of data issued from different analytical methods, and from a careful monitoring of the generation and the fate of degradation by-products.

7
Production

7.1
Producers

The main manufacturers of PGA and glycolic acid-rich PLA_xGA_y polymers in the world are Purac (Purasorb®) in The Netherlands, Boeringher Ingelheim (Resomer®) in Germany, and Birmingham Polymers Inc. and Alkermes in the USA. Information can be found through the Internet.

7.2
World Market

The world market of PLA_xGA_y polymers other that PLA_{100} and PLA_x stereocopolymers offered for environmental applications is rather small – the present applications being limited to small market/high-added value domains such as drug delivery systems.

7.3
Applications

Whereas PLA_{100} and PLA_x stereocopolymers and lactic acid-rich PLA_xGA_y copolymers are preferred to make implants and devices for bone surgery (bone plates, screws, pins, etc.) because they last for a long time in the human body, glycolic acid-rich PLA_xGA_y

copolymers are generally selected for drug-delivery systems, especially the amorphous compounds, because they lead to longer sustained release than semi-crystalline compounds (Mauduit et al., 1993a,b).

The interest of PGA and PLA_xGA_y copolymers for bone surgery was recognized very early, although clinical and industrial developments came much later when controlled polymerization, processing, sterilization, and storage conditions became under good control. Nowadays, several devices are on the market for the temporary internal fixation of bone fractures (screws, plates, pins) and anterior cruciate ligament autografts (interference screws) from several companies, i.e. Biomet Inc. ($PLA_{82}GA_{18}$), Bionx Implants, Inc. (self-reinforced PGA), Instrument Makar, Inc. ($PLA_{42.5}GA_{15}$), Smith & Nephew Endoscopy (PGA; $PLA_{82}GA_{18}$), and Instrument Makar, Inc. ($PLA_{42.5}GA_{15}$) (Barber, 1998; Bostman et al., 1998). As early as 1978 a totally bioresorbale composite system composed of PGA fibers reinforcing PLA_{100} matrices, the long-lasting matrix protecting the fastly degraded fibers, was patented (Vert et al., 1978; Christel et al., 1982, 1985). Later on other composite systems were proposed such as self-reinforced PGA plates where PGA fibers were embedded in a PGA matrix (Törmälä et al., 1991), and went up to clinical and industrial applications. PLA_xGA_y are also worthwhile compounds for tissue reconstruction (Sedel et al., 1978; Hollinger, 1983)

In drug delivery, the literature is full of examples of PLA_xGA_y-based drug formulations, and devices such as implants, microparticles (Anderson and Shive, 1997), and nanoparticles (Brannon-Peppas, 1995) aimed at controlling the delivery of drugs of any kind. Several reviews have been issued that cover well a very rich field (Holland et al., 1985; Li and Vert, 1999; Brannon-Peppas and Vert, 2000). A few systems are commercialized that are used to treat prostate cancer,

i.e. Decapeptyl® by Ipsen-Beaufour in France and Enantone® by Takeda in Japan.

Nowadays, one of the most stimulating areas of potential applications is tissue engineering, a technique that consists of culturing cells on a degradable or bio-degradable scaffold for the sake of implantation for tissue reconstruction. Tissues like skin, bone, and cartilage are the favorite targets of scientists (Agrawal et al., 1995). Among the various degradable and bio-degradable scaffolds presently under investigation, those made of PLA_xGA_y three-dimensional porous devices seem to be the most attractive (Ishaug et al., 1994; Agrawal et al., 1995; Holder et al., 1997; Athanasiou et al., 1998).

7.4
Patents

Many patents involving PLA_xGA_y polymers have been delivered during the last 30 years.

The early patents dealt with polymer and copolymer synthesis: Condensation polymers of hydroxy acetic acid, US2676945; Preparation of high molecular weight poly-hydroxy acetic ester, US2668162.

Later on patents dealing with PGA and glycolic acid-rich PLA_xGA_y crystalline sutures and devices were issued: Absorbale polyglycolic sutures, US3297033; Polyglycolic acid prosthetic devices, US3739773; Method for heat setting of strecht oriented polyglycolic acid filament, US3422181; Process for polymerizing a glycolide, US3442871; Absorbable poly(glycolic acid) suture of enhanced *in vivo* strength retention and a method and apparatus for preparing same, US3626942; Preparation of glycolide polymerizable into polyglycolic acid of consistently high molecular weight, US3597450; Reducing capillarity of polyglycolic acid sutures, US3867190; Coating composition for sutures, US4185637.

More recent patents are: Yttrium and rare earth compounds catalyzed lactone polymerization, US5028667; Verfahren zur Herstellung von Polyestern auf der Basis von Hydroxycarbosäuren, EP0443542A2; Biodegradable chewing gum bases including plasticized poly(DL-lactic acid) and copolymers thereof, WO0019837; Polymerization process and product, US4273920; Method for producing poly(glycolic acid) with excellent heat resistance, KR9502607.

Besides these patents dealing with polymeric matter and surgical devices, a number of patents were delivered for drug delivery systems of different kinds, namely implants, microspheres and nanospheres. The list of recent patents can be easily found using free data banks like http://ep.espacenet.com/

8

Outlook and Perspectives

PLA_xGA_y polymers constitute a large family of aliphatic polyesters that can be synthesized by different routes leading to compounds with different molecular weights, unit distributions, and terminal groups. All the members of this family are hydrolytically degradable. They are not biodegradable under normal conditions either in the animal body or in the environment, although some enzymes are able to biodegrade some members under *in vitro* model conditions. Among the whole family, PGA and glycolic acid-rich polymers are finding applications in the biomedical and pharmacological fields. They are being currently used to make commercial sutures. However, it is in the field of controlled drug delivery via implants, microspheres, and nanospheres that amorphous PLA_xGA_y polymers are showing the greatest potential for applications, and a few formulations being already on the market. Some lactic acid-rich PLA_x-GA_y polymers are being used in orthopedic surgery, especially for fixing anterior cruciate ligament autografts. In contrast to PLA_x stereocopolymers, none of the glycolic acid-containing compounds have been considered as candidates or applications as environmentally degradable polymer, so far.

9
References

Adamus, G., Sikorska, W., Kowalczuk, M., Montaudo, M., Scandola, M. (2000) Sequence distribution and fragmentation studies of bacterial copolyester macromolecules: characterization of PHBV macroinitiator by electrospray ion-trap multistage mass spectrometry, *Macromolecules* **33**, 5797–5802.

Agrawal, C. M., Niedersuer, C. C., Micallef, D. M., Athanasiou, K. A. (1995) The use of PLA–PGA polymers in orthopedics, in: *Encyclopedic Handbook of Biomaterials and Bioengineering*, New York: Marcel Dekker, 2081–2115.

Anderson, J. M., Shive, M. S. (1997) Biodegradation and biocompatibility of PLA and PLGA microspheres, *Adv. Drug Del. Rev.* **28**, 5–24.

Ashley, S. L., McGinity, J. W. (1989) Enzyme-mediated drug release from poly(D,L-lactide) matrices, *Congr. Int. Technol. Pharm.* **5**, 195–204.

Athanasiou, K. A., Schmitz, J. P., Agrawal, C. M. (1998) The effects of porosity on *in vitro* degradation of polylactic acid–polyglycolic acid implants used in repair of articular cartilage. *Tissue Eng.* **4**, 53–63.

Barber, A. F. (1998) Resorbable fixation devices: a product guide, *Orthopaedic Special Edition* **4**, 11–17.

Bostman, O., Hirvensalo, E., Vainionpaa, S., Vihtonen, K., Tormala, P., Rokkanen, P. (1990) Degradable polyglycolide rods for the internal fixation of displaced bimalleolar fractures, *Int. Orthop.* **14**, 1–8.

Brannon-Peppas, L. (1995) Recent advances on the use of biodegradable microparticles and nanoparticles in controlled drug delivery, *Int. J. Pharm.* **116**, 1–9.

Brannon-Peppas, L., Vert, M. (2000) *Polylactic and Polyglycolic Acids as Drug Delivery Carriers, in Handbook of Pharmaceutical Controlled Release Technology* (Wise, L. R., Klibanov, D. L., Mikos, A., Brannon-Peppas, L., Peppas, N. A., Trantalo, D.

J., Wnek, G. E., Michael, J., Yaszemski, M. J., Eds.), New York: Marcel Dekker, pp. 90–130.

Braud, C., Devarieux; R., Garreau, H., Vert, M. (1996) Capillary electrophoresis to analyze water-soluble oligo(hydroxy acids) issued from degraded or biodegraded aliphatic polyesters, *J. Environ. Polym. Degrad.* **4**, 135–148.

Chabot, F., Vert, M., Chapelle, S., Granger, P. (1983) Configurational structires of lactic acid stereo-copolymers as determined by ^{13}C-^{1}H-NMR, *Polymer* **24**, 53–59.

Chen, J., Lee, J. W., Hernandez de Gatica, N. L., Burkhart, C. A., Hercules, D. M., Gardella, J. A. (2000) Time-of-flight secondary ion mass spectrometry studies of hydrolytic degradation kinetics at the surface of poly(glycolic acid), *Macromolecules* **33**, 4726–4732.

Christel, P., Chabot, F., Leray, J. C., Morin, C., Vert, M. (1982) Biodegradable composites for internal fixation, in: *Biomaterials 1980: Advances in Biomaterials* (Winter, G. D., Gibbons, D. F., Plenk, H., Eds.), New York: John Wiley & Sons, Vol. 3, 271–280.

Christel, P., Vert, M., Chabot, F., Garreau, H., Audion, M. (1985) PGA (polyglycolic acid) fiber-reinforced PLA(polylactic acid) as an implant material for bone surgery, Proceedings, Plastic and Rubber Institute, London, p. 11.

Chu, C. C. (1981) The *in-vitro* degradation of poly(glycolic acid) sutures—effect of pH, *J. Biomed. Mater. Res.* **15**, 795–804.

Chu, C. C., Browning, A. (1988) The study of thermal and gross morphologic properties of polyglycolic acid upon annealing and degradation treatments, *J. Biomed. Mater. Res.* **22**, 699–712.

Chu, C. C., Williams, D. F. (1983) The effect of gamma irradiation on the enzymatic degradation of polyglycolic acid absorbable sutures, *J. Biomed. Mater. Res.* **17**, 1029–40.

Cordwener, F. W., Rozema, F. R., Bos, R. R. R., Boering, G. (1995) Material properties and tissue reaction during degradation of poly(96L/4D-lactide)—a study *in vitro* and in rats, *J. Mater. Sci.: Mater. Med.* **6**, 211–218.

Cutright, D. E., Perez, B., Beasley, J. D., Larson, W. J., Rosey, W. R. (1974) Degradation rates of polymers and copolymer of polylactic and polyglycolic acids, *Oral Surg.* **37**, 142–152.

Dos Santos, I., Morgat, J. L., Vert, M. (1998) *J. Labelled Cpd Radiopharm.,* **41** 1005–1012.

Dos Santos, I., Morgat, J. L., Vert, M. (1999) Glycolide deuteration by hydrogen isotope exchange using the HSCIE method, *J. Labelled Cpd Radiopharm.* **42**, 1093–1101.

Dubois, P., Jacobs, C., Jérôme, R., Teyssié, P. (1991) Macromolecular engineering of polylactones and polylactides. IV, Mechanism and kinetics of lactide homopolymerization by aluminum isopropoxide, *Macromolecules* **24**, 2266–2270.

Dunsing, R., Kricheldorf, H. R. (1985) Polylactones. 5. Polymerization of LL-lactide by means of magnesium salts, *Polym. Bull.* **14**, 491–496.

Frazza, E. J., Schmitt, E. E. (1971) A new absorbable suture, *J. Biomed. Mater. Sci.* **1**, 43–58.

Fukuzaki, H., Yoshida, M., Asano, M., Aiba, Y., Kumakura, M. (1990) Direct copolymerization of glycolic acid with lactones in the absence of catalysts, *Eur. Polym. J.* **26**, 457–461.

Gilbert, R. D., Stannett, V., Pitt, C. G., Schindler, A. (1992) The design of biodegradable polymers, in: *Development in Polymer Degradation* (Grassie, N., Ed.), London: Applied Science, Vol. IV, 259–293

Granger, P., Chabot, F., Chapelle, S., Vert, M., Unpublished data.

Gogolewski, S., Jovanovic, M., Perren, S. M., Dillon, J. G., Hughes, M. K. (1993) Tissue response and *in vivo* degradation of selected polyhydroxyacids: Polylactides (PLA), poly(3-hydroxybutyrate-*co*-3-hydroxyvalerate) (PHB/VA), *J. Biomed. Mater. Res.,* **27**, 1135–1148.

Grizzi, I., Garreau, H., Li, S., Vert, M. (1995) Biodegradation of devices based on poly(DL-lactic acid): size dependence, *Biomaterials* **16**, 305–311.

Guilding, D. K., Reed, A. M. (1979) Biodegradable polymers for use in surgery – Polyglycolic/poly(lactic acid) homo- and copolymers: 1, *Polymer* **20**, 1459–1464.

Hakkarainen, M., Karlsson, S., Albertsson, A.-C. (2000) Rapid (bio)degradation of polylactide by mixed culture of compost microorganisms—low molecular weight products and matrix changes, *Polymer* **41**, 2331–2338.

Herrman, J. B., Kelly, R. J., Higgins, G. A. (1970) Polyglycolic acid sutures, *Arch. Surg.* **100**, 486–490.

Higgins N. A. (1950) Condensation polymers of hydroxy-acetic acid, US patent 2676945, Appl. October 18.

Holder, Jr., W. D., Gruber, H. E., Moore, A. L., Culberson, C. R., Anderson, W., Burg, K. J. L., Mooney, D. J. (1998) Cellular in growth and thickness changes in poly-L-lactide and polyglycolide matrices implanted subcutaneously in the rat, *J. Biomed. Mater. Res.* **41**, 412–421.

Holland, S. J., Tighe, B. J., Gould, P. L. (1986) Polymers for biodegradable medical devices. 1. The potential of polyesters as controlled macromolecular release systems, *J. Control. Rel.* **4**, 155–180.

Hollinger, J. O. (1983) Preliminary report on the osteogenic potential of a biodegradable copolymer of polylactide (PLA) and polyglycolide (PGA), *J. Biomed. Mater. Res.* **17**, 71–82.

Holten, C. H. (1971) Intermolecular esters, in: *Lactic acid* (Holton, C. H., Ed.), Weinheim: Verlag Chemie, 221–224.

Ignatius, A. A., Claes, L. E. (1996) In vitro biocompatibility of bioresorbable polymers: poly(L,DL-lactide) and poly(L-lactide-*co*-glycolide), *Biomaterials* **17**, 831–839.

Ishaug, S. L., Yashemski, M. J., Bizios, R., Mikos, A. G. (1994) Osteoblast function on synthetic biodegradable polymers, *J. Biomed. Mater. Res.* **28**, 1445–1453.

Kenley, R. A., Ott Lee, M., Mahoney, T. R., Sanders, L. (1987) Poly(lactide-*co*-glycolide) decomposition kinetics *in vivo* and *in vitro*, *Macromolecules* **20**, 2398–2403.

Kleine, J. and Kleine, H. (1959) Über hochmolekulare, insbesondere optisch aktive Polyester der Milchsäure, ein Beitrag zur Stereochemie makromolekularer Verbindungen, *Makromol. Chem.* **30**, 23–38.

Kohn, F. E., van Ommen, J. G., Feijen, J. (1983) The mechanism of the ring-opening polymerization of lactide and glycolide, *Eur. Polym. J.* **12**, 1081–1088.

Kowalski, A., Duda, A, Penczek, S. (2000) Kinetics and mechanism of cyclic esters polymerization initiated with tin(II) octoate. 3. Polymerization of L,L-dilactide, *Macromolecules* **33**, 7359–7370.

Kricheldorf, H. R., Kreiser, I. (1987) Polylactones. 13. Transesterification of poly(L-lactide with poly(glycolide), poly(β-propiolactone) and poly(ε-caprolactone) *J. Macromol. Sci. Chem.* **A24**, 1345–1358.

Kricheldorf, H. R., Kreiser-Saunders, I., Stricker, A. (2000) Polylactones 48. Sn-Oct₂-initiated poly-merizations of lactide: a mechanistic study, *Macromolecules* **33**, 702–709.

Kronenthal, R. L. (1974) Biodegradable polymers in medicine and surgery, in: *Polymers in Medicine and Surgery* (Kronenthal, R. L., User, Z., Martin, E., Eds.), New York: Plenum Press, 119–137.

Kulkarni, R. K., Moore, E. G., Hegyeli, A. F., Leonard, F. (1971) Biodegradable poly(lactic acid) polymers. *J. Biomed. Mater. Res.* **5**, 169–181.

Kulkarni, R. K., Pani, K. C., Neuman, C., Leonard, F. (1966) Polylactic acid for surgical implants, *Arch. Surg.* **93**, 839–843.

Lee, K. H., Chu, C. C. (2000) The role of superoxide ions in the degradation of synthetic absorbable sutures, *J. Biomed. Mater. Res.* **49**, 25–35.

Leenslag, J. W., Pennings, A. J. (1987) Synthesis of high molecular weight poly(L-lactide) initiated by tin-2-ethyl hexanoate, *Makromol. Chem.* **188**, 1809–1814.

Leray, J., Vert, M., Blanquaert, D. (1976) Nouveau matériau de prothèse osseuse et son application, FR7628629.

Li, S. (1999) Hydrolytic degradation characteristics of aliphatic polyesters derived from lactic and glycolic acids, *J. Biomed. Mater. Res. (Appl. Biomater.)* **48**, 342–353.

Li, S., Vert, M. (1995) Biodegradation of aliphatic polyesters, in: *Biodegradable Polymers, Principles and Applications* (Scott, G. and Gilead, D., Eds.), London: Chapman & Hall, 43–87.

Li, S. M., Vert, M. (1996) Hydrolytic degradation of coral/poly(DL-lactic acid) bioresorbable materials, *J. Biomater. Sci.: Polym. Ed.* **7**, 817–827.

Li., S., Vert., M. (1999) Biodegradable polymers: polyesters, in: *The Encyclopedia of Controlled Drug Delivery* (Mathiowitz, E., Ed.), New York: John Wiley & Sons, 71–93.

Li, S., Garreau, H., Vert, M. (1990a) Structure–property relationships in the case of degradation of solid aliphatic poly(α-hydroxy acids) in aque-ous media: 3. Amorphous and semi-crystalline PLA 100, *J. Mater. Sci. Mater. Med.* **1**, 198–206.

Li, S., Garreau, H., Vert, M. (1990b) Structure–property relationships in the case of degradation of solid aliphatic poly(α-hydroxy acids) in aque-ous media: 2. PLA37.5GA25 and PLA75GA25 copolymers, *J. Mater. Sci. Mater. Med.* **1**, 131–139.

Li, S. M., Garreau, H., Vert, M. (1990c) Structure–property relationships in the case of the degra-dation of massive aliphatic poly-(alphahydroxy acids) in aqueous media: part 1: poly(DL-lactic acid), *J. Mater. Sci. Mater. Med.* **1**, 123–130.

Li, S. M., Girod-Holland, S., Vert, M. (1996) Hydro-lytic degradation of poly(DL-lactic acid) in the presence of caffeine base, *J. Control. Rel.* **40**, 41–53.

Lillie, E., Schulz, R. C. (1975) ¹H and ¹³C-{¹H} NMR spectra of stereocopolymers of lactide, *Makromol. Chem.* **176**, 1901–1906.

Lowe C. H. (1954) Preparation of high molecular weight polyhydroxyacetic ester, US patent 2668162, Appl. March 24, 1952.

Mainil-Varlet, P., Curtis, R., Gogolewski, S. (1997) Effect of *in vivo* and *in vitro* degradation on molecular and mechanical properties of various low-molecular weight polylactides, *J. Biomed. Mater. Res.*, **36**, 360–380.

Mauduit, J., Bukh, N., Vert, M. (1993a) Gentamy-cin/poly(lactic acid) mixtures aimed at sustained release local antibiotic therapy administered per-operatively: II. The case of gentamycin sulfate in high molecular weight poly(DL-lactic acid) and poly(L-lactic acid), *J. Control. Rel.* **23**, 221–230.

Mauduit, J., Bukh, N., Vert, M. (1993b) Gentamy-cin/poly(lactic acid) blends aimed at sustained release local antibiotic therapy administered per-operatively. III. The case of gentamycin sulfate in films prepared from high and low molecular weight poly(DL-lactic acids), *J. Control. Rel.* **25**, 43–49.

Mauduit, J., Perouse, E., Vert, M. (1996) Hydrolytic degradation of films prepared from blends of high and low molecular weight poly(DL-lactic acid)s. *J. Biomed. Mater. Res.* **30**, 201–207.

McDonald, R. T., McCarthy, S. P., Gross, R. A. (1996) Enzymatic degradability of poly(lactide): effects of chain stereochemistry and material crystallinity, *Macromolecules* **29**, 7356–7361.

McNeil, I. C., Leiper, H. A. (1985) Degradation studies of some polyesters and polycarbonates— 1. Polylactide: general features of the degradation under programmed heating conditions, *Polym. Degrad. Stab.* **11**, 267–285.

Miller, R. A., Brady, J. M., Cutright, D. E. (1977) Degradation rates of oral resorbable polyesters for orthopaedic surgery, *J. Biomed. Mater. Res.* **11**, 711–719.

Nieuwenhuis, J. (1992) Synthesis of polylactides, polyglycolides and their copolymers, *Clin. Mater.* **10**, 59–67.

Nishida, H., Yamashita, M., Nagashima, M., Hat-tori, N., Endo, T., Tokiwa, Y. (2000) Theoretical prediction of molecular weight on autocatalytic random hydrolysis of aliphatic polyesters, *Mac-romolecules* **33**, 6595–6601.

Pitt, C. G., Gratzel, M. M., Kimmel, G. L., Surles, J., Schindler, A. (1981) Aliphatic polyesters. 2. The degradation of poly(DL-lactide), poly (ε-caprolactone) and their complexes *in vivo*. *Biomaterials* **2**, 215–220.

Reed, A. M. (1978) *In vivo* and *in vitro* studies of biodegradable polymers for use in medicine, PhD Thesis, Liverpool University.

Reed, A. M., Guilding, D. K. (1981) Biodegradable polymers for use in surgery—poly(glycolic)/poly(lactic acid) homo and copolymers, *Polymer* **22**, 494–498.

Reeve, M. S., McCarthy, S. P., Downey, M. J., Gross, R. A. (1994) Polylactide stereochemistry: effect on enzymatic degradability, *Macromolecules* **27**, 825–831.

Salthouse, T. N., Matlaga, B. F. (1976) Polyglactin 910 suture absorption and the role of cellular enzymes, *Surg. Gynecol. Obstet.* **142**, 544–550.

Schakenraad, J. M., Hardink, M. J., Feijen, J., Molenaar, I., Nienwenhuis, P. (1990) Enzymatic activity toward poly(L-lactic acid) implants, *J. Biomater. Med. Res.* **24**, 529–545.

Schmitt, E. A., Flanagan, D. R., Lindhart, R. J. (1994) Importance of distinct water environments in the hydrolysis of poly(DL-lactide-*co*-glycolide), *Macromolecules* **27**, 743–748.

Schmitt, E. E., Polistina R. A. (1967) Surgical sutures, US3297033.

Schmitt, E. E., Polistina R. A. (1973) Polyglycolic acid prosthetic devices, US patent 3297033.

Schwach, G., Vert, M. (2000) *In vitro* and *in vivo* degradation of lactic acid-based interference screws used in cruciate ligament reconstruction, *Int. J. Biol. Macromol.* **25**, 283–291.

Schwach, G., Coudane, J., Engel, R., Vert, M. (1997) More about the initiation mechanic of lactide polymerization in the presence of stannous octoate, *J. Polym. Sci.: Polym. Chem.* **35**, 3431–3440.

Schwach, G., Coudane, J., Engel, R., Vert, M. (1998a) Ring opening polymerization of DL-lactide in the presence of zinc-metal and zinc-lactate. *Polym. Int.* **46**, 177–182.

Schwach, G., Coudane, J., Engel, R., Vert, M. (1998b) Something new in the field of PGA/LA bioresorbable polymers, *J. Control. Rel.* **53**, 85–92.

Schwach, G., Coudane, J., Engel, R., Vert, M. (1996a) Zn lactate as initiator of DL-lactide ring opening polymerization and comparison with Sn octoate, *Polym. Bull.* **37**, 771–776.

Schwach, G., Coudane, J., Vert, M., Huet-Olivier, J. (1996b) Catalyseur et composition catalytique pour la fabrication d'un polymère biocompatible résorbable, et procédés les mettant en œuvre, FR9602140.

Schwach, G., Coudane, J., Vert, M., *Biomaterials*, in press.

Schwach, G., Engel, R., Coudane, J., Vert, M. (1994) Stannous octoate versus zinc-initiated polymerization of racemic lactide: effect of configurational structures, *Polym. Bull.* **32**, 617–623.

Sedel, L., Chabot, F., Christel, P., de Charantenay, F. X., Leray, J., Vert, M. (1978) Les implants biodégradables en chirurgie orthopédique, *Rev. Chir. Orthop.* **64**, 92–101.

Sharma, C. P., Williams, D. F. (1981) The effects of lipids on the mechanical properties of polyglycolic acid structures, *Eng. Med.* **10**, 8–10.

Tabata, Y., Ikada, Y. (1988) Macrophage phagocytosis of biodegradable microspheres composed of L-lactic acid/glycolic acid homo and copolymers, *J. Biomed. Mater. Res.* **22**, 837–858.

Tanzi, M. C., Verderio, P., Lampugnani, M. G., Resnati, M., Dejana, E., Sturanie E. (1994) Cytotoxicity of some catalysts commonly used in the synthesis of copolymers for medical use, *J. Mater. Sci.: Mater. Med.* **5**, 393–401.

Therin, M., Christel, P., Li, S., Garreau, H., Vert, M. (1992) Degradation of massive poly(α-hydroxy acids) in aqueous living medium: *in vivo* validation of *in vitro* findings, *Biomaterials* **13**, 594–600.

Törmälä, P., Vasenius, J., Vainionpää, S., Laiho, J., Pohjonen, T., Rokkanen, P. (1991) Ultra high strength absorbable self-reinforced polyglycolide (SR-PGA) composite rods for internal fixation of bone fractures: *in vitro* and *in vivo* study, *J. Biomed. Mater. Res.* **25**, 1–22.

Torres, A., Li, S., M., Roussos, S., Vert, M. (1996a) Degradation of L- and DL-lactic acid oligomers in the presence of *Fusarum moniliforme*, *J. Environ. Polym. Degrad.* **4**, 213–223.

Torres, A., Li, S., M., Roussos, S., Vert, M. (1996b) Poly(lactic acid) degradation in soil or under controlled conditions, *J. Appl. Polym. Sci.* **62**, 2295–2302.

Torres, A., Roussos, S., Li, S., M., Vert, M. (1996c) Screening of microorganisms for biodegradation of poly(lactic acid) and lactic acid-containing polymers, *Appl. Environ. Microbiol.* **62**, 2393–2397.

Vert, M. (1990) Degradation of polymeric biomaterials with respect to temporary applications, in: *Degradable Materials*, Boca Raton, FL: CRC Press, 11–37.

Vert, M. (1998) Bioresorbable synthetic polymers and their operation field, in: *Biomaterials in Surgery* (Walenkamp, G., Ed.), Stuttgart: Thieme, 97–101.

Vert., M., Chabot, F., Leray, J., Christel, P. (1978) Nouvelles pièces d'ostéosynthèse, leur préparation et leur application, FR7829878.

Vert, M., Chabot, F., Leray, J., Chrsitel, P. (1981) Bioresorbable polyesters for bone surgery. *Makromol. Chem. Suppl.* **5**, 30−41.

Vert, M., Christel, P., Chabot, F., Leray, J. (1984) Bioresorbable plastic materials for bone surgery, in: *Macromolecular Biomaterials* (Hastings G. W., Ducheyne, P., Eds.), Boca Raton, FL: CRC Press, 119−141.

Vert, M., Li, S. M., Spenlehauer, G., Guerin, P. (1992) Bioresorbability and biocompatibility of aliphatic polyesters. *J. Mat. Sci. Mat. Med.* **3**, 432−446.

Vert, M., Mauduit, J., Li, S. M. (1994a) Biodegradation of PLA/GA polymers: increasing complexity, *Biomaterials* 15, 1209−1213.

Vert, M., Li, S., Garreau, H. (1994b) Attempts to map structure and degradation characteristics of aliphatic polyesters derived from lactic and glycolic acids, *J. Biomater. Sci.: Polym. Ed.* **6**, 639−649.

Vert, M., Torres, A., Li, S. M., Roussos, S., Garreau, H. (1994c) The complexity of the biodegradation of poly(2-hydroxy acid)-type aliphatic polyesters, in: *Biodegradable Plastics and Polymers* (Doi, Y. and Fukuda, K., Eds.), Amsterdam: Elsevier, pp. 11−23

Vert, M., Li. S., Garreau, H. (1995) Recent advances in the field of lactic-acid/glycolic acid polymer-based therapeutic systems *Macromol. Symp.* **98**, 633−642.

Vert, M., Schwach, G., Engel, R., Coudane, J. (1998) Something new in the field of PLA/GA bioresorbable polymers. *J. Control. Rel.* **53**, 85−92.

Williams, D. F. (1981) Enzyme hydrolysis of polylactic acid, *Eng. Med.* **10**, 5−7.

Williams, D. F., Mort, E. (1977) Enzyme accelerated hydrolysis of polyglycolic acid, *J. Bioeng.* **1**, 231−238.

Zaikov, G. E. (1985) Quantitative aspects of polymer degradation in the living body, *J. Macromol. Sci., Rev. Macromol. Chem. Phys.* **C25**, 551−597.

Zhang, X., MacDonald, D. A., Goosen, M. F. A., Mcauley, K. B. (1994) Mechanism of lactide polymerization in the presence of stannous octoate: the effect of hydroxy and carboxylic acid substances, *J. Polym. Sci.: Polym. Chem.* **32**, 2965−2970.

7
Polyanhydrides

Dr. Neeraj Kumar[1], Prof. Ann-Christine Albertsson[2], Dr. Ulrica Edlund[3],
Dr. Doron Teomim[4], Aliza Rasiel[5], Prof. Abraham J. Domb[6]

[1] Department of Medicinal Chemistry and Natural Products, School of Pharmacy,
The Hebrew University of Jerusalem, Jerusalem-91120, Israel;
Tel.: +972-2-6757573; Fax: +972-2-6757629; E-mail: nk31@mailcity.com

[2] Department of Polymer Technology, Royal Institute of Technology,
S-100 44 Stockholm, Sweden; Tel.: +46-8-7908274; Fax: +46-8-100775;
E-mail: aila@polymer.kth.se

[3] Department of Polymer Technology, Royal Institute of Technology,
S-100 44 Stockholm, Sweden; Tel.: +46-8-7908274; Fax: +46-8-100775;
E-mail: aila@polymer.kth.se

[4] Department of Medicinal Chemistry and Natural Products, School of Pharmacy,
The Hebrew University of Jerusalem, Jerusalem-91120, Israel;
Tel.: +972-2-6757573; Fax: +972-2-6757629; E-mail: doront@shellcase.com

[5] Police Headquarters, Forensic Department, Jerusalem;
E-mail: magoraia@012.net.il

[6] Department of Medicinal Chemistry and Natural Products, School of Pharmacy,
The Hebrew University of Jerusalem, Jerusalem-91120, Israel;
Tel.: +972-2-6757573; Fax: +972-2-6758959; E-mail: adomb@cc.huji.ac.il

ACDA	acetylenedicarboxylic acid
a_N	Hyperfine splitting constant
BTC	1,3,5-benzenetricarboxylic acid
Co	drug loading in g mL^{-1}
CPH	1,3-bis(p-carboxyphenoxy)hexane
CPP	1,3-bis(p-carboxyphenoxy)propane
PEG	polyethylene glycol
DMF	N,N- dimethylformamide
DSC	differential scanning calorimetry
EPR	electron paramagnetic resonance
EPRI	electron paramagnetic resonance imaging
EAD	euracic acid dimer
FA	fumaric acid
FAD	fatty acid dimer
FDA	Food and Drug Administration
g-value	Zeeman splitting constant for free electrons
Gelfoam®	absorbable gelatin sponge
Gliadel™	polyanhydride brain tumor implant containing BCNU
GPC	gel permeation chromatography
IPA	isophthalic acid
L_n	average length of sequence

MIT	Massachusetts Institute of Technology
M_w	weight average molecular weight
Mn	number average molecular weight
MRI	magnetic resonance imaging
MTX	methotrexate
NMR	nuclear magnetic resonance
PLA	poly(lactic acid)
PCL	poly(caprolactone)
PHB	poly(hydroxybutyrate)
PSA	poly(sebacic acid)
PTMC	poly(trimethylene carbonate)
PAA	poly(adipic acid)
P(RAS)	poly(ricinoleic acid succinate)
SA	sebacic acid
SEM	scanning electron microscopy
Septicin™	polyanhydride antibacterial bone implant
STDA	4,4'-stilbendicarboxylic acid
Surgicel®	oxidized cellulose absorbable hemostat
TA	terephthalic acid
T_g	glass transition temperature
TMA-gly	trimellitimide-glycine
ToF-SIMS	time-of-flight secondary ion mass spectroscopy
Vicryl®	synthetic absorbable suture
Xc	degree of crystallinity
XPS	X-ray photoelectron spectroscopy

1
Introduction

Polyanhydrides are emerging as a new class of biodegradable polymers for drug delivery, and have been successfully utilized clinically as a carrier vehicle for a number of drugs (Domb et al., 1999; Teomim et al., 1999; Stephens et al., 2000). Polyanhydrides have a hydrophobic backbone with a hydrolytically labile anhydride linkage, such that hydrolytic degradation can be controlled by manipulation of the polymer composition. They are of major interest because they show no evidence of inflammatory reactions, and are hydrolytically unstable and degraded both *in vitro* and *in vivo*, with release of nonmutagenic and noncytotoxic products (Leong et al., 1986a,b). Polyanhydrides are biocompatible, and have excellent controlled release characteristics (Tamargo et al., 1989; Chasin et al., 1990).

Pharmaceutical research has been focused on polyanhydrides derived from sebacic acid (SA), 1,3-bis(*p*-carboxyphenoxy) propane (CPP) and fatty acid dimer (FAD). Recently, the Food and Drug Administration (FDA) has approved the use of the polyanhydride poly[sebacic acid-*co*-1,3-bis(*p*-carboxyphenoxy) propane] to deliver the chemotherapeutic agent for treatment of brain cancer (Dang et al., 1996). The introduction of an imide group into polyanhydrides enhances the mechanical properties of polymers, thus enabling them to be used in orthopedics applications (Uhrich et al., 1995, 1997; Muggli et al., 1999; Young et al., 2000), while the presence of PEG groups in polyanhydrides increases their hydrophilicity and sets up the polymer for rapid drug release applications (Jiang and Zhu, 2000).

The main limitation of polyanhydrides is their storage stability, as they must be stored under refrigeration. Several reviews have been published on polyanhydrides for controlled drug delivery applications (Leong et al., 1989; Domb et al., 1993, 1997; Laurencin et al., 1995). This chapter reviews the chemistry, degradation behavior, biocompatibility, and medical uses of polyanhydrides.

2
Historical Outline

Bucher and Slade (1909) were the first to report on the synthesis of polyanhydrides, and some years later, Hill and Carothers (1930, 1932) synthesized aliphatic polyanhydrides and studied the behavior of diacids towards anhydride formation. These authors prepared super-anhydrides from homologous aliphatic dicarboxylic acid and used them to spin fibers with good mechanical strength. Conix (1958) reported polyanhydrides derived from aromatic acids, and found that these compounds were hydrolytically stable and had excellent film- and fiber-forming properties. Conix proposed a general method for the preparation of aromatic polyanhydrides as shown below:

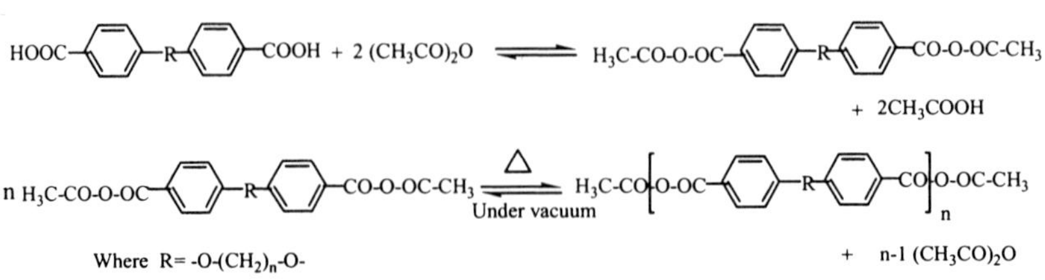

Conix subsequently studied a large number of aromatic polyanhydrides and found that they had glass transition temperatures (T_g) in the range of 50 to 100 °C, were transformed into opaque porcelain-like solids, and had a resistance to hydrolysis even on exposure to alkaline solution. Yoda (1962, 1963) introduced a new class of heterocyclic crystalline compounds into the polyanhydride family. He synthesized various types of five-membered heterocyclic dibasic acids, and polymerized these compounds with acetic anhydride at 200–300 °C under vacuum and nitrogen atmosphere. The heterocyclic polymers thus obtained had melting points in the range 70–190 °C, and good fiber- and film-forming properties. Consequently, extensive research was undertaken to enhance the stability against hydrolysis and to increase the fiber- and film-forming properties. Aliphatic polyanhydrides were considered as the least important compounds due to their unstable nature against hydrolysis. In 1980, Langer was the first to exploit the hydrolytically unstable nature of the polyanhydrides for the sustained release of drugs in controlled drug delivery applications (Rosen et al., 1983), and used these compounds as biodegradable carriers in various medical devices.

3
Synthesis

Polyanhydrides have been synthesized by melt condensation of activated diacids (Domb and Langer, 1987), ring-opening polymerization (ROP), dehydrochlorina-

HOOC—⬡—R—⬡—COOH + 2 (CH₃CO)₂O ⇌ H₃C-CO-O-OC—⬡—R—⬡—CO-O-OC-CH₃

+ 2CH₃COOH

n H₃C-CO-O-OC—⬡—R—⬡—CO-O-OC-CH₃ ⇌ (Δ, Under vacuum) H₃C-CO-[O-OC—⬡—R—⬡—CO]ₙ-O-OC-CH₃

+ n-1 (CH₃CO)₂O

Where R= -O-(CH₂)ₙ-O-tion, and dehydrative coupling agents (Leong et al., 1987; Domb et al., 1988). In general, solution polymerization yielded low-molecular weight polymers. The most widely used method is the melt condensation of dicarboxylic acids treated with acetic anhydride:

$$HOOC\text{-}R\text{-}COOH + (CH_3\text{-}CO)_2O \xrightarrow{\text{Reflux}}$$
$$CH_3\text{-}CO\text{-}(O\text{-}CO\text{-}R\text{-}CO\text{-})_mO\text{-}CO\text{-}CH_3$$
$$(I)$$

$$(I) \xrightarrow{180° C/<1 \text{ mmHg}}$$
$$CH_3\text{-}CO\text{-}(O\text{-}CO\text{-}R\text{-}CO\text{-})_nO\text{-}CO\text{-}CH_3$$

$$m = 1 - 20; n = 100 - 1000$$

The polycondensation takes place in two steps. In the first step, dicarboxylic acid monomers are reacted with excess acetic anhydride to form acetyl-terminated anhydride prepolymers with a degree of polymerization (Dp) ranging from 1 to 20; these are then polymerized at elevated temperature under vacuum to yield polymers with Dp ranging from 100 to over 1000. Acetic acid mixed anhydride prepolymers were also prepared from the reaction of diacid monomers with ketene or acetyl chloride.

The condensation reaction of diacetyl mixed anhydrides of aromatic or aliphatic diacids is carried out in the temperature range of 150 to 200 °C. Recently, we reported the synthesis of polyanhydride from ricinoleic acid half-esters with maleic and succinic anhydride (Teomim et al., 1999). We prepared ricinoleic acid maleate or succinate diacid half-esters which were polymerized by melt condensation (Figure 1).

A variety of catalysts has been used in the synthesis of a range of polyanhydrides. Significantly higher molecular weights, with shorter reaction time, were achieved by utilizing cadmium acetate, earth metal oxides, and $ZnEt_2\text{-}H_2O$. Except for calcium carbonate, which is a safe natural material, the use of these catalysts for the production of medical-grade polymers is limited because of their potential toxicity.

Polyanhydrides can be synthesized by melt condensation of trimethylsilyl dicarboxylates and diacid chlorides to yield polymers with an intrinsic viscosity 0.43 dl g^{-1} (Gupta, 1988). Direct polycondensation of adipic acid at a high temperature under vacuum resulted in the low-molecular weight oligomers (Knobloch and Ramirez, 1975).

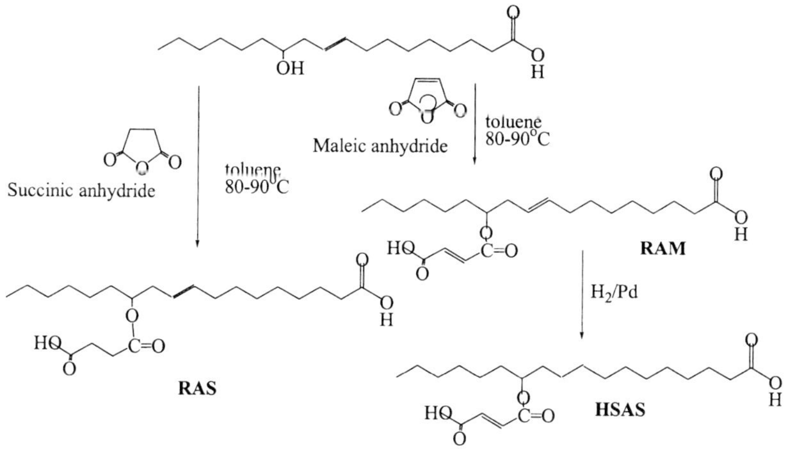

Fig. 1 Synthesis of ricinoleic acid-based monomers.

ROP offers an alternate approach to the synthesis of polyanhydrides used for medical applications. Albertsson and coworkers prepared adipic acid polyanhydride from cyclic adipic anhydride (oxepane-2,7-dione) using cationic (e.g.$AlCl_3$ and $BF_3.(C_2H_5)_2O$), anionic (e.g. $CH_3COO^- K^+$ and NaH), and coordination-type inhibitors such as stannous-2-ethylhexanoate and dibutyltin oxide (Albertsson and Lundmark, 1988, 1990; Lundmark et al., 1991). ROP takes place in two steps: (1) preparation of the cyclic monomer; and (2) polymerization of the cyclic monomers. These authors synthesized oxipane-2,7-dione (Figure 2), a cyclic monomer of adipic acid by heating the reaction mixture containing 15 g (0.1026 mol) adipic acid in 150 mL acetic anhydride for 4 h in a nitrogen atmosphere, the excess of acetic anhydride and acetic acid formed during the course of the reaction being removed by vacuum distillation. The residue was transferred to a Claisen flask, followed by addition of depolymerization catalyst $[(CH_3COO)^{2-}$ Zn. $2H_2O]$ and heated under vacuum. After removal of residual acetic anhydride at 0.9 mbar and 250 °C, $ZnCl_2$ (1 wt.%) was added at room temperature to a solution of 6 g (0.05 mol) adipic anhydride in 50 mL methylene chloride. After 6 h, the reaction was terminated by pouring the reaction mixture into dry ether to precipitate the polymer. The white precipitate thus obtained was filtered, washed with dry ether, and dried.

A variety of solution polymerizations at ambient temperature has been reported (Leong et al., 1987; Domb et al., 1988). Partial hydrolysis of terephthalic acid chloride in the presence of pyridine as an acid acceptor, yielded a polymer of *molecular weight* 2100. The use of N,N-bis(2-oxo-3-oxazolidinyl)phospharamido chloride, dicyclohexylcarbodiimide, chlorosulfonyl isocyanate, and phosgene as coupling agents produced low-molecular weight polymers. Furthermore, homo- and copolyanhydrides have also been synthesized via aqueous and nonaqueous interfacial reaction conditions, albeit with limited success. Various aromatic polymers were prepared by the phase transfer reaction having equimolar amounts of acid dissolved in aqueous base, and corresponding diacid chlorides dissolved in organic solvent (Subramanyam and Pinkus, 1985; Leong et al., 1987; Domb et al., 1988). The copolymers thus obtained present a regularly alternating structure. In the reaction between sebacoyl chloride in chloroform and sodium salt of isophthalic acid in water, a copolymer with large number of SA units was obtained. This result can be explained on the basis of a side reaction that occurs between SA and the product formed in the immiscible organic phase; the result was a polymer having a high SA content.

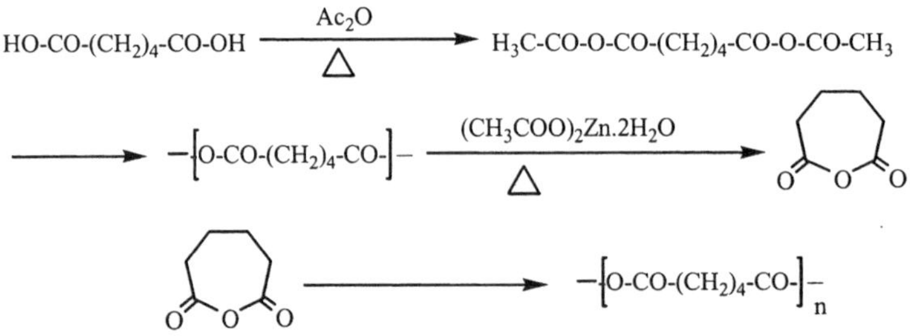

Fig. 2   Ring-opening polymerization mechanism of poly(adipic acid)anhydride.

4
Polyanhydride Structures

Since the introduction of polyanhydrides into the regime of polymers, hundreds of polyanhydride structures have been reported (Mark et al., 1969). A representative list of polymers is shown in Table 1. Polyanhydrides developed for intended use in medicine are described below.

Tab. 1 Representative polyanhydrides synthesized after 1909

Polymer structure	Melting point (°C)
$[-OC-\langle\bigcirc\rangle-COO-]$	400
$[-OC-\langle\bigcirc\rangle-COO-]$	256
$[OC-H_2C-\langle\bigcirc\rangle-CH_2-COO-]$	91
$[-C(=O)-\langle\bigcirc\rangle-C(=O)-O-(CH_2)_2-O-C(=O)-\langle\bigcirc\rangle-C(=O)-O-]$	250
$[-C(=O)-\langle\bigcirc\rangle-C(=O)-NH(CH_2)-NH-C(=O)-\langle\bigcirc\rangle-C(=O)-O-]$	330
$[-C(=O)-\langle\bigcirc\rangle-(CH_2)_6-\langle\bigcirc\rangle-C(=O)-O-]$	151
$[-OC-CH_2-CH_2-S-\langle\bigcirc\rangle-S-CH_2-CH_2-COO-]$	91
$[-OC-H_2C-O-\langle\bigcirc\rangle-O-CH_2-COO-]$	160
$[-OC-CH_2-CH_2-\langle O \rangle-CH_2-CH_2-COO-]$	67
$[-OC-CH_2-CH_2-\langle S \rangle-CH_2-CH_2-COO-]$	78
$[-OC-CH_2-CH_2-\langle N-CH_3 \rangle-CH_2-CH_2-COO-]$	188
$[OC-\langle\bigcirc\rangle-C(CH_3)_2-\langle\bigcirc\rangle-COO-]$	230
$[-OC-\langle anthracene \rangle-COO-]$	> 300
$[-OC-\langle\bigcirc\rangle-C(=O)-\langle\bigcirc\rangle-COO-]$	338

Tab. 1 (cont.)

Polymer structure	Melting point (°C)
[-OC—⬡—O—⬡—COO-]	295
[-OC—⬡—CH₂—⬡—COO-]	332
[-OC—⬡—O-(CH₂)₂-O—⬡—COO-]	237
[-OC—⬡—OCH₂—⬡—CH₂-O—⬡—COO-]	140
[-OC—(naphthalene)—COO-]	450
[-OC-(CH₂)₄ CONH—C(C₆H₅)₂—NH-CO-(CH₂)₄-COO-]	> 300
[-OC-(CH₂)₄ CONH—CH₂—NH-CO-(CH₂)₄-COO-]	285
[OC-(CH₂)₂-SO₂-(CH₂)₂-SO₂-(CH₂)₂-COO]	185
[OC-(CH₂)₂-S-(CH₂)₂-S-(CH₂)₂-COO]	81
[OC-P-COO]	
[OC-(CH₂)$_x$-COO] $x = 4-16$	60–100
[—⬡—C·O·C—O- CH₂-CH₂-O-C·O·C—]	155
{[OC—⬡—COO]$_x$[OC—(CH₂)₈-COO]$_y$}$_n$	220
[C=C, C·O, —C=O]$_n$ and [C≡C, C·O, —C=O]$_n$	–
[C(=O)—⬡—O-(CH₂)x—C(=O)-O]$_n$ $x = 1-10$	< 100
[(C(=O)—⬡—C(=O)-O)$_x$-(C(=O)—⬡—C(=O)-O)$_y$]$_n$	> 200

Tab. 1 (cont.)

Polymer structure	Melting point (°C)
TA IPA	98–176
	–
	59.3
	70.4
	61.1
	71.2–77.8
{R = (CH₂)₃; (CH₂)₂-O-(CH₂)₈-O-(CH₂)₂}G	58–177
N-(CH₂)ₓ—C-O—	
—C— (CH₂)ₓ—N ... N—(CH₂)ₓ —C-O x = 1–5	
N-CH-C-O— R R = -CH₂-CH-(CH₃)₂ R = -CH2-C₆H₅-OH R = -CH₂-S-CH₂-C₆H₅	
N—(CH₂)ₓ—C-O C-(CH₂)₈—C	45 (1:1, x = 1)

4.1
Unsaturated Polymers

A series of unsaturated polyanhydrides were prepared by melt or solution polymerization of fumaric acid (FA), acetylenedicarboxylic acid (ACDA), and 4,4′-stilbendicarboxylic acid (STDA) (Domb et al., 1991). The double bonds remained intact throughout the polymerization process, and were available for a secondary reaction to form a cross-linked matrix. The unsaturated homopolymers were crystalline and insoluble in common organic solvents, whereas copolymers with aliphatic diacids were less crystalline and soluble in chlorinated hydrocarbons (Domb et al., 1991):

$$\left[CO-\!\!\!\bigcirc\!\!\!-CH=CH-\!\!\!\bigcirc\!\!\!-CO-O \right]_n$$

4.2
Aliphatic-aromatic Homopolymers

Polyanhydrides of diacid monomers containing aliphatic and aromatic moieties, poly-[(p-carboxyphenoxy)alkanoic anhydride], were synthesized by either melt or solution polymerization with *molecular weight* of up to 44,600 (Domb and Langer, 1989; Domb et al., 1989):

$$\left[\overset{O}{\underset{\|}{C}}-\!\!\!\bigcirc\!\!\!-O-(CH_2)x-\overset{O}{\underset{\|}{C}}-O \right]_n$$

$x = 1 - 10$

The polymers of carboxyphenoxy alkanoic acid having methylene groups (n = 3, 5, and 7) were soluble in chlorinated hydrocarbons, and melted at temperatures below 100 °C. These polymers displayed a zero-order hydrolytic degradation profile for 2 to 10 weeks.

The length of the alkanoic chain dictated the degradation time wherein an increasing degradation time was observed with an increasing chain length.

4.3
Soluble Aromatic Copolymers

Aromatic homopolyanhydrides are insoluble in common organic solvents, and melt at temperatures above 200 °C (Domb, 1992). These properties limit the use of purely aromatic polyanhydrides, since they cannot be fabricated into films or microspheres using solvent or melt techniques. Fully aromatic polymers that are soluble in chlorinated hydrocarbons and melt at temperatures below 100 °C were obtained by copolymerization of aromatic diacids such as isophthalic acid (IPA), terephthalic acid (TA), 1,3-bis(carboxyphenoxy)-propane (CPP), or 1,3-bis(carboxyphenoxy)-hexane (CPH).

$$\left[\left(CO-\!\!\!\bigcirc\!\!\!-CO-O \right)_n \left(CO-\!\!\!\bigcirc\!\!\!-O-CH_2CH_2CH_2-O-\!\!\!\bigcirc\!\!\!-CO-O \right)_m \right]_z$$

4.4
Poly(ester-anhydrides)

4,4′-Alkane- and oxa-alkanedioxydibenzoic acids were used for the synthesis of polyanhydrides (McIntyre, 1994). Polymers melted at a temperature range of 98 to 176 °C, and had a *molecular weight* of up to 12,900. Di- and tri-block copolymers of poly(caprolactone), poly(lactic acid), and poly(hydroxybutyrate) have been prepared from carboxylic acid-terminated low-molecular weight polymers copolymerized with SA prepolymers by melt condensation (Abuganima, 1996). Similarly, di-, tri- and brush copolymers of poly(ethylene glycol) with poly(sebacic anhydride) have been prepared by melt copolymerization of carboxylic acid-terminated PEG (Gref et al., 1995).

$$\left[\text{C} \underset{\text{O}}{\|} \bigcirc \text{-O-} \underset{\text{O}}{\overset{\|}{\text{C}}} \text{-R-} \underset{\text{O}}{\overset{\|}{\text{C}}} \text{-O-} \bigcirc \text{-} \underset{\text{O}}{\overset{\|}{\text{C}}} \text{-O} \right]_n$$

R = (CH$_2$)$_{2-8}$; (CH$_2$)$_2$-O-(CH$_2$-CH$_2$-O)$_2$-(CH$_2$)$_2$

4.5
Fatty Acid-based Polyanhydrides

Polyanhydrides were synthesized from dimer and trimer of unsaturated fatty acids; the details of this are described in Section 3 (Domb, 1993; Domb and Maniar, 1993a,b; Teomim and Domb, 1999). The dimers of oleic acid and eurecic acid are liquid oils containing two carboxylic acids available for anhydride polymerization. The homopolymers are viscous liquids, and copolymerization with increasing amounts of SA forms solid polymers with increasing melting points as a function of the SA content. The polymers are soluble in chlorinated hydrocarbons, tetrahydrofuran, 2-butanone, and acetone. Polyanhydrides synthesized from nonlinear hydrophobic fatty acid esters, based on ricinoleic, maleic acid, and SA, possessed desired physicochemical properties such as low melting point and hydrophobicity and flexibility to the polymer formed, in addition to biocompatibility and biodegradability. The polymers were synthesized by melt condensation to yield film-forming polymers with *molecular weight* > 100,000 (Domb and Nudelman, 1995a)

The properties of polyanhydrides were modified by the incorporation of long-chain fatty acid terminals such as stearic acid in the polymer composition; this alters the polyanhydride's

hydrophobicity and decreases its degradation rate (Teomim and Domb, 1999; Teomim et al., 1999b; Domb and Maniar, 1993b).

Since natural fatty acids are monofunctional, they would act as chain terminators in the polymerization process to control the molecular weight of a polymer. A detailed analysis of the polymerization reaction shows that for up to about 10 mol.% content of stearic acid, the final product is essentially a stearic acid-terminated polymer; this also applies in the case of octanoic acid and lauric acid. In contrast, the higher concentration of acetyl stearate in the reaction mixture resulted in the formation of an increasing amount of stearic anhydride byproduct, with minimal effect on the polymer molecular weight – which remains in the range of 5000. Physical mixtures of polyanhydrides with triglycerides and fatty acids or alcohols did not form uniform blends.

4.6
Amino Acid-based Polymers

General methods for the synthesis of poly(amide-anhydrides) and poly(amide-esters) based on naturally occurring amino acids have been described (Domb, 1990). The polymers were synthesized by amidation of the amino group of an amino acid with a cyclic anhydride, or by the amide coupling of two amino acids with a diacid chloride. Low-molecular weight polymers from methylene bis(*p*-carboxybenzamide) were synthesized by melt condensation (Hartmann and Schultz, 1989). A series of amido-containing polyanhydrides based on *p*-aminobenzoic acid were synthesized by melt condensation; these polymers melted at 58 to 177 °C and had *molecular weight* in the range 2500 to 12,400.

Poly(anhydride-*co*-imides) were synthesized by melt condensation polymerization (Uhrich et al., 1995; Staubli et al., 1991a,b). The trimellitic-amino acid polymers and its copolymers were extensively studied for use as drug carriers (Uhrich et al., 1995; Staubli et al., 1990, 1991a,b). The following amino acids were incorporated in a cyclic imide structure to form a diacid monomer: glycine, β-alanine, γ-aminobutyric acid, ʟ-leucine, ʟ-tyrosine, 11-aminododecanoic acid, and 12-aminododecanoic acid. During the course of the reaction, trimellitic anhydride and pyromellitic anhydride were acetylated by treating imide acid with excess of acetic anhydride, and heating at a reflux temperature until the starting material was dissolved (1 h); this was followed by cooling and drying. The reaction mixture was precipitated in cold, dry diethyl ether. The white precipitate thus obtained was re-washed with diethyl ether, dried and forwarded for melt polymerization to obtain the polymer (Domb and Langer, 1987). The homopolymers of all *N*-trimellitylimido acids containing amino acids were rigid and brittle, with *molecular weight* below 10,000 (Staubli et al., 1991a,b). Higher-molecular weight polymers were obtained by the incorporation of flexible segments, i.e., copolymers with aliphatic diacids, in the polymer backbone. Copolymers of *N*-trimellitylimido-glycine or aminodecanoic acid with either SA or 1,6-bis(*p*-carboxyphenoxy)hexane (CPH) were prepared in defined ratios. High-molecular weight copolymers ($M_w > 100,000$) were generally obtained with an increasing content of the SA or CPH comonomer.

4.7
Modified Polyanhydrides and Blends

The physical and mechanical properties of polyanhydrides can be altered by modification of the polymer structure, with a minor change in the polymer composition. Several such modifications include the formation of polymer blends, branched and cross-linked polymers, partial hydrogenation and reaction with epoxides.

Biodegradable polymer blends of polyanhydrides and polyesters have been investigated as drug carriers (Domb, 1993; Abuganima, 1996). Albertsson and coworkers blended poly(trimethylene carbonate) (PTMC) with poly(adipic anhydride) (PAA), and the matrix of PTMC-PAA blend was found biocompatible in *in vitro* and *in vivo* experiments, as well as being a promising candidate for controlled drug delivery (Edlund and Albertsson, 1999; Edlund et al., 2000). These authors suggested that the erosion rate of the blend-matrix could be controlled by varying the proportions of PTMC and PAA. In general, polyanhydrides of different structures form uniform blends with a single melting temperature. Low-molecular weight poly(lactic acid) (PLA), poly(hydroxybutyrate) (PHB), and poly(caprolactone) (PCL) are miscible with polyanhydrides, while high-molecular weight polyesters ($M_w > 10,000$) are not compatible with polyanhydrides. Uniform blends of PCL with 10 to 90% by weight of poly(dodecanedioic anhydride) (PDD) were prepared by melt mixing, and the product exhibited good mechanical strength. Differential scanning calorimetry (DSC) thermograms showed two separate peaks (at 55 °C and 75–90 °C for PCL and PDD, respectively) for all compositions. Infra-red (IR) and weight-loss measurements of the blends during hydrolysis indicated a rapid degradation of the anhydride component. After 20 days, the blends contained only PCL with some diacid degradation products, and no anhydride polymer. These studies indicated that the anhydride component degraded and released from the blend composition, without affecting the PCL degradation.

Branched and cross-linked polyanhydrides were synthesized in the reaction of diacid monomers with tri- or polycarboxylic acid branching monomers (Maniar et al., 1990). Muggli *et al.* (1999) synthesized highly cross-linked, surface-eroding polymers by the process of photopolymerization. They used anhydride monomers (e.g., SA, CPP, CPH) end-capped with methacrylate functionalities, and observed that the degradation time scale could be controlled by variation in the network composition, from about 2 days for poly(MSA) to 1 year for poly(MCPH). SA was polymerized with 1,3,5 benzenetricarboxylic acid (BTC) and poly(acrylic acid) to yield random and graft-type branched polyanhydrides. The molecular weights of the branched polymers were significantly higher (M_w 250,000) than the molecular weight of the respective linear polymer (M_w 80,000). The specific viscosities of the branched polymers were lower than those of the linear polyanhydrides, albeit with similar molecular weights. Except for the differences in molecular weight, there were no noticeable changes in the physico-chemical or thermal properties of the branched polymers and the linear polymers. Drug release was faster from the branched polymers as compared with the respective linear polymer of a comparable molecular weight.

5
Characterization

The characterization of polyanhydrides, together with data obtained for their chemical composition, structure, crystallinity and thermal properties, mechanical properties, and thermodynamic and hydrolytic stability, are summarized in this section.

5.1
Composition by ¹H-NMR

The following copolymer characteristics have been studied by ¹H-NMR (Ron et al., 1991): the degree of randomness that suggests whether the polyanhydride is either a random or block copolymer; the average length of sequence (L_n); and the frequency of occurrence of specific comonomer sequences. Copolymers of CPP and SA were used as model polymer. The protons on the aromatic ring close to the anhydride groups experience a low shielding electron density, and absorb at low frequency; in contrast, the protons next to aliphatic comonomers absorb at higher frequency. Accordingly, the CPP-CPP and CPP-SA diads were represented by peaks at 8.1 and 8.0 respectively, and the triplets at 2.6 and 2.4 represent the SA-CPP and SA-SA diads, respectively. The degree of randomness, average block length, and the probability of finding the diad SA-SA or SA-CPP were calculated by integrating the ¹H-NMR spectral peaks of poly(CPP-SA) of various compositions. Similar data analysis was also applied to aromatic-aliphatic homopolymers (Domb et al., 1989) and to copolymers of SA with FA (Domb et al., 1991), 1,3-bis(*p*-carboxyphenoxy)hexane (CPH) (Uhrich et al., 1995), and trimellitimide derivatives (Staubli et al., 1991b).

Kim et al. (2000) used a ¹H NMR spectral technique for the estimation of the percent conversion of PEG-dicarboxylate to the methacrylated derivative. This conversion was calculated by using integral ¹H-NMR intensities of the protons adjacent to anhydride (A), and those adjacent to carboxylic acid (B), in the following manner:

Conversion $\% = A/(A+B) \times 100$

These authors compared the ^{1}H-NMR spectra of PEG-succinate with dimethacrylated macromer, and observed that the chemical shifts for ethylene protons adjacent to the anhydride bond in succinate were shifted from 2.62–2.69 ppm to 2.69–2.84 ppm, corresponding to the protons in unconverted succinate. The ratio of the peak integral of two methacrylate protons to methylene protons adjacent to the ester bond in the PEG backbone was approximately 1, which was in accordance with the theoretical value.

5.2
Molecular Weight

Molecular weights of the polyanhydrides were determined by viscosity measurements and gel permeation chromatography (GPC) (Ron et al., 1991) Weight average molecular weight (M_w) of polyanhydrides were in the range of 5000 to 300,000, and a polydispersity of 2 to 15 which increased as M_w increased. The intrinsic viscosity [η] increases with the increase in M_w. The Mark-Houwink relationship for poly(CPP-SA) was calculated from the viscosity data and the M_w values, which were determined by universal calibration of the GPC data using polystyrene standards.

$$[\eta]_{CHCl_3}^{23\,°C} = 3.88 \times 10^{-7} \ M_w^{0.658}$$

The acetic acid end group determination for molecular weight estimation was not used because the polymer may contain cyclic macromolecules with no acetate end groups (Domb et al., 1991).

5.3
Crystallinity

Since crystallinity is an important factor to control the polymer erosion, an analysis was made of the effect of polymer composition on crystallinity (Ron et al., 1991; Staubli et al., 1991b; Uhrich et al., 1995). Polymers based on SA, CPP, CPH, and FA were investigated. Crystallinity was determined by X-ray diffraction, a combination of X-ray and DSC, and data generated from ^{1}H-NMR spectroscopy and Flory's equilibrium theory. Homopolyanhydrides of aromatic and aliphatic diacids were crystalline (> 50% crystallinity). The copolymers possess a high degree of crystallinity at high molar ratios of either aliphatic or aromatic diacids. The heat of fusion values for the polymers demonstrated a sharp decrease as CPP was added to SA, or vice versa. After adding one monomer, a decreasing trend in crystallinity appeared in the results of X-ray and DSC analysis, which was a direct result of the random presence of other units in the polymer chain. A detailed analysis of the copolymers of SA with the aromatic and unsaturated monomers, CPP, CPH, FA, and trimellitic-amino acid derivative was reported (Staubli et al., 1991). Copolymers with high ratios of SA and CPP, TMA-gly, or CPH were crystalline, while copolymers of equal ratios of SA and CPP or CPH were amorphous. The poly(FA-SA) series displayed high crystallinity, regardless of comonomer ratio.

5.4
Infra-red and Raman Analysis

Anhydrides present characteristic peaks in the IR and Raman spectra. In general, the carbonyl of aliphatic polymers absorb at 1740 and 1810 cm^{-1}, and aromatic polymers at 1720 and 1780 cm^{-1}. Typical IR spectra of aliphatic and aromatic polymers that contain aliphatic and aromatic anhydride bonds may present three distinct peaks, in which the aromatic peak appears at 1780 cm^{-1} and the peaks at 1720–1740 cm^{-1} show a general overlap. The presence of carboxylic acid groups in the polymer can be determined from the presence of a peak at 1700 cm^{-1}. The degradation of polyanhydrides can be estimated by the ratio of the anhydride peaks at 1810 cm^{-1} and 1700 cm^{-1}. The significance of this analysis is that one can measure the degradation of the anhydride bonds without any dissolution of the degradation products, this being dependent on the solubility of the latter.

Fourier transform (FT-IR) spectra have also been used for calculating the relative concentration of free acid to anhydride released during degradation of the polyanhydride matrix (Santos et al., 1999). In this study, the ratio of the peak area of anhydride (1860–1775 cm^{-1}) and carboxylic acid (1775–1675 cm^{-1}) bonds was plotted with time, and the concentrations of free acid were then calculated.

FT-Raman spectroscopy (FTRS) was used to characterize an homologous series of aliphatic polyanhydrides, poly(carboxyphenoxy)alkanes, and copolymers of carboxyphenoxy propane (CPP) and SA. All anhydrides show two diagnostic carbonyl bands: the aliphatic polymers have the carbonyl pairing at 1803/1739 cm^{-1}, and the aromatic polymers have the band pair at 1764 and 1712 cm^{-1} (Davies et al., 1991; Tudor et al., 1991). All the homopolymers and copolymers show different methylene bands due to deformation, stretching, rocking and twisting; the spectra for the aromatic polyanhydrides such as PCPP also have diagnostic benzene para-substitution bands. Thus, it is possible to differentiate between aromatic and aliphatic anhydrides bonding and, in conjunction with other diagnostic bands, to monitor the change in individual monomer composition within a copolymer mixture.

FTRS was also used to study the hydrolytic degradation of polyanhydrides (Davies et al., 1991). PSA rods exposed to water for 15 days were analyzed daily by FTRS. The intensity of the carbonyl anhydride band pair (1803/1739 cm^{-1}) disappeared from day 0 to 15, with the emergence of the complementary acid carbonyl band (1640 cm^{-1}) which increased in intensity over the same period. Similarly, an increase in the intensity of the C–C deformation at 907 cm^{-1} with hydrolysis reflects the gain in freedom of the methylene chain of low-molecular weight oligomers.

5.5
Surface and Bulk Analysis

The morphology of polyanhydrides was studied by scanning electron microscopy (SEM) to elucidate the mechanism of polymer degradation and drug release from polyanhydrides (Mathiowitz et al., 1990). The surface chemical structures of aliphatic polyanhydride films have been examined using time-of-flight secondary ion mass spectroscopy (ToF-SIMS) and X-ray photoelectron spectroscopy (XPS) (Davies et al., 1996). The main peak observed at 285 eV was due to the C–H bond, and the peak at 289.5 eV appeared from O–C=O. The XPS data confirmed the purity of the surface, and the elemental ratios of the experimental surface were in good agreement with the known stoichiometry of the examined polyanhydrides. ToF-SIMS spectra of the polyanhydrides show the reflection of polymer structure. Both negative- and positive-ion SIMS spectra, occurring throughout the entire series of the polyanhydrides, were examined as a result of confirmation on systematic fragmentation where radical cations were observed in the positive-ion spectra. The ion at m/z 71 may arise from fragmentation of the anhydride unit; thus, the appearance of $CH_2=CHCOO^-$ is observed for all the polyanhydrides. The combined use of ToF-SIMS and XPS provides a detailed insight into the interfacial chemical structure of polyanhydrides. Atomic force microscopy (AFM) was used to follow the degradation of poly(sebacic acid) and its blends with poly(lactide) (Shakesheff et al., 1994, 1995). These AFM studies reveal the surface polymer morphology to a resolution comparable with that achieved with vacuum-based SEM, and demonstrate the influence of a variety of factors, including polymer crystallinity and the pH of the aqueous environment on the kinetics of the degradation of biodegradable polymers and their blends.

5.6
Stability

The stability of polyanhydrides in solid state and in dry chloroform solution was reported (Domb and Langer, 1989). Aromatic polymers such as poly(CPP) and poly(CPM) maintained their original molecular weight for at least one year in the solid state. In contrast, aliphatic polyanhydrides such as PSA decreased in molecular weight over time. The decrease in molecular weight shows first-order kinetics, with activation energies of 7.5 Kcal mol^{-1}°K. The decrease in molecular weight was explained by an internal anhydride interchange mechanism, as revealed from elemental and spectral analysis. This mechanism was supported by the fact that the decrease in molecular weight was reversible, and heating of the depolymerized polymer at 180 °C for 20 min yielded the original high-molecular weight polymers. However, under similar conditions the hydrolyzed polymer did not increase in molecular weight. It was also found that, in many cases, the stability of polymers in the solid state or in organic solutions did not correlate with their hydrolytic stability. A similar decrease in molecular weight as a function of time was also observed among the aliphatic-aromatic-*co*-polyanhydrides

and imide-*co*-polyanhydrides (Domb et al., 1989, 1991; Staubli, 1991a,b).

Mader et al. (1996) have utilized the γ-radiation technique for sterilization of polyanhydrides. In this technique, aliphatic and aromatic homo- and copolymers were irradiated at 2.5 Mrad dose under dry ice, with changes in properties being monitored before and after irradiation. The mechanical and physical properties of the polymer were found to be the same before and after irradiation, as were those monitored by ^{1}H-NMR and FTIR spectra. Using the same concept, these studies were extended for the saturated and unsaturated polyanhydrides (Teomim et al., 2001). Ricinoleic acid-based copolymers with SA and P(CPP-SA) were irradiated under dry ice and at room temperature, while P(FA-SA) was irradiated only at room temperature, under the same conditions as described earlier. It was concluded that saturated polyanhydrides are sufficiently stable during γ-irradiation, while the presence of the double bond conjugated to an anhydride bond makes it unstable and leads to the formation of free radicals. These free radical polyanhydrides transform via an anhydride interchange process into less conjugated polyanhydrides, as shown in Scheme 1.

The outcome of this process was self-depolymerization via inter- and/or intramolecular anhydride interchange to form lower-molecular weight polymers, as shown in Figure 3.

In general, polymers with high melting points and high crystallinity provided a high yield of observable radicals at room temperature. These endogenous free radicals were used to study the processes of water penetration and polymer degradation *in vivo* (Mader et al., 1996). The detection of γ-irradiation sterilization-induced free radicals *in vivo* (using EPR) might be of significance in that the changes in the mobility of the

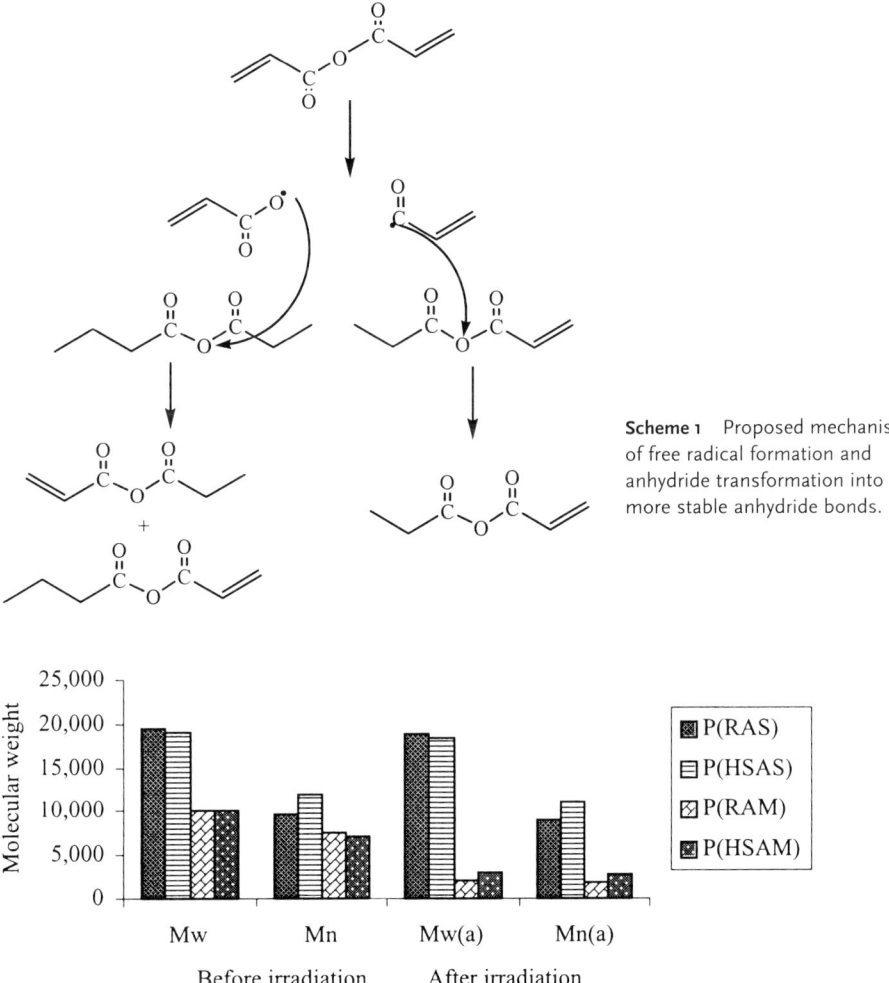

Scheme 1 Proposed mechanism of free radical formation and anhydride transformation into more stable anhydride bonds.

Fig. 3 Molecular weight changes of irradiated ricinoleic acid-based polymers.

radicals could be used as a tool to study drug-release kinetics in a noninvasive and continuous fashion, without any need to introduce paramagnetic species.

6
Fabrication of Delivery Systems

Polymers with low melting points and good solubility in common organic solvents (e.g., methylene chloride) allow easy dispersion of a drug into their matrix. Drugs can also be incorporated using either compression or melt-molding processes. For example, drugs can be incorporated into a slab either by melt mixing the drug into the melted polymer, or by solvent casting. Polymer slabs loaded with drug can also be prepared by compression-molding a powder that contains the drug. Similarly, the drug–polymer formulation can be molded into beads or rods. Polymer films can be prepared by casting the polymer solution containing the drug onto a Teflon-coated dish, followed by evaporation of solvent. Microsphere-based delivery systems can be formulated using a variety of common techniques, including solvent removal, hot-melt encapsulation, and spray drying (Mathiowitz and Langer, 1987; Bindschaedler et al., 1988; Pekarek et al., 1994). However, it is essential that all processes be performed under anhydrous conditions in order to avoid hydrolysis of the polymer.

7
In vitro Degradation and Drug Release

The degradation of polyanhydride matrices depends on many factors, including the chemical nature and hydrophobicity of the monomers used in formation of the polymer, the shape and geometry of the implants (as erosion rate depends on the surface area of the polymer), accessibility of the implants to water (as porous materials degrade rapidly in comparison with non-porous materials), the level of drug loading in the polymer matrix, and the pH of the surroundings (polymers degrade rapidly in alkaline media). However, the porosity of the implant can be changed by altering the method of fabrication. For example, a compression-molded device will degrade faster than an injection-molded device due to the former having a higher porosity in the polymer mass. There are many examples of the degradation rate of various types of polyanhydrides. Among all the polyanhydrides, the clinically tested members – poly(FAD-SA), poly(CPP-SA), poly(FA-SA) and poly(EAD-SA) – have been the focus of these studies, although other examples are also available. In general, during the initial 10 to 24 h of water incubation in aqueous medium, the molecular weight fell rapidly, with

no loss in wafer mass. However, this was followed by a rapid decrease in wafer mass, accompanied by a small change in polymer molecular weight. The period of extensive mass loss starts when the number average molecular weight (M_n) of polymer approaches 2000, regardless of the initial molecular weight of the polymer. During this period (which lasts for about one week), SA – a relatively water-soluble comonomer – is released from the wafer, leaving the less soluble comonomer, CPP or FAD, which is slow to solubilize (Dang et al., 1996). Increasing the SA content in the copolymer increases the hydrophilicity of the copolymer, which in turn results in a higher erosion rate and hence a higher drug release rate. In a similar mechanism, poly(adipic anhydride) (PAA) in PMTC/PAA blends, when incubated in an aqueous medium, degrades rapidly into its diacid monomers and diffuses out of the matrix. Typical scanning electron micrographs of PTMC/PAA (Edlund and Albertsson, 1999) blends which contain different ratios of monomers and different exposure times in water are shown in Figure 4. This situation could be explained by the fact that the anhydride linkages in the polymer are hydrolyzed subsequent to penetration of water into the polymer. The water uptake depends on the hydrophobicity of the polymer, and therefore the hydrophobic polymers that prevent water uptake have slower erosion rates and lower drug release rates. This is valuable information, as the hydrophobicity of the polymer can be altered by changing the structure and/or the content of the copolymer, thereby being able to alter the drug release rate. An increasing release of methotrexate (MTX) by increasing the SA content in the fatty acid-terminated polyanhydrides (Figure 5) was recently reported (Teomim and Domb, 1999). In the P(CPP-SA) and P(FAD-SA) series of copolymers, a 10-fold increase in

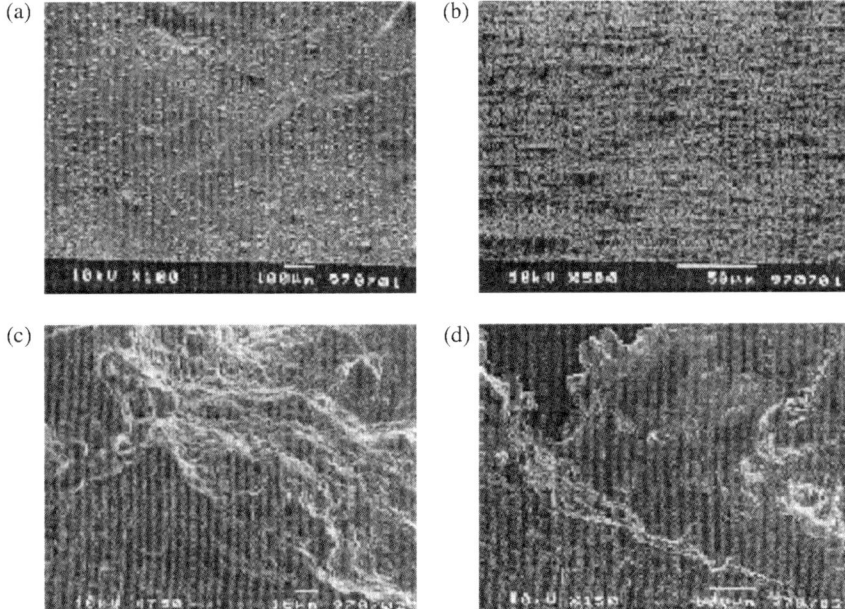

Fig. 4 Scanning electron microscopy micrographs showing the topology of PTMC-PAA films at different stages of erosion: (a) blend of 50% PTMC (medium molecular weight, MMW) and 50% PAA before incubation; (b) the same material after 2 days of water exposure; (c) the same material after 3 weeks; (d) blend of 20% PTMC (low molecular weight, LMW) and 80% PAA after 3 weeks (taken from Edlund and Albertsson, 1999).

drug release rate was achieved by alteration of the ratio of the monomers; thus, both polymers can be used to deliver drugs over a wide range of release rates.

As mentioned earlier, there is no correlation between the rate of drug release and polymer degradation expressed as % decrease in the molecular weight. At first glance, this might appear to be contradictory (D'Emanuele et al., 1994), but on closer examination it appears that a drug dispersed in the polymer matrix is released when the eroding polymer brings the drug with it into solution. Thus, the release rate would depend on the rate of erosion expressed as volume of the matrix dissolved per unit time, multiplied by the drug load, rather than the rate of polymer degradation. The implication is that drug release should correlate with weight loss, which is a more appropriate

indicator of the erosion rate than the decrease in molecular weight. Another feature of surface erosion is that the molecular weight of the polymer at the surface may decrease, while the interior of the device may retain the same molecular weight. Then, the lower-molecular weight fragments so formed may not diffuse out or dissolve into the release medium. Therefore, it is not the decrease in molecular weight but the subsequent weight loss to the diffusion and erosion of molecular weight fragments, which should correlate with drug release. This also explains why drug release from these polymer devices was independent of the initial molecular weight of the copolymer (D'Emanuele et al., 1994).

Thus, the factors that affect drug release from the polyanhydride matrix include polymer composition, fabrication method,

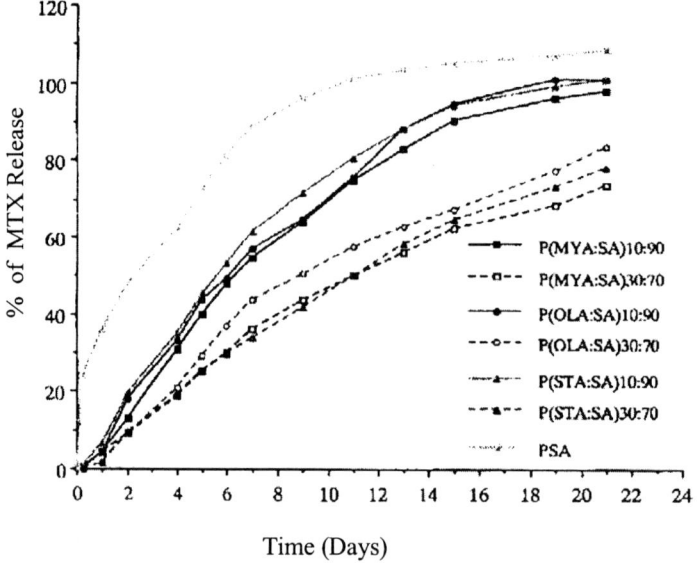

Fig. 5 *In vitro* release of methotrexate (MTX) from C_{14}-C_{18} fatty acid-terminated polymers. MTX release was conducted in phosphate buffer, pH 7.4, at 37 °C and determined by UV absorbance at 303 nm (from Teomim and Domb, 1999).

size and geometry, particle size of incorporated drug, drug solubility, and drug loading. Recently, heparin release from thin flexible sheets composed of PLA laminated polyanhydride P(EAD-SA) was reported (Teomim et al., 1999a). Drug release was observed to follow first-order kinetics, and ~50% of the drug was released within 24 h from incubated poly (EAD-SA) film. The polymer matrix coated with PLA and PLA-PEG sheets controlled the release rate, as heparin release was observed continuously for 2–3 weeks. Polyanhydride matrix coated with PLA-PEG block copolymer was more hydrophilic than the coated with PLA, and showed a faster release (Figure 6). It is observed that hydrophilic drugs are released much more rapidly than the hydrophobic drugs and suffer from a significant initial burst effect, which is a function of drug particle size. Moreover, the correlation between drug release and polymer degradation is better for injection or melt-molded devices than with compression-molded devices (Domb et al., 1994a). In a recent report, Stephens et al. (2000) investigated the *in vitro* release behavior of gentamicin from a poly(EAD-SA) matrix, and found that release rate of the drug was faster in water than in phosphate buffer, pH 7.4. The beads of polyanhydride matrix were cracked in water, and drug release was more rapid due to hydrophilicity of the drug and its diffusion through the cracked matrix. In contrast, drug release in buffer at pH 7.4 was diffusion-controlled, as no crack was observed in the bead dipped in phosphate buffer. The release of indomethacin from poly(CPP-SA) and poly(FAD-SA) was studied and found to be independent of drug loading (Gopferich and Langer, 1993). A simple model that takes into account the following kinetic steps, namely the spontaneous degradation of polymer to crystallized monomer, the creation of pores, the dissolution of monomers inside the pores, and the final release of monomer via

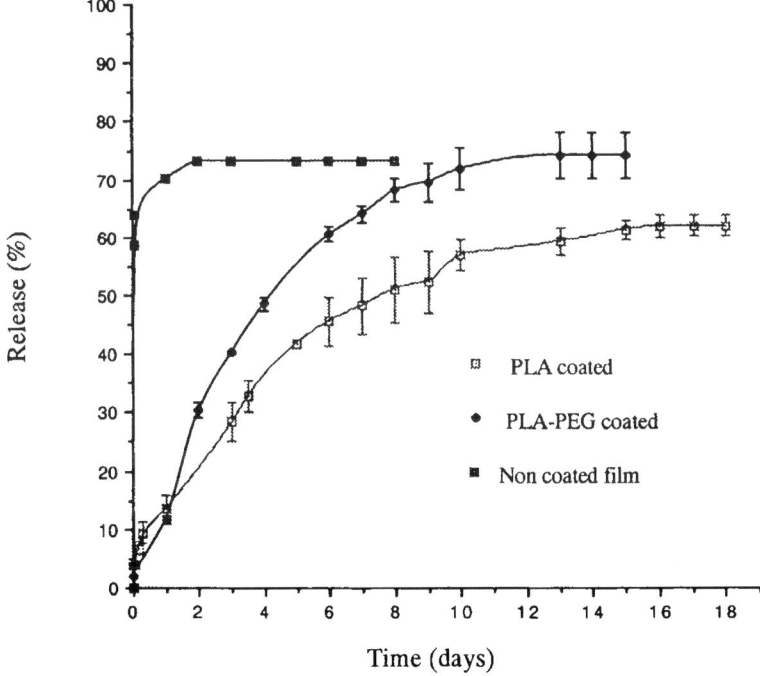

Fig. 6 *In vitro* release of heparin from laminated and coated polyanhydride films [2% (w/v) 10 x 5 x 0.1 mm³]. The release was conducted in phosphate buffer, pH 7.4, at 37 °C (from Teomim et al., 1999a).

diffusion through the pore network was proposed by Gopferich and Langer to explain this unusual release behavior (Gopferich and Langer, 1993, 1995a,b; Gopferich et al., 1995). Erosion was then simulated using a Monte Carlo method that describes these morphological changes during erosion.

Noninvasive *in vivo* and *in vitro* monitoring of drug release and polymer degradation using EPR spectroscopy was carried out by Mader et al. (1997a,b). By incorporating a nitro-oxide radical probe such as 2,2,5,5-tetramethyl-3-carboxyl-pyrrollidine-1-oxyl (PCA) at 5 mmol kg^{-1}, the microviscosity and drug mobility, as well as the pH within a device, can be monitored both *in vivo* and *in vitro* using EPR methods. Low-frequency EPR spectrometry was used to study the *in vivo* and *in vitro* degradation of polyanhydrides. Tablets loaded with PCA were ex-

posed to 0.1 M phosphate buffer pH 7.4 at 37 °C, or implanted subcutaneously into the back of rats or mice. Measurements were carried out using a standard 9.4-Ghz EPR spectrometer or a 1.1-GHz spectrometer equipped with a surface coil. EPR measurements of pH are based on the effect of protonation/deprotonation of groups located in close proximity to the radical moiety, which induces changes in the hyperfine splitting constant, a_N and the g value (the Zeeman splitting constant for free electrons). The measurements of pH values in

g = 2.0054; a_N = 1.557 mT   g = 2.0057; a_N = 1.43 mT

the range between 0 and 9 are possible by using probes with different pK_a values.

In vitro and *in vivo* studies have demonstrated the environment within the polyanhydride tablets as acidic (pH ≤ 4) when the degradation is carried out in 0.1 M pH 7.4 phosphate-buffered solution. One approach that has been taken to counteract the acidity is the incorporation of buffering substances into the polymer matrix during the device fabrication. This has been shown to cause an increase in pH inside the delivery system. The microviscosity and drug mobility, as well as the formation of radicals, within a device were monitored both *in vivo* and *in vitro* using EPR methods. EPR imaging (EPRI) introduces a special dimension by means of additional gradients, and is applied to characterize the degradation front and the microenvironment of polyanhydride disks loaded with a pH-sensitive nitrooxide (Mader et al., 1997a). Exposure to buffer (pH 7.4) resulted in the formation of a front of degraded polymer from outside to inside. A pH gradient was found to exist within the polymer matrix, and the pH rises with time from 4.7 to 7.4. The issue of a possible chemical reaction between amine and hydroxyl group-containing drug moieties with the anhydride bonds in the polymer during drug incorporation and release has also been investigated (Mader et al., 1997a).

8
Biocompatibility and Elimination

The Food and Drug Administration (FDA), in testing and evaluating new biomaterials, established the biocompatibility and safety of polyanhydrides following the 1986 guidelines. Several accepted criteria and tests to evaluate new biomedical materials were used to assess the safety of polyanhydrides (Braun et al., 1982; Leong et al., 1986a; Laurencin et al., 1990). In these studies,

poly[bis(*p*-carboxy-phenoxy)propane anhydride (PCPP) and its copolymers with SA were tested. Neither mutagenicity nor cytotoxicity or teratogenicity was associated with the polymers or their degradation products, as evaluated by mutation assays. The tissue response of these polyanhydrides was studied by subcutaneous implantation in rats, and in the cornea of rabbits. The polymers did not provoke inflammatory responses in the tissues over a 6-week implantation period, and histologic evaluation indicated relatively minimal tissue irritation, with no evidence of local or systemic toxicity (Laurencin et al., 1990). Systemic response to the polymer was evaluated by monitoring of blood chemistry and hematologic values, and by comprehensive examination of organ tissues. Neither method revealed any significant response to the polymer. Recently, a report on the biocompatability of ricinoleic acid-based polymer (published by our laboratory; Teomim and Domb, 1999) in which biocompatability of poly(RAS-SA) was tested in rats and compared with Vicryl™ surgical suture and sham surgery. It was observed that no any abnormal gross pathological finding occurred at the implantation site after 3, 7 and 21 days of implantation. Blood chemistry and cell counts were similar for treated and untreated rats. The polymer was degraded into ricinoleic acid with constant release of incorporated drug, while the inflammatory response after subcutaneous implantation was minimal to mild, and comparable with that seen clinically with Vicryl™ absorbable sutures.

Since the CPP-SA copolymer was designed to be used clinically to deliver an anticancer agent directly into the brain for the treatment of brain neoplasm, *in vivo* safety evaluations and brain biocompatibility were assessed in rats (Tamargo et al. 1989), rabbits (Brem et al., 1989) and monkeys (Brem et al., 1988). In the rat brain study, the tissue reaction of the polymer (PCPP-SA 20:80) was compared with the reaction observed with two standard materials used in surgery, which have been extensively studied, namely Gelfoam® (an absorbable gelatin sponge) and Surgicel® (an oxidized cellulose absorbable hemostat commonly used in brain surgery). Histologic evaluation of the tissue demonstrated a small rim of necrosis around the implant, and a mild to marked cellular inflammatory reaction limited to the area immediately adjacent to the implantation site. The pathologic response associated with poly(CPP-SA) copolymer was slightly more pronounced than with Surgicel® at the earlier time points, but noticeably less marked than with Surgicel® at the later times. The reaction to Gelfoam® was essentially equivalent to that observed in control rats. In a similar brain biocompatibility study carried out in monkeys, no tissue abnormalities were noted either in computed tomography or magnetic resonance imaging (MRI) scans. Furthermore, no abnormalities were observed either in the blood chemistry or hematologic evaluations (Brem et al., 1988). There appeared to be no adverse systemic effects due to the implants as assessed by histologic evaluation of the tissue tested. Overall, no unexpected or untoward reaction to treatment was observed. Copolymers of SA with several aliphatic comonomers such as dimer of erucic acid (FAD), FA acid and isophthalic acid were also tested both subcutaneously and in the rat brain, and were also found to be biocompatible (Rock et al., 1991). The hydrolysis and elimination processes of polyanhydrides have been studied using a series of polyanhydrides derived from different linear aliphatic diacids (Domb and Nudelman, 1995b). These polymers degrade into their monomer or oligomer units at about the same rate, but differ in the water solubility of their degradation products. Polymers based

on natural diacids of the general structure $-[OOC-(CH_2)_x-CO]-$ (where x is between 4 and 12) were implanted subcutaneously in rats, and the elimination of polymers from the implantation site was studied. The in vitro hydrolysis of this polymer series was studied by monitoring the weight loss, release of monomer degradation products, and changes in content of anhydride bonds of polymer as a function of time. It was observed that, both in vitro and in vivo, the rate of polymer elimination was a function of monomer solubility. The elimination time for polymers based on soluble monomers (x = 4–8) was 7–14 days, while the polymers based on monomers with lower solubility (x = 10–12) were eliminated only after 8 weeks. All polymers were found to be biocompatible and useful as carriers of drugs.

The elimination of poly(CPP-SA)-based implant (Gliadel™), which is currently in clinical use for the treatment of brain cancer, was studied in rabbit and rat brains using radioactive polymer and drug (Domb et al., 1994b, 1995a). The implant is composed of N,N-bis(2-chloroethyl)-N-nitrosourea (BCNU) dispersed in a co-polyanhydride matrix of CPP and SA. Four groups of rabbits were implanted with wafers loaded with BCNU, one in a ^{14}C-SA-labeled polymer, another in a ^{14}C-CPP-labeled polymer, and two groups with ^{14}C-BCNU in a non-labeled polymer in which one was for a BCNU disposition study and one for a residual drug study. In the rabbits implanted with the ^{14}C-SA-labeled polymer, approximately 10% of the radioactivity was found in the urine and 2% in the feces, and about 10% remained in the device after 7 days of implantation. In contrast, only 4% of the radioactivity associated with the ^{14}C-CPP labeled polymer was found in urine and feces during this period. However, a drastic increase in CPP excretion was found after 9 days; moreover, after 21 days 64% of the implanted ^{14}C-CPP was recovered in the urine and feces, and 29% was still in the recovered wafers. Studies with radiolabeled BCNU in rabbit brain revealed that approximately 50% of the BCNU in the wafers was released in 3 days, and over 95% was released after 6 days in the rabbit brain. Excretion of this polymer after implantation in the rat brain using radiolabeled polymers showed that over 70% of the SA comonomer was excreted in 7 days, with about 40% of the SA metabolized to CO_2 (Domb et al., 1995). The elimination of poly(FAD-SA) rods loaded with 0, 10, and 20 wt.% of gentamicin sulfate after implantation in the femoral muscle and bone of dogs was studied as part of the preclinical studies for Septacin® bone implant (Domb and Amselem, 1994). Most of the polymer implant was gradually eliminated from bone and muscle within 4 to 8 weeks post implantation, with the elimination from bone being faster and leading to new bone formation in the implant site, without any polymer entrapment. The elimination rate was dependent mainly on the amount of polymer implanted. Gentamicin was released for a period of about 3 weeks, with no residual drug being detected in the polymer remnants 8 weeks post implantation. In all experiments, no local or systemic toxicity was observed.

Mader et al. (1997b) utilized the noninvasive technique of MRI to visualize both polymer erosion (*in vivo* and *in vitro*) and the physiologic response (edema, encapsulation) to the implant. MRI enables monitoring, *in vivo* and in a noninvasive manner, of the water content, implant shape, and response of the biological system to an implant in real time, without stopping the experiment. MRI images were taken during the course of degradation of slabs of PSA – a rapidly degrading polyanhydride – placed in physiologic buffer solution and implanted subcutaneously in rats. However, a water penetration front was clearly observed at day 21 *in vitro* and at day 32; thus, the bright image inside the MRI images indicated that the entire polymer matrix was filled with water. This brightness was never noticed *in vivo*. Instead, a deformation of the implant was observed, starting with the rounding of the corners (day 9), which progressively increased (days 16 and 20) and was completed at day 28.

9
Applications

Polyanhydrides have been investigated as a candidate for controlled release devices of drugs for treating eye disorders (Albertsson et al., 1996), chemotherapeutic agents (Wu et al., 1994), local anesthetics (Masters et al., 1993a,b; Maniar et al., 1994), anticoagulants (Chickering et al., 1996), neuroactive drugs (Kubek et al., 1998), and anticancer agents (Domb and Ringel, 1994). Polyanhydride poly(CPP-SA) were loaded with carmustine and implanted into the brains of rats, rabbits, and monkeys, and were found to be biocompatible and efficacious against brain tumors (Tamargo et al., 1993; Fung et al., 1998). Based on the findings of these laboratory studies, a phase I–II clinical trial was completed which showed treatment to be well tolerated in all patients, apparently without the production of any systemic side effects (Brem et al., 1991). Phase III human clinical trials have demonstrated that the site-specific delivery of BCNU (carmustine) from a poly(CPP-SA)20:80 wafer (Gliadel®) in patients with recurring brain cancer (glioblastoma multiforme) significantly prolongs patient survival (Brem et al., 1995). On the basis of the results of clinical trials, the FDA has given its first approval during the past 23 years for a new treatment for brain tumors (Brem and Lawson, 1999). Gliadel®

was subsequently approved for the treatment of brain tumors in Canada, South America, Israel, South Korea and Europe. Masters et al. (1993a,b) used a polyanhydride cylinder for the delivery of local anesthetics in close proximity to the sciatic nerve to produce a neuronal block for several days. The use of polyanhydrides in the oral delivery of insulin and plasmid DNA has also been investigated (Mathiowitz et al., 1997). During the past five years, investigations have expanded to newer polymers and other drugs such as 4-hydroperoxy cyclophosphamide (4HC), cisplatin, carboplatin, paclitaxel, as well as several alkaloid drugs, in an effort to develop a better system for treating brain tumors (Laurencin et al., 1993; Brem et al., 1994; Judy et al., 1995; Olivi et al., 1996). Carboplatin incorporated into poly(FAD-SA), prepared by mixing the drug in the melted polymer, has been evaluated for the treatment of brain tumors in laboratory animals, and has shown promising results (Olivi et al., 1996). Poly(CPP-SA) has also been used to develop a delivery system for gentamicin sulfate for the treatment of osteomyelitis (Laurencin et al., 1993; Domb and Amselem, 1994). A sustained release of gentamicin sulfate over a period of few weeks was obtained both *in vivo* and *in vitro* using this system. This delivery device, which is in the form of a chain of beads, has been undergoing human clinical trials in the USA. The effect of long-term glutamic acid stimulation of trigeminal motorneurons, using poly(FAD-SA) microspheres, has also been explored. This study was undertaken to determine the role of glutamate in possible growth disorders of the craniofacial skeleton. Pronounced skeletal changes in the snout region were observed in growing rats receiving glutamate, indicating that a sustained release of glutamic acid *in vivo* can affect the development of skeletal tissue (Hamilton-Byrd

et al., 1992). Recently, the use of polyanhydrides for the delivery of heparin and the treatment of osteomyelitis have also been reported (Teomim et al., 1999a; Stephens et al., 2000).

10
Outlook and Perspectives

Polyanhydrides have now been investigated as a 'smart' biomaterial used for the short-term release of drugs for more than two decades. Over these years, intensive research has been conducted in both academia and industry which has yielded hundreds of publications and patents describing new polymer structures, studies on chemical and physical characterization of these polymers, degradation and stability properties, toxicity studies, and applications of these polymers mainly for the controlled delivery of bioactive agents. Such research also yielded a device (Gliadel®) which can be used clinically in the treatment of brain cancer. Due to the rapid degradation and limited mechanical properties, the main application for this class of polymers is for the short-term controlled delivery of bioactive agents.

The advantages and disadvantages of these polymers relate to the hydrolytic instability of the anhydride bond that degrades rapidly, thereby changing the polyanhydride into its nontoxic monomers. The main advantages of this class of polymers are:

- They may be prepared from readily available, low-cost resources that are generally considered as safe dicarboxylic acid building blocks, many of which are either body constituents or metabolites.
- They are prepared in a one-step synthesis, with no need for purification steps.
- They have a well-defined polymer structure, with controlled molecular weight;

moreover, they degrade hydrolytically into their building blocks at a predictable rate.
- They can be manipulated accordingly to release bioactive agents in a predictable rate for periods of weeks.
- They are processable, either by low melting injection molding or extrusion for mass production, and have versatile properties which can be obtained by monomer selection, composition, surface area, and additives.
- They degrade to their respective diacids, and are completely eliminated from the body within periods of weeks to months.
- They can be sterilized by terminal γ-irradiation, without there being any adverse effect on polymer properties.

It should be noted that, although these polymers have so many advantages, they also have certain disadvantages, the main ones being their short-term degradation and release periods (which may extend from weeks to a few months), and the need for specialized storage conditions (e.g., refrigeration, dry). Ironically, however, the high rate at which these polymers degrade renders them most suitable for the short-term, i.e., days to weeks, controlled delivery of bioactive agents in the form of solid implants or injectable microspheres.

On this basis, it is expected that these polymers will be developed in the coming years into implantable devices for the local and systemic delivery of drugs.

11
Patents

Patents relating to the application of polyanhydrides are listed in Table 2.

Tab. 2 Patents for the application of polyanhydrides

S. No.	Patent No./ Issue Date	Title	Author(s)	Applicant(s)
1	US04886870 12/12/1989	Bioerodible articles useful as implants and prostheses having predictable degradation rates	D'Amore, P.; Leong, K. W.; Langer, R.S.	Massachusetts Institute of Technology (USA)
2	US04898734 02/06/1990	Polymer composite for controlled release or membrane formation	Mathiowitz, E.; Langer, R. S.; Warshawsky, A.; Edelman, E.	Massachusetts Institute of Technology(USA)
3	US04906473 03/06/1990	Biodegradable poly(hydroxy-alkyl)amino dicarboxylic acid) derivatives, a process for their preparation, and the use thereof for depot formulations with controlled delivery of active ingredient	Ruppel, D.; Walch, A.	Hoechst Aktiengesellschaft (Germany)
4	US04999417 03/12/1991	Biodegradable polymer compositions	Domb, A. J.	Nova Pharmaceutical Corporation (USA)
5	US05019379 05/28/1991	Unsaturated polyanhydrides	Domb, A. J.; Langer, R. S.	Massachusetts Institute of Technology (USA)
6	US05175235 12/29/1992	Branched polyanhydrides	Domb, A. J.; Maniar, M.	Nova Pharmaceutical Corporation (USA)
7	US05179189 01/12/1993	Fatty acid-terminated polyanhydrides	Domb, A. J.; Maniar, M.	Nova Pharmaceutical Corporation (USA)

Tab. 2 (cont.)

S. No.	Patent No./ Issue Date	Title	Author(s)	Applicant(s)
8	US05197466 03/30/1993	Method and apparatus for volumetric interstitial conductive hyperthermia	Marchosky, J. A.; Moran, C. J.; Fearnot,N. E.	MED Institute Inc. West Lafayette, IN
9	US05395916 03/07/1995	Biodegradable copolymer from hydroxy proline	Mochizuki, S.; Nawata, K.; Suzuki, Y.	Teijin Limited (Japan)
10	US05459258 10/17/1995	Polysaccharide-based biodegradable thermoplastic materials	Merrill, E. W.; Sagar, A.	Massachusetts Institute of Technology (USA)
11	US05473103 12/05/1995	Biopolymers derived from hydrolyzable diacid fats	Domb, A. J.; Nudelman, R.	Yissum Research Development Co. of the Hebrew University of Jerusalem (Israel)
12	US05522895 06/04/1996	Biodegradable bone templates	Mikos, A. G.	Rice University (USA)
13	US05545409 08/13/1996	Delivery system for controlled release of bioactive factors	Laurencin, C. T.; Lucas, P. A.; Syftestad, G.T. ; Domb, A. J.; Glowacki, J.; Langer, R. S.	Massachusetts Institute of Technology (USA)
14	US05618563 04/08/1997	Biodegradable polymer matrices for sustained delivery of local anesthetic agents	Berde, C. B.; Langer, R. S.	Children's Medical Center Corporation (USA)
15	US05626862 05/06/1997	Controlled local delivery of chemotherapeutic agents for treating solid tumors	Brem, H.; Langer, R. S.; Domb, A. J.	Massachusetts Institute of Technology (USA)
16	US05660851 08/26/1997	Ocular inserts	Domb, A. J.	Yissum Research Development Company of the Hebrew University of Jerusalem (ISRAEL)
17	US05716404 02/10/1998	Breast tissue engineering	Vacanti, J. P.; Atala, A.; Mooney, D. J.; Langer, R. S.	Massachusetts Institute of Technology (USA)
18	US05756652 05/26/1998	Poly (ester-anhydrides) and intermediates therefore	Storey, R. F.; Deng, Z. D.; Peterson, D. R.; Glancy, T. P.	DePuy Orthopedics, Inc. (USA)
19	US05855913 01/05/1999	Particles incorporating surfactants for pulmonary drug delivery	Hanes, J.; Edwards, D. A.; Evora, C.; Langer, R. S.	Massachusetts Institute of Technology (USA)
20	US06046187 04/04/2000	Formulations and methods for providing prolonged local anesthesia	Berde, C. B.; Langer, R. S.; Curley, J.; Castillo, J	Children's Medical Center Corporation (USA)

12
References

Abuganima, E. (1996) *Synthesis and characterization of copolymers and blends of polyanhydrides and polyesters*, M.Sc. Thesis, The Hebrew University of Jerusalem.

Albertsson, A.-C., Lundmark, S. (1988) Synthesis of poly(adipic anhydride) by use of ketenes, *J. Macromol. Sci.- Chem. A* **25**, 247–258.

Albertsson, A.-C., Lundmark, S. (1990) Synthesis of poly(adipic) anhydride by use of ketene, *J. Macromol. Sci.- Chem. A* **27**, 397–412.

Albertsson, A.-C., Carlfors, J., Sturesson, C. (1996) Preparation and characterization of poly(adipic anhydride) microspheres for ocular drug delivery, *J. Appl. Polym. Sci.* **62**, 695–705.

Bindschaedler, C., Leong, K., Mathiowitz, E., Langer, R. (1988) Poly(anhydride) microspheres formulation by solvent extraction, *J. Pharm. Sci.* **77**, 696–698.

Braun, A. G., Buckner, C. A., Emerson, D. J., Nichinson, B. B. (1982) Quantitative correspondence between the in vivo and *in vitro* activity of teratogenic agents, *Proc. Natl. Acad. Sci. USA* **79**, 2056.

Brem, H., Lawson, H. C. (1999) The development of new brain tumor therapy utilizing the local and sustained delivery of chemotherapeutic agents from biodegradable polymers, *Cancer* **86**, 197–199.

Brem, H., Tamargo, R. J., Pinn, M., Chasin, M. (1988) Biocompatibility of a BCNU-loaded biodegradable polymer: a toxicity study in primates. *Am. Assoc. Neurol. Surg.* **24**, 381.

Brem, H., Kader, A., Epstein, J. I., Tamargo, R. J., Domb, A. J., Langer, R., Leong, K. W. (1989) Biocompatibility of bioerodible controlled release polymer in the rabbit brain, *Sel. Cancer Ther.* **5**, 55–65.

Brem, H., Mahley, M. S., Vick, N. A. et al. (1991) Interstitial chemotherapy with drug polymer implants for the treatment of recurrent gliomas, *J. Neurosurg.* **74**, 441–446.

Brem, H., Walter, K. A., Tamargo, R. J., Olivi, A., Langer, R. (1994) Drug delivery to the brain, in: *Polymeric Site-Specific Pharmacotherapy* (Domb, A., Ed.), John Wiley & Sons: Chichester, 117–140.

Brem, H., Piantadosi, S., Burger, P. C. et al. (1995) Placebo-controlled trial of safety and efficacy of intraoperative controlled delivery by biodegradable polymers of chemotherapy for recurrent gliomas. The Polymer-brain Tumor Treatment Group, *Lancet* **345**, 1008–1012.

Bucher, J. E., Slade, W. C. (1909) The anhydrides of isophthalic and terephthalic acids, *J. Am. Chem. Soc.* **31**, 1319–1321.

Chasin, M., Domb, A., Ron, E., Mathiowitz, E., Leong, K., Laurencin, C., Brem, H., Grossman, S., Langer, R. (1990) Polyanhydrides as drug delivery systems, in: *Biodegradable Polymers as Drug Delivery Systems* (Chsin, M., Langer, R., Eds.), New York: Marcel Dekker, 43–70.

Chickering, D., Jacob, J., Mathiowitz, E. (1996) Poly(fumaric-co-sebacic) microspheres as oral drug delivery systems, *Biotechnol. Bioeng.* **52**, 96–101.

Conix, A. (1958) Aromatic poly(anhydrides): a new class of high melting fiber-forming polymers, *J. Polym. Sci.* **29**, 343–353.

Dang, W. B., Daviau, T., Nowotnik, D. (1996) *In vitro* erosion kinetics of implantable polyanhydride Gliadel™, *Proc. Int. Symp. Control. Rel. Bioact. Mater.* **23**, 731–732.

Davies, M. C., Khan, M. A., Domb, A. J., Langer, R., Watts, J. F., Paul, A. (1991) The analysis of the surface chemical structure of biomedical aliphatic poly(anhydrides) using XPS and ToF-SIMS, *J. Appl. Polym. Sci.* **42**, 1597–1605.

Davies, M. C., Shakesheff, K. M., Shard, K. M., Domb, A. J., Roberts, C. J., Tendler, S. J. B.,

Williams, P. M., Tendler, S. J. B. and Williams, P. M. (1996) Surface analysis of biodegradable polymer blends of poly(sebacic anhydride) and poly-(DL-lactic acid), *Macromolecules* **29**, 2205–2212.

D'Emanuele, A., Hill, J., Tamada, J. A., Domb, A. J., Langer, R. (1994) Molecular weight changes in polymer erosion, *Pharmaceut. Res.* **9**, 1279–1283.

Domb, A. J. (1990) Biodegradable polymers derived from amino acids, *Biomaterials* **11**, 686–689.

Domb, A. J. (1992) Synthesis and characterization of bioerodible aromatic anhydride copolymers, *Macromolecules* **25**, 12–17.

Domb, A. J. (1993) Biodegradable polymer blends: screening for miscible polymers, *J. Polym. Sci.: Polym. Chem.* **31**, 1973–1981.

Domb, A. J., Amselem, S. (1994) Antibiotic delivery systems for the treatment of bone infections, in: *Polymeric Site-Specific Pharmacotherapy* (Domb, A.J., Ed.), Chichester: John Wiley & Sons, 242–265.

Domb, A. J., Langer, R. (1987) Poly(anhydrides). I. Preparation of high molecular weight polyanhydrides, *J. Polym. Sci. Polym. Chem.* **25**, 3373–3386.

Domb, A. J., Langer, R. (1989) Solid-state and solution stability of poly(anhydrides) and poly(esters), *Macromolecules* **22**, 2117–2122.

Domb, A. J., Maniar, M. (1993a) Absorbable biopolymers derived from dimers fatty acids, *J. Poly. Sci.: Polym. Chem.* **31**, 1275–1285.

Domb, A. J., Maniar, M. (1993b) Fatty acid terminated polyanhydrides, US Patent 5,179,189.

Domb, A. J., Nudelman, R. (1995a) Biodegradable polymers derived from natural fatty acids, *J. Polym. Sci.* **33**, 717–725.

Domb, A. J., Nudelman, R. (1995b) *In vivo* and *in vitro* elimination of aliphatic polyanhydrides, *Biomaterials* **16**, 319–323.

Domb, A. J., Ringel, I. (1994) Polymeric Drug Carrier Systems in the Brain, in: *Providing Pharmaceutical Access to the Brain, Methods in Neuroscience* (Flanaga, T. R., Emerich, D. F., Winn, S. R. Eds.) CRC Press: Boca Raton, FL, 169–183, Vol. 21.

Domb, A. J., Ron, E., Langer, R. (1988) Poly(anhydrides). II. One step polymerization using phosgene or diphosgene as coupling agents, *Macromolecules* **21**, 1925–1929.

Domb, A. J., Gallardo, C. F., Langer, R. (1989) Poly(anhydrides). 3. Poly(anhydrides) based on aliphatic-aromatic diacids, *Macromolecules* **22**, 3200–3204.

Domb, A. J., Mathiowitz, E., Ron, E., Giannos, S., Langer, R. (1991) Polyanhydrides IV. Unsaturated and cross-linked poly(anhydrides), *J. Polym. Sci. Part A: Polym. Chem.* **29**, 571–579.

Domb, A. J., Amselem, S., Shah, J., Maniar, M. (1993) Polyanhydrides: Synthesis and characterization, in: *Advances in Polymer Sciences* (Peppas, N. A., Langer, R., Eds.), Heidelberg: Springer-Verlag, 93–141.

Domb, A. J., Amselem, S., Langer, R., Maniar, M. (1994a) Polyanhydrides as carriers of drugs, in: *Designed to Degrade Biomedical Polymers* (Shalaby, S., Ed.), Munich: Carl Hanser Verlag, 69–96.

Domb, A. J., Rock, M., Perkin, C., Proxap, B., Villemure, J. G. (1994b) Metabolic disposition and elimination studies of a radiolabelled biodegradable polymeric implant in the rat brain, *Biomaterials* **15**, 681–688.

Domb, A. J., Rock, M., Perkin, C., Proxap, B., Villemure, J. G. (1995) Excretion of a radiolabelled biodegradable polymeric implant in the rabbit brain, *Biomaterials* **16**, 1069–1072.

Domb, A. J., Elmalak, O., Shastri, V. R., Ta-Shma, Z., Masters, D. M., Ringel, I., Teomim, D., Langer, R. (1997) Polyanhydrides, in: *Handbook of Biodegradable Polymers* (Domb, A. J., Kost, J., Weiseman, D. M., Eds.), Amsterdam: Harwood Academic Publishers, 135–159.

Domb, A. J., Israel, J. H., Elmalak, O., Teomim, D., Bentolila, A. (1999) Preparation and characterization of Carmustine loaded biodegradable disc for treating brain tumors, *Pharm. Res.* **16**, 762–765.

Edlund, U., Albertsson, A.-C. (1999) Copolymerization and polymer blending of trimethylene carbonate and adipic anhydride for tailored drug delivery, *J. Appl. Polym. Sci.* **72**, 227–239.

Edlund, U., Albertsson, A.-C., Singh, S. K., Fogelberg, I. (2000) Sterilization, storage, stability and *in vivo* biocompatibility of poly(trimethylene carbonate)/poly(adipic anhydride) blends, *Biomaterials* **21**, 945–955.

Fung, L. K., Ewend, M., Sills, A. et al. (1998) Pharmacokinetics of interstitial delivery of carmustine, 4-hydroperoxycyclophosphamide and paclitaxel from a biodegradable polymer implant in the monkey brain, *Cancer Res.* **58**, 672–684.

Gopferich, A., Langer, R. (1993) The influence of microstructure and monomer properties on the erosion mechanism of a class of polyanhydrides, *J. Polym. Sci.* **31**, 1445–1458.

Gopferich, A., Langer, R. (1995a) Modeling of polymer erosion in three dimensions: rotationally symmetric devices, *AIChE J.* **41**, 2292–2299.

Gopferich, A., Langer, R. J (1995b) Modeling monomer release from bioerodible polymers, *J. Control. Rel.* **33**, 55–69.

Gopferich, A., Karydas, D., Langer, R. (1995) Predicting drug release from cylindrical polyanhydride matrix discs, *Eur. J. Pharm. Biopharm.* **41**, 81–87.

Gref, R., Minamitake, Y., Peracchia, M. T., Domb, A. J., Trubetskoy, V., Torchilin, V., Langer, R. (1995) Poly(ethylene glycol) coated nanospheres, *Adv. Drug Deliv. Rev.* **16**, 215–233.

Gupta, B. (1988) Polyanhydride process from bis(trimethylsilyl) ester of dicarboxylic acid, US Patent 4,868,265.

Hamilton-Byrd, E. L., Sokoloff, A. J., Domb, A. J., Terr, L., Byrd, K. E. (1992) L-Glutamate microsphere stimulation of the trigeminal motor nucleus in growing rats, *Polym. Adv. Technol.* **3**, 337–344.

Hartmann, M., Schultz, V. (1989) Synthesis of poly(anhydride) containing amido groups, *Macromol. Chem.* **190**, 2133.

Hill, J. W. (1930) Studies on polymerization and ring formation. VI. Adipic anhydride. *J. Am. Chem. Soc.* **52**, 4110–4114.

Hill, J. W., Carothers, H. W. (1932) Studies on polymerization and ring formation. XIV. A linear superpolyanhydride and a cyclic dimeric anhydride from sebacic acid, *J. Am. Chem. Soc.* **54**, 5169.

Jiang, H. L., Zhu, K. J. (2000) Pulsatile protein release from a laminated device comprising of polyanhydrides and pH-sensitive complexes, *Int. J. Pharm.* **194**, 51–60.

Judy, K. D., Olivi, A., Buahin, K. G., Domb, A. J., Epstein, J. I., Colvin, O. M., Brem, H. (1995) Effectiveness of controlled release of a cyclophosphamide derivative with polymers against rat gliomas, *J. Neurosurg.* **82** 481–486.

Kim, B. S., Hrkach, J. S., Langer, R. (2000) Synthesis and characterization of novel degradable photocrosslinked poly(ester-anhydride) networks, *J. Polym. Sci. A: Polym. Chem.* **38**, 1277–1282.

Knobloch, J. O., Ramirez, F. (1975) *J. Org. Chem.* **40**, 1101–1106.

Kubek, M. J., Liang, D., Byrd, K. E., Domb, A. J. (1998) Prolonged seizure suppression by a single implantable polymeric TRH microdisk preparation, *Brain. Res.* **809**, 189–197.

Laurencin, C. T., Domb, A. J., Morris, C., Brown, V., Chasin, M., McConnell, R., Lange, N., Langer, R. (1990) Poly(anhydrides) administration in high doses in vivo: studies of biocompatibility and toxicology, *J. Biomed. Mater. Res.* **24**, 1463–1481.

Laurencin, C., Gerhart, T., Witschger, P., Satcher, R., Domb, A. J., Hanff, P., Edsberg, L., Hayes, W., Langer, R. (1993) Biodegradable polyanhydrides for antibiotic drug delivery, *J. Orthop. Res.* **11**, 256–262.

Laurencin, C. T., Ibim, S. E. M., Langer, R. (1995) Poly(anhydrides) in: *Biomedical Applications of Synthetic Biodegradable Polymers* (Hollinger, J. O., Ed.), Boca Raton, FL: CRC Press, 59–102.

Leong, K. W., D'Amore, P., Langer, R. (1986a) Bioerodible poly(anhydrides) as drug carrier matrices. II. Biocompatibility and chemical reactivity, *J. Biomed. Mater. Res.* **20**, 51–64.

Leong, K.W., Kost, J., Mathiowitz, E., Langer, R. (1986b) Poly(anhydrides) for controlled release of bioactive agents, *Biomaterials* **7**, 364.

Leong, K. W., Simonte, V., Langer, R. (1987) Synthesis of poly(anhydrides): melt-polycondensation, dehydrochlorination, and dehydrative coupling, *Macromolecules* **20**, 705–712.

Leong, K. W., Domb, A., Langer, R. (1989) Poly(anhydrides), in: *Encyclopedia of Polymer Science and Engineering*, 2nd Edition, New York: John Wiley & Sons.

Lundmark, S., Sjoling, M., Albertsson, A.-C. (1991) Polymerization of oxipane-2,7-dione in solution and synthesis of block copolymer of oxipane-2,7-dione and 2-oxipanone, *J. Macromol. Sci.- Chem. A* **28**, 15–29.

Mader, K., Domb, A. J., Swartz, H. M. (1996) Gamma sterilization induced radicals in biodegradable drug delivery systems, *J. Appl. Rad. Isotop.* **47**, 1669–1674.

Mader, K., Bacic, G., Domb, A. J., Elmalak, O., Langer, R., Swartz, H. M. (1997a) Noninvasive in vivo monitoring of drug release and polymer erosion from biodegradable polymers by EPR spectroscopy and NMR imaging, *J. Pharm. Sci.* **86**, 126–134.

Mader, K., Cremmilleleux, Y., Domb, A. J., Dunn, J. F., Swartz, H. M. (1997b) *In vitro/in vivo* comparison of drug release and polymer erosion from biodegradable P(FAD-SA) polyanhydrides – a noninvasive approach by the combined use of Electron Paramagnetic Resonance Spectroscopy and Nuclear Magnetic Resonance Imaging, *Pharm. Res.* **14**, 820–826.

Maniar, M., Xie, X., Domb, A. J. (1990) Poly(anhydrides). V. Branched poly(anhydrides), *Biomaterials* **11**, 690–694.

Maniar, M., Domb, A., Haffer, A., Shah, J. (1994) Controlled release of local anesthetics from fatty acid dimer based polyanhydrides, *J. Control. Rel.* **30**, 233–239.

Mark, H.E., et al. (Eds.) (1969) Poly(anhydrides), in: *Encyclopedia of Polymer Science and Technology* New York: John Wiley & Sons, 630, Vol. 10.

Masters, D. B., Berde, C. B., Dutta, S. K., Griggs, C. T., Hu, D., Kupsky, W., Langer, R. (1993a)

Prolonged regional nerve blockade by controlled release of local anesthetic from a biodegradable polymer matrix, *Anesthesiology* **79**, 340–346.

Masters, D. B., Berde, C. B., Dutta, S., Turek, T., Langer, R. (1993b) Sustained local anesthetic release from bioerodible polymer matrices: a potential method for prolonged regional anesthesia, *Pharm. Res.* **10**, 1527–1532.

Mathiowitz, E., Langer, R. (1987) Poly(anhydride) microspheres as drug carriers. I. Hot-melt microencapsulation, *J. Control. Rel.* **5**, 13–22.

Mathiowitz, E., Kline, D., Langer, R. (1990) Morphology of poly(anhydride) microspheres delivery systems, *J. Scan. Microsc.* **4**, 329.

Mathiowitz, E., Jacob, J. S., Jong, Y. S., Carino, G. P., Chickering, D. E., Chaturvedi, P., Santos, C. A., Vijyaraghavan, K., Montgomery, S., Basset, M., Morrell, C. (1997) Biologically erodible microspheres as potential oral drug delivery systems, *Nature* **386**, 410–414.

McIntyre, J. E. (1994) British Patent 978,669.

Muggli, D. S., Burkoth, A. K., Anseth, K. S. (1999) Crosslinked polyanhydrides for use in orthopedic applications: degradation behavior and mechanics, *J. Biomed. Mater. Res.* **46**, 271–278.

Olivi, A., Awend, M.G., Utsuki, T., Tyler, B., Domb, A. J., Brat, D. J., Brem, H. (1996) Interstitial delivery of carboplatin via biodegradable polymers is effective against experimental glioma in the rat. *Cancer, Chemother. Pharmacol.* **39**, 90–96.

Pekarek, K. J., Jacob, J. S., Mathiowitz, E. (1994) One-step preparation of double-walled microspheres *Nature* **357**, 258–260.

Rock, M., Green, M., Fait, C., Gell, R., Myer, J., Maniar, M., Domb, A. (1991) Evaluation and comparison of biocompatibility of various classes of polyanhydrides, *Polym. Preprints* **32**, 221.

Ron, E., Mathiowitz, E., Mathiowitz, G., Domb, A., Langer, R. (1991) NMR characterization of erodible copolymers, *Macromolecules* **24**, 2278–2282.

Rosen, H. B., Chang, J., Wnek, C. E., Lindhardt, R J., Langer, R. (1983) Biodegradable poly(anhydrides) for controlled drug delivery, *Biomaterials* **4**, 131–133.

Santos, S. A., Freedman, B. D., Leach, K. J., Press, D. L., Scarpulla, M., Mathiowitz, E. (1999) Poly(fumaric-co-sebasis anhydride) – A degradation study as evaluated by FTIR, DSC, GPC and X-ray diffraction, *J. Control. Rel.* **60**, 11–22.

Shakesheff, K. M., Davies, M. C., Roberts, C. J., Tendler, S. J. B., Shard, K. M., Domb, A. J. (1994) In situ AFM imaging of polymer degradation in an aqueous environment, *Langmuir* **10**, 4417–4419.

Shakesheff, K. M., Chen, X., Davies, M. C., Domb, A. J., Roberts, M. C., Tendler, S. J. B., Williams, P. M. (1995) Relating the phase morphology of a biodegradable polymer blend to erosion kinetics using simultaneous *in situ* atomic force microscopy and surface plasmon resonance analysis, *Langmuir* **11**, 3921–3927.

Staubli, A., Ron, E., Langer, R. (1990) Hydrolytically degradable amino acid containing polymers, *J. Am. Chem. Soc.* **112**, 4419–4424.

Staubli, A., Mathiowitz, E., Lucarelli, M., Langer, R. (1991a) Characterization of hydrolytically degradable amino acid containing poly(anhydride-co-imides), *Macromolecules* **24**, 2283–2290.

Staubli, A., Mathiowitz, E., Langer, R. (1991b) Sequence distribution and its effects on glass transition temperatures of poly(anhydrides-co-imides) containing asymmetric monomers, *Macromolecules* **24**, 2291–2298.

Stephens, D., Li, L., Robinson, D., Chen, S., Chang, H. C., Liu, R. M., Tian, Y. Q., Ginsberg, E. J., Gao, X. Y., Stultz, T. (2000) Investigation of the *in vitro* release of gentamicin from a polyanhydride matrix, *J. Control. Rel.* **63**, 305–317.

Subramanyam, R., Pinkus, A. G. (1985) Synthesis of poly(terephthalic anhydride) by hydrolysis of terephthaloyl chloride triethylamine intermediate adduct: characterization of intermediate adduct, *J. Macromol. Sci. Chem.* **A22**, 23.

Tabata, Y., Domb, A. J., Langer, R. (1994) Polyanhydride granules provide controlled release of water-soluble drugs with reduced initial burst, *J. Pharm. Sci.* **83**, 5–11.

Tamargo, R. J., Epstein, J. I., Reinhard, C. S., Chasin, M., Brem, H. (1989) Brain biocompatibility of a biodegradable controlled-release polymer in rats, *J. Biomed. Mater. Res.* **23**, 253–266.

Tamargo, R. J., Myseros, J. S., Epstein, J. I., Yang, M. B., Chasin, M., Brem, H. (1993) Interstitial chemotherapy of the 9L-glyosarcoma – Controlled release polymers for drug delivery in the brain, *Cancer Res.* **53**, 329–333.

Teomim, D., Domb, A. J. (1999) Fatty acid terminated polyanhydrides, *J. Polym. Sci. A: Polym. Chem.* **37**, 3337–3344.

Teomim, D., Fishbien, I., Golomb, G., Orloff, L., Mayberg, M., Domb, A. J. (1999a) Perivascular delivery of heparin for the reduction of smooth muscle cell proliferation after endothelial injury, *J. Control. Rel.* **60**, 129–142.

Teomim, D., Nyska, A., Domb, A. J. (1999b) Ricinoleic acid based biopolymers, *J. Biomed. Mater. Res.* **45**, 258–267.

Teomim, D., Mader, K., Bentolila, A., Magora, A., Domb, A. J. (2001) *Macromolecules* (in press).

Tudor, A. M., Melia, C. D., Davies, M. C., Hendra, P. J., Church, S., Domb, A. J., Langer, R. (1991) The application of the Fourier-Transform Raman spectroscopy to the analysis of poly(anhydride) home and copolymers, *Spectrochim. Acta* **47A**, 1335–1343.

Uhrich, K. E., Gupta, A., Thomas, T. T., Laurencin, C. T., Langer, R. (1995) Synthesis and characterization of degradable poly(anhydride-co-imides), *Macromolecules* **28**, 2184–2193.

Uhrich, K. E., Thomas, T. T., Laurencin, C. T., Langer, R. (1997) *In vitro* degradation characteristics of poly(anhydride-co-imides) containing trimellityimidoglycine, *J. Appl. Polym. Sci.* **63**, 1401–1411.

Wu, M. P., Tamada, J. A., Brem, H., Langer, R. (1994) *In vivo* versus *in vitro* degradation of controlled release polymers for intercranial surgery therapy, *J. Biomed. Mater. Res.* **28**, 387–395.

Yoda, N. (1962) Synthesis of poly(anhydrides). Poly(anhydrides) of five-membered heterocyclic dibasic acids, *Makromol. Chem.* **55**, 174–190.

Yoda, N. (1963) Synthesis of poly(anhydrides). Crystalline and high melting poly(amide) poly(anhydrides) of methylene bis(*p*-carboxyphenyl) amide, *J. Polym. Sci., Part A* **1**, 1323–1338.

Young, J. S., Gonzales, K. D., Anseth, K. S. (2000) Photopolymers in orthopedics: Characterization of novel crosslinked polyanhydrides, *Biomaterials* **110**, 1181–1188.

8
Polylactides
"NatureWorks™ PLA"

Dr. Patrick Gruber[1], Mr. Micheal O'Brien[2]
[1] Cargill Dow LLC, 15305 Minnetonka Blvd, Minnetonka, Minnesota 55345, USA;
 Tel.: +01 952 742 0444; Fax: +01 952 742 0477;
 E-mail: pat_gruber@cargilldow.com
[2] Cargill Dow LLC, 15305 Minnetonka Blvd, Minnetonka, Minnesota 55345, USA;
 Tel.: +01 952 742 0523; Fax: +01 952 742 0477;
 E-mail: michael_O'brien@cargilldow.com

AATCC	American Association of Textile Chemists and Colorists
BON	biaxially oriented Nylon
CD	cross-direction
COF	coefficient of friction
Den	denier
EVA	ethylene vinyl acetate
GPPS	general-purpose polystyrene
GRAS	generally recommended as safe
HDPE	high-density polyethylene
HIPS	high-impact polystyrene
LDPE	low-density polyethylene
MD	machine direction
Mn	number average molecular weight
OPET	oriented polyethylene terephthalate
OPLA	oriented polylactic acid
OPP	oriented polypropylene
PET	polyethylene terephthalate
PLA	polylactic acid (or polylactide)
PP	polypropylene
PS	polystyrene
ROP	ring-opening polymerization
T_g	glass transition temperature
VA EVA	vinyl acetate/ethylene vinyl acetate copolymer
WVTR	water vapor transmission rate

1

Introduction

Polylactic acid (PLA) is a highly versatile aliphatic, compostable polymer derived from 100% annually renewable resources. Because such resources replace oil as the feedstock, PLA requires 20–50% less fossil resources than comparable petroleum-based plastics. With PLA, CO_2 is removed from the atmosphere when growing the feedstock crop, and returned to the Earth when PLA is degraded. Since the process recycles the Earth's carbon, PLA has the potential to reduce atmospheric CO_2 levels. Disposal of PLA fits with existing systems, including the additional option of composting. Long term, with the correct infrastructure, PLA products could be recycled back to a monomer and into polymers.

The land mass necessary for feedstock production is minimal. Producing 500,000 tonnes of PLA requires less than 0.5% of the annual US corn crop. Since corn is a cheap dextrose source, the current feedstock supply is more than adequate to meet foreseeable demand.

In spite of PLA's excellent balance of properties and environmental benefits, its commercial viability has been limited traditionally by high production costs (more than $4 per kg). Until recently, PLA has enjoyed little success in replacing petroleum-based plastics outside of biomedical applications such as sutures (Lipinsky and Sinclair, 1986). The recent formation of Cargill Dow LLC in 1997 has brought focus to the development and production of Nature-Works™ PLA. Although the development of PLA is clearly in its early stages compared with conventional plastic, this new company is significantly reducing production costs while expanding the use of PLA beyond biomedical applications.

2
Historical Outline

Somewhat surprisingly, PLA has been the subject of many investigations for over a century. In 1845, Pelouze condensed lactic acid by distillation of water to form low molecular-weight PLA and the cyclic dimer of lactic acid, lactide (Pelouze, 1845). About 50 years later, an attempt was made to prepare PLA from lactide, but without success (Bischoff and Walden, 1894). In 1932, although another group was able to polymerize lactide to PLA, the method was unsuitable for practical use (Carothers et al., 1932). Later, Watson published a review on the possible uses of PLA for coatings and as a constituent in resins (Watson, 1948). Even though PLA has been known for over 100 years, its commercial viability or usefulness had not been practical. In 1986, PLA was described as having potential as a commodity plastic (Lipinsky and Sinclair, 1986).

In 1988, Cargill Incorporated began an investigation into lactic acid, lactide, and PLA and concluded that although PLA was an interesting material, it was not practical with the then-known technology. Consequently, Cargill began to address the manufacturing, melt processing and cost issues, and in 1994 commenced operation of a semi-works with 4000 tonnes annual capacity, together with large-scale trials on conventional polymer processing equipment. In 1997, Cargill and The Dow Chemical Company formed Cargill Dow LLC in order to develop and bring to full commercialization the PLA technology and products. It is planned that Cargill Dow will commence full commercial manufacture of PLA in November 2001.

3
Production of PLA

PLA can be prepared by both direct condensation of lactic acid and by the ring-opening polymerization (ROP) of the cyclic lactide dimer (Figure 1). Because the direct condensation route is an equilibrium reaction, difficulties in removing trace amounts of water during the late stages of polymerization generally limit the ultimate molecular weight achievable by this approach. Most work has focused on the ROP, although Mitsui Toatsu Chemicals has patented an azeotropic distillation process using a high-boiling solvent to drive the removal of water in the direct esterification process to obtain high molecular-weight PLA (Enomoto et al., 1995; Ichikawa et al., 1995; Kashima et al., 1995; Ohta et al., 1995).

Cargill Dow LLC has developed a patented, low-cost continuous process for the production of lactic acid-based polymers (Gruber et al., 1992a,b, 1993a,b,c, 1994, 1996). The process combines the substantial environ-

H₂O

HO—⟨CH₃, H⟩—OH

L-Lactic acid

O
⟨O⟩
H CH₃
L-PLA

H₂O

O
H
CH₃
L-Lactide

Fig. 1 Polymerization routes to polylactic acid.

mental and economic benefits of synthesizing both lactide and PLA in the melt rather than in solution and, for the first time, provides a commercially viable compostable commodity polymer made from annually renewable resources. The process starts with a continuous condensation reaction of aqueous lactic acid to produce low molecular-weight PLA prepolymer (Figure 2). Next, the prepolymer is converted into a mixture of lactide stereoisomers using tin catalysis to enhance the rate and selectivity of the intramolecular cyclization reaction. The molten lactide mixture is then purified by vacuum distillation. Finally, PLA high polymer is produced using a tin-catalyzed, ring-opening lactide polymerization in the melt, completely eliminating the use of costly and environmentally unfriendly solvents. After the polymerization is complete, any remaining monomer is removed under vacuum and recycled to the beginning of the process (Figure 3). This process is currently in operation at a 6000 tonnes per year market development facility in Minnesota. The construction of a 140,000 tonnes per year commercial-scale PLA plant in North America was recently announced by Cargill Dow

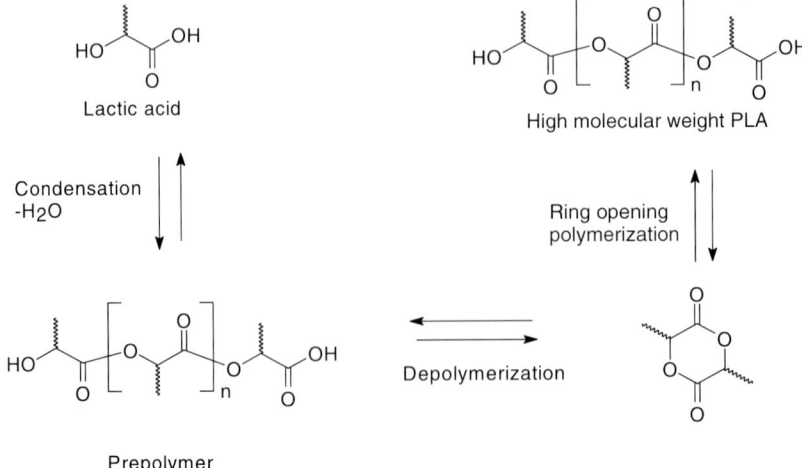

Fig. 2 Schematic of PLA production via prepolymer and lactide.

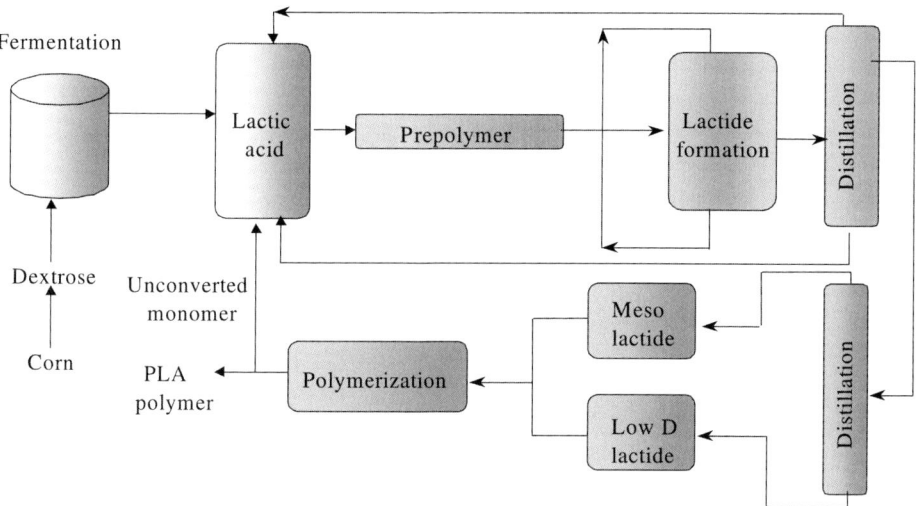

Fig. 3 Non-solvent process to prepare polylactic acid.

LLC for start-up in the fourth quarter of 2001, with plans to construct an addition plant in Europe in the near future.

3.1
Polymerization of Lactide

Many catalyst systems have been evaluated for the polymerization of lactide, including complexes of aluminum, zinc, tin, and lanthanides. Even strong bases such as metal alkoxides have been used with some success. Depending on the catalyst system and reaction conditions, almost all conceivable mechanisms [cationic (Kricheldorf and Krieser, 1986), anionic (Kurcok et al., 1992), coordination (Kricheldorf et al., 1995; Nijenhuis et al., 1995), etc.] have been

proposed to explain the kinetics, side reactions, and nature of the end groups observed in lactide polymerization. Tin compounds, especially tin(II) bis-2-ethylhexanoic acid (tin octoate), are preferred for the bulk polymerization of lactide due to their solubility in molten lactide, high catalytic activity, and low rate of racemization of the polymer. Conversions of >90% and less than 1% racemization can be obtained while providing polymer of high molecular weight.

The polymerization of lactide using tin octoate is generally thought to occur via a coordination–insertion mechanism (Kricheldorf et al., 1995; Nijenhuis et al., 1995), with ring opening of the lactide to add two lactic acid molecules to the growing end of the polymer chain (Figure 4).

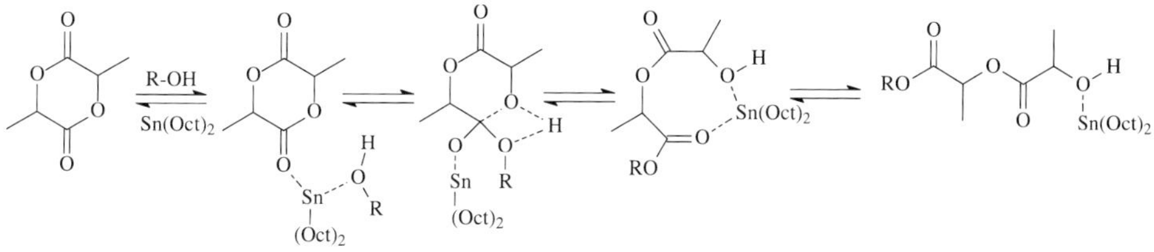

Fig. 4 Copolymerization of lactide and caprolactone.

High molecular-weight polymer, a good reaction rate, and low levels of racemization are observed with tin octoate-catalyzed polymerization of lactide. Typical conditions for polymerization are 180–210 °C, tin octoate concentrations of 100–1000 ppm, and 2–5 h to reach ca. 95% conversion. The polymerization is first order in both catalyst and lactide. Frequently, hydroxyl-containing initiators, such as 1-octanol, are used both to control molecular weight and to accelerate the reaction.

Copolymers of lactide with other cyclic monomers such as caprolactone (Sinclair, 1977a,b) can be prepared using similar reaction conditions (Figure 5). These monomers can be used to prepare random copolymers or block polymers because of the end growth polymerization mechanism. Cyclic carbonates, epoxides and morpholine-diones have also been copolymerized with lactide.

4
PLA Packaging Applications and Performance

PLA represents a real breakthrough, and a significant step towards more sustainable packaging applications. PLA polymers exhibit a balance of performance properties that are comparable with – and in certain cases superior to – those of traditional thermoplastics. Initially, two specific packaging areas have received close attention, namely high-value films and rigid thermoformed containers. The functional properties and benefits of PLA in these areas are listed in Table 1.

4.1
Deadfold and Twist Retention

One feature that is inherent to paper and foil, but is usually lacking in plastic films, is the ability to hold a crease or fold, or the ability to

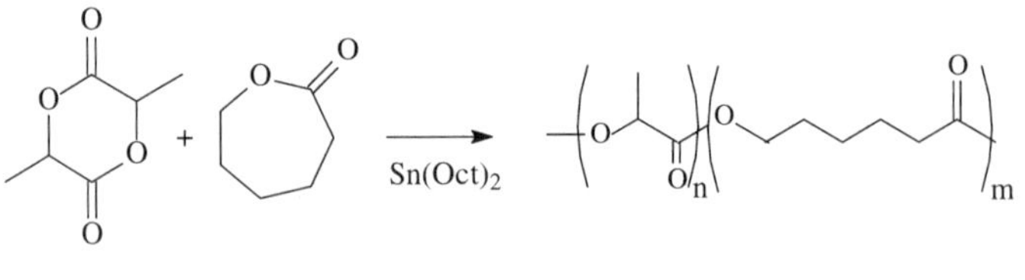

Lactide   ε-Caprolactone

Fig. 5 Generalized coordination–insertion chain growth mechanism of lactide to PLA. R = growing polymer chain.

Tab. 1 PLA Functional Properties for Packaging

Functional Property	Packaging Improvement
Deadfold, twist and crimp	Improved folding and sealing
High Gloss & Clarity	Package aesthetics
Barrier Properties	Grease and oil resistance
Renewable Resource	Made from CO_2
Flavor and Aroma Prop.	Reduced taste/odor issues
Low temp. Heat Seal	Stronger seals at lower temperatures
High Tensile and Modulus	Wet paper strength, ability to downguage coating
Low COF, Polarity	Printability
GRAS Status	Food contact approved

retain a twist that is imparted in order to close the edges of the film around a small object (such as a candy lozenge). Oriented PLA film is shown to have excellent deadfold and twist retention. The combination of high modulus, high clarity, and characteristics of excellent deadfold and twist retention found in oriented PLA film are more comparable with those of cellophane films than many other plastic films.

4.2
High Gloss and Clarity

PLA is of relatively high modulus, this being comparable with that of cellophane and oriented polyethylene terephthalate (PET), two to three times that of oriented Nylon 6,6 or polypropylene, and more than 10 times that of low-density polyethylene (LDPE). The optical properties of PLA are known to be sensitive to additive and fabrication effects.

4.3
Barrier Properties

PLA is an inherently polar material due to the basic repeat unit of lactic acid. This high polarity leads to a number of unique attributes such as a high critical surface energy that yields excellent printability. Another benefit of this polar polyester polymer is the resistance to aliphatic molecules such as oils and terpenes. Further details are provided in Tables 2 and 3.

Tab. 2   Grease Resistance of Coated Paper 55 °C

Polymer Coating	Hours to Failure at 55 °C
PLA	> 120 hrs, no failure
LDPE	10 hrs
Fluorinated Paper	96 hrs

Tappi T507, ASTM F119

Tab. 3   Grease Resistance of PLA Coated Paper

Grease or Oil	23 °C no press.	55 °C with press.
Mineral Oil	> 120 hrs	> 24 hrs
Olive Oil	> 120 hrs	> 24 hrs
Oleic Acid	> 120 hrs	> 24 hrs
Butter	> 120 hrs	> 24 hrs

Tappi T507

4.4
Low Temperature Heat Seal

The heat seal initiation temperatures of the amorphous PLA sealant and the 18% vinyl acetate/ethylene vinyl acetate copolymer (VA EVA) formulation were equivalent and between 80 and 85 °C. The ultimate seal strength of the EVA formulation in this type of high-density polyethylene (HDPE) coextrusion has been optimized over time to have approximately 360 g cm^{-1} peel strength over a broad temperature range. Approximately 540 g cm^{-1} peel strength was observed with natural PLA in this initial coextrusion structure. The combination of a delamination failure mode with this peel strength value provides interesting potential for PLA as a component of an "easy-open" package that utilizes a flavor and aroma barrier polymer as the high-performance sealant. Features leading to the descriptor of "high performance" would include low seal initiation temperatures, high hot tack strength, and the addition of attributes such as resistance to flavor and odor scalping, or the addition of easy-open seal characteristics. Hot tack comparisons of the EVA formulation and the PLA coextrusions show enhanced hot tack strengths for PLA, with a peak hot tack strength observation of 450 g cm^{-1} with the PLA coextrusion and a peak hot tack strength of 130 g cm^{-1} for the EVA formulation.

4.5
High-tensile Modulus Strength

PLA is considered a high-modulus polymer, and has a tensile modulus of 3500–4000 MPa that is comparable with that of PET or cellophane and about double that of polyolefins such as HDPE or polypropylene (PP). This increased modulus and tensile strength give the excellent deadfold and crimp similar to cellophane; in addition, when replacing polyolefins, it provides the ability to downgauge either the paper/board substrate or the coating thickness while retaining package strength.

4.6
Film Characteristics

PLA is a glassy polymer at room temperature, with a T_g of about 55–65 °C. PLA yields enhanced physical properties when oriented, leading to interest in biaxially oriented films produced via tentering operations, and thermoformed rigid containers. PLA with D-isomer levels below 8% can be semi-crystalline if nucleated, annealed, or subjected to strain-induced orientation processes. Enhanced toughness and heat resistance accompanies crystallinity development. The melting point of PLA increases with decreasing content of the D-isomer, and ranges from about 130 to 175 °C.

PLA has critical surface energy of 38 dynes cm², making it receptive to holding corona treatment well, and therefore easily printable and metallizable. The relatively high critical surface energy also leads to films with inherent anti-fogging behavior.

4.7
Packaging Conclusions

Polylactide, or PLA, is a new thermoplastic for packaging applications that is derived from annually renewable agricultural resources. PLA can be fabricated in a variety of familiar processes and brings a new combination of attributes to packaging, including stiffness, clarity, deadfold and twist retention, low-temperature heat sealability, as well as an interesting combination of barrier properties including flavor and aroma barrier characteristics. The combined attributes of this new thermoplastic make PLA a performance polymer that is environmentally and economically appealing. The physical properties of PLA relating to its packaging applications are summarized in Table 4.

5
PLA for Fibers and Nonwovens

Some of the key performance features that distinguish PLA from other fiber materials are described in Table 5. These features are fueling interest in using PLA for several fiber and nonwoven applications.

5.1
Applications and Properties

Many fiber manufacturers and end-users are developing market applications for PLA fibers. Some of the more promising areas include clothing, home furnishings, binder fibers and nonwoven applications for use in wipes, hygiene and agriculture. The following sections discuss various applications using PLA and the specific properties of greatest interest for that application.

5.2
Clothing

Properties of particular interest to the clothing industry are highlighted in Table 6.

Tab. 4 PLA Physical Properties Summary

Film Type	*MD, 1% Secant Modulus, MPa*	*CD, 1% Secant Modulus, MPa*	*Dead fold*	*Twist Retention*	*Haze [%]*	*20 Gloss*	*O_2P*	*WVTR*
PLA blown natural	3000	3000	–	–	0.7	110	570	340
PLA blown with COF additives	–	–	–	–	11.8	30	710	370
OPLA with COF additives	3400	5200	95	280	2.8	60	570	320
Cellophane	3900	2500	71	310	–	–	–	–
OPET	4100	4100	36	190	4.1	80	50	25
BON	1600	1500	15	16	–	–	20	160
OPP	1500	1500	12	150	–	–	1500	5
LDPE (0.6 MI)	200[a]	200[a]	–	–	5.7	20	6500	20

[a]LDPE data was 2% secant modulus.

Test procedures: Physical testing included ASTM test methods for secant modulus (ASTM D 882), haze (ASTM D 1003), gloss (ASTM D 2457), moisture transmission rate (ASTM F 1249), and oxygen permeability rate (ASTM D 3985). Heat seal and hot tack data were generated on a DTC hot tack tester using test conditions of 0.275 N/mm2 psi pressure and a 0.5 s dwell time with samples 25.4 microns wide. The heat seal samples were pulled open 24 hours after preparation using a tensiometer and a 10 in/min crosshead speed. The hot tack samples were pulled open 0.2 seconds after preparation using a pull rate of 200 mm/min. The permeability of flavor and aroma molecules was measured using a Dow designed instrument which has been described in the literature[4]. Deadfold was measured by clamping the film in place along one end, folding and creasing the film using 20 psi pressure for 0.5 seconds, and then releasing the film and measuring the angle retained. For the dead fold retention test, a larger value indicates that the film had improved dead fold characteristics. Twist retention was measured by clamping the film in place along one end, twisting the free end 360 °, and releasing. The retained angle was measured at 1 minute. A larger value indicates that the film had improved twist retention.

Tab. 5 PLA Key Performance Features – Fibers and Nonwovens

More hydrophilic than PET	Lower density than PET
Excellent hand, drape and feel	Dyeable with dispersion dyes
Good resilience	Outstanding processability
Excellent crimp and crimp retention	Controllable thermal bonding temperature
Controllable shrinkage	Grades offer range of crystalline melt temps from 120–170 °C
Tenacity up to 7 g/den^{-1}	Low flammability and smoke generation
Unaffected by UV light	

PLA is an aliphatic polyester and therefore contains no aromatic ring structures. The moisture regain and wicking properties are superior to those of PET, and garments made from 100% PLA or blends of PLA with wool and cotton feel more comfortable. In addition, the lower modulus leads to better drape and hand-feel, whilst the elastic recovery and crimp retention properties lead to excellent shape retention and crease resistance. Although not a nonflammable polymer, the fiber shows improved self-extinguishing characteristics.

Kanebo, Inc. introduced a PLA fiber under the trade name LACTRON™ fiber at the February 1998 Nagano Olympics under the theme of “Fashion for the Earth.” Kanebo exhibited several garments from PLA or

Tab. 6 Clothing: Key Property Comparisons

	PET	*PLA*
Moisture management	Good wicking, cos. contact angle = 0.135, 0.2%–0.4% moisture regain	Better wicking cos. contact angle = 0.254, 0.4%–0.6% moisture regain
Flammability	Burns 6 min. after flame removed	Burns 2 minutes after flame removed
Resilience	51% recovery at 10% strain	64% recovery at 10% strain
Renewable resource	Petroleum-based	Dextrose based (corn)
Drape/Hand-feel	Poor	Good
Luster	Medium to low	Very high to low
Crease Resistance	Good	Excellent

Tab. 7 Laundering Test

Simulated conditions	AATCC test[a]	Burst strength (psi)	Lightfastness (AATCC Gray Scale Rating)*	% Dimensional change (width/length)	Mn	Mw
Control		83	–	–	57694	117970
Cold hand wash(40⁰C)	–	79	5.0	0/ − 3.82	56343	107835
Same, no bleach	1A	82	4.7	0/ − 3.13	52123	108115
Cold machine wash(40⁰C)	–	75	5.0	0/ − 4.17	56281	111206
Same, no bleach	–	78	4.8	0/ − 3.82		
Warm machine wash(50⁰C)	5A	74	5.0	+ 6.25/ − 7.98	57190	112086
Same, no bleach	2A	78	4.5	+ 7.64/ − 7.98		
Hot machine wash(70⁰C)	4A	74	4.8	+ 2.08/ − 6.25	58085	112510
Same, no bleach	3A	76	4.9	+ 2.08/ − 6.25		

*AATCC gray scale for color change rating where: 5 denotes "no" change, 4 denotes "slight" change, 3 denotes "noticeable" change, 2 denotes "considerable" change, and 1 denotes "heavy" change in color
[a]AATCC Test Method 61–1994 (35% PLA/ 65% Cotton Blend Knitted Shirt, Simulates five Washings)

Tab. 8 Skin Modeling Test

Test:	PLA/Cotton Comparison Rating:
Thermal Resistance	Superior
Water Vapor Permeability Index	Equivalent
Water Vapor Absorbency	Superior
Buffering Index	Equivalent
Moisture Permeability	Equivalent
Sweat Transport	Equivalent
Sweat Uptake	Equivalent
Thermal Resistance	Superior
Drying Time	Equivalent

PLA/natural fiber blends. In autumn 2000, Woolmark and Cargill Dow announced a joint-marketing agreement for wool and NatureWorks™ fibers (Cargill Dow's trade name for PLA fibers).

Fabrics containing PLA are also being considered for sportswear as well as outerwear applications. The durability in laundering of fabrics from PLA has been tested under simulated laundry studies (Table 7) in accordance with American Association of Textile Chemists and Colorists (AATCC) standards.

There is considerable interest in PLA either as an inner wicking layer or as an intimate blend with other natural fibers. In terms of comfort under normal and active wear conditions, independent laboratory testing by the Hoehenstein Research Institute demonstrated that PLA fibers perform better than polyester (PET) fibers when combined with cotton in the plaited fabric that was tested.

The results of the Hoehenstein testing concluded that wearers of PLA/cotton fabric will experience improved physiological comfort versus equivalent polyester faced fabric (Forchungsinstitut Hohenstein, 2000). Tables 8 9 10 summarize the results of the skin model, sensorial testing and comfort votes of PLA/cotton fabric are summarized in Tables 8, 9 and 10.

Tab. 9 Sensorial Comfort Testing

Test	PLA/Cotton Comparison Rating
Wet cling index	Equivalent
Surface Index	Superior
Stiffness	Superior

Tab. 10 Overall rating of PLA/cotton versus PET/cotton fabric

Test	PLA/Cotton Comparison Rating
Sensorial wear comfort	Equivalent
Total wear comfort	Superior

5.3
Carpeting and Furnishings

The superior UV resistance and reduced smoke characteristics are two attractive properties, which differentiate PLA fabrics (Table 11). PLA's superior resilience is an added feature that brings promise to this potential application area.

5.4
Other Fiber Application Areas

There are many other potential applications that might benefit from these new PLA polymers (Table 12).

5.5
Summary

This new fiber, derived entirely from annually renewable resources (e.g., corn) combines the comfort and feel of natural fibers with the performance of synthetics (Table 13). The unique property spectrum of PLA fibers allows the creation of products

Tab. 11 Furnishings key property comparisons

	PET	PLA
Resilience	65% recovery at 5% strain	93% recovery at 5% strain
Fading resistance	30% loss in elong 100 h xenon arc	0% loss in elong. 100 h xenon arc.
Flammability	Burns 6 min. after flame removed	Burns 2 min. after flame removed
Smoke generation	394 m^2 kg^{-1}	63 m^2 kg^{-1}
Heat liberation	Peak release rate 38 W m^{-2}	Peak release rate 22 W m^{-2}
Sustainability	Petroleum-based	Dextrose-based (corn)
Density	1.34 g mL^{-1}	1.25 g mL^{-1}

Tab. 12 Other fiber application areas

Wipes (household and industrial)	Disposable incontinence (diaper top sheets, acquisition layers)
Feminine hygiene	Disposable garments (sms constructions)
Filtration and high loft nonwovens	UV resistant fabrics for exterior use (awnings, ground cover, etc)
Soil erosion control (geotextiles)	General agriculture (plant pots)
Self crimping/fibrillatable/microdenier – bicomponents	

Tab. 13 Fibers properties summary

Fiber property	Synthetics			PLA	Natural fibers			
Specific gravity	1.14	1.39	1.18	1.25	1.52	1.52	1.34	1.31
Tenacity (g/d)	5.5	6.0	4.0	6.0	2.5	4.0	4.0	1.6
Moisture regain %	4.1	0.2–0.4	1.0–2.0	0.4–0.6	11	7.5	10	14–18
Elastic recovery (5% strain)	89	65	50	93	32	52	52	69
Flammability	Medium	High smoke	Medium	Low smoke	Burns	Burns	Burns	Burns slowly
UV resistance	Poor	Fair	Excellent	Excellent	Poor	Fair-poor	Fair-poor	Fair
Wicking (L-W slope; higher slope, more wicking)	–	0.7–0.8 (no finish)	–	6.3–7.5 (no finish) 19–26 (with finish)	–			

with superior hand-feel and touch, drape, comfort, moisture management, UV resistance and resilience. Combining these performance features with the comfort of natural fibers enables PLA to be used in a wide spectrum of products including clothing, carpets, nonwoven/fiberfill and household and industrial markets.

6
Environmental Sustainability

Conventional synthetic polymers rely on oil reserves for their feedstock source, the main drawbacks being that fossil resources not only take millions of years to regenerate but that their use also has a harmful impact on the environment. In contrast, the monomer in PLA is derived from renewable resources such as agricultural crops that are grown and harvested annually, harnessing energy from the sun. However, as is the case for all polymers, fossil fuels are used in the processing of raw materials and resin production for PLA. The total petroleum use (feedstock + process energy) of pellet production for conventional plastics compared with the projected total petroleum use for the production of PLA pellets is shown in Figure 6. PLA year 1 data represent the production facility currently under construc-

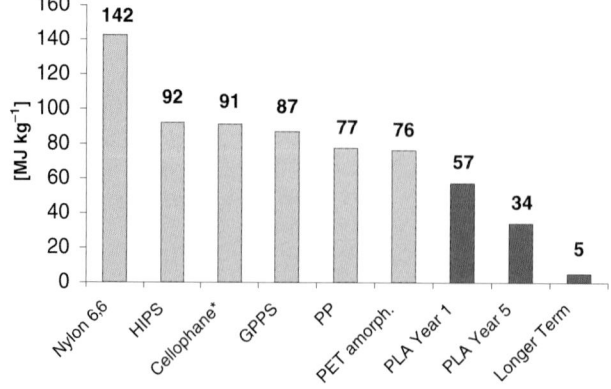

Fig. 6 Total fossil resource use in common plastics, measured in energy.

tion by Cargill Dow. PLA year 5 and long-term data assume new product technology and alternative energy sources further maximize production efficiencies for PLA.

6.1
CO₂ Emissions

With PLA, CO_2 is removed from the atmosphere when growing the feedstock crop and returned to the Earth when PLA is degraded. As the process recycles the Earth's carbon, PLA has the potential to reduce CO_2 levels in the atmosphere. In Figure 7, PLA year 1 data represent the production facility currently under construction by Cargill Dow but projected to be on-line in late 2001. PLA year 5 and long-term data assume that new process technology and alternative energy sources further reduce CO_2 contributions by PLA. Figure 7 compares net CO_2 emissions between conventional plastics and PLA (from cradle to pellets).

6.2
Degradation and Hydrolysis

The primary mechanism of degradation is hydrolysis, followed by bacterial attack on the fragmented residues. The environmental degradation of PLA occurs by a two-step process. During the initial phases of degradation, the high molecular-weight polyester chains hydrolyze to lower molecular-weight oligomers, this reaction being accelerated by acids or bases and being affected by both temperature and moisture levels. Embrittlement of PLA occurs during this step at a point where the Mn decreases to less than ~40,000. At about this same Mn, microorganisms in the environment continue the degradation process by converting these lower molecular-weight components to CO_2, water, and humus (Narayan, 1993). The structural integrity of molded PLA articles decreases as the molecular weight falls, and eventually the article disintegrates. A typical degradation curve of PLA under composting conditions is shown in Figure 8.

During typical use and storage conditions, PLA products are extremely stable. In addition, certain additives can be used to retard

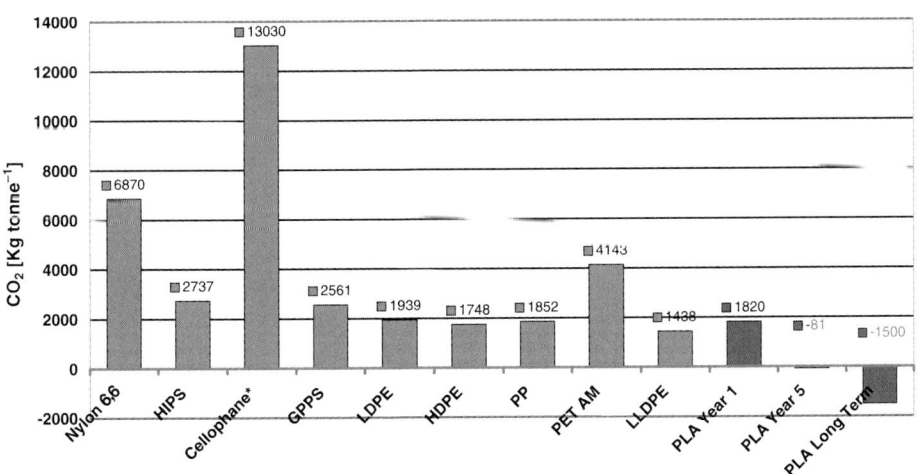

Fig. 7 Net CO_2 emissions: cradle to pellets.

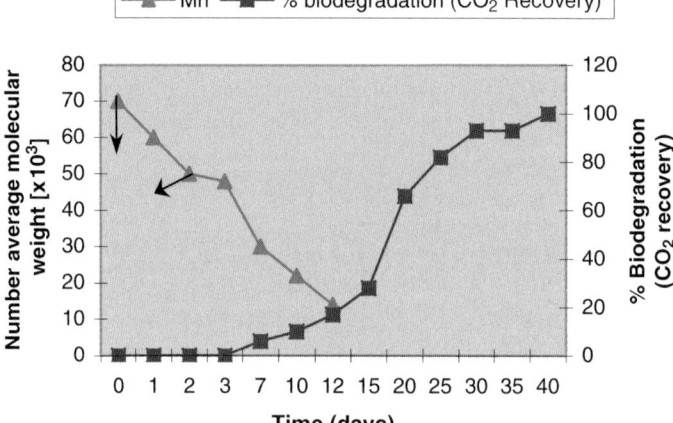

Fig. 8 Biodegradation of PLA in compost at 60 °C.

hydrolysis, though continuing studies in this area will lead to increased PLA stability for more extreme applications.

6.3
Disposal Options

PLA fits all disposal routes available to conventional plastics, including recycling and incineration. However, because of PLA's ability to be degraded under controlled conditions, composting and recycling back to the monomer represent two new disposal options available to the public. Where an infrastructure exists, composting can be especially useful in reducing the landfill space that is needed in the more densely populated continents, such as Europe. Figure 9 demonstrates how the use of annually renewable resources in the production of PLA and composting can aid in the recycling of the Earth's CO_2 content.

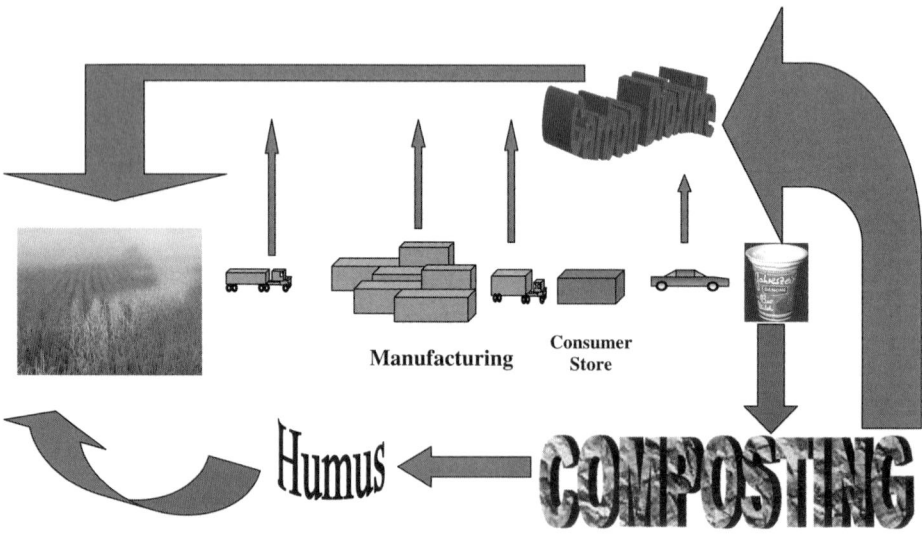

Fig. 9 Composting of PLA.

7
Outlook and Perspectives

Low-cost PLA products are finding uses in many applications, including packaging films, fibers, nonwovens and a host of molded articles. PLA is not being used in these applications solely because of its degradability, nor because it is made from renewable resources – it is being used because it functions very well and provides excellent properties at a competitive price. The use of PLA as a cost-effective alternative to commodity petrochemical-based plastics will in turn increase the demand for agricultural products such as corn and sugar beet, and lessen the dependence of plastics on oil. Although PLA is in its infancy as a material, technological improvements to modify its ductility, heat resistance and hydrolytic stability in order to widen its material performance are already underway at a variety of institutes and companies. Finally, PLA provides the first example in the polymer industry where environmentally friendly bioprocessing and chemical processing technologies are brought together with biomass feedstocks to create a major commercial opportunity. The techniques learned with PLA will translate to other renewable resource-based chemicals and polymers, thereby establishing a new industry in the long term.

Acknowledgments

The authors thank Mark Hartmann, Nicole Whitemann, Erwin Vink, James Lunt, Kevin McCarthy, Lang Phommahaxay, Ray Drumright, David E. Henton, Patrick Smith, Jed Randall and Tom Bremel for assistance in these studies. NatureWorks™ PLA is a registered trademark of Cargill Dow LLC

8
References

Nijenhuis, A. J., Du, Y. J., Van Aert, H. A. M., Bastiaansen, C. (1995) *Macromolecules* **28**, 2124.

Bischoff, C. A., Walden, P. (1894) Ueber Derivate der Milchsäure. *Liebigs Ann. Chem.* **279**, 71–99.

Carothers, W. H., Dorough, G. L., Van Natta, F. J. (1932) Studies of polymerization and ring formation, X. The reversible polymerization of six-membered cyclic esters. *J. Am. Chem. Soc.* **54**, 761–772.

Chemical Week, May 2000, **162**(18), 24E.

Ichikawa, F., Kobayashi, M., Ohta, M., Yoshida, Y., Obuchi, S., Itoh, H. (1995) U.S. Patent 5 440 008.

Forschungsinstitut Hohenstein (2000) Test Report no. ZO.4.3805. Comparison of Physiological Comfort of polyester (PET)/cotton and Nature-Works™ fibers/cotton fabrics, 24 May 2000.

Kricheldorf, H. R. and Krieser, I. (1986) *Makromol. Chem.* **187**, 1861.

Kricheldorf, H. R., Kreiser-Saunders, I., Boettcher, C. (1995) *Polymer* **36**(6), 1253.

Enomoto, K., Ajioka, M., Yamaguchi, A. (1995) U.S. Patent 5 310 865.

Lipinsky, E. S., Sinclair, R.G. (1986) *Chemical Engineering Progress*, August.

Ohta, M., Yoshida, Y., Obuchi, S., Yoshida, Y. (1995) U.S. Patent 5 440 143.

Kurcok, P., Penczek, J., Franek, J., Kedlinski, Z. (1992) *Macromolecules* **25**(9), 2285.

Gruber, P. R., Hall, E. S., Kolstad, J. J., Iwen, M. L., Benson, R. D., Borchardt, R.L. (1992a) U.S. Patent 5 142 023.

Gruber, P. R., Hall, E. S., Kolstad, J. J., Iwen, M. L., Benson, R. D., Borchardt, R.L. (1992b) U.S. Patent 5 247 058.

Gruber, P. R., Hall, E. S., Kolstad, J. J., Iwen, M. L., Benson, R. D., Borchardt, R.L. (1993a) U.S. Patent 5 247 059a.

Gruber, P. R., Hall, E. S., Kolstad, J. J., Iwen, M. L., Benson, R. D., Borchardt, R.L. (1993b) U.S. Patent 5 258 488.

Gruber, P. R., Hall, E. S., Kolstad, J. J., Iwen, M. L., Benson, R. D., Borchardt, R.L. (1993c) U.S. Patent 5 274 073.

Gruber, P. R., Hall, E. S., Kolstad, J. J., Iwen, M. L., Benson, R. D., Borchardt, R.L. (1994) U.S. Patent 5 357 035.

Gruber, P. R., Hall, E. S., Kolstad, J. J., Iwen, M. L., Benson, R. D., Borchardt, R.L. (1996) U.S. Patent 5 484 881.

Pelouze, J. (1845) Mémoire sure l'acide lactique, *Ann. chimie.* **13**, 257–268.

Plastics Technology, January 1998, pp. 13–15.

Narayan, R. (1993) Degradation of polymeric materials, in: *Science and Engineering of Composting: Design, Environmental, Microbiol* Narayan, R. (1993) Degradation of polymeric materials, in: *Science and Engineering of Composting: Design, Environmental, Microbiological and Utilization Aspects* (Hoitink, H. A., Keener, H. N., Eds.), Ohio: Renaissance Publications, 00–00.

Ogical and Utilization Aspects (Hoitink, H. A., Keener, H. N., Eds.), Ohio: Renaissance Publications, 00–00.

Narayan, R. (1993) Degradation of polymeric materials, in: *Science and Engineering of Composting: Design, Environmental, Microbiological and Utilization Aspects* (Hoitink, H. A., Keener, H. N., Eds.), Ohio: Renaissance Publications, 00–00.

Sinclair, R. G. (1977a) U.S. Patent 4 045 418.

Sinclair, R. G. (1977b) U.S. Patent 4 057 537.

Kashima, T., Kameoka, T., Ajioka, M., Yamaguchi, A. (1995) U.S. Patent 5 428 126.

Watson, P. D. (1948) Lactic acid polymers as constituents of synthetic resins and coatings. *Ind. Eng. Chem.* **40**, 1393–1397.

9
Polylactic acid "LACEA"

Nobuyuki Kawashima[1], Dr. Shinji Ogawa[2], Shoji Obuchi[3], Mitsunori Matsuo[4],
Tadashi Yagi[5]

[1] LACEA Business Development Unit, Mitsui Chemicals, Inc., 3-2-5 Kasumigaseki,
Chiyoda-ku, Tokyo 100-6070, Japan; Tel.: +81-3-3592-4479; Fax: +81-3-3592-4268;
E-mail: nobuyuki.kawashima@mitsui-chem.co.jp

[2] PLA Group Process Technology Laboratory R&D Center, Mitsui Chemicals, Inc., 30
Asamuta-machi, Omuta-shi, Fukuoka 836-8610, Japan; Tel.: +81-944-51-8261;
Fax: +81-944-52-4725; E-mail: sinzi.ogawa@mitsui-chem.co.jp

[3] PLA Group Process Technology Laboratory R&D Center, Mitsui Chemicals, Inc.,
580-32 nagaura, sodegaura-shi, Chiba 299-0265, Japan; Tel.: +81-438-64-2300;
Fax: +81-438-64-2356; E-mail: shoji.obuchi@mitsui-chem.co.jp

[4] LACEA Business Development Unit, Mitsui Chemicals, Inc., 3-2-5 Kasumigaseki,
Chiyoda-ku, Tokyo 100-6070, Japan; Tel.: +81-3-3592-4479; Fax: +81-3-3592-4268;
E-mail: mitsunori.matsuo@mitsui-chem.co.jp

[5] Chemical Safety Research Laboratory, Mitsui Chemical Analysis, & Consulting
Services, Inc.144 Togo, Mobara-shi, Chiba 297-0017, Japan; Tel.: +81-475-25-6732;
Fax:+81-475-25-6750; E-mail: tadashi.yagi@mitsui-chem.co.jp

BPS	Biodegradable Plastics Society in Japan
D_0	on-set temperature of weight loss
D_1	1% weight loss temperature
GreenPla	a nick-name of biodegradable plastics which was selected publicly under the sponsorship of METI and BPS in Japan
JORA	Japan Organics Recycling Association
LACEA	trade name of PLA production developed by Mitsui Chemicals, Inc.
LCA	life cycle assessment
METI	The Ministry of Economy, Trade and Industry
OECD	Organization for Economic Cooperation and Development
OPP	oriented polypropylene
PET	polyethylene terephthalate
PLA	polylactic acid
POY	partially oriented yarns
PP	polypropylene
PS	polystyrene
PVC	polyvinylchloride
ROP	ring-opening polymerization
T_c	crystallization temperature
T_g	glass transition temperature
T_m	melting point

1
Introduction

Polylactic acid (PLA) is a novel material that is endowed with stability and functions similar to those of conventional plastics, and is safely biodegradable in the composting process or in soil or water environments. Unlike other plastics that are derived from dwindling fossil resources, PLA is available from plant resources, and is expected to contribute to reducing various problems associated with environment, waste disposal and resources.

Mitsui Chemicals is conducting PLA business development under the tradename of LACEA, and in this chapter the raw materials for PLA production, together with details of the biodegradability, production and processing technologies for PLA will be described. In addition, the marketing activities for LACEA, the identification and labeling system for GreenPla (the brand-name for biodegradable plastics which was selected publicly under the sponsorship of METI and BPS in Japan) and its laws and regulations in Japan, and challenges to the market expansion for GreenPla and LACEA will be discussed.

2
Historical Outline

Although the existence of PLA has been recognized widely since the 1960s, its application has in the past been limited to medical devices, mainly because of its high cost. Later, during the late 1980s, Battelle and Dupont each conducted a study of PLA production technology and processing technology aimed mainly at applications for commodity plastics. During the mid-1990s, Cargill commenced semi-commercial production of PLA using a lactide polymeri-

zation process, and subsequently announced the construction of plant to produce 140,000 tonnes of PLA.

Mitsui Chemicals has now developed a direct condensation polymerization process of lactic acid for the production of PLA, and has commenced the operation of a pilot plant. This company has also recently successfully developed a solid-state polymerization process for PLA production which is similar, in principle, to the production process for polyethylene terephthalate (PET).

In parallel with the production process development, Mitsui Chemicals is conducting PLA-based material development, processing technology development and market development.

3
Raw Materials

The lactic acid from which PLA is produced is generated by the fermentation of starch and saccharides available from corn, sugar beet, sugar cane, potatoes, and other biomass (Figure 1). If about 5% (equal to 6.8 million tons of the worldwide annual plastics production were to be replaced by PLA, the theoretical requirement for saccharides would be 9.9 million tons. The annual output of major plant resources in the world that can produce saccharides is about 3,900 million tons, and the quantity of saccharides obtained from them is estimated

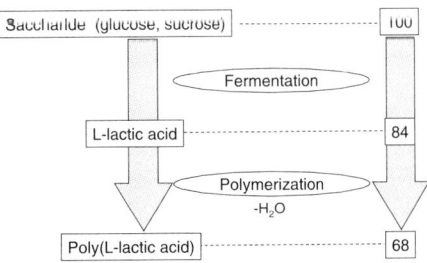

Fig. 1 Polylactic acid available from saccharide.

to be about 1,400 million tons. This suggests that 5% of the world's annual output of plastics can be replaced with PLA by utilizing about 0.7% of the major plant resources (Kawashima, 2000) (Tables 1 and 2). Consequently, the consumption of fossil resources can be reduced by an equivalent amount. A life cycle assessment (LCA) report has shown that PLA and starch-based materials (starch/aliphatic polyester-based materials) will not strain fossil resources as much as conventional plastics when the complete life cycle from raw material to disposal after use is considered, and will in fact reduce environmental loadings (Abstracts of Biodegradable Plastics Conference, 2000).

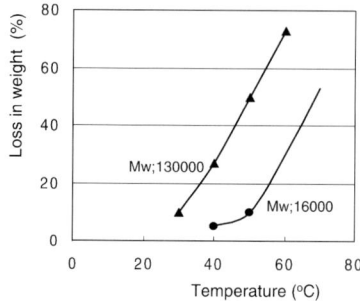

Fig. 2 Temperature dependence of PLA hydrolysis.

4

Biodegradability

The hydrolysis of PLA has been studied extensively, and details of the pH dependence of hydrolysis have been well documented (Makino et al., 1985). PLA hydrolysis is known to be accelerated by either acidic or alkaline conditions, and is also temperature-dependent (Yagi et al., 1997a) in association with the glass transition temperature (T_g) (Figure 2). It is known that, whilst hydrolysis at room temperature under neutral conditions produces an oligomer of PLA (Li et al., 1990), hydrolysis under the conditions specified in Figure 2 (pH 9, 60 °C) sharply

Tab. 1 Major plant resources and PLA

Major plant resources (as converted into saccharide)	PLA
9.9 million tons	6.8 million tons
0.7% of potential production in the world	Approx. 5% of plastics production in the world

Tab. 2 Major plant resources, and their availability in terms of saccharide (Units: 100 million tons year⁻¹, %)

Major plant resources	Worldwide crop[a]	Theoretical yield of saccharide	Theoretical quantity of saccharide
Wheat	6.10	60[b]	3.66
Corn	5.86	60[b]	3.52
Rice	5.73	60[b]	3.44
Potato	2.95	15[b]	0.44
Cassava	1.64	20[b]	0.33
Sweet potato	1.38	25[b]	0.34
(Subtotal)	(23.66)		(11.73)
Sugar cane	12.4	15[c]	1.86
Sugar beet	2.63	15[c]	0.39
Total	38.7		14.0

[a]FAO Production Yearbook 1998, Geographical Statistics Handbook. [b]Summation of starch yield and glucose yield. [c]Calculated from sucrose yield.

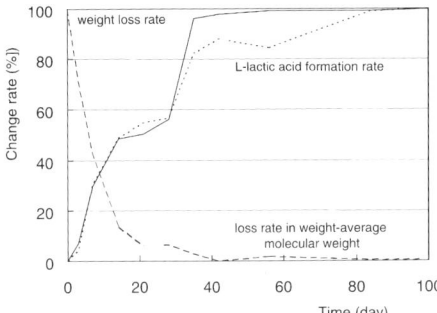

Fig. 3 PLA hydrolysis curve (conditions: pH 9, 60 °C).

reduces the molecular weight to an extent that the temporarily generated oligomer is finally degraded into monomers (Figure 3) (Yagi et al., 1997a).

According to the latest reports, a number of microorganisms have been discovered that are capable of degrading PLA. It is interesting to note that high molecular-weight PLA can easily be degraded by specific microorganisms which can degrade silk, the primary structure of which is similar to that of PLA (Tokiwa et al., 1996). We have studied the molecular weight dependence of biodegradability through quantitative analysis of degradability (OECD301C) using standard activated sludge of a typical microbial flora (Figure 4)

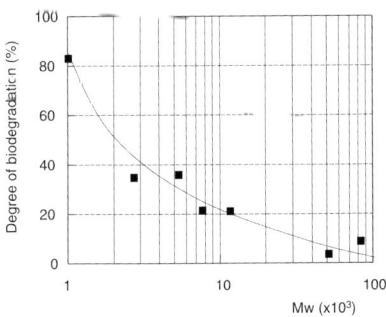

Fig. 4 Molecular weight dependence of PLA bio-degradation (condition: standard activated sludge, 25 °C, 20 days).

(Yagi et al., 1997a). Whereas high molecular-weight PLA is difficult to degrade using ordinary microorganisms, PLA oligomers with a lower molecular weight can easily penetrate microbial membranes, thereby showing good water solubility, and hence can easily be biodegraded. It has also been found that biodegradation does not lead to residual oligomers, which was observed during hydrolysis (see above).

All these findings suggest that high molecular-weight PLA will be degraded rapidly into low molecular-weight PLA under high-temperature, alkaline conditions as are found in compost, and that low molecular-weight PLA will then simply be degraded into CO_2 and water by microorganisms (Table 3) (Yagi et al., 2000). This has in fact been demonstrated in a biodegradability study using compost (Lunt, 1998). A later report (Department of Drinking Water, 1999) has corroborated that PLA can quickly be biodegraded in a municipal landfill. In contrast, in an environment held at neutral pH and room temperature, as represented by soil and water, PLA is degraded at a slow rate.

Bearing in mind the facts discussed above, it can be concluded that PLA has a high degree of stability under normal conditions of use and storage, and can be degraded rapidly after use (Figure 5).

In order to validate the suitability of PLA, LACEA, for composting, Mitsui Chemicals has conducted an inherent biodegradability test, a biodegradability test under actual composting conditions, and a quality test of the final compost. The results are shown below. In accordance with ISO 14855, LACEA has been verified to be degraded by more than 60% in 60 days in mature and neutral compost with a low microbial activity, which attests to its inherent biodegradability. When LACEA labeled with a radio-isotope (^{14}C) was tested in actual compost held at an active fermentation stage, it

Tab. 3 Characteristics of PLA degradation: logarithmic 4-steps degradation

Molecular weight (Mw)	Materials	Degradation mechanism		Degradation rate	
		Hydrolysis	Biodegradation	Loss in Mw	Loss in weight
10^5	Polymer (ex.molded articles)				
⇓	⇓	Main	Secondary	Fast	Slow
10^4			only PLA degradable microorganism		
10^3	⇓		Both occurred □	Slow	Fast
	Increase in water solubility and membrane penetration	Secondary	Main	Fast	Slow
	⇓				
10^2□	Monomer: Lactic acid				
	⇓	none	Only occurred	Fast	Fast
10	Carbon dioxide, water, etc.				

showed the same degree of biodegradability as organic waste in an actual composting process: 60% of LACEA was converted into inorganic substances in about 10 days, and >90% was degraded in about 20 days (Figure 6) (Yagi et al., 1997b). The final compost into which LACEA was degraded was subjected to plant growth tests in accordance with DIN 54900, and no harmful effects were observed. In 1997, LACEA was certified as compostable material in Germany. Since autumn 1999, LACEA garbage bags with excellent biodegradability and

compostability have been used by Kosaka Town, Akita Prefecture, and this has contributed towards an integrated bio-recycling system from the disposal of waste to organic farming.

LACEA, which has been developed with emphasis on biodegradability and safety, has been certified in accordance with strict standards set forth by the Biodegradable Plastics Society's Biodegradable Plastics (GreenPla) Identification and Labeling System in Japan which was launched in June, 2000.

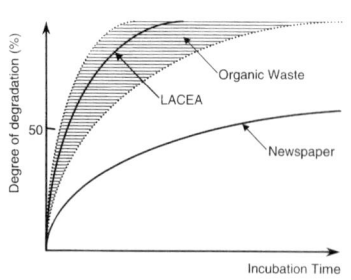

Fig. 5 Degradation rate at composting facilities.

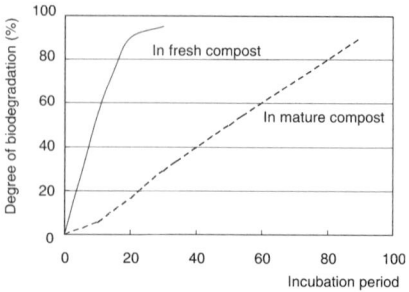

Fig. 6 Curve of degradation of PLA in compost (conditions: 58 °C, based on ISO14855).

5
Production Technologies

In order to obtain PLA with sufficient molecular weight or strength for practical purposes, there are two polymerization routes available (Figure 7): 1) ring-opening polymerization (ROP) (Jamshidi, 1987; US Patent 5,357,035) of L-lactide, which is a cyclic dimer of L-lactic acid; and 2) direct polycondensation (Ajioka, 1995) of L-lactic acid. For both routes, extreme effort has been made to streamline their commercial implementation.

5.1
Ring-opening Polymerization

Figure 8 provides a schematic view of the PLA production process based on the ROP of L-lactide. L-Lactic acid is condensed to PLA oligomers with low molecular weight (several hundreds to several thousands), and these are then depolymerized in the presence of a catalyst and subjected to a backbiting reaction for the production of L-lactide. Having been purified, L-lactide can easily be subjected to ROP in the presence of a catalyst and an initiator, to yield high molecular-weight PLA.

The crude L-lactide synthesized as above contains water, lactic acid and lactic acid oligomers as impurities, as well as its optical

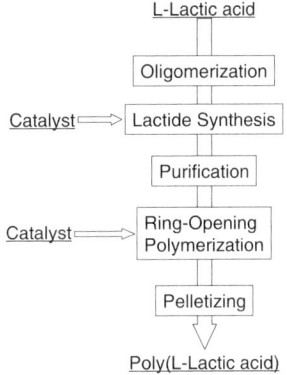

Fig. 8 Schematic diagram of the ring-opening polymerization process.

isomers, D-lactide and meso-lactide. Even with a high-purity feedstock of L-lactic acid, racemization may occur in the process of lactide synthesis, and this will lead to the production of such optical isomers. In order to apply the ROP process of L-lactide to the production of PLA with desired molecular weight and physical properties, it is necessary to use highly refined L-lactide. In view of this, many refining processes have been studied, including a recrystallization process with an organic solvent (US Patent 5,463,086), a process using solvent extraction and crystallization in combination (WO Patent 93/31494), and a melt crystallization process (Canada Patent CA2,115,472). Recently, Cargill Dow established a simple distillation refining technology for the pro-

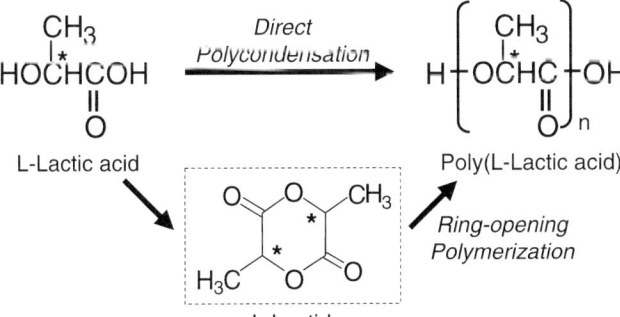

Fig. 7 Polymerization routes of PLA.

duction of high-purity L-lactide on an industrial scale, without the use of organic solvents. Cargill Dow have also recently announced a plan to build a 140,000 tonnes per year plant for the commercial production of PLA (*Wall Street Journal*, 2000).

For the ROP of high-purity L-lactide, a solvent-free melt polymerization process (Gruber et al., 1996) or bulk polymerization process is employed. In these processes, the polymerization rates are controlled by temperature and the amount of catalyst. In the melt polymerization process, high molecular-weight PLA can be obtained simply, but it is characterized by the presence of a small percentage of L-lactide that invariably remains in the PLA produced due to the reaction equilibrium involved (Figure 9). This residual amount of L-lactide can be reduced by lowering the polymerization temperature, however (Gruber et al., 1996).

5.2
Direct Polycondensation

A method of synthesizing PLA by polycondensation of L-lactic acid was first introduced by Carothers and Van Natta in 1933. Since then, many research groups have endeavored to implement this scheme, but have encountered difficulties in obtaining a high molecular-weight PLA with sufficient strength for practical use. In 1994, Mitsui Chemicals established a process for the direct polycondensation of L-lactic acid (US patent 5,310,865; Ajioka et al., 1995). This process, which is a solution polymerization, is outlined in Figure 10 and features the use of a highly active catalyst and an organic solvent. Water, which is generated as a byproduct of the polycondensation reaction, can be removed with an azeotropic solvent; the solvent is then dried and recycled. In this way, dehydration is efficiently performed without affecting the equilibrium of the condensation reaction. This solution reaction allows a reaction temperature to be chosen which is below the melting point of PLA, and this efficiently prevents depolymerization and racemization. It is reported that a high-purity PLA with a molecular weight of up to 300,000 daltons was successfully produced using this process (Ajioka et al., 1995).

These solution polymerization techniques have been studied not only for the production of PLA, but also for the copolymeriza-

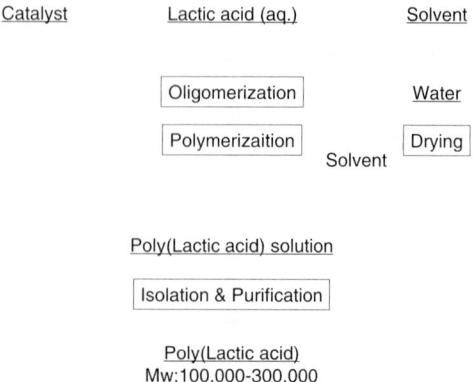

Fig. 9 Relationship between residual lactide and temperature.

Fig. 10 Schematic diagram of the direct polycondensation process.

tion with other polyesters or the cross-linking reaction of PLA to produce new materials, as well as for the polymerization of aliphatic polyesters (US Patent 5,637,631; US Patent 5,914,381; US Patent 5,428,126).

It should be noted, however, that the direct polymerization process illustrated in Figure 10 has several disadvantages. For example, there is a need for a relatively large reactor because of a low volume efficiency for a production scale-up well comparable with the commercial system for conventional plastics. There is also a need for evaporation and recovery of the solvent. Bearing these points in mind, the fine-tuning of the polymerization process based on the direct polycondensation has been studied in various ways. Although research and development relating to polycondensation catalysts and reaction mechanisms is still in progress (Kimura et al., 1999; Otera et al., 1996), further research is being directed by the extruder manufacturer to simplify the process with the use of a twin-screw extruder as a reactor (US Patent 5,574,129).

Recently, Mitsui Chemicals developed a new process based on direct polycondensation of L-lactic acid to enable the production of high molecular-weight PLA, without the use of an organic solvent (US Patent 5,574,129; Ogawa et al., 1999). As shown in Figure 11, this process consists of liquid-state polycondensation and solid-state poly-condensation. PLA oligomer directly synthesized from L-lactic acid by a liquid-state polycondensation in the presence of a catalyst is pelletized. The pellets thus obtained are crystallized and then subjected to solid-state polycondensation in the stream of an inert gas for producing PLA with a high molecular weight. This process is capable of producing high molecular-weight polymers without any use of solvent, and it dispenses with the evaporation operation for principal chemical components, thus minimizing the specific energy consumption required for production. It is a simple process, and very similar to that widely used for the production of PET for bottles.

In solid-state polycondensation, PLA is produced at relatively low reaction temperatures below the melting point of the polymer, and therefore has a higher optical purity than when produced by the ROP process; moreover, the polymer can be easily produced with a high crystallinity. It has also been reported that virtually no lactide or other low molecular-weight compounds are formed. The basic physical, mechanical and molding properties of this PLA are equivalent to those of the polymer produced by the solution polymerization process, and efforts are being made to develop technologies for implementation of direct polycondensation as applicable to large-scale industrial production (Ogawa et al., 1999).

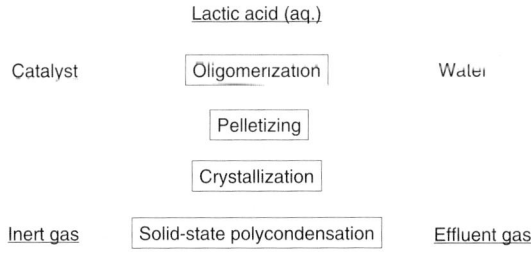

Fig. 11   Schematic diagram of the solid-state polycondensation based on direct polycondensation.

5.3
Other Polymerization Processes

Mitsui Chemicals have also developed a process for producing high molecular-weight PLA at near-room temperature (US Patent 5,719,256). In this process, the oligomer of L-lactic acid is reacted with haloiminium compounds to produce an acid halide, which is then reacted with a base for the production of a high molecular-weight product. It is said that this process can realize a very high reaction rate at relatively low temperatures, and the PLA thus obtained features an excellent hue.

6
Processing Technologies

Although PLA may be processed using a variety of methods, the optimization of processing technology and improvement of material properties are necessary in order to achieve a commercially feasible process. In this respect, many investigations have been performed during the past 15 years, and for a variety of applications.

6.1
General Physical Properties

PLA is a crystalline thermoplastic polymer with a comparatively high degree of thermal stability, as indicated by a T_g of ~60 °C, a crystallization temperature (T_c) of ~125 °C, a melting point (T_m) of ~170 °C, an on-set temperature of weight loss (D_0) of ~285 °C, and a weight loss temperature (D_1) at ~295 °C of 1%. PLA is an aliphatic polyester having ester bonds, and its melt viscosity falls as a result of hydrolysis when molten in the presence of water, just as with PET. However, in dry conditions its melt viscosity is stable. The tolerance is less than about

100 p.p.m. Under dry conditions, PLA is only minimally subject to deterioration, even when held at high temperatures, and remains almost unchanged from the melting point up to about 250 °C (Figure 12).

The melt viscosity of PLA shows a higher dependence on temperature in comparison with that of polypropylene (PP), but is little dependent on the shear rate in the low shearing range, where it demonstrates Newtonian behavior (Figure 13).

A temperature range of about 170–250 °C is applicable to the processing of PLA. As the melting conditions may be widely adjusted from high viscosity ($>10^4$ Poise) to low viscosity ($\sim 10^2$ Poise), it is possible to apply any type of forming (injection molding, film

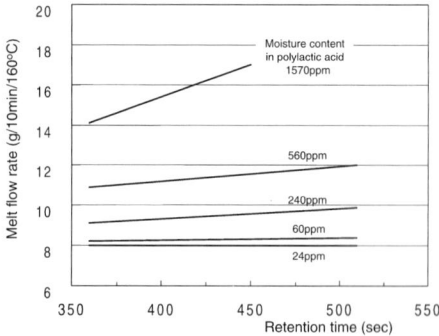

Fig. 12 Influence of moisture content on melt flow rate.

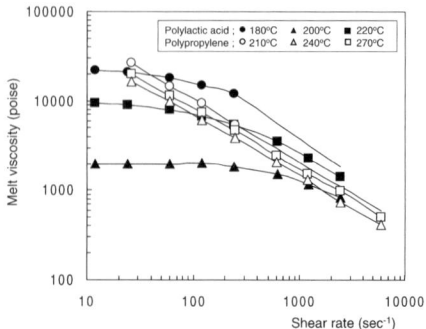

Fig. 13 Relationships between shear rate and melt viscosity.

forming, spinning, blow molding, foaming, blown film, etc.) merely by controlling the resin temperature.

All these processing technologies have been confirmed as being successfully applied to LACEA.

6.2
Injection Molding

The molding conditions for PLA acid lie in a comparatively narrow range, mainly because its melt viscosity is heavily dependent upon temperature. It is therefore preferable to mold PLA at lower pressures and lower rates than would be used for polypropylene.

The molding shrinkage is 0.2–0.4% of the mold size, or almost the same level as for polystyrene (PS). The mold is therefore recommended to be designed as same as that for PS. The molded products obtainable under ordinary molding conditions are amorphous and show a transparency equivalent to, or higher than, PS and PET. However, their applications are limited because their impact resistance and thermal resistance are not sufficiently high (Table 4).

For the purpose of improving impact resistance and thermal resistance, studies have been made to blend formulas with nucleating agents, inorganic fillers and other biodegradable polymers, etc. While there are several reported cases of improvement in impact resistance and thermal resistance at the neglect of transparency, improvements in physical properties without impairing transparency are being studied.

6.3
Film Forming

A transparent film or sheet of PLA can be produced by utilizing a T-die extruder (US Patent 5,076,983). In addition, by drawing, orientation and crystallization, it is possible to improve both the thermal resistance and impact resistance of the film or sheet to the same level of strength and stiffness as oriented polypropylene (OPP) or PET, while maintaining its high transparency (Table 5).

Tab. 4 Physical properties measured by ASTM for LACEA

		LACEA		Commodity plastics	
		Standard	Impact resistant grade	GPPS	PET
Tensile strength	[MPa]	68	44	45	57
Elongation at break	[%]	4	3	3	300
Flexural strength	[MPa]	98	76	76	88
Flexural modulus	[MPa]	3,700	4,700	3,000	2700
Izod impact	[J m^{-1}]	29	43	21	59
Vicat softening point	[°C]	58	114	98	79
Heat distortion temperature	[°C]	55	66	75	67
Under 0.45 MPa load				(1.82)	
Density	[kg m^{-3}]	1.26	1.48	1.05	1.4
Light transmission	[%]	94	opaque	90	–
Refractive index		1.45	–	–	1.58
Volume resistivity	[$\Omega \cdot$cm$\cdot 10^{16}$]	1$\leq$	1$\leq$	1$\leq$	1$\leq$
Flame retardancy	[UL94]	HB	HB	HB	–

Tab. 5 Physical properties of LACEA film

		LACEA		Commodity Plastics	
		Nonoriented	Oriented	OPP	PET
		250 µm	25 µm MD/TD	25 µm MD/TD	25 µm MD/TD
Tensile strength	[MPa]	70	110	130/300	240
Elongation at break	[%]	5	140	170/45	130
Flexural modulus	[MPa]	2.0	3.9	2.0/3.6	4.0
Flexural strength	[MPa]	–	27/21	9/5	20/5
Heat contraction	[%]	–	2.4/0.9	1.5/0.2	–
HAZE	[%]	1.0	1.3	2.2	3.0
Gas permeability					
O_2	[$cm^3 \, m^{-2} \, day^{-1} \, MPa^{-1}$]	550	4,400	15,000	620
N_2		37	830	–	–
CO_2		1200	17,000	51,000	2,400
H_2O permeability	[$g \, m^{-2} \, day^{-1}$]	31	160	5	23

A transparent nonoriented film is obtained under melting temperature (180–250 °C) and cooling roll temperature (10–40 °C) conditions. An oriented film will be obtained by stretching it to two to ten times its original length at 60–80 °C, and further annealing it at temperatures between the stretching temperature and melting point. An oriented film may be processed for dry lamination, printing and heat seal or other applications, and it can be used for various types of packaging.

By optimizing the annealing conditions during stretching, it is possible to modify the film into a shrinkable form when it is reheated. An application for shrinkable film is also expected.

A nonoriented sheet can easily be molded by a vacuum or pneumatic process into transparent containers and trays (Figure 14). The molding drawdown is relatively small compared with that for polyvinyl chloride (PVC). The features are that the obtained moldings are high in transparency, and that

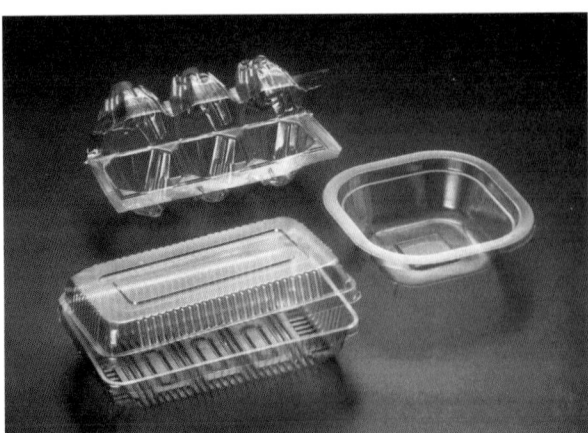

Fig. 14 Containers and tray.

a high dimensional accuracy of the mold can be maintained.

6.4
Spinning

Multifilaments and staple fibers may easily be produced by melt spinning of PLA. Thermal-resistant, high-strength fibers are obtained by stretching partially oriented yarns (POY) spun at a temperature of about 210–230 °C with a speed of about 2500–3500 m min⁻¹, followed by further re-stretching at a temperature of 100–140 °C (Figure 15). The textiles thus produced have physical properties well comparable with those of PET and nylon, and show strength of about 5–6 g dyne⁻¹ and an elongation of about 10–30%. Those fabrics are silky to the touch, and feature excellent dye affinity and light stability (Ikado, 1999).

For manufacturing a nonwoven material, various processes such as spun-bonding, melt-blowing, slitting, etc., are available, and any of these are suitable for PLA (see Figure 15). In order to produce a heat-resistant nonwoven when using the spun-bonding process, it is necessary to complete orientation and crystallization while quench-

ing and solidifying the molten filaments. Orientation and crystallization progress with the increase in spinning speed; indeed, it has been reported that 40% crystallinity is obtained at a spinning speed of about 2000 m min⁻¹ (Mezghani and Spruiell, 1998).

Many reports have been published in recent years about composite filaments of PLA, other aliphatic polyesters and their copolymers, including various types of core shell filaments (Japan Patent H09-302925).

6.5
Blow-molding, Extrusion-foaming and Expansion-molding

The melt tension of PLA is comparatively low, and a proper forming process should be selected in order to meet a specific geometry of moldings. For forming large containers, it is desirable to use an injection-blowing process (Figure 16). PLA containers obtained in this way show a high degree of transparency and a resistance that compares well with that of PET containers. Their water vapor permeability is about 10-fold higher than that of PET, and hence appropriate applications should be selected.

Fig. 15 Film, nonwoven cloth, pellet and fiber.

Fig. 16 Blow-molded bottles.

In the low extrusion-foaming process based on a chemical expanding agent, sheets expanded two- to three-fold with cell sizes of 100–200 μm are available. By contrast, in the high extrusion-foaming process using a physical expanding agent, studies are in progress to change the expanding agents from chlorofluorocarbons and hydrocarbons to CO_2. At present, a foaming expanded 30–40 times the original volume is available, but further improvements are necessary for practical use because of comparatively large cell sizes (Figure 17).

Recently, expansion moldings expanded 30–40 times, with cell sizes of ~100 μm have been developed. Compared with PS expansion moldings, these are somewhat inferior in restitution characteristics, but pose no practical problems. Their implementation is anticipated (see Figure 17).

6.6
Blown Film

PLA is rigid and hence difficult to process for blown film. Copolymerization technology

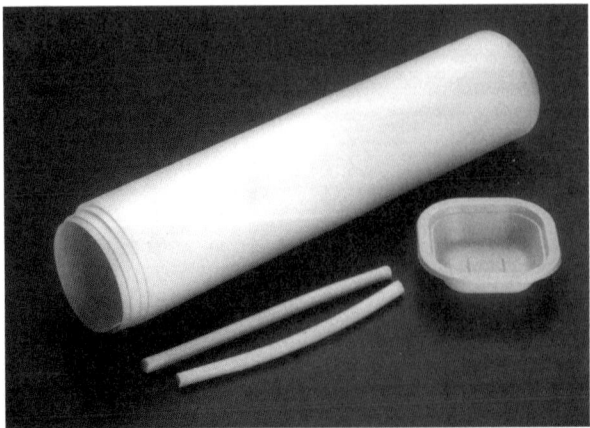

Fig. 17 Expansion-molded articles.

applied in the lactic acid polymerization process, coupled with blending technology to mix PLA with flexible polymers or plasticizers, have been reported to yield PLA-based flexible materials. Mitsui Chemicals successfully developed the PLA-based flexible material as a grade of LACEA. By using such technology, this grade of LACEA shows physical properties well comparable with those of $CaCO_3$-added polyethylene that is currently available commercially (Table 6). Compared with garbage disposal bags made of polybutylenesuccinate, polycaprolactone and other aliphatic polyesters, this grade of LACEA is stable during garbage storage, but can be biodegraded quickly during composting.

7

Marketing Activities for LACEA

In order to take full advantage of the excellent features of PLA, including the utilization of abundant plant resources and biodegradability (Table 7), Mitsui Chemicals is attempting to develop markets primarily for packaging and containers, agricultural and civil engineering materials and composting materials (Table 8). Supported by many years of wide practical knowledge, experience and know-how in material and processing technology regarding PLA, Mitsui Chemicals has provided expert assistance to respond to the needs of customers in various areas.

7.1
Packaging and Containers

PLA is the only material among the Green-Pla products that has a high transparency, and is an excellent material for packaging. As compared with conventional plastics, PLA has a low heat of combustion (as low as paper), and can be composted together

Tab. 6 Properties of LACEA blown film

Item	LACEA blown film MD/TD	PE (CaCO₃-added) MD/TD
Thickness [μm]	30	30
Tensile strength [MPa]	20/20	40/30
Elongation [%]	200/200	300/500
Tensile elastic modulus [MPa]	4.0/3.5	5.5/6.5
Tear strength [g/16 sheets]	70/70	40/540

Tab. 7 Characteristics of PLA "LACEA"

Raw material	• Lactic acid derived from renewable resources (plant resources)
Production	• Polymerization of lactic acid by direct condensation polymerization • Transparent thermoplastic resin having high rigidity
Processing	• Mechanical properties and processability (extrusion, injection molding, vacuum molding, melt spinning, coating, lamination, dyeability and printability) which are similar to those of PS and PET
Use	• Suitable for fungal resistance test (JIS Z 2911) and food sanitation test (Notification No. 20)
Disposal after use	• Fast degradation in compost and slow degradation in the environment • Hydrolysis in alkaline condition • Low heat of combustion (4500 Kcal kg⁻¹, about the same as that of paper) • No toxic gas emission

Tab. 8 Market Development for PLA "LACEA"

Area	Characteristics of LACEA	Effects	Commercial Examples
Packaging and Containers	• High transparency • Plant resources	• The contents can be seen • The containers are in harmony with the contents (natural materials) in concept.	• Organic vegetable packaging • Containers for cosmetics/repellent containing natural extract
	• Plant resources • Low heat of combustion • Alkaline degradability	• Product design not requiring separation from paper wastes. • Paper recycling is not obstructed.	• Cutting blade for wraps • Envelopes with window • Coated paper box and containers
Agricultural and Civil engineering materials	• Degradability in environment (in soil) • Plant resources	• Waste collection is not required. • Labor saving • Bio-recycling	• Mulch film • Nonwoven • Nursery pots • Sandbags
	• Alkaline degradability (in concrete)	• Efficiency in construction work is improved.	• Containers for water-impermeability agent for concrete
Composting materials	• Compostability	• Separation from organic waste is not required. • Labor saving • Biorecycling	• Organic waste collection bags
Other	• Plant resource • Plant resources	• Product design based on natural material is possible.	• Kitchen net • Cards
	• High rigidity • Pri ntability	• Product design based on natural material is possible.	Fiber products
	• Plant resources • Silky touch • Dyeability		

with paper waste. It may also be subjected to alkaline degradation, and therefore does not affect the recycling process of waste paper. In addition, PLA is compatible with various other waste disposal processes. Accordingly, it is possible to use composite materials of PLA and paper (including coatings and laminates) for the merchandising of products, thereby avoiding the troublesome separation of paper and plastics for recycling, this being inevitable for polyolefin-paper composite materials.

PLA can be used to replace the metal cutting blade on the roll box for food wrapping film; a blade made from PLA produces equally satisfactory cuts as its metal counterpart, and has achieved an excellent market reputation since it can be disposed of without being separated from the roll box. In the spring of 2000, the Tokyo Metropolitan Government agreed that wrapping film roll boxes with a LACEA cutting blade could be disposed of as combustible waste, considering that LACEA is derived from plant material.

Market appraisals of LACEA applications to window envelopes, fancy boxes and other products are under way. Since PLA can be said to be a natural material similar to paper and wood, the merchandising and test marketing of this new packaging is currently promoted in harmony with the contents. For example, LACEA packaging and containers are used for organic fruits and vegetables, and for cosmetic items made from natural ingredients.

7.2
Agricultural and Civil Engineering Materials

The open-field burning of mulch films and plastic netting is banned by law in Japan, making it difficult to recover or dispose of them, or to prevent them from being left in the environment. However, PLA is not only environmentally compatible, it also it dispenses with any need for recovery and disposal, as well as reducing associated labor costs. At present, field tests of mulch films and nonwovens made from LACEA are being conducted at various facilities, including national agricultural experiment stations.

7.3
Composting Materials

The Organic Waste Recycling Study Committee established by the Ministry of Agriculture, Forestry and Fisheries in fiscal year 1998 proposed in its report (Organic Waste Recycling Study Committee, 1999) that the government should make efforts to promote the composting as a way of organic waste disposal. The number of local governments studying garbage collection bags made of LACEA has been increasing, and Kosaka Town, Akita Prefecture has decided to adopt this concept. This LACEA product is also being studied for the disposal of organic waste generated by convenience stores, supermarkets, fast food chain stores, schools and hospitals. Applications in this area are expected to expand significantly since the recent introduction of the Food Recycling Law in Japan which is designed to promote the composting of organic waste.

7.4
Other Applications

In December 1999, prepaid cards describing the rigidity, environmental friendliness and naturalness of PLA were marketed for a new cellular phone system developed by a major telephone service company.

PLA can also be used as a new natural textile material with a silky touch, and developmental efforts are being made for its application in the textile industry.

7.5

Labeling System and Regulations

In Japan, the Container and Packaging Recycling Law, the Food Recycling Law, the Law on Promoting Green Purchasing, and various other laws and regulations have been established under the Basic Law for Establishing the Recycling-based Society in which the fiscal year 2000 is set as the starting year for the recycling-oriented (in other words, sustainable or closed-cycle) society (Table 9). All these laws and related actions are expected to boost the marketing and promotion of GreenPla products that have been developed so far to solve the problems associated with resources, environment and waste disposal.

For example, the Container and Packaging Recycling Law (which is already in force) promotes the recovery and recycling of containers and packaging which account for 60% of refuse in volumetric ratio; all "other" plastics in addition to PET are required to be recycled starting from April 2000. At present, GreenPla products are classified as "other plastics", but it is said that in future, a system enabling the spent GreenPla products to be composted together with organic waste such as garbage will be established, with the growth of their consumption. This will make GreenPla products exempt from legal disposal charges in accordance with the Container and Packaging Recycling Law. The Food Recycling Law requires that fast food restaurants and other food manufacturing/processing/retail industries recover and recycle food waste (organic waste) for feedstuff and organic fertilizer. The food processing industry, convenience stores and major fast food chains are planning to introduce GreenPla products as compostable materials which can be composted together with food waste,

Tab. 9 Statutory regulations relating to waste management and recycling (Japan)

The Basic Law for Establishing the Recycling-based Society	
Waste Management and Public Cleansing Law * (as amended)	● Securing restriction of the generation of wastes and appropriate recycling and disposal of wastes.
Law for Promotion of Effective Utilization of Resources (as amended)	● Promoting restriction of the generation of wastes and reuse and recycling of wastes.
Container and Packaging Recycling Law*	● Recycling of containers and packaging that have been sorted and collected, was made obligatory of business parties that produce / use containers and packaging, and other related parties.
Electric Household Appliance Recycling Law	● Collection and recycling of waste electric household appliances was made obligatory of business parties that produce / sell electric household appliances, and other related parties.
Construction Material Recycling Act	● Sorted dismantling of a building and recycling of construction wastes were made obligatory of parties who receive orders for building construction, and other related parties.
Food Recycling Law*	● Restriction of the generation of leftover foods and recycling of foods were made obligatory of business parties that produce / sell foods, restaurants, etc.
Law on Promoting Green Purchasing*	● The National Government, etc., takes initiative in promoting procurement of reproduced products, etc.

* Regulations which influence biodegradable plastics development

or as accelerators for food waste composting. According to the Waste Management and Public Cleansing Law, companies generating waste are responsible for its disposal, and since 1997 it has been illegal to burn spent agricultural materials such as mulch films and nonwovens on open fields. As an open-field burning alternative, farmers are beginning to express considerable interest in GreenPla products which can be ploughed in or composted after use, at low cost and without affecting the environment. The central government agencies and local governments are required by the Law on Promoting Green Purchasing to give environmentally friendly products preference when procuring equipments and materials, and are also studying the feasibility of including GreenPla products in their designated list of purchases.

For the purpose of increasing the marketing volume of GreenPla products, it is necessary to provide measures to ensure that their biodegradability and safety features are brought to the attention of all consumers. To this end, the Biodegradable Plastics Society launched the GreenPla Identification and Labeling System in June 2000. This system is one of materialized examples out of 18 suggestions described in a report titled "The Age of New Plastics" (Ministry of Economy, Trade and Industry, 1995). This report was prepared in 1995 by the Study Committee for the Practical Use of Biodegradable Plastics after series of discus-sions among scholars, government and industry for the purpose of accelerating the implementation of GreenPla.

Biodegradability testing methods adapted to various environmental conditions have been proposed and established in the form of JIS (Japan), ASTM (USA), DIN (Germany), CEN (EU), and ISO, and a system for the certification of compostable materials has already been implemented in Europe (Table 10). In contrast to Europe and the USA, where the system has been developed to focus on composting as a preferable Green-Pla disposal environment, Japan's system is organized with an eye on biodegradability and safety of all ingredients of GreenPla products considering a variety of needs in the application of the products to food packaging, agricultural equipments and materials, composting, etc. A certification system for compostable materials in Japan which would be similar to European system is expected to be proposed by JORA (Japan Organics Recycling Association), which was founded in July, 2000. The fundamental idea of the system would be applied to specifically GreenPla compostability certification. A similar system is also being realized in the USA.

As of December 2000, 34 GreenPla products – including the packaging for organic vegetables and garbage bags – have been certified and authorized to bear the GreenPla Symbol (Figure 18). Many of PLA products, including LACEA products, are among

Tab. 10   Biodegradability testing methods incorporated in ISO and JIS

Dis. No.	Title
14851	Evaluation of the ultimate aerobic biodegradability of plastic materials in an aqueous medium-Method bydetermining the oxygen demand in a closed respirometer
14852	Evaluation of the ultimate aerobic biodegradability of plastic materials in an aqueous medium-Method by analysis of released carbon dioxide
14855	Evaluation of the ultimate aerobic biodegradability and disintegration of plastics under controlled composting conditions-Method by analysis of released carbon dioxide

Fig. 18 German and Japanese symbol marks.

these. GreenPla packages are currently coming on to the market, having been adopted by major electric appliance manufacturers for distribution of their hot-selling products, and Japan's Labeling system will soon be recognized by the general public.

8
Challenges to the Market Development for GreenPla and LACEA

In line with some 18 suggestions proposed by the Study Committee mentioned above, various measures are being implemented for the promotion of GreenPla products, including standardization of biodegradability testing methods, launching of a composting model project, designation of brand-name and symbol, and assurance of safety of the entire GreenPla products, including intermediates. The major challenges are: 1) the establishment and management of a scheme for identification, labeling and certification to draw a clear line between GreenPla and nondegradable plastics; 2) the formulation of technical standards for food packaging applications and acquisition of official approvals; and 3) the development of an infrastructure, including the construction of composting facilities, that is indispensable for the promotion of GreenPla products. With the support of the Ministry of Economy, Trade and Industry and other competent authorities, the Biodegradable Plastics Society (chairperson: Dr. Hiroyuki Nakanishi,

President of Mitsui Chemicals) is committed to responding efficiently to these challenges. The identification and labeling system has been launched in June 2000, as described previously.

In the whirl of these activities, Mitsui Chemicals is promoting LACEA development while pressing ahead with various projects, including: 1) the development of markets to support the business start-up and future expansion with various ranges of application; 2) the configuration of material processing technologies to facilitate and expand the use of LACEA; and 3) the optimization of mass production technologies for cost reduction.

9
Conclusions

In October 1999, the Closed-Cycle (or Sustainable) Industrial Technology Study Committee (Chairperson: Prof. Hiroyuki Yoshikawa, Chancellor of the University on the Air) of the Ministry of Economy, Trade and Industry defined the technologies to support the realization of a Closed-Cycle oriented society as Closed-Cycle Technologies and released a final report (Ministry of Economy, Trade and Industry, 1999) summarizing the guidelines for promoting the development of those technologies and implementing future measures (Figure 19). Closed-Cycle Technologies are considered to contribute to: 1) a minimization of the input of resources

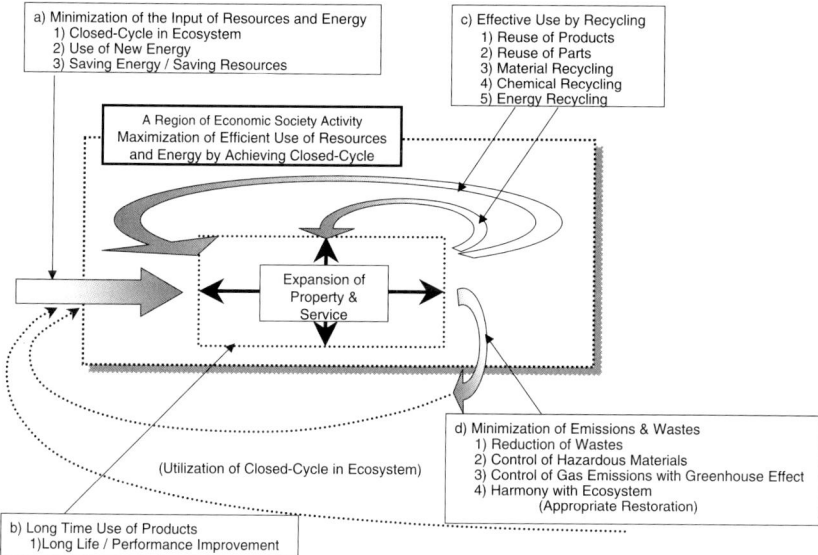

a) Minimization of the Input of Resources and Energy
  1) Closed-Cycle in Ecosystem
  2) Use of New Energy
  3) Saving Energy / Saving Resources

c) Effective Use by Recycling
  1) Reuse of Products
  2) Reuse of Parts
  3) Material Recycling
  4) Chemical Recycling
  5) Energy Recycling

A Region of Economic Society Activity
Maximization of Efficient Use of Resources
and Energy by Achieving Closed-Cycle

Expansion of
Property &
Service

(Utilization of Closed-Cycle in Ecosystem)

d) Minimization of Emissions & Wastes
  1) Reduction of Wastes
  2) Control of Hazardous Materials
  3) Control of Gas Emissions with Greenhouse Effect
  4) Harmony with Ecosystem
       (Appropriate Restoration)

b) Long Time Use of Products
  1)Long Life / Performance Improvement

Fig. 19   The position of the closed-cycle (or sustainable) technologies.

and energy considering ecosystem balance, for example by increasing use of biomes; 2) a long-term use of products; 3) their efficient use by recycling; and 4) a minimization of emissions and wastes considering ecosystem balance as key technologies. It is evident that the GreenPla products contribute towards reducing the waste referred to in item (4). As discussed in this chapter, PLA is a contributor to items (1) and (4), and is considered to offer a new breed of technology and material interlinking between disposal (biodegradation) and raw material (renewable resources).

As the role to be played by PLA and other GreenPla products is made clear, it is expected that the market for PLA and the GreenPla products will expand dramatically.

10
Patents

Patents referred to in this chapter are summarized in Tables 11 and Table 12. Mitsui Chemicals has filed a variety of patents for production, material technology, processing technology to end-use applications; over 300 patents are filed in Japan, and over 40 patents have been granted in the USA. Cargill Dow and Shimadzu as PLA manufacturers also file PLA-related patents.

Tab. 11 Patent list

	Patent No. Publication No.	Applicant	Inventors	Title	Date of patent	Date of publication
1	US5357035	Cargill, Incorporated, (Minnetonka, MN)	Gruber; Patrick R. (St. Paul, MN); Hall; Eric S. (Crystal, MN); Kolstad; Jeffrey J. (Wayzata, MN); Iwen; Matthew L. (Minneapolis, MN); Benson; Richard D. (Maple Plain, MN); Borchardt; Ronald. L. (Eden Prairie, MN)	Continuous process for manufacture of lactide polymers with purification by distillation	1994/10/18	
2	US5463086	Dainippon Ink and Chemicals, Inc. (Tokyo, JP)	Kubota; Kazuomi (Chiba, JP); Murakami; Yoichi (Chiba, JP)	Process for producing actides and process for purifying crude lactides	1995/10/31	
3	WO9631494	CHRONOPOL INC (US)	EGGEMAN TIMOTHY J; MIAO FUDU	METHOD TO PRODUCE AND PURIFY CYCLIC ESTERS		1996/10/10
4	WO9319058	BIOPAK TECHNOLOGY LTD (US)	CREMEANS GEORGE E (US); LIPINSKY EDWARDS (US); CHEUNG ALEX (US); BENECKE HERMAN P (US); HILLMAN MELVILLE E D (US); MARKLE RICHARD A (US); SINCLAIR RICHARD G (US)	PROCESS FOR THE PRODUCTION OF CYCLIC ESTERS FROM HYDROXY ACIDS AND DERIVATIVES THEREOF		1993/9/30
5	CA2115472	DU PONT (E. I.) DE NEMOURS AND COMPANY (United States)	Oapos;Brien, William George (United States); Sloan, Gilbert Jacob (United States)	MELT CRYSTALLIZATION PURIFICATION OF LACTIDES		1994/8/18
6	US5310865	Mitsui Toatsu Chemicals, Incorporated (Tokyo, JP)	Enomoto; Katashi (Kanagawa, JP); Ajioka; Masanobu (Kanagawa, JP); Yamaguchi; Akihiro (Kanagawa, JP)	Polyhydroxycarboxylic acid and preparation process thereof	1994/5/10	
7	US5637631	Mitsui Toatsu Chemicals, Inc. (Tokyo, JP)	Kitada; Ikumi (Kanagawa-ken, JP); Higuchi; Chojiro (Kanagawa-ken, JP); Ajioka: Masanobu (Kanagawa-ken, JP); Yamaguchi; Akihiro (Kanagawa-ken, JP)	Preparation process of degradable polymer	1997/6/10	
8	US5914381	Mitsui Chemicals, Inc. (Tokyo, JP)	Terado; Yuji (Kanagawa-ken, JP); Suizu; Hiroshi (Kanagawa-ken, JP); Higuchi; Chojiro (Kanagawa-ken, JP); Ajioka; Masanobu (Kanagawa-ken, JP)	Degradable polymer and preparation process of the same	1999/6/22	

Tab. 11 (cont.)

	Patent No. Publication No.	Applicant	Inventors	Title	Date of patent	Date of publication
9	US5428126	Mitsui Toatsu Chemicals, Inc. (Tokyo, JP)	Kashima; Takeshi (Kanagawa, JP); Kameoka; Taiji (Kanagawa, JP); Higuchi; Chojiro (Kanagawa, JP); Ajioka; Masanobu (Kanagawa, JP); Yamaguchi; Akihiro (Kanagawa, JP)	Aliphatic polyester and preparation process thereof	1995/6/27	
10	US5574129	The Japan Steel Works, Ltd. (Tokyo, JP)	Miyoshi; Rika (Hiroshima, JP); Sakai; Tadamoto (Hiroshima, JP); Hashimoto; Noriaki (Hiroshima, JP); Sumihiro; Yukihiro (Hiroshima, JP); Yokota; Kayoko (Hiroshima, JP); Koyanagi; Kunihiko (Hiroshima, JP)	Process for producing lactic acid polymers and a process for the direct production of shaped articles from lactic acid polymers	1996/11/12	
11	US5719256	Mitsu Toatsu Chemicals, Inc. (Tokyo, JP)	Tamai; Shoji (Fukuoka-ken, JP); Mori; Yukiko (Fukuoka-ken, JP); Goto; Kenichi (Fukuoka-ken, JP); Watanabe; Katsuji (Fukuoka-ken, JP); Kohgo: Osamu (Fukuoka-ken, JP); Shimizu; Kotaro (Fukuoka-ken, JP); Kataoka; Toshiyuki(Fukuoka-ken, JP); Kuroki; Takashi (Kanagawaken, JP); Yamashita; Wataru (Fukuoka-ken, JP); Mizuta; Hideki (Fukuoka-ken, JP); Nagata; Teruyuki (Fukuoka-ken, JP)	Process for preparing polycondensation polymer compound	1998/2/17	
12	US5076983	E. I. Du Pont de Nemours and Company (Wilmington, DE)	Loomis; Gary L. (Drexel Hill, PA); Ostapchenko; George J. (Wilmington)	Polyhydroxy acid films	1991/12/31	
13	tokkaihei H09-302925	Sega Industries, Co.	Yoshio Saga (JP); Mitsuo kitamura (JP	unknown		1997/11/25
14	US5539081	Cargill, Incorporated	Gruber; Patrick R. (St. Paul, MN); Kolstad; Jeffrey J. (Wayzata, MN); Hall; Eric S. (Crystal, MN); Eichen Conn; Robin S. (Minneapolis, MN); Ryan; Christopher M. (Chisago City, MN)	Melt-stable lactide polymer composition and process for manufacture thereof	1996/7/23	

11
References

Abstracts of Biodegradable Plastics (2000) Conference held in Frankfurt.

Ajioka, M. (1995) Biodegradable aliphatic polyesters, *Petrotechnology* **18**, 202–206.

Ajioka, M., Enomoto, K., Suzuki, K., Yamaguchi, A. (1995) Basic properties of polylactic acid produced by the direct condensation polymerization of lactic acid, *Bull. Chem. Soc. Jpn.* **68**, 2125–2131.

Carothers, W. H., Van Natta, F. J. (1933) Studies of polymerization and ring formation. XVIII, Polyesters from omega-hydroxydecanoic acid, *J. Am. Chem. Soc.* **55**, 4714.

Closed-Cycle Industrial Technology Study Committee (1999) *Technical Issues to Clear for the Development of Closed-Cycle Society*, Environmental Administrative Guidance Office, Environmental Siting Bureau, Ministry of Economy, Trade and Industry, October.

Department of Drinking Water, Domestic Sanitation Bureau, Ministry of Health and Welfare (1999) FY1999 Report on the Environmental Impact of Landfilled Biodegradable Plastics, Environmental Impact Study Committee on the Landfilling of Biodegradable Plastics.

Gruber, P. R., Kolstad, J. J., Hall, E. S., Eichem Conn, R. S., Ryan, C.M. (1996) U. S. Patent 5,539,081.

Ikado, S. (1999) LACEA, *Plastics Age* **45**, 122.

Jamshidi, K., Eberhart, R. C., Hyon, S. H., Ikada, Y. (1987), *Polym. Prep.* **28**, 236.

Kawashima, N. (2000) *Chemical Economics*, June 2000.

Kimura, Y. et al. (1999) 45th Koubunshi kenkyu-happyoukai, Kobe.

Li, S., Garreau, H., Vert, M. (1990) Structure-property relationships in the case of the degradation of massive poly(alpha-hydroxy acids) in aqueous media. Part 3. Influence of the morphology of poly(ʟ-lactic acid), *J. Mater. Sci.: Mater. Med.* **1**, 198–206.

Lunt, J. Properties of Polylactic Acid Polymers – presented at Techtextile, March 22–24, 2000.

Makino, K., Arakawa, M., Kondo, T. (1985) Preparation and *in vitro* degradation properties of polylactide microcapsules, *Chem. Pharmaceut. Bull.* **33**, 1195–1201.

Mezghani, K, Spruiell, J. E. (1988) High-speed melt spinning of poly(ʟ-lactic acid) filaments, *J. Polym. Sci.: Part B Polym. Phys.* **36**, 1005–1012.

Ogawa, S., Hiraoka, S., Obuchi, S., Ajioka, M., Kawashima, N. (1999) Ninth BEDPS Meeting, Hawaii.

Organic Waste Recycling Study Committee (1999) *Current Status and Issues of Organics Wastes Recycling*, February.

Otera, J., Kazuko, K., Yano, T. (1996) Direct condensation polymerization of ʟ-lactic acid catalyzed by distannoxane, *Chem. Lett.* 225–226.

Study Committee for the Practical Use of Biodegradable Plastics (1995) *The Age of a New Plastics*, Biochemical Industry Division, Ministry of Economy, Trade and Industry.

Tokiwa, Y. et al. (1996), *Appl. Environ. Microbiol.* **62**, 2393.

Wall Street Journal, January 11, 2000.

Yagi, T. et al. (1997a) Manuscript on 97–2 meeting of Ecomaterial Society, Tokyo, Japan, 1997.

Yagi, T. et al. (1997b) Manuscript on Organic Recovery and Biological Treatment, 97 International Conference, Harrogate, UK.

Yagi, T. et al. (2000) Proceeding on ICS-UNIDO International Workshop, Seoul, Korea.

10
Aliphatic Polyesters: "Bionolle"

Dr. Ryoji Ishioka[1], E. Kitakuni[2], Y. Ichikawa[3]
[1] Showa Highpolymer Co. Ltd., Bionolle Project, 3-20 Kanda Nishiki-cho 3-Chome Chiyoda-ku, Tokyo, Japan; Tel.: +81-3-3293-8411; Fax: +81-3-3293-8995; E-mail: ryoji.ishioka@shp.co.jp
[2] Showa Denko K. K., Central Research Laboratory, 1-1 Onodai 1-Chmoe Midori-ku, Chiba-Si, Japan; Tel.: +81-43-226-5257; Fax: +81-43-3293-5262; E-mail: kitakuni@ctl.sdk.co.jp
[3] Showa Highpolymer Co., Ltd., Bionolle Project, 3-20 Kanda Nishiki-cho 3-Chome Chiyoda-ku, Tokyo, Japan; Tel.: +81-791-67-2092; Fax: +81-791-67-1302; E-mail: Yasushi Ichikawa@sdk.co.jp

BOD	biological oxygen demand
DSC	differential scanning calorimetry
GPC-MALLS	gel-permeation chromatography-MALLS
HDPE	high-density poly(ethylene)
HPLC/MS	high-performance liquid chromatography/mass spectrometry
HV	3-hydroxyvalerate
LLDPE	low-density poly(ethylene)
Mn	number average molecular weight
Mw	weight average molecular weight
NMR	nuclear magnetic resonance
P(3HB)	poly(3-hydroxybutyrate)
PBA	polytetramethylene adipate
PBS	polytetramethylene succinate
PBSA	poly(butylene succinate adipate)
PBSE	polytetramethylene sebacate
PCL	poly(caprolactone)
PEA	poly(ethylene adipate)
PEAz	polyethylene azelate
PEDe	polyethylene decamethylate
PES	polyethylene succinate
PESE	polyethylene sebacate
PESu	polyethylene suberate
PHB	poly(3-hydroxybutyrate)
PHSE	polyhexamethylene sebacate
PPL	polypropiolactim
ThOD	theoretical oxygen demand

1
Introduction

Only within the past 30–40 years has it been shown that aliphatic polyesters, which form solid polymers that are insoluble in water, can be degraded by microorganisms. In 1968, Darby and Kaplan first reported that a low-molecular-weight polyester could support the growth of filamentous fungi (Darby and Kaplan, 1968). Some years later, Tokiwa and co-workers reported that high molecular-weight polyesters such as poly(ethylene adipate) (PEA; molecular weight 3000 Daltons) and poly(caprolactone) (PCL; molecular weight of 25,000 Daltons) were also biodegradable (Tokiwa and Suzuki, 1974; Tokiwa et al., 1976).

2
Historical Outline

Research on polymerization *per se* has a long history. For example, a report by Carothers in the 1930s disclosed that PEA had an average molecular weight of 3000, but when the degree of polymerization of aliphatic polyester was increased, then depolymerization was found to occur in parallel. Thus, great difficulties were encountered in attaining a molecular weight of $\geq 10^4$ which is required for structural materials.

In recent years, the handling of plastic waste has become a controversial issue, and there has been a growing recognition of a need for biodegradable plastics that would exhibit performance similar to that of conventional plastics during use, but which would be biodegradable. This in turn has prompted a plethora of studies to be carried out on these plastics since the late 1980s, and has led to major technical advances in the field of biodegradable plastics.

In 1994, Showa High Polymer Co., Ltd. successfully produced an aliphatic polyester of high molecular weight, and as a result marketed the "Bionolle" series. These compounds are aliphatic polyesters that have excellent practical properties and are also biodegradable.

3
Composition of "Bionolle"

The selection of glycols and dicarboxylic acids is a key issue in the production of biodegradable plastics from a glycol-dicarboxylic acid polyester. Such selection greatly influences the ease of polymerization, the physical properties and the biodegradability of the polymer produced, as well as its production costs.

When developing biodegradable plastics, if the plastics are to be used as general structural materials, then their component polymers must be heat-resistant, and in particular they should have a melting point which is higher than about 80 °C. The melting points of aliphatic polyesters have been investigated in great detail (Korshak and Vloogradora, 1953), and consequently glycols and dicarboxylic acids can be selected on the basis of such properties for the production of aliphatic polyesters. Candidates for polyester production should have melting points of ≥ 80 °C, and these include both acids (e.g., oxalic acid and succinic acid) and glycols (e.g., ethylene glycol, 1,4-butanediol, neopentyl glycol, and C_{20} glycols) (Table 1).

Another criterion to determine the composition of a polyester is to monitor its degradability by microorganisms. Such biodegradation generally involves two steps: 1) a primary stage which is enzyme-catalyzed to form a lower molecular-weight compound; and 2) degradation of the lower molecular-weight compound within the cells of the microorganisms. In the natural environ-

Tab. 1 Melting points of a variety of aliphatic polyesters

Glycols HOROH R	Dicarboxylic acid HOOC(CH₂)ₙCOOH								
	$n=0$	1	2	3	4	5	6	7	8
$-(CH_2)-$	159	-2	102	-19	47	25	63	44	72
CH_3 $-CHCH_2-$	–	–	-2	-25	-25	-37	-41	-46	-34
$-(CH_2)_3-$	66	-25	43	35	36	41	47	46	49
CH_3 $-CH(CH_2)_2-$	-4	-20	-15	-32	-36	-43	–	-52	-44
$-(CH_2)_4-$	103	-24	113	36	58	38	–	49	54
$-(CH_2)_5-$	49	-26	32	22	37	39	43	46	53
$-(CH_2)_6-$	70	-48	52	28	55	52	61	52	65
$-(CH_2)_{10}-$	76	29	71	55	70	63	70	67	71
$-(CH_2)_{20}-$	88	67	86	77	85	82	86	34	87
$-(CH_2)_2-O-(CH_2)_2-$	5	-18	-11	-30	-29	-32	28	-36	44
$-(CH_2)_2-O-(CH_2)_2-O-(CH_2)_2-$	-14	-34	-24	-36	-39	-42	-41	-43	28
CH_3 $-CH_2-C-CH_2-$ CH_3	111	67	36	–	37	–	17	0	26

ment, the primary stage is the rate-determining step of the biodegradation, and therefore the overall biodegradation rate of a polyester can be estimated with reasonable accuracy from the rate at which it is hydrolyzed enzymatically. When the hydrolysis of an aliphatic polyester by lipase, or by an enzyme which hydrolyzes PEA was studied, a negative correlation was found between hydrolysis rate and polyester's melting point (Tokiwa et al., 1988) (Figure 1).

From the above data, it appears that the biodegradation rate of polyester derived from ethylene glycol decreases in the following order of carboxylic acids:

adipic acid > azelaic acid > succinic acid.

Similarly, the biodegradation rate of polyester derived from adipic acid is considered to decrease in the following order of glycols:

ethylene glycol > 1,4-butanediol.

The biodegradation rate of plastics also changes when the materials comprise a copolymer of polyester or blended polymers. When the hydrolysis characteristics of homo-polyesters (poly(hexamethylene succinate) and PCL and co-polyesters were investigated, the latter compounds exhibited a considerably higher hydrolytic rate (Doi, 1995). This suggests that the biodegradation of polyester may depend not only on its primary structure but also on the higher-order structure, and this is important in the control of biodegradability.

The combination of sources for producing polyester is determined by considering the physical properties and cost of the sources, as well as their heat resistance and biodegradability. Structural repeating units of Bionolle products which are under pre-marketing development on the basis of the above criteria are shown in Table 2.

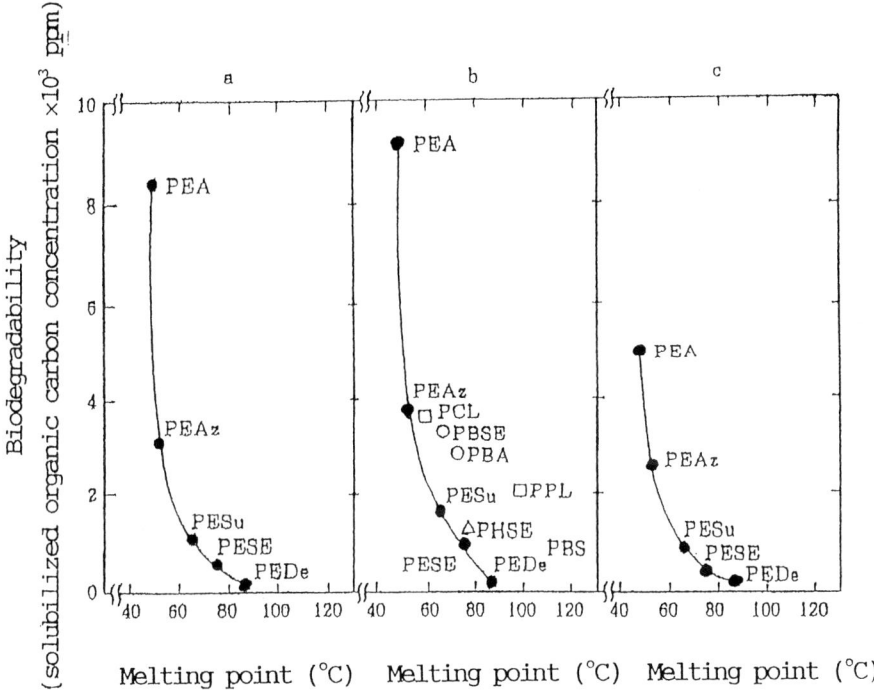

Fig. 1 Hydrolysis rates and melting points of aliphatic polyesters. (a) *R. delemar* lipase; (b) *R. arrhizus* lipase; (c) *Penicillium* sp. 14-3PEA hydrolytic enzyme. PEA: polyethylene adipate; PEAz: polyethylene azelate; PEDe: polyethylene decamethylate; PBA: polytetramethylene adipate; PHSE: polyhexamethylene sebacate; PCL: polycaprolactone; PESu: polyethylene suberate; PESE: polyethylene sebacate; PBS: polytetramethylene succinate; PBSE: polytetramethylene sebacate; PPL: polypropiolactim.

Tab. 2 Structural formulas and abbreviations of Bionolle products

Poly(butylene succinate) (PBS)	$\text{-(O-CH}_2\text{-CH}_2\text{-CH}_2\text{-CH}_2\text{-O-}\overset{\overset{\displaystyle O}{\|}}{C}\text{-CH}_2\text{-CH}_2\text{-}\overset{\overset{\displaystyle O}{\|}}{C}\text{)-}$	Bionolle #1000
poly(butylene succinate adipate) (PBSA)	$\text{-(O-CH}_2\text{-CH}_2\text{-CH}_2\text{-CH}_2\text{-O-}\overset{\overset{\displaystyle O}{\|}}{C}\text{-(CH}_2)_x\text{-}\overset{\overset{\displaystyle O}{\|}}{C}\text{)-}$	Bionolle #3000

$$X: \ 2 \ \text{and} \ 4$$

4
Grades of "Bionolle"

Bionolle products consist predominantly of two types of polyesters, i.e., poly(butylene succinate) (PBS) and poly(butylene succi-

nate adipate) (PBSA). The former is referred to as "Bionolle #1000" and the latter as "Bionolle # 3000".

The physical properties of typical Bionolle grades and general-purpose resins are listed in Table 3. When a polyester is used as a

Tab. 3 Characteristics of typical Bionolle grades (comparison with general purpose resins)

Properties	Unit	PBS (#1000 series)				PBSA (#3000 series)		General purpose resins		
Grade		#1001	#1903	#1020	#1020X30	#3001	#3020	PP	HDPE	LDPE
MFR (190 °C) 2.16 kg-load	g 10 min^{-1}	1.5	4.5	25	22	1.4	25	4 (230 °C)	2	2
Density	G cm^{-3}	1.26			1.48	1.23		0.90	0.95	0.92
m.p. (crystalline)	°C	114	115	115	115	94	95	164	130	108
T_g	°C	−32				−45		+5	−120	−120
Crystallization temp.	°C	75	88	76	81	50	53	120	104	80
HDT	°C	97			100	69		110	82	49
Flexural modulus	kgf cm^{-2}	6,700	7,000	7,000	23,000	3,300	3,500	14,000	11,000	1,800
Tensile yield strength	kgf cm^{-2}	330	400	350	340	190	195	320	280	120
Tensile break strength	kgf cm^{-2}	580	360	210	-	480	350	450	400	360
Tensile elongation	%	700	50	320	10	900	400	800	650	400
Crystallization degree	%	35-45				20-30		56	69	49
Chain structure		Linear	Long-chain branched	Linear	Linear	Linear		Linear		Long-chain branched
Relative biodegradability in environment		Slower				Faster		Nonbiodegradable		
ISO/FDIS 14855		O				O				
ISO/FDIS 14851		O				O				
Heat of combustion	Cal kg^{-1}	5,640				5,720		10,500	11,000	

*Physical property data are representative, and are not to be construed as quality assurance values.

structural material, it must have a melting point of $\geq 80\,°C$. Thus, the combination of diols and dicarboxylic acid has been investigated so as to provide a polyester having the above-described melting point, and as a result, the Bionolle #1000 series – which has a composition containing butanediol and succinic acid – has been created.

Investigation has shown that reducing crystallinity through copolymerization, thereby increasing the rate of biodegradation, is an effective means of enhancing biodegradability, and use of this approach has resulted in the creation of the Bionolle #3000 series.

As shown in Table 3, Bionolle products have high tensile strength, but low flexural modulus. In other words, they display both toughness and flexibility. One grade of Bionolle, #1020X30, is reinforced by using fillers so as to enhance the elastic modulus, and this results in a flexural modulus of $\sim 30{,}000\ \mathrm{kg\ cm^{-2}}$.

Typical Bionolle products are linear-chain polymers which have a low melt tension, though this leads to some disadvantages, for example a high degree of neck-in during molding into sheets, as well as difficulty in forming sheets that exhibit a high expansion ratio. In order to solve these problems, a long-chain branched structure is incorporated into a typical Bionolle chain structure, and as a result, molding processability is greatly improved. The product obtained in this way is designated the #1903 grade. The long-chain branched structure also effectively elevates the crystallization temperature, thereby increasing the efficiency of molding cycles (see Table 3).

Bionolle has excellent processability and can be molded into a variety of products, such as films, injection-molded products, filaments, and nonwoven fabrics. Several grades differing in melt flow indices are available so that molding can be carried out in straightforward manner using a typical molding apparatus (Table 4).

5
Structure and Properties of Bionolle

5.1
Chemical Structure

Bionolle is available in two grades, #1000 and #3000. These are aliphatic polyesters comprising butanediol and succinic acid (#1000), and butanediol, succinic acid, and adipic acid (#3000). #3000 Bionolle has improved biodegradability compared with #1000.

5.2
Molecular Weight and Molecular Weight Distribution

The available MFR (melt flow rate: 2.16 kgf at 190 °C) range of Bionolle is 1.0–30 g per 10 min, depending on the application.

Tab. 4 Grades and processability of Bionolle

MFR standards	PBS	PBSA	Molding methods and use
0.8–3	#1001	#3001	Tubular film, monofilament, blown
3–9	#1003, #3003	#1903	T-die cast film, foamed film, laminate
9–15	#1010	#3010	Mono-filament
20–34	#1020	#3020	Multi-filament, staple, injection
35–45	#1030	-	Non-woven fabric

Figure 2 shows MFR as a function of weight average molecular weight (Mw). The molecular weight distribution of Bionolle as defined by a Mw/Mn ratio of ~3.0, was very close to the theoretical value for polyesters.

5.3
Branching Structure

In the field of polyolefins, the introduction of long-chain branches into the main chain has long been known as a means of controlling rheological characteristics such as elongation viscosity, which in turn improves processability. In #1903 Bionolle, a long-chain branched structure was introduced, this being designed to facilitate processability during extrusion coating and foaming. Yoshikawa et al. (1996) reported the molecular weight and branched structure of Bionolle as determined by use of gel-permeation chromatography-MALLS (GPC-MALLS). The Mw/Mn ratios of Bionolle #1001 and #1903 were found to be 3.0 and 5.0, respectively, and #1903 was found to contain two long-chain branches in a single molecule. In other words, #1903 has an "H"-shaped structure.

5.4
Crystal Structure

The physical properties, such as modulus and heat distortion temperature, of semi-crystalline polymers are closely related to their crystal structure and crystallinity. Since Bionolle is a semi-crystalline polymer, its crystal structure and crystallinity must be examined in order to elucidate its physical properties. The crystal structure of #3000 is the same as that of #1000, because adipic acid is a minor component, although #3000 exhibits less crystallinity than #1000. Bionolle #1000 has two types of crystal modifications (Ichikawa et al., 1994): crystal transition between the two modifications takes place reversibly under the application and removal of stress. The crystal transition mechanism is found to be a first-order thermodynamic phase transition (Ichikawa et al., 1995). Similar crystal transition is reported for poly(butylene terephthalate) (Tashiro et al., 1980). The crystal modifications which appear under zero stress and under the application of stress are denoted by α- and β-forms, respectively. Figure 3 shows the two crystal structures (Ichikawa et al., 2000).

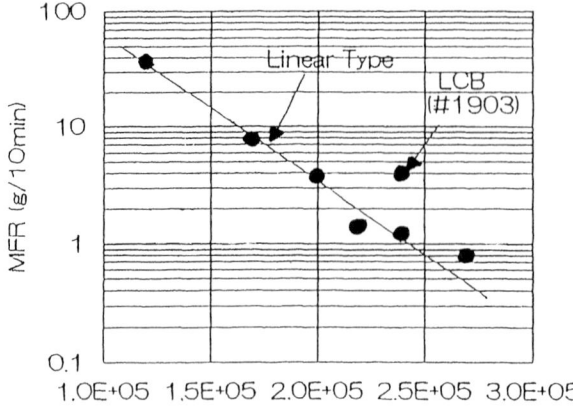

Wt. av. mol. wt.

Fig. 2 Relationship between melt flow ration (MFR) and molecular weight.

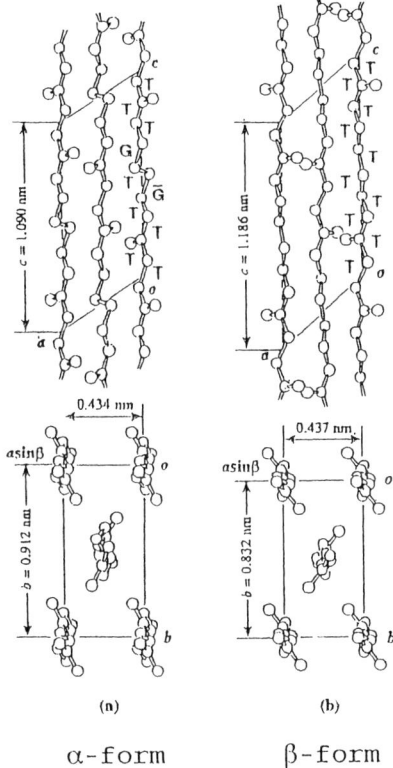

(n) (b)

α-form β-form

Fig. 3 Crystal structures of Bionolle.

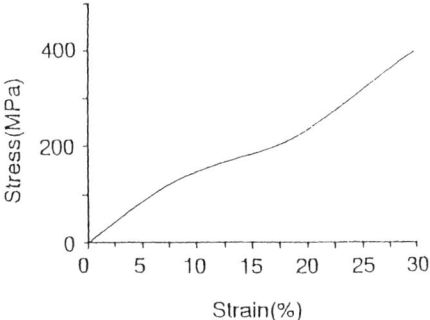

Fig. 4 Stress–strain curve.

point where the crystal transition between α- and β-forms takes place. In the third region, where only the β-form exists, the modulus is governed by the modulus of the β-form. Note that the modulus of the β-form is slightly greater than that of the α-form. Only the α-form can be considered for application in conventional environments (below the critical stress), since the β-form does not exist.

5.5
Melting and Crystallization

A significant difference between the two crystal forms is seen in terms of conformation in the main chain, despite the crystal forms having identical molecular packing and symmetry. The difference in conformation results in a difference in modulus between the α- and β-forms, however. Figure 4 shows a stress–strain curve of a uniaxially-oriented fiber specimen of Bionolle #1000, where the effect of the crystal structure can be clearly seen. The stress–strain curve has three regions. In the low strain region, where only the α-form exists, the modulus of a fiber is governed by the modulus of the α crystal. In the second region, where the α and β-forms co-exist, a plateau can be seen. The stress at the plateau is defined as being "critical stress" at the

Bionolle shows eccentric melting behavior, as shown by differential scanning calorimetry (DSC). Figure 5 shows DSC curves of Bionolle #1000 under various conditions: the samples were maintained at 160 °C for 5 min to initiate melting, cooled to 30 °C at various cooling rates, and then maintained at this temperature for 5 min. The samples were then heated up to 160 °C at 10 °C min^{-1}. At a very high cooling rate of 200 °C min^{-1}, the sample shows an exothermic peak just below the endothermic peak, corresponding to the melting point. In contrast, the sample shows two melting peaks when the cooling rate is low (5 °C min^{-1}). In addition, the ratio of the peak areas depends on the cooling rate. Samples prepared by cooling at 200 °C min^{-1}, annealing between the exothermic and endothermic peak tempera-

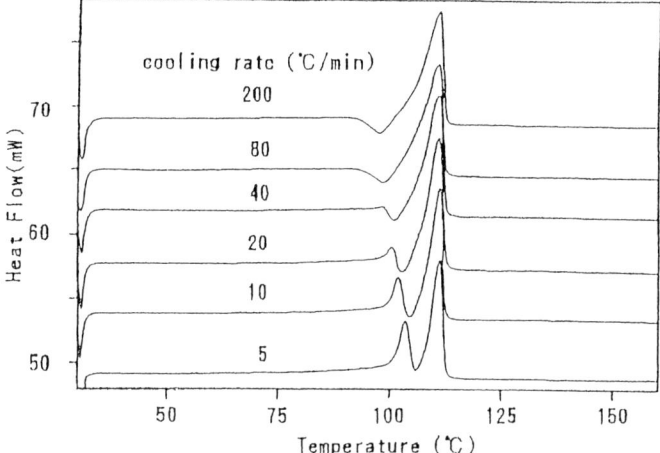

Fig. 5 Change in differential scanning calorimetry melting curve of Bionolle #1000 under various cooling conditions.

tures, and then cooling to room temperature, show a melting behavior similar to that of samples prepared at a cooling rate of 200 °C min^{-1} (Figure 6), except that no exothermic peak characteristic is seen. In other words, no difference in melting enthalpy is observed. These findings indicate that the small exothermic peak can be attributed to a reduction in crystal defects, and not to crystallization. Double peaks that

appear at a low cooling rate (5 °C min^{-1}) are due to the size distribution of α-form crystals; no β-form is identified in the sample.

The crystallization temperature of Bionolle is somewhat low. In particular, the crystallization temperature of #3000 is around 50 °C, and therefore crystallization may not be complete in summer when ambient temperatures are high. A useful way

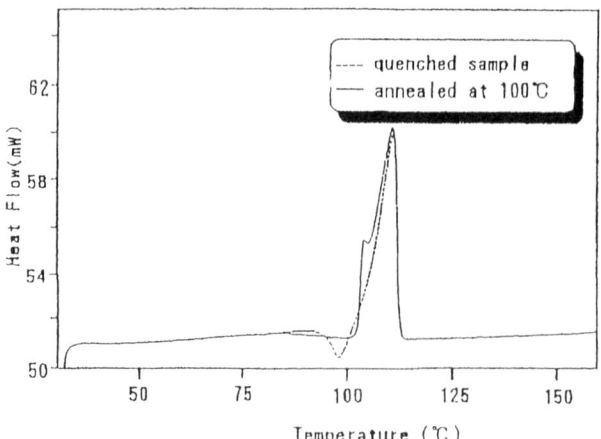

Fig. 6 Change in differential scanning calorimetry melting curve of Bionolle #1000 after quenching and annealing.

to overcome this problem is to add Bionolle with a long-chain branch, such as #1903. Figure 7 shows how the addition of such a branched component improves the crystallization temperature by 20 °C in the case of blends #3003 and #1903. Miyata and Masuko (1998) reported that the equilibrium melting point and melting enthalpy of a perfect crystal are 132 °C and 200 J g^{-1}, respectively; the amorphous phase density is 1.18 g cm^{-3}.

Crystallinity can be determined by X-ray diffraction, high-resolution solid-state nuclear magnetic resonance (NMR), DSC, and density, all of which detect the α-form. Kitsukawa et al. (1995) reported that the crystallinity obtained by NMR correlates well with that shown by the X-ray method, though the absolute values are different. The NMR method seems preferable to the X-ray method when the absolute value is to be determined, because the latter method introduces some ambiguity into the peak/halo separation. When DSC is used, crystallinity can be obtained by the ratio of "melting enthalpy of sample" to "enthalpy of a perfect crystal (200 J g^{-1})". It should be noted,

however, that the baseline must be determined carefully. When the method relying on density is used, crystallinity can be determined by measuring the density of a sample, based on knowledge that the densities of a crystal and amorphous phase are 1.33 g cm^{-3} and 1.18 g cm^{-3}, respectively.

6
Processability

Any conventional process that is applicable to polyolefins can also be applied to Bionolle. However, Bionolle must be treated with the same special care that is taken with other polyesters. First, a larger torque is needed for extrusion due to the characteristic narrow molecular weight distribution (note: Mw/Mn ~ 3). Second, the cooling time should be lengthened due to the relatively slow crystallization rate of Bionolle. Third, moisture content must be strictly controlled in order to prevent chain scission caused by hydrolysis during processing. The conditions are not as strict as those required for polyamides, unless Bionolle is used at high temperature

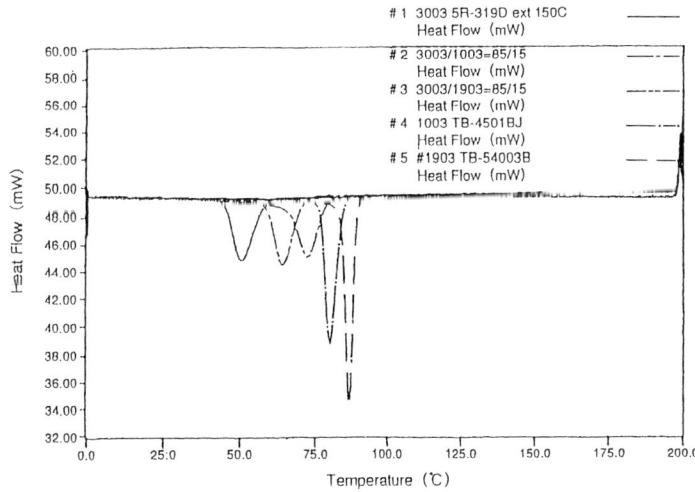

Fig. 7 Crystallization curves of Bionolle.

and humidity. Finally, the processing temperature should be 140–230 °C. Above 230 °C, chain scission would occur, resulting in weaker mechanical strength. However, extrusion coating and foaming would be typical exceptions. Details for blown film are discussed in the next section.

7
Blown Film

Bionolle #1001 and #3001 are both suitable for blown film applications; these are high molecular-weight grades having an MFR of ~1.2 g 10 min^{-1}. Desirable processing temperatures for #1001 and #3001 are 180–190 °C and 170–180 °C, respectively. The machine should have a high-torque motor, due to the narrow molecular weight distribution of Bionolle. In order to avoid overtorque, a blend of polyethylene and Bionolle (50/50) can be used during the conversion of polyethylene into Bionolle.

The most critical factor in the processing of Bionolle is the cooling method. Dual Lip type air rings (vertical blow type) for linear low-density polyethylene (LLDPE) and a large-capacity blower are needed, especially for Bionolle #3001. The air ring for high-density polyethylene (HDPE) is not applicable to Bionolle, because, due to its poor cooling characteristics, the bubble would not be stable or the frost line level would be high. In order to keep the physical properties of the film in balance, the blow-up ratio should be larger than that for polyolefins; for instance, a blow-up ratio >3 is desirable for Bionolle. If the blow-up ratio is <2, the impact strength of the resultant film is lowered dramatically. In addition, the lip-gap should be 1.0–1.5 mm, as impact strength and tear strength decrease when the lip-gap is increased.

The heat-seal strength (a measure of heat-seal capability) of Bionolle can be made quite high by optimizing the heat-seal temperature (Figure 8). In addition, since the printability of Bionolle is better than that of polyolefins, less or no corona treatment would be required.

The characteristics of blown film are summarized in Table 5; Bionolle shows physical properties comparable with those of polyolefins, with the exception of tear strength. An attempt has been made to solve

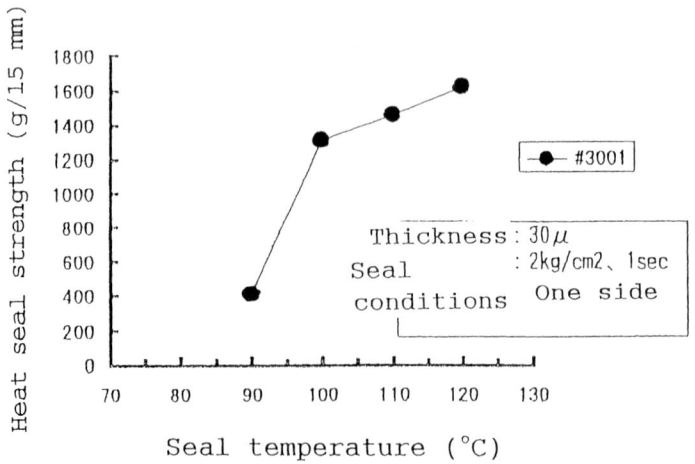

Fig. 8 Heat-seal strength of Bionolle #3001.

Tab. 5   Characteristics of blown film with thickness of 30 μm

Item	Direction	Unit	Test Method	#1001	#3001	#3001F25	L-LDPE
Density	---	g/l	JIS K 6760	1260	1230	1420	920
MFR (190 °C)	---	g/10min.	JIS K 6760	2	1.4	1.4	1.6
Tensile Strength			ASTM-D638				
σ y at yield	MD	MPa		31.1	17.9	19.2	12.7
σ b at yield	TD	MPa		30.6	18.1	19.6	10.8
σ b at break	MD	MPa		62.2	39.9	38.5	43.1
σ b at break	TD	MPa		59.3	44.7	44.7	21.6
Elongation at break	MD	%		660	780	570	370
	TD	%		710	970	760	420
Young's Modulus	MD	MPa	ASTM	4800	3300	754	3400
	TD	MPa	D882	5500	3500	784	3800
Tear strength	MD	N/mm	ASTM	3.6	4.4	11.5	9.8
	TD	N/mm	D1922	11	23	35.3	196
Film impact strength	–	KJ/m	ASTM D781	23.5	29.4	66	21.6
Haze	–	%	JIS-K7105	42	20	84	7

*These physical properties are representative values, and provide no guarantee in quality (Data from Showa Denko Soken, Kawasaki Laboratory).

this problem by upgrading the material or by optimizing processing conditions.

of polyester products, the mechanism of decomposition must be elucidated in detail in advance.

8
Stability of "Bionolle"

Polyester undergoes hydrolysis in the presence of water, thereby lowering its molecular weight. It is well known that, at high temperature, polyester decomposes in a nonradical manner to form an acid and a low molecular-weight compound having an unsaturated bond. Decomposition proceeds via a six-membered ring intermediate. As such decomposition is related to the stability

8.1
Hydrolysis

When polytetramethylene succinate (PBS) film, PBSA film, and polyethylene succinate (PES) film were placed at 65 °C under high-moisture conditions (in sterilized sawdust powder, water content 65%), the molecular weight of each sample decreases over time, with the rate of decrease being considerably high. Table 6 shows time-lapse changes in the molecular weight of each sample. Hence,

Tab. 6   Change in molecular weight under high-temperature and high-moisture conditions

	0 weeks			2 weeks			8 weeks		
Samples	$Mp \times 10^{-4*}$	$Mw \times 10^{-4}$	Mw/Mn	$Mp \times 10^{-4*}$	$Mw \times 10^{-4*}$	Mw/Mn	$Mp \times 10^{-4*}$	$Mw \times 10^{-4}$	Mw/Mn
PBS	12.1	10.8	2.46	2.89	3.16	2.15	0.411	0.596	1.33
PBSA	19.1	23.4	2.97	2.79	3.57	2.15	0.452	0.610	1.33
PES	8.98	16.7	4.35	1.61	2.55	2.07	0.340	0.432	1.18

*Mp represents peak-top molecular weight; each molecular weight is reduced from that of PMMA.

the conditions of use must be checked before such film is used at high temperature.

The hydrolysis rates of PBSA in water at a pH of 7 and at several temperatures are shown in Figure 9, and the effect of pH on hydrolysis rate is shown in Figure 10. Rate constants of PBSA obtained from these data are listed in Table 7.

Hydrolysis is accelerated with increasing temperature. The Mw of PBSA decreases from 200,000 to 120,000 Daltons during hydrolysis at 35 °C for 150 days, and deterioration of physical properties can be considered. Thus, when PBSA is in contact with water its use is limited to a short time period, and to temperatures that are not low.

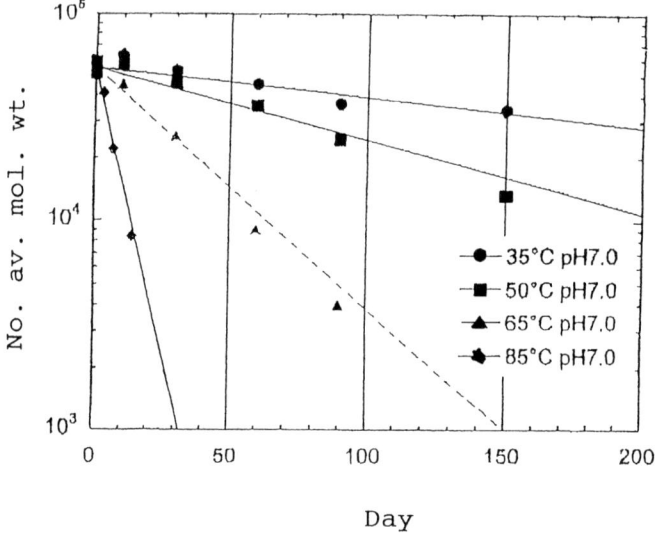

Fig. 9 Hydrolysis rate of PBSA.

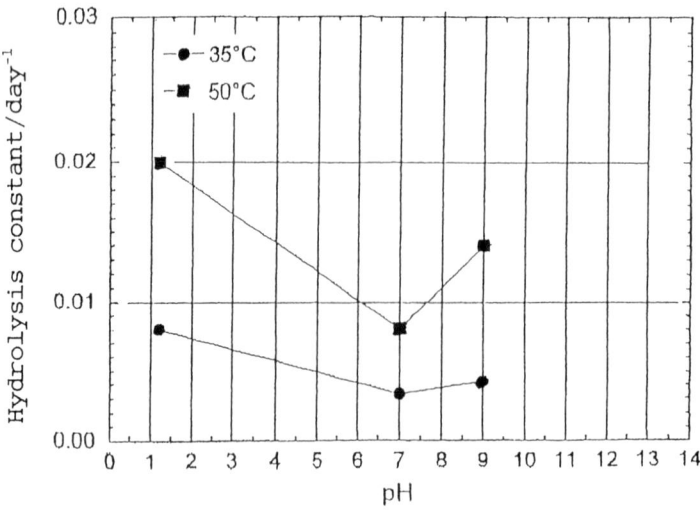

Fig. 10 Effect of pH on hydrolysis rate.

Tab. 7 Hydrolysis rate constants (K) of PBSA

	Temperature °C	PH	KdAy[a]
PBSA	35	7.0	0.003
	50	7.0	0.008
	65	7.0	0.027
	85	7.0	0.12
PLLA	23	7.0	0.02[a]
	37	7.0	0.024[a]
PDLLA	23	7.ß	0.018[a]
	37	7.0	0.051[a]

[a] M.Hilpinen –Vainio, *J.Appl.Polym.Sci*, **59**, 1281 ('96).

Reference values of hydrolysis rate constants of lactic acid polymers are listed in Table 8.

Tab. 8 Hydrolysis rate constants (K) of poly(butylene succinate adipate)

	Temperature °C	PH	KdAy^{-1}
PBSA	35	7.0	0.003
	50	7.0	0.008
	65	7.0	0.027
	85	7.0	0.12
PLLA	23	7.0	0.020[1]
	37	7.0	0.024[1]
PDLLA	23	7.0	0.018[1]
	37	7.0	0.051[1]

[1] M.Hiljanen-Vainio, J. Appl. Polam. Sci **59** 1281 ('96)

8.2
Pyrolysis

Polyester undergoes pyrolysis at > 200 °C. Pyrolysis proceeds in a nonradical manner and is known to be accelerated by a polymer-end carboxyl group. Even though Bionolle is a hydroxyl-terminated polymer, it exhibits a high pyrolysis rate at 250 °C or higher. Some examples of pyrolysis features are listed in Table 9.

For temperatures of ≤ 200 °C, the molecular weight decreases slowly, and heat deterioration for a short period of time can be negligible. When the temperature is ≥ 200 °C, the pyrolysis rate increases gradually. When the temperature is ≥ 250 °C, a considerable decrease in molecular weight occurs. However, when polyester is subjected to heating for a short period of time, such as heating experienced during a typical molding process, then heat deterioration is not significant, i.e., the polyester remains stable, until the temperature reaches approximately 280 °C.

During pyrolysis, the formation of a double bond occurs at a polymer end portion. The amount of acid formed is relatively smaller than the amount of double

Tab. 9 Pyrolysis of PBS

Heating conditions	Mu (×10⁴)	Mw (×10⁴)	Double bond terminal[a]	Acid terminal[a]	OH terminal[a]
Initial	10.0	22.8		ND	2.2
190 °C					
1 h	9.9	22.8	–	ND	1.8
4 h	9.7	21.6	0.1	ND	2.1
220 °C					
1 h	9.7	20.5	0.1	ND	1.8
4 h	9.4	18.7	0.1	ND	2.5
250 °C					
1 h	5.4	10.8	0.5	ND	3.2
4 h	3.3	5.8	1.9	1.0	4.5

[a] The number of C atoms per ester main chain of 1000 C atoms (measured by ^{13}C-NMR).

bonds. One possible reason for this might be elimination, to the outside the system, of the acid in the form of acid anhydride.

9
Biodegradability

9.1
Aerobic Biodegradation

The potential biodegradability of Bionolle is determined by using aerobic microorganisms in an aqueous system according to ISO14851 (International Organization for Standardization, 1999a). Bionolle is reduced to powder by a freeze mill, attaining a particle size of 125–250 μm, after which it is incubated in a suspension of active soil in a closed flask in a respirometer (Coulometer,

Ohkura Electric, Tokyo, Japan). The consumption of oxygen (biological oxygen demand; BOD) is automatically determined by measuring the change in pressure. The degree of biodegradability is determined by expressing the BOD as a percentage of the theoretical oxygen demand (ThOD). The soil used in the biodegradation is obtained from the surface of a cultivated field in Hiroshima Prefecture. Figure 11 shows the results, which reveal that Bionolle is as readily biodegradable as cellulose (microcrystalline cellulose powder), and potential biodegradability is estimated to be about 70%. The difference in biodegradability between Bionolle #1000 and #3000 can be attributed to the time required to attain a constant level of BOD, at which point no further biodegradation is expected. The results also show that Bionolle #1000 and #3000 take about 20 and

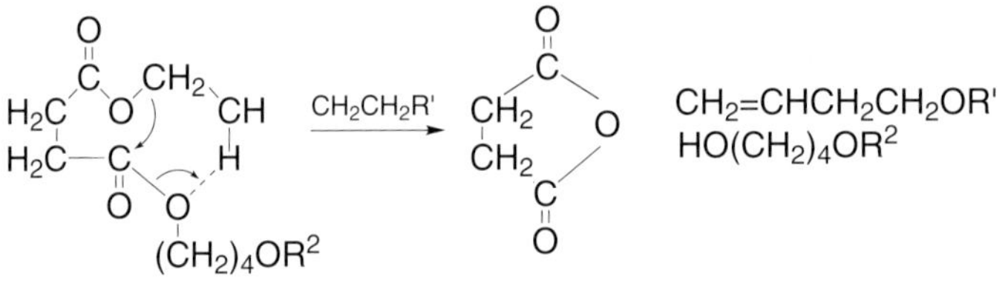

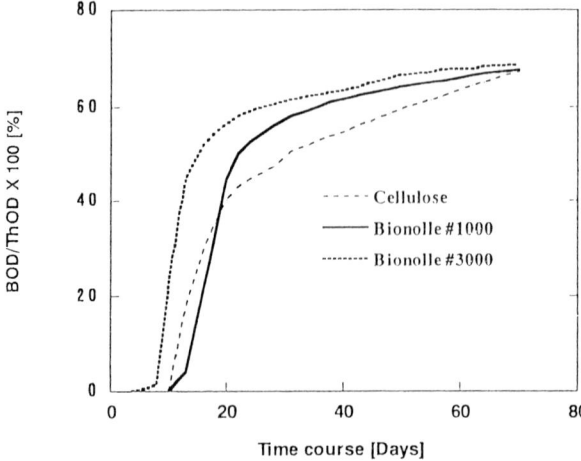

Scheme 1 Mechanism of Pyrolysis of PBS

Fig. 11 Aerobic biodegradability of Bionolle.

40 days, respectively, to achieve a percentage biodegradation value of 60%. The lifetime of Bionolle #1000 can be said to be double that of Bionolle #3000.

9.2
Compostability

The organic recovery of packaging materials, which includes aerobic composting in municipal or industrial biological waste treatment facilities, has become an important option for reducing and recycling packaging waste. During its early development, Bionolle was expected to offer real compostability, and was certified by "OK Compost" in Belgium in 1997. A controlled aerobic composting test, which is technically identical to that of ISO14855 (International Organization for Standardization, 1999b), should provide unequivocal information on the inherent and ultimate biodegradability of Bionolle as a packaging material. Based on results for day 40, the final product (mature compost) after aerobic composting shows Bionolle's excellent compostability (Figure 12). The resultant compost is checked for residual amounts of heavy metals and

other hazardous substances listed in prEN13432 (European Committee for Standardization, 2000), and also ability to attain normal seed germination and plant growth. Bionolle was found to have no negative effect on the quality of resultant compost and therefore be suitable for organic recovery.

9.3
Degradation in the Environment

Bionolle in film form has been examined in different environments for practical degradation, and also for biodegradability by measuring the loss in weight (Figure 13). Bionolle undergoes degradation as readily in seawater or fresh water as it does in farm soil, showing a prerequisite of ability to degrade in any environment. The Biodegradable Plastic Society of Japan carried out biodegradation tests on five different biodegradable polymers (copolymers of 3-hydroxybutyrate and 3-hydroxyvalerate (PHB/HV), poly(ε-caprolactone) (PCL), Bionolle #1000, Bionolle #3000, and poly(lactic acid)) at 22 locations in Japan and one location in the USA over a 12-month period

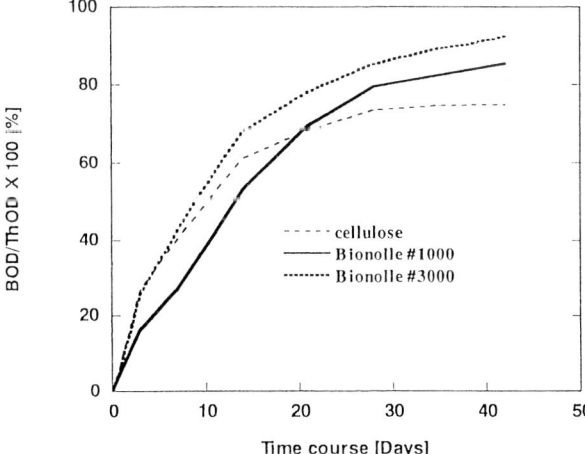

Fig. 12 Compostability of Bionolle films.

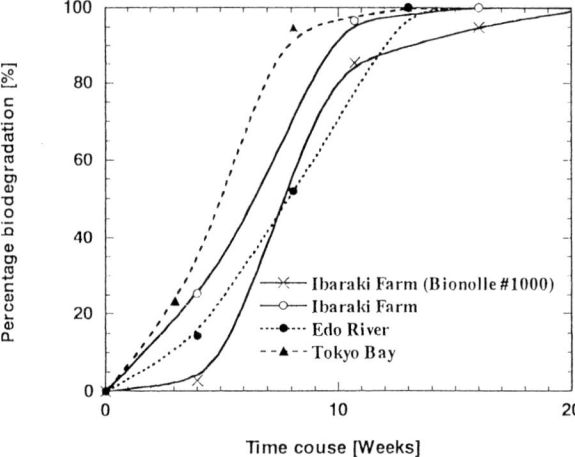

Fig. 13 Environmental degradation of Bionolle.

starting in November 1995, and compared the change in mechanical properties, i.e., tensile strength, elongation at break, specific viscosity, and average molecular weight (Hoshino et al., in press). The test results concluded that polymer degradation is accompanied by losses in mechanical properties and a reduction in molecular weight, and this could be largely accounted for by the state of the inoculum at the site where the polymer is exposed. Moreover, the polymer disposition for degradation (general tendency toward degradation) is determined by the degree of overall weight loss. Bionolle, therefore, ranks with PHB/HV and PCL in terms of biodegradability.

9.4
Microbial Degradation

Fungi and bacteria produce a variety of extracellular enzymes that function as catalysts in the hydrolytic degradation of polymers. Several microbial strains have been isolated from soil which can utilize Bionolle as sole carbon source. *Paecylomyces lilacinus*, *Aspergillus* sp., *Malbranchea sulfureum* (Figure 14), *Alcaligenes* sp., *Pseudomonas* sp., *Bacillus* sp., *Corynebacteria* and *Streptomyces* sp. are examples (Nishioka et al., 1994) of common strains found worldwide. These strains are thought to degrade the polymer structure into small fragments via enzymatic

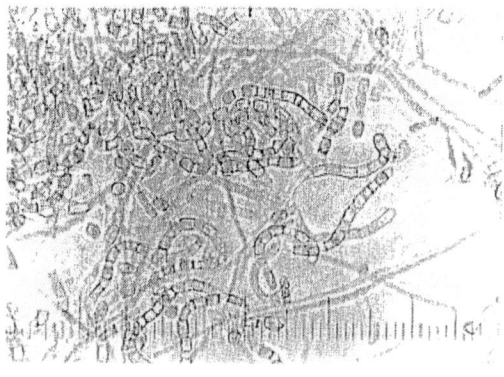

Fig. 14 *Mabranchea sulfureum* BN-432.

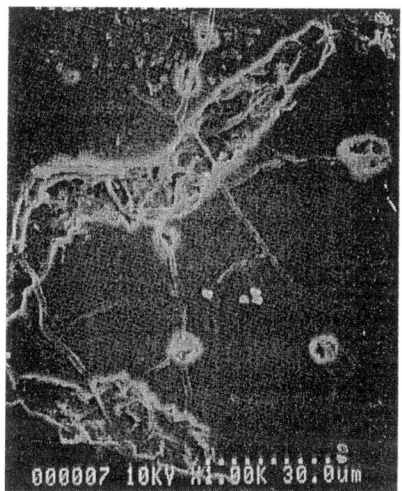

Fig. 15 Scanning electron micrograph of the degrading film surface.

hydrolysis (Ando et al., 1998), and to use these fragments as nutrients. Figure 15 shows an electron micrograph of the surface of a rotted film.

10
Environmental Friendliness

It follows that there is a need to define and evaluate Bionolle to ensure that it will indeed degrade and mineralize in an environmentally acceptable manner in natural and/or microbially active environments, without leaving harmful residues. The primary products of degradation were obtained experimentally by enzymatic hydrolysis of Bionolle in an aqueous solution, and the chemical structures determined analytically using high-performance liquid chromatography/ mass spectrometry (HPLC/MS) and ^{1}H-NMR. The results indicated that polyester-based compounds of monomer units were produced predominantly, but that small amounts of diurethane compounds (derived from diisocyanate serving as a coupling agent) were also formed. The resultant products were assessed for biodegradability in an activated sludge according to OECD 301C, and found to have been completely mineralized within 2 weeks (Figure 16) (Kitakuni et al., 2000). Moreover, the 50% lethal dose for either compound was estimated at $> 5,000$ mg kg^{-1}, while the Ames mutagenicity test proved negative; this suggested that, during degradation, these materials had no harmful effect on life within the environment.

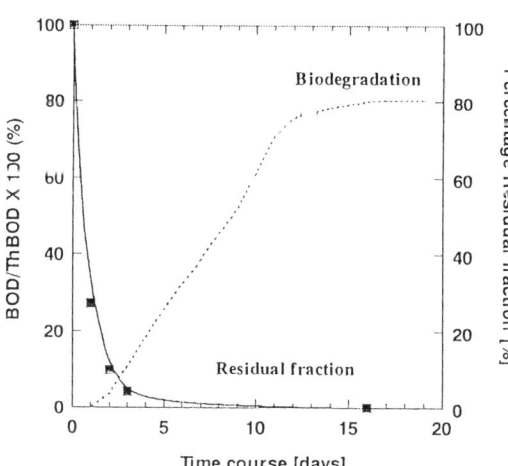

Fig. 16 Biodegradation of enzymatic hydrolysis products of Bionolle.

11

Fitness for Sustainable Development

The need to understand the biological reactivity of polymers has increased in response to the integration of biological waste treatment systems (e.g., composting and soil application) into the waste management plans of communities worldwide. Since 1994 – an early stage of development – we have been aware of the need to ensure that Bionolle products are safe and compatible with the environment. Our efforts include clarifying the fate of Bionolle materials after disposal and designing its packaging to better integrate with the existing waste disposal infrastructure. From a point of view of environmental friendliness, Bionolle is ideal for rapid biodegradation in a microbially active aerobic environment, i.e., compost or sewage sludge, which requires the maximum rate and greatest extent of mineralization. The reason why Bionolle itself is so conducive to biodegradation is probably related to the degree of crystallinity (Figure 17). The mechanical properties of crystalline polymers depend strongly on their crystal structures, which can be altered by pressure, temperature, and stress, as well

as on the crystallinity of the polymer (Ichikawa et al., 2000). Figure 17 shows clearly that a change in degree of crystallinity from 41% to 54% reduces the biodegradation rate by one-third, suggesting that the predominant action is one whereby Bionolle chains in an amorphous state and exposed on the surface are susceptible to degradation. This is supported by the observation that the rate of enzymatic hydrolysis of poly(3-hydroxybutyrate) (poly(3HB)) film by PHB depolymerase (*Alcaligenes faecalis*) depends strongly on the degree of crystallinity of P(3HB) film, but is independent of the size of spherulites, and why the rate of enzymatic erosion for poly(3HB) chains in an amorphous state occurs is much more rapid than for chains in a crystalline state (Koyama and Doi, 1997). In order to be accepted for a wider range of applications, degradable plastic products must possess certain key characteristics. These include a "Fit for Purpose" degradation rate with respect to different applications under various environmental conditions while maintaining acceptable physical product performance properties during actual use that are comparable with those of conventional, nondegradable plastics. To obtain these desirable character-

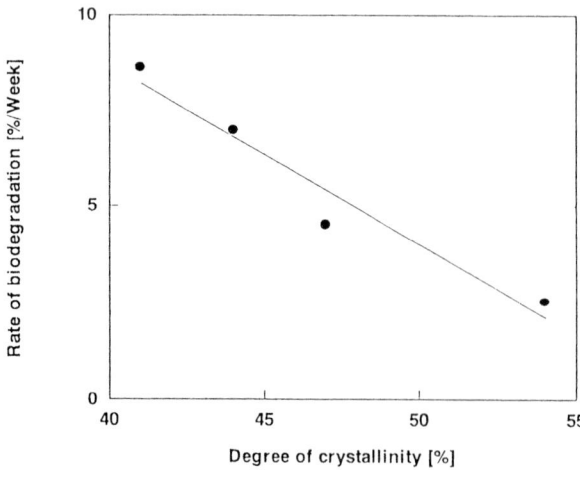

Fig. 17 The effect of crystallinity on biodegradation rate.

istics, copolymerization of different monomeric units and blending with other degradable polymers can be implemented (Abe et al., 1995; Doi et al., 1995; Koyama and Doi, 1996). Bionolle shows good compatibility with other polyesters, e.g., PLA and PCL. Therefore, the range of Bionolle formulations feasible today will make Bionolle widely applicable and able to meet the requirements of numerous packaging, health, and agricultural needs. Bionolle products can fit nicely into the natural biocycle and thereby qualify as environmentally friendly plastics, fulfilling the worldwide objectives of sustainable development.

12
Outlook and Perspectives

Surrounded by an enormous variety of products designed for all manners of functions, we currently enjoy unprecedented material affluence. This affluence is a fruit of technical innovation, the roots of which can be traced to the 1950s. However, this innovation has also brought about a world of unprecedented scales of production, consumption, and waste. The sustainment of social progress requires our development as a circulation society that takes into consideration the capacity of nature's intrinsic purifying function, because the resources and energy available to us are not infinite.

Although plastics support modern society through their excellent properties and adaptability to mass production technology, their ubiquitous use and long service life raise serious environmental problems. Currently, plastics are consumed in a great variety and in vast amounts. Accordingly, waste plastics find multifarious uses and statuses complicating the solution of environmental problems. A multilateral approach for solving

these problems may include the following measures:

- reducing the total amount of plastics we use by avoiding excessive use;
- promoting effective use of plastics through, e.g., recycling; and
- applying nature's intrinsic purifying mechanisms to solving problems.

Many of these measures – including prolonging the service life of plastic products, recycling of plastic materials and chemicals for use as carbon resources, and developing biodegradable plastics – must be considered at the design stage of polymer materials.

Our approach is to impart biodegradability to general use so as to make use of the power of nature, and we hope that our Bionolle line of biodegradable plastics contributes to more effective solutions to environmental problems.

Bionolle is now increasingly applied to situations where plastic products are used in the natural environment and involve uneconomically high waste collection costs, as well as situations where users seek to suppress the environmental load imposed by plastic products left unintentionally in the natural world.

13
Patents

1. Takiyama, E., et al., USP 5,436,056 (1995/07/25)
2. Takiyama, E., et al., USP 5,530,058 (1996/06/25)
3. Takiyama, E., et al., EPC 0565235 (1999/12/22)
4. Takiyama, E., et al., EPC 0569153 (1999/10/06)
5. Takiyama, E., et al., EPC 0572682 (1997/10/01)

14
References

Abe, H., Doi, Y., Aoki, H., Akehata, T., Hori, Y., Yamaguchi, A. (1995) Physical properties and enzymatic degradability of copolymers of (R)-3-hydroxybutyric and 6-hydroxyhexanoic acids, *Macromolecules* **28**, 7630–7637.

Ando, Y., Yoshikawa, K., Yoshikawa, T., Nishioka, M., Ishioka, R., Yakabe, Y. (1998) Biodegradability of poly(tetramethylene succinate-co-tetramethylene adipate): I. Enzymatic hydrolysis, *Polym. Degrad. Stab.* **61**, 129–137.

Darby, R. T., Kaplan, A. M. (1968) *Appl. Microbiol.* **16**, 900.

Doi, Y. (1995) Report on Development Marino Forum 21, of Biodegradable Fishing Tackle 1994: 193.

Doi, Y., Kitamura, S., Abe, H. (1995) Microbial synthesis and characterization of poly(3-hydroxybutyrate-*co*-3-hydroxyhexanoate), *Macromolecules* **28**, 4822–4828.

European Committee for Standardization (2000) Packaging – Requirements for packaging recoverable through composting and biodegradation – Test scheme and evaluation criteria for the final acceptance of packaging. prEn13432, rue de Stassart, 36, B-1050 Brussels, Belgium.

Hoshino, A., Sawada, H., Yokota, M., Tsuji, M., Fukuda, K., Kimura, M. Influence of weather conditions and soil properties on degradation of biodegradable plastics in soil, *Soil Sci. Plant Nutr.* (in press).

Ichikawa, Y., Suzuki, J., Washiyama, J., Moteki, Y., Noguchi, K., Okuyama, K. (1994) Strain-induced crystal modification in poly(tetrametylene succinate), *Polymer* **35**, 3338.

Ichikawa, Y., Washiyama, J., Moteki, Y., Noguchi, K., Okuyama, K. (1995) Crystal transition mechanisms in poly(tetrametylene succinate), *Polym. J.* **27**, 1230.

Ichikawa, Y., Kondo, H., Igarashi, Y., Noguchi, K., Okuyama, K., Washiyama, J. (2000) Crystal structures of α and β forms of poly(tetramethylene succinate). *Polymer* **41**, 4719–4727.

International Organization for Standardization (1999a) Determination of the ultimate aerobic biodegradability of plastic materials in an aqueous medium - Method by measuring the oxygen demand in a closed respirometer. ISO 14851, Geneve 20, Switzerland.

International Organization for Standardization (1999b) Determination of the ultimate aerobic biodegradability of plastic materials in an aqueous medium - Method by measuring the oxygen demand in a closed respirometer. ISO 14855, Geneve 20, Switzerland.

Kitakuni, E., Yoshikawa, K., Nakano, K., Sasuga, J., Nobiki, M., Naoi, H., Yokota, Y., Ishioka, R., Yakabe, Y. (2000) The biodegradation of poly(tetramethylene succinate-*co*-tetramethylene adipate) and poly(tetramethylene succinate) through water-soluble products, *Environ. Toxicol. Chem.* **20**, 941–946.

Kitsukawa, M., Yosikawa, K., Nishioka, M., Moteki, Y., Ichikawa, Y. (1995) Proceedings of the Society of Solid-State NMR for Materials, No. 18.

Korshak, V. V., Vloogradova, S. V. (1953) *DoAd. Akad. Nauk. SSSR* 1017.

Koyama, N., Doi, Y. (1996) Miscibility, thermal properties, and enzymatic degradability of binary blends of poly[(R)-3-hydroxybutyric acid] with poly(ε-caprolactone-*co*-lactide), *Macromolecules* **29**, 5843–5851.

Koyama, N., Doi, Y. (1997) Effects of solid-state structures on the enzymatic degradability of bacterial poly(hydroxyalkanoic acids), *Macromolecules* **30**, 826–832.

Miyata, T., Masuko, T. (1998) *Polymer* **39**, 1399.

Nishioka, M., Tuzuki, T., Wanajyo, Y., Oonami, H., Horiuchi, T. (1994) Biodegradation of BIO-NOLLE, in: Doi, Y., Fukuda, K., Eds), *Biodegradable Plastics and Polymers*, Studies in Polymer Science 12. Amsterdam, The Netherlands: Elsevier Science B.V., 584–590.

Tashiro, K., Nakai, Y., Kobayashi, M., Tadokoro, H. (1980) Solid-State transition of poly(butylene terephthalate), *Macromolecules* **13**, 137.

Tokiwa, Y., Ando, T., Suzuki, T. (1976) *J. Ferment. Technol.* **54**, 603.

Tokiwa, Y., Suzuki, T. (1974) *J. Ferment. Technol.* **52**, 393.

Tokiwa, Y., Suzuki, T., Takeda, K. (1988) *Agric. Biol. Chem.* **52**, 1937.

Yoshikawa, K., Ofuji, N., Imaizumi, M., Moteki, Y., Fujimaki, T. (1996) Molecular weight distribution and branched structure of biodegradable aliphatic polyesters determined by s.e.c-MALLS, *Polymer*, **37**, 1281.

11
Biodegradable Aliphatic-
Aromatic Polyesters:
"Ecoflex®"

MSc. Motonori Yamamoto[1], Dr. Uwe Witt[2], DI. Gabriel Skupin[3],
Dr. Dieter Beimborn[4], Dr. Rolf-Joachim Müller[5]

[1] BASF Aktiengesellschaft, ZKT/U-B1, Polymer Laboratory, D-67056 Ludwigshafen,
Germany; Tel.: +49-621-60-43080; Fax: +49-621-60-20313;
E-mail: motonori.yamamoto@basf-ag.de

[2] BASF Aktiengesellschaft, KSS/BP, Business Unit Biodegradable Polyester,
D-67056 Ludwigshafen, Germany; Tel.: +49-621-60-47825;
Fax: +49-621-60-93303; E-mail: uwe.witt@basf-ag.de

[3] BASF Aktiengesellschaft, KSS/BP, Business Unit Biodegradable Polyester,
D-67056 Ludwigshafen, Germany; Tel.: +49-621-60-41912;
Fax: +49-621-60-93384; E-mail: gabriel.skupin@basf-ag.de

[4] BASF Aktiengesellschaft, ZH/TC-Z570„ Experimental Toxicology and Ecology,
D-67056 Ludwigshafen, Germany; Tel.: +49-621-60-58224;
Fax: +49-621-60-58043; E-mail: dieter.beimborn@basf-ag.de

[5] Gesellschaft für Biotechnologische Forschung mbH, Mascheroder Weg 1,
D-38124 Braunschweig, Germany; Tel.: +49-531-6181-610;
Fax: +49-531-6181-175; E-mail: rmu@GBF.de

ASTM	American Society for Testing and Materials
BPS	Biodegradable Plastics Society, Japan
CEN	Comitée Européen de Normalisation
DIN	Deutsches Institut für Normung
DIN CERTCO	Gesellschaft für Konformitätsbewertung mbH
GC-MS	gas chromatography coupled with mass spectrometry
GPC	gel permeation chromatography
ISO	International Organisation for Standardization
JIS	Japanese Industrial Standards
MALDI-MS	matrix-assisted laser desorption/ionization mass spectrometry
MS	mass spectrometry
NMR	nuclear magnetic resonance
PE-LD	polyethylene, low density
PBT	poly(butylene terephthalate)
PET	poly(ethylene terephthalate)
PS	Pruefsubstanz (test substance)
T. fusca	*Thermomonospora fusca*

1
Introduction

1.1
Perspectives for Biodegradable Plastics

The market for biodegradable materials is growing significantly every year. A positive reaction can be expected from those applications, where the biodegradability offers a clear advantage for customers and the environment, with typical examples being packaging, compost bags, agricultural films, and coatings. However, in order to gain wider acceptance in the market for such innovative solutions, the following requirements must be fulfilled:

- Standard test methods (DIN 54900, CEN, ISO, ASTM, JIS).
- Good properties and processability comparable with that of conventional plastics.
- Competitive prices and supply in sufficient quantities, taking advantage of economies of scale.

BASF has risen to this challenge, and in 1998 a special type of biodegradable copolyester based on synthetic raw materials was commercialized by the company under the tradename Ecoflex® With Ecoflex®, BASF aims to make an active contribution to the development of the market for biodegradable polymers. Following analysis of the existing market and promising applications, it is estimated that good opportunities exist especially for the successful development of large volumes in materials such as compost bags, fast food/disposables, agricultural films, hygiene films, and paper coatings.

1.2
BASF's Integrated Manufacturing System Offers Advantages

For the production of Ecoflex®, advantage is taken of BASF polymer technology and economies of scale. The key success factor for Ecoflex®, notably with regard to availability and price, is the integration into the "BASF Verbund". The use of monomers from BASF's integrated manufacturing system and the production in existing facilities in Ludwigshafen enable favorable cost structures.

2
Historical Outline

In 1990, at the request of the German state government, BASF began a feasibility study of biodegradable and compostable plastics for use of packaging. This study revealed that the most important factors for the biodegradable plastics for commercial uses are price, performance, availability of monomers/polymers, and use of existing plant. On the basis of these prerequisites, BASF began screening suitable monomers on the laboratory scale. Pilot plant production and sampling of Ecoflex® were started in 1994 and, in 1997, following a good response from several customers, BASF began to produce Ecoflex® commercially at an existing poly(butylene terephthalate) (PBT) plant.

3
Structure of Ecoflex®:
The BASF Concept of Modular Units

Ecoflex® is an aliphatic-aromatic copolyester based on terephthalic acid, adipic acid, 1,4-butanediol and modular units.

The product properties of Ecoflex® are designed to meet the requirements of a biodegradable plastic: ideally, a combination of processability, utilization properties, and biodegrabability. This is achieved by the synthesis of tailor-made molecular structures, obtained through modular units by which the statistical copolyester units, including 1,4-butanediol and the dicarbonic acids, adipic acid and terephthalic acid, are linked (Figure 1).

This modular system involves the incorporation of hydrophilic components of monomers with branching, leading to chain-lengthening, and thereby increasing the molecular weight to yield tailor-made products with totally different material properties.

density (PE-LD) (Tables 1 and 2). The films are tear-resistant and flexible, and also resistant to both water and fluctuations in humidity. The extreme strength and failure energy are clear product characteristics of Ecoflex®, and these significantly exceed the respective properties of PE-LD films. The barrier properties differ from those of PE-LD, however; Ecoflex® films are breathable because of their moderate water vapor permeability, and this can be adjusted within different Ecoflex®-batches (Tables 3 and 4).

4
Properties, Processing, and Application of Ecoflex®

Examples of the mechanical properties, processing (rheological) properties and applications of Ecoflex® are outlined in the following sections.

4.1
Property Profile of Ecoflex®

The mechanical properties of Ecoflex® are comparable with those of poly(ethylene)-low

4.2
Processing of Ecoflex®

4.2.1
Film Extrusion
Ecoflex® can be processed by conventional blown film lines for PE-LD. The excellent draw-down ability of Ecoflex® (Figures 2 and 3) leads to interesting applications in the thin film segment (<20 μm) and, depending on the processing equipment available, 10 μm films can be obtained.

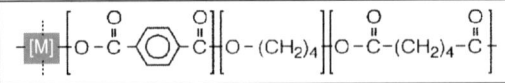

Fig. 1 The BASF concept of modular systems. M: e.g., monomers with a branching and chain extension effect.

Tab. 1 Typical basic material properties of Ecoflex®

Property	Unit	Test method	Ecoflex®	Lupolen® 2420 F (PE-LD)
Mass density	g cm^{-3}	ISO 1183	1.25–1.27	0.922–0.925
Melt flow rate	mL 10 min^{-1}	ISO 1133	3–8	–
MVR 190 °C, 2,16 kg	g 10 min^{-1}		–	0.6–0.9
MFR 190 °C, 2,16 kg				
T_m	°C	DSC	110–115	111
T_g	°C	DSC	– 30	
Shore D hardness	–	ISO 868	32	48
Vicat VST A/50	°C	ISO 306	80	96

T_m, melting point; T_g, glass transition temperature.

Tab. 2 Typical properties of blown film (50 µm) of Ecoflex®

Property	Unit	Test method	Ecoflex®	Lupolen® 2420 F (PE-LD)
Transparency	%	ASTM D 1003	82	89
Tensile strength	N mm^{-2}	ISO 527	32/36	26/20
Ultimate strength	N mm^{-2}	ISO 527	32/36	–
Ultimate elongation	%	ISO 527	580/820	300/600
Failure energy (Dyna-Test)	J mm^{-1}	DIN 53373	14.3	5.5
Tear propagation resistance	N mm^{-1}	DIN 53363	236/124	–
Permeation rates:				
Oxygen	Cm^3 (m^{-2} d · bar)	DIN 53380	1600	2900
Water vapor	g (m^{-2} ·d)	DIN 53122	140	1.7

Tab. 3 Masterbatches of Ecoflex®

Batch type	Description
Ecoflex® Batch SL1	Slip agent (erucamide)
Ecoflex® Batch SL2	Slip agent (wax-type)
Ecoflex® Batch AB1	Antiblock agent (fine talc)
Ecoflex® Batch AB2	Antiblock agent (coarse talc)
Ecoflex® Batch C White	White masterbatch
Ecoflex® Batch C Black	Black masterbatch

Ecoflex® does not require any special handling; neither is the pre-drying usually associated with thermoplastic polyesters necessary with Ecoflex®, and this provides an additional advantage for the converter. Films made from Ecoflex® can be printed and welded with conventional equipment used with PE-LD.

Hence, Ecoflex® is an outstanding material for applications such as compost bags, films in the agricultural sector, or hygiene films.

4.2.2
Master Batches of Ecoflex®

For extrusion processing, the production of colored films, adjustment of the water vapor barrier, antiblock and slip properties, and transparency, master batches based on Ecoflex® were developed to fulfil the different customers needs (Table 3).

The barrier properties of Ecoflex® films can be adjusted by using special additives. Water vapor permeabilities of Ecoflex® films produced with different master batches are listed in Table 4.

Tab. 4 Water vapor permeability of Ecoflex® films

	Film thickness [µm]	Water vapor permeability [g m^{-2} · d]	Water vapor permeability [g (100 µm) · m^{-2} · d]
Ecoflex®	15	> 500	140
Ecoflex® + SL2	16	200	32
Ecoflex® + SL2 + AB1	14	170	24

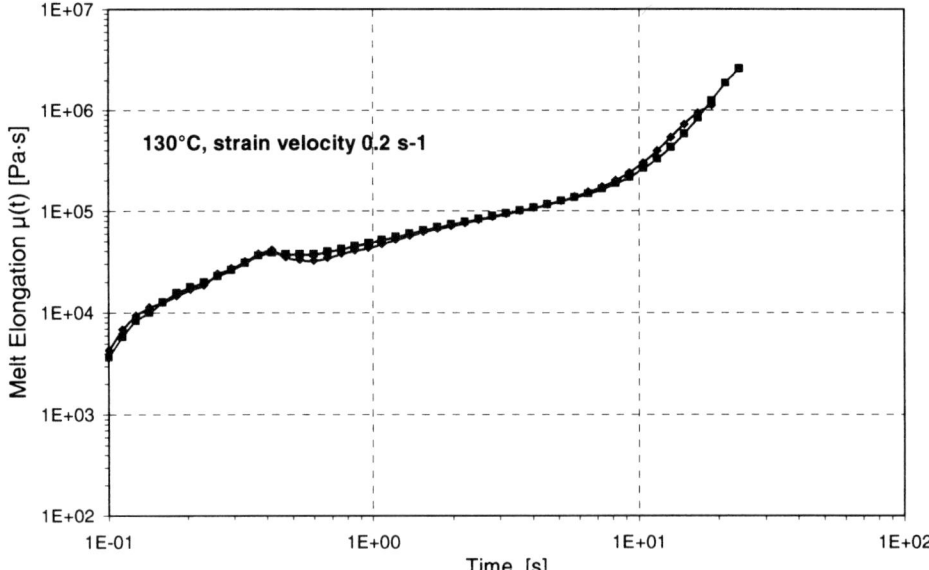

Fig. 2 Melt elongation of Ecoflex® at 130 °C.

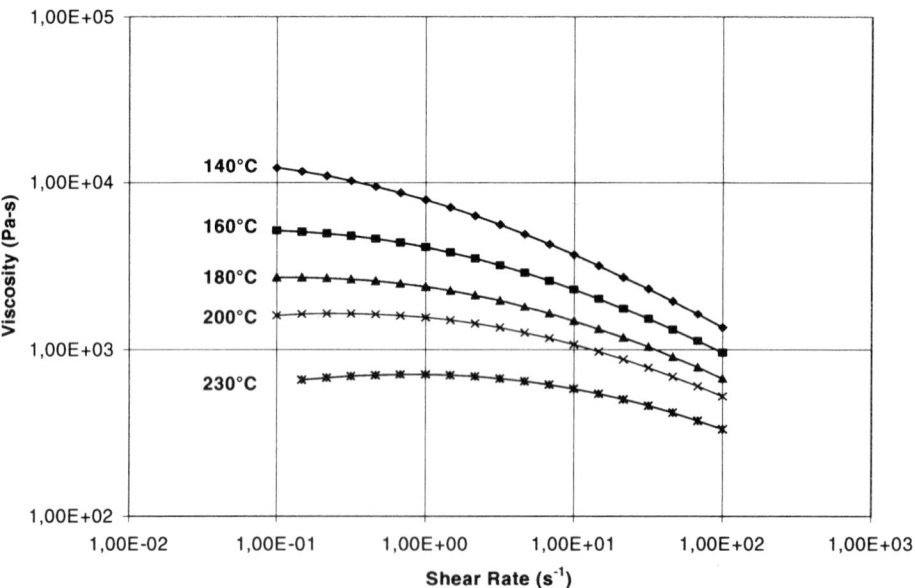

Fig. 3 Viscosity functions of Ecoflex® at different temperatures.

4.3
Applications

Typical applications of Ecoflex® are compost bags for organic waste, films in the agricultural sector, household films, and coating or lamination (e.g., paper), and coating materials for starch-based products (e.g., plates, cups) within the fast food and catering industries.

4.3.1
Compost Bags for Organic Waste

Kitchen waste can be collected hygienically and composted together with a biodegradable bag. Ecoflex® meets the requirements of compost bags; these include in particular the wet strength of the film and the time over which it remains stable to the organic waste, as well as problem-free processing in the compost facility.

4.3.2
Mulch Films

Uses of mulch covers include earlier harvesting, higher yields, and better crop quality. The advantage of mulch films made from biodegradable plastics such as Ecoflex® is that, following the harvest they can simply be ploughed together with the plant residue into the soil, where they fully degrade.

4.3.3
Coated or Laminated Materials

Coating or lamination of, for example, paper is used when there is a need for high wet strength and fat resistance. The use of Ecoflex® as a biodegradable laminate offers the additional advantage of problem-free disposal by composting. Applications of particular interest are therefore packaging materials soiled with food residues, such as paper wraps, drinking cups, fast-food packaging, or boxes and containers for frozen food.

4.3.4
Orientated Films

Like other thermoplastic polyesters, Ecoflex® shows a high orientability-enhanced mechanical properties. Consequently, orientated films and monofilaments are interesting applications for Ecoflex® where biodegradability is an advantage (e.g., knitted nets).

4.3.5
Transparent Films for Food Wrapping (Cling Films)

By adding special additives and optimizing the processing conditions, transparent films can be obtained using a blown film process. These films can be used for the wrapping of foodstuffs, including meats, vegetables and fruits sold in supermarkets.

4.4
Starch Ecoflex® Blends

Starch is an inexpensive raw material that is available in large quantities. The disadvantage, however, is that native starch cannot be processed like thermoplastic materials without the need for additives. Moreover, starch materials have a somewhat limited range of application due to their relatively high water absorbing properties. Blending starch with a hydrophobic polymer makes it possible to overcome these disadvantages, but clearly in order to create completely biodegradable starch blends it is crucial to use biodegradable hydrophobic polymers.

The use of Ecoflex® for starch blends leads to the targeted hydrophobization of the starch, and achieves the water-resistance that such blends require in many applications. Films made from this material show good mechanical properties such as high ultimate strength and elongation at break, they are antistatic, permeable to oxygen and water vapor, can be printed on and sealed, and they are very soft to the touch. Thus, Ecoflex® is an essential component for the processing of renewable materials such as starch; consequently, BASF is supplying Ecoflex® to several starch blend producers in Europe.

5
Biodegradation and Ecotoxicity of Ecoflex®

The copolyester Ecoflex® combines the good material properties of aromatic polyesters such as poly(ethylene terephthalate) (PET) with the biodegradability observed with many aliphatic polyesters, for example poly-(ε-caprolactone). Due to the complex structure of copolyesters like Ecoflex®, their biodegradability has been evaluated in a number of basic investigations and standard tests.

The compostability of Ecoflex® was tested according to German standard DIN V 54900. Further detailed investigations were targeted at the analysis of the intermediates, especially aromatic ester sequences, formed during the degradation. For some tests, a pure strain of *Thermomonospora fusca*, which occurs naturally in compost (Kleeberg et al., 1998), in addition to a currently isolated extracellular lipase from *T. fusca* (Kleeberg, 1999), was used in laboratory tests under defined conditions.

5.1
General Biodegradability of Aromatic Polyesters and Aliphatic-Aromatic Copolyesters

Pure aromatic polyesters such as PET or PBT have shown to be quite insensitive to hydrolytic degradation under physiological conditions. For example, Edge et al. (1991) and Allen et al. (1994) estimated the lifetime of PET in the environment to be up to 50 years. No significant direct microbial attack has been observed on these homopolyesters to date (Tokiwa and Susuki, 1977; Lefevre et al., 1999).

Based on results from early investigations on the biologically induced degradation of aliphatic-aromatic copolyesters, it was concluded that only at relative low fractions of the aromatic component would the copolyesters be attacked biologically (Tokiwa et al., 1990; Jun et al., 1994a,b).

The first report of any microbial attack on a block copolyester with 50 mol.% of terephthalic acid was made in the early 1990s (Witt

et al., 1994). However, one year later the same authors published results showing that aliphatic-aromatic copolyesters in which adipic acid (aliphatic) and terphthalic acid (aromatic) were distributed statistically were also degraded in a compost simulation test with a terephthalic acid content of up to 50 mol.% (Witt et al., 1995). A detailed overview of attempts to synthesize biodegradable aliphatic-aromatic copolyesters has been provided by Müller et al. (2001).

The first studies to focus on the fate of the aromatic sequences of such copolyesters were published in 1996 (Witt et. al., 1996a), and showed that sequences containing one or two terephthalic acids were rapidly metabolized in different environments. The degradation of longer aromatic oligomeric intermediates was also demonstrated in subsequent studies (Witt, 1996b).

5.2
Evaluation of the Biodegradation of Ecoflex® with Standard Tests

Recently, a German technical standard (DIN V 54900) became available for the determination of compostability of polymeric materials (Norm, DIN V 54900, 1998).

The standard consists of three parts (Table 5):

1) A compilation of relevant chemical data must be performed, for example to check the heavy metal content.
2) An evaluation is performed as to whether a polymeric material is biodegraded (mineralized) under defined laboratory conditions, by measuring the carbon dioxide evolved during degradation.
3) Acceptance in parts 1 and 2 is a prerequisite for part 3, where disintegration of the material under conditions of practical relevance, including a compost quality assessment, is tested.

Tab. 5 DIN V 54900 – Investigation of compostability of polymeric materials

	Description
Part 1	Chemical test
Part 2	Evaluation of biodegradability of polymeric materials in laboratory tests
Part 3	Testing of compostability under conditions of practical relevance including compost quality assessment

Based on the result of the DIN V 54900 testing, a material can be certified and labeled by DIN CERTCO to be "compostable".

Further biodegradability testing of Ecoflex® in addition to standard tests was focused on the proof of complete biodegradation of this material, as well as additional ecotoxicity tests, though these are not yet a mandatory part of the standards procedure.

5.2.1
Aerobic Biodegradation Test under Controlled Composting Conditions (ISO14855) and Analysis of the Residual Polymer

The biodegradability of Ecoflex® was investigated according to the "aerobic biodegradation test under controlled composting conditions" (Norm, ISO14855, 1999), which is analogous to DIN V 54900 Part 2. In this test, a defined quantity of polymer is mixed with compost, and the amount of carbon dioxide generated by microbial conversion is measured. As a control, the same amount of compost with cellulose instead of the polymeric material, and a sample of the same compost without any polymer, are also tested.

The results obtained for Ecoflex® are shown in Figure 4. The degradation rates (percentage of theoretical CO_2) were measured in three parallel reactors, and after 124 days were 93% (PS1), 95% (PS2), and 96% (PS3). The mean value of the degradation curves was 95% at the end of the test.

After the test, the residual polymer was extracted from the compost and analyzed. The amounts of polymer, determined by gel permeation chromatography (GPC) measurements were 2.3, 1.3 and 3.4%, respectively (Table 6). No aromatic oligomers intermediates were detected. The extracted residual polymer was analyzed with respect to its stoichiometric composition, using ^{13}C-NMR. A significant increase in the aromatic acid content would point to a slow or no degradation of aromatic block components. Within the accuracy of the method employed, no change in stoichiometric compo-

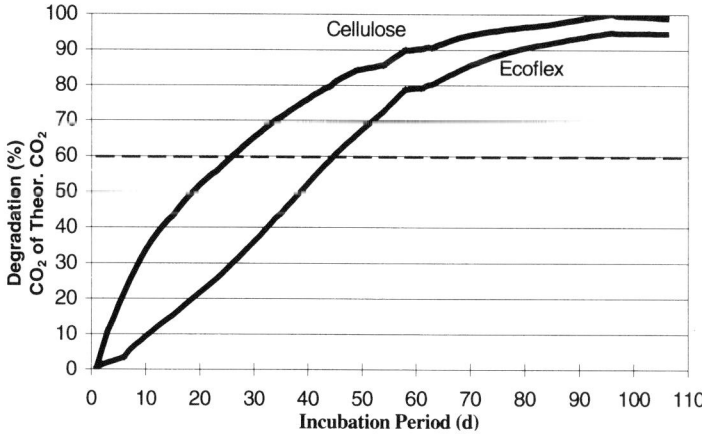

Fig. 4 Degradation of Ecoflex® in a controlled composting test.

Tab. 6 Analysis of residual polymer from 100 g of a dried mixture of compost and polymer for the investigation of compostability of Ecoflex® according to DIN V 54900, part 2

Sample	Initial amount of polymer [g 100 g⁻¹ mixture]	Residual polymer [g 100 g⁻¹ mixture] Measurement 1	Measurement 2	Residual polymer [%] Average	Average
PS1	15	0,39	0,30	0,35	2,3
PS2	15	0,15	0,22	0,19	1,3
PS3	15	0,57	0,44	0,51	3,4

sition was observed. Further analysis of the extracted residual polymer by mass spectrometry (MS) and matrix-assisted laser desorption/ionization mass spectrometry (MALDI-MS) has revealed that all types of aliphatic and aromatic polymer fragments are present in the residual polymer (BASF, 1997). These results suggest that aliphatic and aromatic ester bonds in the polymer chain are degraded statistically during the composting process, and no accumulation of long aromatic oligomeric intermediates occurs.

Inhomogeneities within the solid compost matrix, for example if zones are too wet, too dry, or not well ventilated, or if polymer particles are adhering to the wall of the reactor, represent plausible explanations for the existence of residual polymer.

5.2.2
Degradation under Conditions of Practical Relevance

The results of the composting tests of Ecoflex® films according to DIN V 54900, part 3 (Disintegration of the material under practical composting conditions), are shown in Figure 5.

After 12 weeks of composting, no visible Ecoflex® fragments of the initial film (thickness 120 μm) were found, and the rotting degrees of the compost were similar with or without Ecoflex®. The mature compost showed no negative effect in both a plant

compatibility test with summer barley (PlanCoTec, 1998) and in an earthworm acute toxicity test (Krieg, 1998) compared with a compost produced without Ecoflex®.

5.2.3
Certification of Ecoflex®

With the results presented here, Ecoflex® is registered as a compostable material in the positive list of DIN CERTCO, as well as at BPS in Japan as "GreenPla".

5.3
Detailed Biodegradation Studies of Ecoflex® in Defined Laboratory Tests

The accuracy of the residual polymer analysis after the composting test described above is limited due to the complexity of the compost matrix. More accurate conclusions about the fate of the different copolyester components, and especially aromatic intermediates, can be drawn from especially optimized laboratory tests with defined synthetic media as the environment. The degradation behavior of the longer aromatic sequences in the copolyester is of special interest, because it is well known that high-molecular weight pure PBT (e.g., Ultradur®) is not biodegradable. Hence, the question arises whether longer aromatic oligomer-intermediates are, ultimately, also totally degraded.

Prior to composting

After 1 week

After 2 weeks

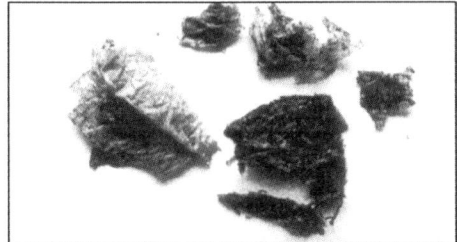

After 3 weeks

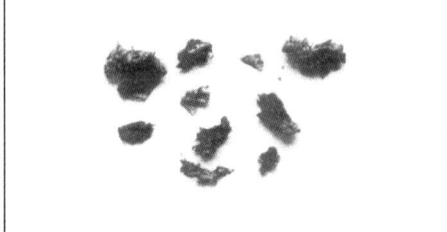

Fig. 5 Compostability of Ecoflex® according to DIN V 54900, part 3.

In order to answer the question of complete biodegradability, a specific naturally occurring microorganism (*T. fusca*) which was shown to depolymerize Ecoflex® very effectively was used (Kleeberg et al., 1998). Surprisingly, it was revealed that this organism is able to cleave the polymer chains of Ecoflex®, but cannot metabolize the degradation products, e.g., the monomers, in significant amounts. Thus, these organisms are ideal for investigating the intermediates. Degradation of Ecoflex® in an aquatic test system with *T. fusca* results in an artificially high concentration of intermediates, which facilitates the precise analysis of the polymeric intermediates and residues.

In a further step, only the extracellular, polyester-degrading enzyme from *T. fusca* was used for additional degradation experiments with Ecoflex®.

5.3.1
Degradation of Ecoflex® with the Thermophilic Actinomycete, *T. fusca*

After 22 days of degradation of Ecoflex® with *T. fusca* at 55 °C in a defined synthetic medium, more than 99.9% of the polymer had depolymerized and dissolved in the medium (residual polymer analyzed by GPC analysis of the solid fraction). With regard to the degradation of the diacid and diol components of Ecoflex®, only the monomers (1,4-butanediol, terephthalate and adipate) could be detected in the medium by GC-MS measurements (Witt et al., 2001). In interrupted degradation experiments, the mono-esters of adipic acid and terephthalic acid with 1,4-butanediol were observed predominantly, in addition to the monomers.

The depolymerization products generated by *T. fusca* in the experiments above were rapidly metabolized when a mixed microbial culture (from a soil eluate) was added to the accumulated intermediates in the degrada-

Tab. 7 Fragments of Ecoflex® after degradation with isolated pure strain (tests 1–3) and with pure and mixed cultures from compost (test 4). X = detected; − = not detected

Test[a]	Monomers[b]			Aliphatic oligomers[b]		Aromatic oligomers[b]	
	B	A	T	BA	ABA	BT	BTB
1	X	X	X	X	X	X	X
2	X	X	X	X	X	−	−
3	X	X	X	−	−	−	−
4	−	−	−	−	−	−	−

[a]Test 1: 1750 mg polyester in 80 mL media. Intermediates from isolated pure culture after 21 days. Enzyme activity stopped by pH-shift (*in situ* building of high amounts of acids during degradation). Tests 2, 3: 350 mg polyester in 80 mL media. Intermediates from isolated pure culture after 7 days (test 2) and after 21 days (test 3). Test 4: Residues after a 7-day inoculation by isolated pure culture, and a 14-day inoculation with compost eluate; [b]A: Adipic acid; B: 1,4-butanediol; T: terephthalic acid. Intermediates from the cleavage of "M" (see Figure 1) are not considered. According to test 4, no intermediates were detectable by GC-analysis.

tion media (Table 7, test 4). After 14 days incubation of these solutions, the intermediates could no longer be detected by GC measurements as they completely had been metabolized by the microorganisms.

These tests also showed that longer aromatic sequences occurring in the polymer chains of Ecoflex® are the subject of total biodegradation under the environmental conditions applied.

5.3.2
Depolymerization of Ecoflex® with an Extracellular Enzyme from *T. fusca*

Compared to the degradation of the copolyester with whole living organisms, depolymerization with only the enzyme responsible for the first step of biodegradation has the advantage of there being fewer substances in the medium, which are not degradation intermediates; this facilitates analytical characterization of the water-soluble monomers and oligomers formed.

Degradation of Ecoflex® with the extracellular hydrolase from *T. fusca* was performed at 55 °C in buffer solution. Within 43 days, 80% of Ecoflex® micronizate (initial quantity, 350 mg) was enzymatically degraded into

monomers (1,4-butanediol, terephthalate, and adipate) and the monoesters of adipic acid and terephthalic acid with 1,4-butanediol. ^{13}C-NMR analysis of the remaining polymer has shown no change in stoichiometric composition, again indicating again the lack of any accumulation of longer aromatic oligomeric intermediates. (BASF and Gesellschaft biotechnologischer Forschung, unpublished data).

In general, the enzymatic degradation test confirms the results obtained from the analysis of residual polymer in compost and the degradation experiment with *T. fusca*.

5.4
Ecotoxicology of the Degradation Products of Ecoflex®, and Risk Assessment

To ensure that the degradation intermediates of Ecoflex® do not have any negative effect on living organisms in the environment, two types of ecotoxicological tests with *Daphnia magna* and *Photobacterium phosphorum* were performed, using the highly concentrated intermediate solutions generated from Ecoflex® degradation with *T. fusca* and the extracellular enzyme, respectively. These ecotoxicological tests go beyond the those demanded in standards for evaluation of the compostability of plastics, where usually only plant growth tests or seed germination tests are required.

In none of the toxicological tests was any significant toxicological effect observed, neither for the monomeric depolymerization products nor for the oligomeric intermediates.

From a risk assessment taking into account very conservative assumptions, it could be concluded that there is no indication for an environmental risk when aliphatic-aromatic copolyesters of the Ecoflex® type are introduced into composting processes (Witt et al., 2001).

6

Outlook and Perspectives

Ecoflex® is a new and innovative class of fully biodegradable plastics, and completes the existing product range of plastic materials of BASF. Ecoflex® offers an outstanding cost:performance ratio, and represents a convincing solution for applications in which biodegradable plastics can be used favorably. Furthermore, Ecoflex® provides opportunities for the utilization and processing of renewable resources such as starch by blending. This brings a further cost advantage – an important factor for the development of the biodegradable polymer market. With Ecoflex®, BASF intends to make an active contribution to the development of the market for biodegradable plastics.

7

Patents

International publication numbers for the biodegradable copolyester of BASF on the basis of aliphatic-aromatic monomers with modular units are summarized in Table 8.

Tab. 8 International publications of Ecoflex®-related copolyester

International Publication Number
WO 96/15173
WO 96/15174
WO 96/21689

8
References

Allen, N. S., Edge, M., Mohammadian, M., Jones, K. (1994) Physicochemical aspects of the environmental degradation of poly(ethylene terephthalate), Polym. Degrad. Stab. **43**, 229–237.

BASF Aktiengesellschaft (1997), unpublished results.

Edge, M., Hayes, M., Mohammadian, M., Allen, N. S., Jewitt, K., Brems, K., Jones, K. (1991) Aspects of poly(ethylene terephthalate) degradation for archival life and environmental degradation, Polym. Degrad. Stab. **32**, 131–153.

Jun, H. S., Kim, B. O., Kim, Y. C., Chang, N. H., Woo, S. I. (1994a) Synthesis of copolyesters containing poly(ethylene terephthalate) and poly(ε-caprolactone) units and their susceptibility to Pseudomonas sp. lipase, J. Environ. Polym. Degrad. **2**, 9–18.

Jun, H. S., Kim, B. O., Kim, Y. C., Chang, N. H., Woo, S. I. (1994b) Synthesis of copolyesters containing poly(ethylene terephthalate) and poly(ε-caprolactone), and its biodegradability, Stud. Polym. Sci. **12**, 498–504.

Kleeberg, I. (1999) Untersuchungen zum mikrobiellen Abbau von aliphatisch-aromatischen Copolyestern sowie Isolierung und Charakterisierung eines polyesterspaltenden Enzyms, PhD Thesis, Technical University Braunschweig, Braunschweig, Germany, Internet: http://opus.tu-bs.de/opus/volltexte/2000/90

Kleeberg, I. Hetz, C., Kroppenstedt, R. M., Müller, R.-J., Deckwer, W.-D. (1998) Biodegradation of aliphatic-aromatic copolyesters by Thermomonospora fusca and other thermophilic compost isolates, Appl. Environ. Microbiol. **64**, 1731–1735.

Krieg, W. (1998) Effects of compost containing the biologically degradable copolyester Ecoflex® on the mortality of the earthworm Eisenia foetida, BASF Aktiengesellschaft internal test result.

Lefevre, C., Mathieu, C., Tidjani, A., Dupret, A., Vander Wauven, C., De Winter, W., David, C. (1999) Comparative degradation by microorganisms of terephthalic acid, 2,6-naphthalene dicarboxylic acid, their esters and polyesters, Polym. Degrad. Stab. **64**, 9–16.

Müller, R.-J., Kleeberg, I., Deckwer, W.-D. (2001) Biodegradation of polyesters containing aromatic constituents – a review, J. Biotechnol. **86**, 87–95.

Norm, DIN – Deutsches Institut für Normung (1998) Testing of the compostability of polymeric materials, German standard DIN V 54900 parts 1–3.

Norm, ISO14855 (1999) Evaluation of the ultimate aerobic biodegradability and disintegration under controlled composting conditions – Method by analysis of released carbon dioxide.

PlanCoTec (1998) Kompostierbarkeit des biologisch abbaubaren Werkstoffes Ecoflex® der Fa. BASF AG nach E DIN 54900.

Tokiwa, Y., Suzuki, T. (1977) Hydrolysis of polyesters by lipases, Nature **270**, 76–78.

Tokiwa, Y., Ando, T., Suzuki, T., Takeda, T. (1990) Biodegradation of synthetic polymers containing ester bonds, Polym. Mater. Sci. Eng. **62**, 988–992.

Witt, U., Müller, R.-J., Augusta, J., Widdecke, H., Deckwer W.-D. (1994) Synthesis and biodegradability of polyesters based on 1,3-propanediol, Macromol. Chem. Phys. **195**, 793–802.

Witt, U., Müller, R.-J., Deckwer, W.-D. (1995) Biodegradation of polyester copolymers containing aromatic compounds, J. Macromol. Sci. - Pure Appl. Chem. **A32**, 851–856.

Witt, U., Müller, R.-J., Deckwer, W.-D. (1996a) Evaluation of the biodegradability of copolyesters containing aromatic compounds by investigations of model oligomers, J. Environ. Polym. Degrad. **4**, 9–20.

Witt, U. (1996b) Synthese, Charakterisierung und Beurteilung der biologischen Abbaubarkeit von anwendungsorientierten biologisch abbaubaren ali-phatisch/aromatischen Copolyestern, PhD Thesis, Technical University Braunschweig, Braun-schweig, Germany.

Witt, U., Einig, T., Yamamoto, M., Kleeberg, I., Deckwer, W.-D., Müller, R.-J. (2001) Biodegra-dation of aliphatic-aromatic copolyesters: evalu-ation of the final biodegradability and ecotoxico-logical impact of degradation intermediates, Chemosphere 44, 289–299.

12
Polyesteramides

Dr. Ralf Timmermann
Bayer AG, KU-FE/TKT, Rheinuferstr. 7-12, 47812 Krefeld, Germany;
Tel.: +49-2151-887364; Fax: +49-2151-885599; E-mail: ralf.timmermann.rt@bayer-ag.de

HDT-B	heat distortion temperature, method B
MVR	melt volume rate

1
Introduction

Discussions about biodegradable polymers and their possible commercial application have led to a renewed interest in the development of this group of compounds. A polymer created from a combination of ester and amide bonds – both of which occur naturally in fatty acid esters and proteins – should (in theory) undergo biodegradation and hence present minimal threat to the environment. In contrast, the polyesters and polyamides currently used for thermoplastics have superior mechanical properties and processing behavior, but are resistant to biodegradation.

2
Historical Outline

Polyesteramides are an historical class of polymers, their first syntheses from diacids, diols and diamines being conducted during the 1930s (Carothers and Hill, 1932; Lederer, 1934; Carothers, 1937). However, due to the poor mechanical and thermal properties (e.g., melting point) of the polyesteramides, it was the pure polyamides such as polyamide-66 (prepared from adipic acid and hexamethylene diamine) and aromatic polyesters such as polyethylene terephthalate that subsequently became far more successful in terms of their commercial application.

Polyesteramides are well known in the field of hot melts, and by using dimeric fatty acids the crystallization tendency of polyesteramides compares well with that of polyesteramides, with a reduction in the number of side chains (Veazey, 1984). As a result, these polymers are low melting and sticky in nature.

Polyesteramides made from natural amino acids are generally synthesized by a complex series of reactions, with the aid of protecting groups (Paredes et al., 1998; Ho and Huang, 1999).

Attempts have also been made to synthesize block-built polyesteramides (Buehler et al., 1996; Stapert et al., 1997, 1999), with preformed polyamide and polyester blocks typically being linked during the last stage of the synthesis. In an alternative synthesis, polyesters and polyamides are heated together in order to produce a degree of amidation and esterification. Once synthesized however, the structure of these block-built polymers is not stable, because the polyesteramides are regarded as "living" systems. This means that the amidation and esterification processes are still occurring under thermal influence until a random distribution of the ester and amide groups is finally achieved. Consequently, these compounds are not sufficiently stable for processing because the ongoing changes that are occurring in their structures will influence their thermal properties such as melting point. As result, it was necessary to develop alternative processing conditions.

Alternating copolyesteramides derived from β-hydroxy acids and α-amino acids or ε-aminocaproic acid have been synthesized in two different ways: (1) polycondensation of N-(β-hydroxyacyl)-amino acids; and (2) ring-opening polymerization (ROP) of cyclic amide-esters (Keul et al., 1999).

The anionic copolymerization of ε-caprolactone and ε-caprolactam was studied intensively and led to the introduction of a range of block-built polyesteramides (Goodman, 1984).

Biodegradable polyesteramides have been widely investigated as suture materials in medicine (Roby and Jiang, 1998), the main advantage (when compared with the well-known polylactic acid and polyglycolic acid alternatives) being the remarkably lower pH of the tissue that is achieved after degradation. This effect is due to the buffering of the amino acid structures, and in turn leads to a reduction in the extent of local tissue inflammation.

All these chemical approaches have led to the development of polyesteramides (mainly lower molecular-weight polymers) with properties that were inadequate for use as thermoplastic polymers. Moreover, from a commercial standpoint they were too expensive. As recently as 1988 it was considered that polymers formed by simple polymerization of ester and amide components had poor mechanical properties, although some short-chain polyesteramides have been dissolved in organic solvents (e.g., BioSol®) and used as coating materials (Azuma, 1997).

In most cases the main problem was the formation of high molecular-weight polymers, and to overcome this, activated derivatives of dicarbonic acids (e.g., acid chlorides) were used. This approach resulted in the formation of high molecular-weight polymers, though the conditions required to achieve polymerization are not technically feasible for commercial production.

Hence, the chemical synthesis of high molecular-weight polymers (polyesteramides) from easily available raw materials and under easily controlled conditions was

the first main step in investigating polyesteramides as biodegradable polymers.

3
Chemistry

3.1
Chemical Structures

High molecular-weight polyesteramides are synthesized from commercially available raw materials such as adipic acid, caprolactam, AH-salt (1:1 salt from adipic acid and 1,6-hexamethylenediamine), 1,4-butanediol and diethylene glycol (Timmermann et al., 1993).

There are two main groups of polyesteramides (Grigat et al., 1998): (1) polyesteramides based on polyamide-6, with caprolactam as the amide-forming component (Figure 1); and (2) those based on polyamide-66, with the AH-salt as the amide-forming component (Figure 2). In all cases the ester-forming components are adipic acid and 1,4-butanediol – in some cases in combination with diethylene glycol. Ethylene glycol has also been used as an alternative diol source, but as well as limiting the production of high molecular-weight polymers, a wide variety of byproducts is formed.

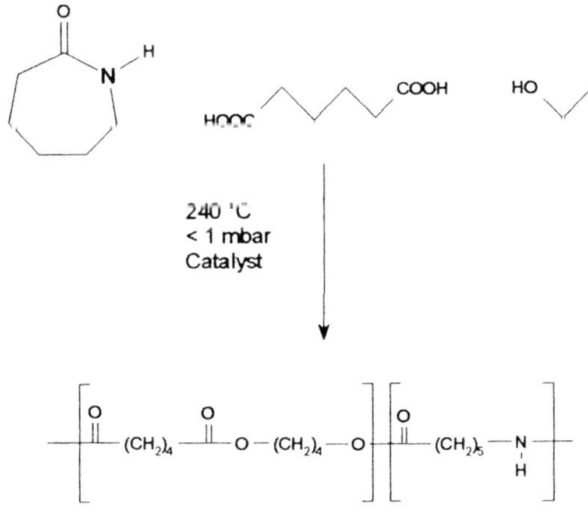

Fig. 1 Synthesis of polyamide-6-based polyesteramides (BAK 403- and 404-grades).

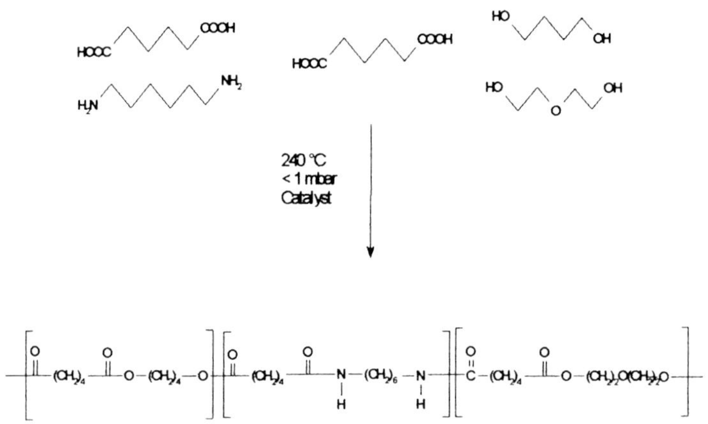

Fig. 2 Synthesis of polyamide-66-based polyesteramides (BAK 402-grades).

3.2
Synthesis

The polymerization reaction is a typical polyesterification process. As in the synthesis of pure polyesters, the diol component is used in an excess of 10–15 mol.%. In the first reaction stage, after mixing of all monomers, esterification of the acidic components with an excess of diols occurs, accompanied by ring-opening of the caprolactam and the initial amidation reactions to form oligomers. In the second stage, polycondensation occurs to form high molecular-weight polyesteramides. The first stage is carried out under normal pressure, while the second stage is performed under vacuum (<1 mbar).

A crucial point in the synthesis of high molecular-weight polyesteramides is that the surface of the reaction melt is maintained in an optimal manner so that any excess of diols and water formed may be removed.

The reaction is catalyzed by esterification catalysts, notably titanium-tetraisopropoxide. This well-known catalyst controls the esterification step, which is recognized as the slowest among the series of different parallel reactions. When the reaction is complete, the catalyst remains in the polymer as titanium dioxide, and so causes no environmental problems.

The manner in which polycondensation of the monomers results in the formation of statistical polyesteramides infers that there is a statistical order of ester and amide groups along the polymer chain. This is especially important in view of the polymer's subsequent biodegradation, as long blocks of polyamides with molecular weights >2000 daltons form highly crystalline units that are stable against biodegradation (hydrolysis) (Table 1).

3.3
Properties

The main difference between the two groups of polyesteramides synthesized using this method are their melting points. The mechanical properties are also changed, together with the required rate of biodegradation. With regard to the effect of amide composition and ester content on biodegradation, ester contents <30 wt.% lead to the formation of products that are stable against biodegradation (Figure 3).

Tab. 1 Overview of BAK resins

		PA-6-based Successor BAK 1095		PA-66-based Successor BAK 2195	
BAK grade		BAK 403–004	BAK 404–004	BAK 402–005	BAK 402–006
Amide content	[wt.%]	60	65	40	40
Main application		Film	Film	Fiber	Injection molding
Melting point	[°C]	116	137	172	172
Crystallization temperature	[°C]	54	65	143	130
Tensile modulus	[MPa]	200	245	535	510
Elongation at break	[%]	345	340	130	220
MVR (200 °C; 1,2 kg)	[cm³ 10 min]⁻¹	7	4	13	49

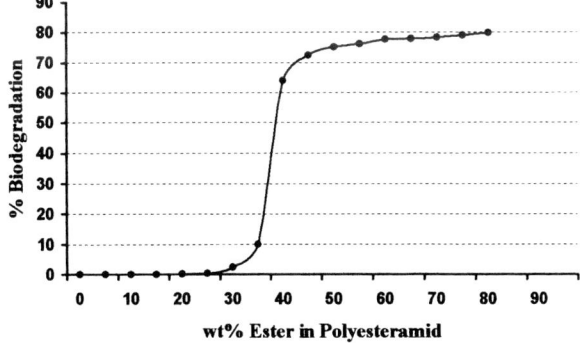

Fig. 3 Biodegradation rate depending on ester content in polyesteramides.

In contrast, the mechanical properties increase in line with a higher amide content; hence, the amide content is maximized to achieve good mechanical properties whilst maintaining rapid and complete biodegradation.

The mechanical properties of polyesteramides can be tailored depending upon their ultimate use, though an acceptable rate of biodegradation must be always be borne in mind. Polyamide-6-based polyesteramides are mainly suited to film applications, with the mechanical properties of films made from BAK 403 or BAK 404 being very similar to those of polyethylene-based films (Table 2).

The preparation of these grades is possible on typical polyethylene processing aggregates, though some minor modifications of the conditions are necessary. In particular,

the temperature profile of the extruder has an inverted structure in comparison with that used for polyethylene – this is because the melting behavior of the polyesteramide granules differs from that of polyethylene granules. The temperature in the die must also be lower than that in the melting zone of the extruder because the statistically ordered polyesteramides have a lower rate of crystallization. By contrast, polyamide-66-based polyesteramides show a much greater degree of crystallinity, and are therefore more suitable for injection molding and fiber extrusion.

As pure resins, polyesteramides are very soft materials; thus, in order to produce a stiffer material with higher tensile properties it is necessary to use fillers. Typical fillers include talc, calcium carbonate, mica, or wollastonite. Grades such as BAK 105 or

Tab. 2 Mechanical properties of BAK grades in comparison with polyethylene

		BAK 403-004	BAK 404-004	BAK 402-005	Polyethylene Standard grade
Tensile modulus	[MPa]	200	245	535	150–500
Yield strength	[MPa]	15	18	20	9–12
Tensile stress at break	[MPa]	26	30	20	15–22
Tensile strain at break	[%]	345	340	130	500–650
Notched impact strength	(23 °C) [kJ m^{-2}]		7.7	10	2.5–14
Melting point	[°C]	116	137	172	105–118

Tab. 3 Mechanical properties of BAK grades in comparison with polypropylene

		BAK 105-007	BAK 106-006	Polypropylene Standard grade
Tensile modulus	[MPa]	1600	1500	700–1800
Elongation	[%]	11	34	25–40
Tensile stress at break	[MPa]	29	29	25–40
Impact strength	(23 °C) [kJ m^{-2}]	71	n.b.	20–n.b.
Notched impact strength	(23 °C) [kJ m^{-2}]	7.7	10	2.5–14
Melting point	[°C]	172	172	160–165
HDT-B	[°C]	120	100	90–130

n.b. = not broken.

Tab. 4 Mechanical properties of BAK grades in comparison with polystyrene

		BAK 105-007	BAK 106-006	Polystyrene Standard grade
Tensile modulus	[MPa]	3500	3200	3000–3500
Tensile stress at break	[MPa]	41	41	30–60
Impact strength	(23 °C) [kJ m^{-2}]	43	99	10–4
Notched impact strength	(23 °C) [kJ m^{-2}]	9,8	13	2
HDT-B	[°C]	140	140	78–102

BAK 106 possess mechanical properties similar to those of polypropylene or even to polystyrene (Tables 3 and 4). However, because of their excellent flow behavior, combined with their good mechanical properties, polyesteramides can be used in the production of thin-walled articles such as flower pots, as well more rigid items such as cutlery (Figure 4).

Legal restrictions (DIN 54900) limit the permitted content of inorganic fillers in different grades of BAK to not more than 47 wt.%.

Fig. 4 Cutlery made from BAK®106-005.

Polyesteramides show superior affinity with organic materials such as wood flour, cellulose, or starch, this property being dependent upon the ability to create hydrogen bonds with the natural component(s). There is no need to optimize the connection between polymer and filler (Jiang and Hinrichsen, 1999; Averous et al., 2000), as the mechanical properties of these compounds are, in general, remarkably good (BAK 200 grades).

4
Applications

Polyesteramides are suitable for many film applications, especially if there is need for rapid and complete degradation after use (e.g., hygiene films, agricultural films) (Figure 5). Further applications include injection-molded components, flower pots and rigid cutlery. The combination of polyesteramides with wood flour has led to a number of valuable compounds for use in the furniture industry, and injection-moldable

Fig. 5 Bags for collecting organic waste.

wood has proved to be very useful when designing and creating unusual shapes.

A combination of polyesteramides with high levels of metal (e.g., iron powder) has resulted in a material which can be used as a replacement for metallic lead. Ultimately, the polyesteramide binder will biodegrade, while the remaining metallic iron will be oxidized an become environmentally neutral.

5
Biodegradation

Unlike pure aliphatic polyesters, which are nearly all biodegradable, pure polyamides are not biodegradable. This is a well-recognized characteristic of polyamides, and is dependent mainly on the crystallinity and melting points of the corresponding polymers (Tokiwa et al., 1979; Nagata and Kiyotsukuri, 1994). Polyesteramides are generated from both components: the ester component is responsible for the rapid biodegradation, whilst the amide component provides the stability and excellent mechanical properties of the polymer. For all grades of BAK, the balance of amide/ester is adjusted such that a maximum rate of biodegradation and a high level of mechanical properties are achieved.

The polyamide-6-based polyesteramides have an ester content of about 35–40 wt.%, whilst the polyamide 66 based compounds (due to their higher crystallinity) have an ester content of ~60 wt.%

The polyesteramides, as a group, have undergone many standardized tests, and all have been certified according to the German DIN 54900 (Figure 6). The test methods incorporated within this DIN cover laboratory tests as Sturm tests, controlled composting, and composting under natural conditions.

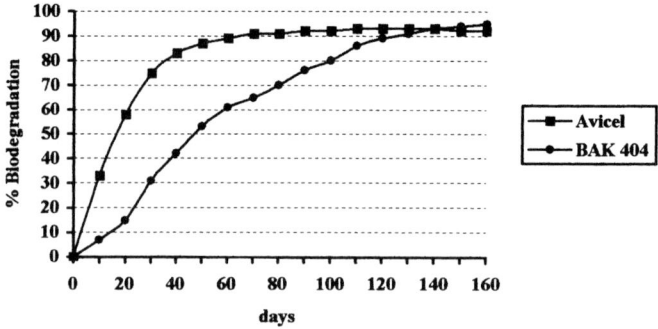

Fig. 6   Biodegradation of polyesteramide under DIN 54900, Part 2 (modified Sturm test).

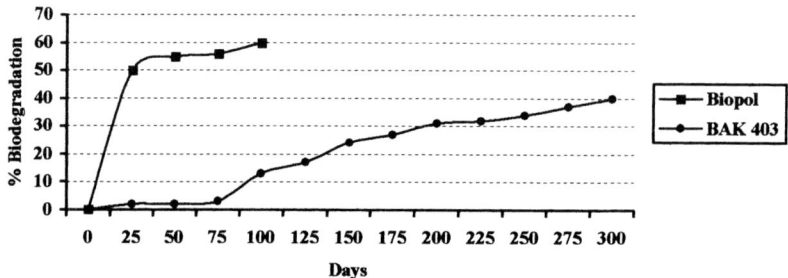

Fig. 7   Biodegradation of polyesteramides in North Sea brackish water.

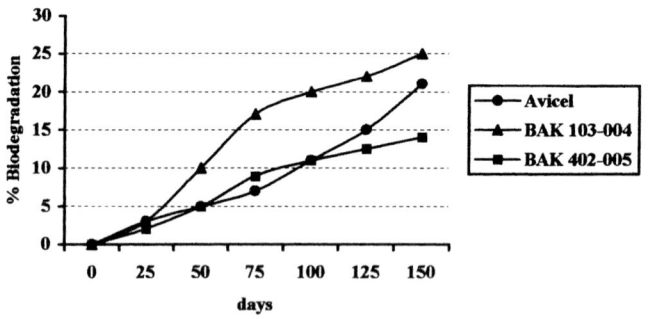

Fig. 8   Biodegradation of polyesteramides in soil according to ASTM D5988.

Polyesteramide degradation has also been monitored in other media, including both sea water (Figure 7) and soil (Figure 8).

The biodegradation of polyesteramides is a combination of hydrolytic and enzyme-catalyzed processes, ultimately to create CO_2 (Koch, 1997; Wiegand et al., 1999). The first step – hydrolysis – occurs at the ester bonds, and this results initially in lower molecular-weight chains of polyesteramides and finally in oligomers and monomers. These monomers can be detected using various analytical methods (Negoro, 1994).

Further degradation of the monomers occurs according to well-documented degradation pathways. In the case of adipic acid, β-oxidation of the carbon chain results in the production of acetic and succinic acids, both of which are widely distributed in nature.

1,4-Butanediol is oxidized twice to succinic acid (the CH_2-OH end-groups will be oxidized to COOH groups, resulting in succinic acid), and is therefore degraded along the same pathway as adipic acid.

Aminocaproic acid (the structure of which is very similar to the natural amino acid lysine) follows the degradation pathways of amino acids, with α-decarboxylation leading to the hexamethylene diamine, after which two-fold oxidation leads to adipic acid, thereafter following the above-mentioned pathway.

A number of microorganisms that are also able to degrade polyesteramides have been investigated (Wiegand et al., 1999). Such studies have shown that polyesteramide-degrading bacteria occur in many ecosystems, the polyesteramides being degraded completely to CO_2, biomass and water.

The carbon balance of the degradation of BAK 403 in liquid media shows that, under these circumstances, almost 65% of the polymer's carbon is degraded to CO_2, and 25% is incorporated into the biomass. The remaining carbon appears as CO_2 dissolved in water (carbonate and hydrogen carbonate)

and as organic carbon in the form of water-soluble, short-chain acids (Figure 9).

The excellent biodegradation of polyester-amides is illustrated in the maximum wall thickness (i.e., the maximum thickness that is completely degraded in composting plants) as determined in DIN 54900, part 3. Depending upon the BAK grade, maximum polyesteramide thicknesses of between 500 µm and > 2000 µm are effectively biodegraded (Table 5).

6
Production

Biodegradable polymers with the tradename BAK® were first introduced in 1995 by Bayer during the trade fair K'95, and commercial production of these polymers was started in about 1997 at a polyester plant producing polyethylene terephthalate. In general, the polymerization process employed was two-staged, as described in Section 3. Only minor changes were necessary to the existing plant, mainly in the pelletizer system. An underwater cutting system provided the best solution to this problem, and the resultant

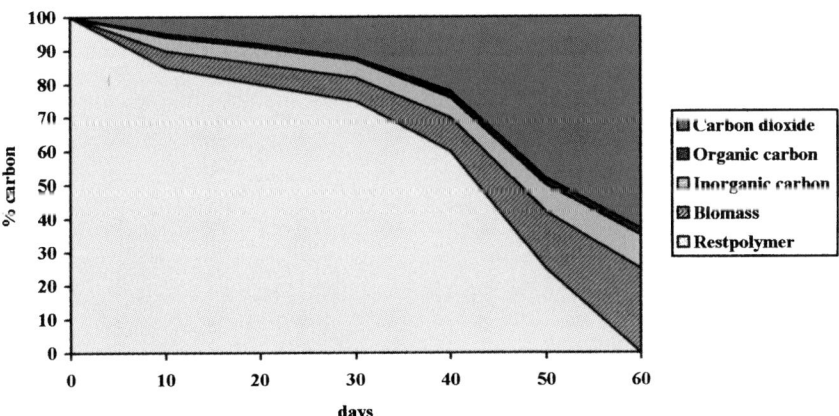

Fig. 9 Carbon-balance of BAK 403 (liquid degradation test).

Tab. 5 Biodegradation, composting of BAK grades according to DIN 54900, Part 3

		BAK 1095 BAK 403-004	BAK 404-004	BAK 402-005 BAK 402-006
Degree of degradation After 12 weeks	Technical test	1000 μm/100% 1600 μm/99.7% 2000 μm/97.7%	500 μm/97.8%	800 μm/96.4% 1000 μm/90.2%
Degree of degradation After 12 weeks	Composting plant	2000 μm/100%	500 μm/100%	500 μm/100% 800 m/100% 1000 μm/100%
Maximum degradable Thickness		> 2000 μm	500 μm	1000 μm

lens-shaped granules proved to be well accepted in the new process.

7
Outlook and Perspectives

The production and sales of BAK was stopped by Bayer in early 2001. At present, no other company sells biodegradable polyesteramides as a thermoplastic material for extrusion and injection molding.

8
Patents

Many patents relating to polyesteramides have been registered, the most important being those outlined in Section 2. The patent applications relating to biodegradable polyesteramides BAK are listed in Table 6.

Tab. 6   Patents covering the biodegradable polyesteramides BAK®

Number of patent	Patent holder	Inventors	Title of patent	Date of publication
EP 0641817	Bayer AG	Timmermann, R., Dujardin, R., Koch, R.	Thermoplastic, biodegradable, workable high molecular weight copolymer – contains aliphatic ester and aliphatic amide units	08.03.1995
EP 0765911	Bayer AG	Grigat, E., Müller, H.P., El Sayed, A., Timmermann, R., Schulz-Schlitte, W.	Reinforced biodegradable thermoplastic molding materials – contain mainly aliphatic polyester, polyester:urethane, polyester:carbonate or polyester:amide and natural fillers such as wood dust, minerals, etc.	02.04.1997
EP 1036107	Bayer AG	Timmermann, R., Schulz-Schlitte, W., Voigt, M.	Biodegradable thermoplastic polyesteramides with good mechanical properties for molding, film and fiber, useful for, e.g. compostable refuse bag	10.06.1999
EP 1098930	Bayer AG	Voigt, M., Schulz-Schlitte, W.	Biodegradable molding composition containing metallic or mineral filler useful for the production of molded articles has a high density and low toxicity	20.01.2000
WO 200063282	Bayer AG	Hashimoto H., Nishikata A., Okuno, H., Sano, S., Satani, S., Schulz-Schlitte, W., Timmermann, R., Voigt, M., Wada, N.	Biodegradable resin composition for multilayered films, comprises a stabilizing additive selected from anti-oxidants, light stabilizers, quenchers, anti-clouding agents, anti-fogging agents and biodegradable polymer	26.10.2000
EP 1099544	Bayer AG	Kaschel, G., Kleemiss, M., Timmermann, R.	Monolayer or multilayer biodegradable thermoplastic films, and their use as packaging film or in cosmetic or personal care articles	16.05.2001

9
References

Averous, L., Fauconnier, N., Moro, L., Fringant, C. (2000), Blends of thermoplastic starch and polyesteramide: processing and properties, *J. Appl. Polym. Sci.* **76**, 1117–1128.

Azuma, T. (1997) Biodegradable aliphatic ester-amide copolymers and their solutions, JP 09227672.

Buehler, F. S., Fanelli, R., Meier, P. Treutlein, R. (1996) Compositions based on biodegradable aliphatic polyesters for manufacture of blown films, EP 708148.

Carothers, W. H. (1937) Linear condensation polymers, U.S. Patent 2 071 250.

Carothers, W. H., Hill, J.W. (1932) Studies of polymerization and ring formation. XIII. Polyamides and mixed polyester-polyamides *J. Am. Chem. Soc.* **54**, 1566–1569.

DIN 54900, Deutsches Institut für Normung, Verlag Beuth, Berlin.

Gonsalves, K. E., Chen, X. (1992) Hydrolytic degradation of nonalternating polyesteramides, *Polym. Prep. (Am. Chem. Soc., Div. Polym. Chem.)*, **33**, 53–54.

Goodman, I. (1984) Copolyesteramides – IV. Anionic copolymers of ε-caprolactone and ε-caprolactame, *Eur. Polym. J.* **20**, 549–557.

Gray, H.W. (1943) Polyesteramide, U.S. Patent 2 333 923.

Grigat, E., Müller, H.P., El Sayed, A., Timmermann, R., Schulz-Schlitte, W. (1997) Reinforced biodegradable thermoplastic moulding materials – contain mainly aliphatic polyester, polyestercarbonate, polyesterurethane or polyesteramide and natural fillers such as wood dust, minerals, etc., EP 0765911.

Grigat, E., Koch, R., Timmermann, R. (1998) BAK 1095 und BAK 2195: completely biodegradable synthetic thermoplastics, *Polym. Degrad. Stabil.* **59**, 223–226.

Hashimoto, H., Nishikata, A., Okuno, H., Sano, S., Satani, S., Schulz-Schlitte, W., Timmermann, R., Voigt, M., Wada, N. (2000) Biodegradable resin composition for multilayered films, comprises a stabilizing additive selected from anti-oxidants, light stabilizers, quenchers, anti-clouding agents, anti-fogging agents and biodegradable polymer, WO 200063282.

Ho, L.-H., Huang, S. J. (1999) Biodegradable polymers derived from amino acids, *Macromol. Symp.* **144**, 7–32.

Jiang, L., Hinrichsen, G. (1999a) Flax and cotton fiber-reinforced biodegradable poly(ester amide composites. Part 1. Manufacture of composites and characterisation of their mechanical properties, *Angew. Makromol. Chem.* **268**, 13–17.

Jiang, L., Hinrichsen, G. (1999b) Flax and cotton fiber-reinforced biodegradable poly(ester amide composites. Part 2. Characterization of biodegradation, *Angew. Makromol. Chem.* **268**, 18–21.

Kaschel, G., Kleemiss, M., Timmermann, R. (2001) Monolayer or multilayer biodegradable thermoplastic films, and their use as packaging film or in cosmetic or personal care articles, EP 1099544.

Keul, H., Robertz, B., Hoecker, H. (1999) New alternating poly(amide-ester)s. Synthesis and properties, *Macromol. Symp.* **144**, 47–61.

Koch, R. (1997) Degradation of biodegradable polyesteramides with enzymes, WO 9743014.

Lederer, H. (1934) Über Faserstoffsynthesen, *Kunstseide*, 43–49.

Nagata, M., Kiyotsukuri, T. (1994) Biodegradability of copolyesteramides from hexamethylene adipate and hexamethyleneadipamide, *Eur. Polym. J.* **30**, 1277–1281.

Negoro, S. (1994) The nylon oligomer biodegradation system of *Flavobacterium* and *Pseudomonas*, *Biodegradation* **5**, 185–194.

Paredes, N., Rodriguez-Galan, A., Puiggali, J., Peraire, C. (1998) Studies on the biodegradation and biocompatibility of a new poly(ester amide) derived from L-alanine, *J. Appl. Polym. Sci.* **69**, 1537–1549.

Roby, M. S., Jiang, Y. (1998) Bioabsorbable poly-esteramide, its preparation and composition and surgical devices therefrom, WO 9832779.

Stapert, H.R., Van der Zee, M., Dijkstra, P.J., Feijen, J. (1997) *Polym. Mat. Eng. Sci.* **76**, 414–415.

Stapert, H.R., Bouwens, A., Dijkstra, P.J., Feijen, J. (1999) Environmentally degradable aliphatic poly(ester amides)s based on short, symmetrical and uniform bisamide-diol blocks. Part 1. Synthesis and interchange reactions, *Macromol. Chem. Phys.* **200**, 1921–1929.

Timmermann, R., Dujardin, R., Koch, R. (1993) Thermoplastic, biodegradable, workable high molecular weight copolymer – contains aliphatic ester and aliphatic amide units, EP 0 641 817.

Timmermann, R., Schulz-Schlitte, W., Voigt, M. (1999) Biodegradable thermoplastic polyestera-mides with good mechanical properties for molding, film and fiber, useful for e.g. compostable refuse bags, EP 1036107.

Tokiwa, Y., Suzuki, T., Ando, T. (1979) Synthesis of copolyamide-esters and some aspects involved in their hydrolysis by lipase, *J. Appl. Polym. Sci.* **24**, 1701–1711.

Veazey, R.L. (1984) Poly(ester-amide) hot melt adhesives, EP 0 156 949.

Voigt, M., Schulz-Schlitte, W. (2000) Biodegradable molding composition containing metallic or mineral filler useful for the production of molded articles has a high density and low toxicity, EP 1098930.

Wiegand, S., Steffen, M., Steger, R., Koch, R. (1999) Isolation and identification of microorganisms able to grow on the polyester amide BAK 1095, *J. Environ. Polym. Degrad.* **7**, 145–156.

13
Chemical Modification
of Natural and Synthetic
Polyesters

Prof. Dr. Emo Chiellini[1], Dr. Ranieri Bizzarri[2], Dr. Federica Chiellini[3]
[1] Chemistry and Industrial Chemistry Department, University of Pisa, Via
   Risorgimento 35, I-56126 Pisa, Italy; Tel.: +39-050-918299; Fax: +39-50-28438;
   E-mail: chlmeo@dcci.unipi.it
[2] Chemistry and Industrial Chemistry Department, University of Pisa, Via
   Risorgimento 35, I-56126 Pisa, Italy; Tel.: +39-050-3139643; Fax: +39-50-28438;
   E-mail: ranieri@dcci.unipi.it
[3] Chemistry and Industrial Chemistry Department, University of Pisa, Via
   Risorgimento 35, I-56126 Pisa, Italy; Tel.: +39-050-3139641; Fax: +39-50-28438;
   E-mail: federica@dcci.unipi.it

ATRP	atom transfer radical polymerization
BOC	*tert*-butoxycarbonyl
Bz	benzyl
CDI	1,1′-carbonyldiimidazole
DBN	1,5-diazabicyclo[4.3.0]non-5-ene
DBU	1,8-diazabicyclo[5.4.0]undec-7-ene
DCC	dicyclohexylcarbodiimide
DIAD	diisopropylazodicarboxylate
DMAD	dimethylazodicarboxylate
DMAP	dimethylaminopyridine
DMF	dimethylformamide
GRGDY	glycine-arginine-glycine-aspartic acid-tyrosine
HOBt	1-hydroxybenzotriazole
MCPBA	methachloroperbenzoic acid
NCA	*N*-carboxyanhydride
NMP	*N*-methyl pyrrolidinone;
PCC	pyridinium chlorochromate
PCL	poly(ε-caprolactone)
PET	poly(ethylene terephthalate)
PGA	poly(glycolic acid)
PHA	poly(hydroxy alkanoate)
PHB	poly(β-hydroxybutyrate)
PHBV	poly(β-hydroxybutyrate-co-β-hydroxyvalerate)
pHEMA	poly(2-hydroxyethyl methacrylate)
PLA	poly(lactic acid)
PMB	pentamethylbenzyl
PMLA	poly(β-malic acid)
RGD	arginine-glycine-aspartic acid
ROH	generic alcohol
ROP	ring-opening polymerization
TEAB	tetraethylammonium benzoate
TFFA	trifluoroacetic anhydride

THP   tetrahydropyranyl
TOSUO   1,4-trioxaspiro[4.6]-9-undecan-9-one
TPP   triphenylphosphine
Z   benzyloxycarbonyl
α-PMA   poly(α-malic acid)
ε-CL   ε-caprolactone

1
Historical Outline and Introductory Remarks

Polyesters are polycondensation polymeric materials characterized by the presence of an ester group in the repeating units. These can be formally derived by the heterocondensation of two complementary components, a diol and a diacid, or by homocondensation of hydroxyacids. These materials have attained a unique position in the field of chemical technology for their valuable physico-chemical characteristics. It is worth noting that polyesters account for almost half of the marketed synthetic fibers in the world (Davis and Hill, 1982). The general procedures for the preparation of synthetic polyesters starting from fossil fuel or renewable resources feedstock are shown diagrammatically in Figure 1.

The history of polyesters began with the pioneering works of W. H. Carothers at DuPont during the 1930s. However, Car-

others investigated mostly linear aliphatic polyesters, whose thermoplastic characteristics did not allow them to be marketed for the fabrication of plastic items. During World War II, scientific interest was devoted to the use of aromatic monomeric precursors for the preparation of polyesters. Consequently, poly(alkylene terephthalate)s with tunable length of the alkyl block in the backbone chain were quickly recognized as the basis for useful melt-spinnable synthetic fibers. In 1953, poly(ethylene terephthalate) (PET) was introduced by DuPont on the North American market under the trade name Dacron®, and this material is still sold extensively today. PET represents one of the most successful polymeric materials in the history of macromolecular technology. Indeed, during the 1990s, the world production of PET fibers exceeded 9 million tons per year. In addition to PET and other poly(alkylene terephthalate)s, liquid-crystalline thermotropic polyesters were also revealed

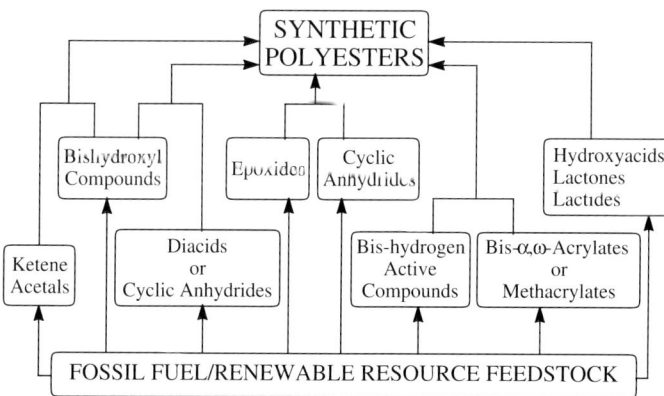

Fig. 1 General procedure for the preparation of synthetic polyesters.

as being important in the preparation of semi-commodity (i.e., between commodity and engineering) "high-tech" plastics. These materials possess a combination of physico-chemical characteristics that makes them ideally suited to the preparation of the complex polymeric systems required by today's microelectronics industry (Kirk and Othmer, 1991).

Following the continuous demand for highly sophisticated materials with active roles in specific environments, polymer chemistry research and industrial production have recently shifted their attention from polymers based on the availability of raw materials and cost-effective production (e.g., commodity plastics) to polymers designed for specific applications. Accordingly, many aliphatic polyesters, such as poly(lactic acid) (PLA), poly(glycolic acid) (PGA), and poly(ε-caprolactone) (PCL), have been extensively investigated for value-added applications, among which medical devices are probably the most prominent. Indeed, such materials have been shown to possess many valuable properties within the biological environment, including hydrolytic degradation, bioresorption, and low toxicity (SaintPierre and Chiellini, 1987; Chiellini and Solaro, 1996a; Chiellini et al., 1998). Nowadays, biocompatible and bioerodible polyesters have begun to replace other common materials as major components of drug delivery systems, degradable sutures, and prostheses, as well as a variety of more specific healing devices (Chiellini and Solaro, 1996b; Kohn and Langer, 1996). A number of problems remain to be solved however, notably with regard to the better tailoring of the materials' characteristics to their final applications. In this respect, the introduction into the polyester aliphatic chains of functional groups that are susceptible to chemical modification represents a convenient option either to modulate the basic properties of the polymer (e.g., hydrophilic–hydrophobic balance and degradation rate), or alternately to provide sites to which bioactive compounds may be bound to varying degrees.

Perhaps surprisingly, the idea of functional polyesters is not new, and in fact the reaction products of polyhydric alcohols and polybasic acids (alkyd resins) have been used widely by the coatings industry since the 1940s. The success of alkyd resins is mainly due to their great versatility and low to moderate production costs. Indeed, relatively inexpensive commodity items provide a large variety of the reaction ingredients, which in turn allows structural modulation of the final resin properties. Beside alkyd resins, a well-known family of functional polyesters is represented by thermosetting resins derived from a mixture of unsaturated dibasic acids, glycols, and unsaturated derivatives such as styrene or methyl methacrylate. Although crosslinked unsaturated polyesters were first obtained in the late 1940s, they are still used today in large amounts (2 million tons per year), mainly as the continuous matrix in glass fiber-reinforced composites.

Despite the straightforward production methods that are used, the physico-chemical characteristics of these two classes of functional polyester do not allow them to be used in many high-tech applications, mostly because neither the molecular weight nor the final macromolecular architecture can be conveniently adjusted. Furthermore, the crosslinked nature of the materials prevents them from being soluble in organic media, and this greatly limits the extent of the chemical modifications that can be carried out on the macromolecular chains. Consequently, during the past few years a great deal of effort has been devoted to the design and preparation of functional polyesters with controlled structure and 'tunable' physico-chemical characteristics.

The potential synthetic pathways for the creation of "functional polyesters" are based on the preparation of functional monomeric precursors, which can in turn be readily transformed into the respective homopolymers, or copolymerized with other functional or nonfunctional molecules (Figure 2). The functional groups in the parent monomers or monomeric precursors must be selected in such a way that they do not interfere with the polymerization mechanism. Although some chemical groups comply with this situation under various polymerization conditions, it is often necessary to install preliminary protection of the functional groups to be incorporated. In these cases, the protecting groups should be chosen so that the deprotection procedure does not modify the characteristics of the parent polymer, as might occur with trans-esterification or chain cleavage reactions. It should be mentioned however that reactions carried out on preformed polymers are subject to certain fundamental drawbacks that often limit the practical applicability of polymer chemical transformation technology (Platè, 1973) used to create new polymeric materials.

The nature of the polymerization process plays an important role in the design of the synthetic strategy to produce suitably protected functional monomers. The most convenient way to prepare polyesters is by ring-opening polymerization (ROP) of cyclic ester monomers (Brunelle, 1993). Therefore, functional monomers for ROP are lactone molecules bearing pendant chemical groups that do not interact specifically with the ionic propagating species. Furthermore, ROP of lactones often proceeds according to a "living" polymerization mechanism, and this permits strict control of the final polymer chain characteristics with regard to the incorporation of functional groups.

The polycondensation of diacids and dialcohols is another important polymerization methodology, but this normally requires very demanding experimental conditions if high molecular-weight polyesters are to be prepared. In this case, the introduction of functional groups is generally difficult, even if protecting groups are used.

Polymerization procedures which allow for selective end-group insertion offer another possibility of obtaining functional polyesters. Although only some cyclic monomers can undergo this type of polymerization, the use of suitable monomers may prove useful to produce structurally ordered polyfunctional polymers and subsequently complex macromolecular architectures.

The most valuable methods for the incorporation of functional groups into a polyester chain are summarized in the next sections, though clearly a fully comprehensive review of the subject is not possible within the scope of this chapter. The preparation of functional monomeric com-

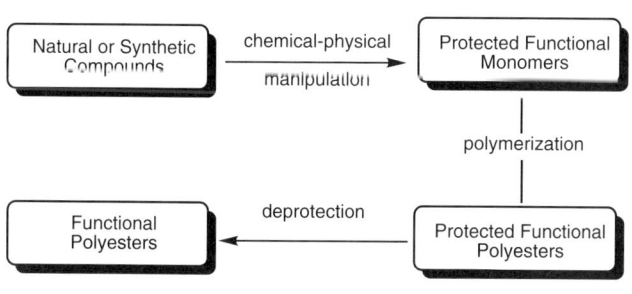

Fig. 2 General procedure for the preparation of polyesters.

pounds and their polymerization and co-polymerization will be detailed in Sections 2 to 6, whilst the introduction of functional groups as chain termini will be outlined in Section 7.

2
Malic Acid-based Polyesters

Malic acid is a natural compound that constitutes the structural unit of a wide class of functional bioerodible-biodegradable polymeric materials (Scheme 1). The fundamental unit, poly(malic acid), is a polyester which carries carboxyl pendant groups that can exist with alternative chain ester linkages of either α or β type. As for PLA, the stereoregularity of the polymeric chain is determined by the effective absolute configuration of the asymmetric carbon atom present in each repeating monomeric unit.

Direct polymerization of malic acid cannot be accomplished, since the trifunctional nature of the parent monomer leads to strongly ramified oligomers. Thus, the selective protection of one of the two carboxyl functions in the molecule is necessary for effective formation of both poly(α-malic acid) and poly(β-malic acid). The protective benzyl group (**Bz**) is generally used for this

purpose, as it can easily be cleaved by hydrogenolysis – a chemical process that does not affect the ester bonds in the main chain in any way.

2.1
Poly(α-Malic Acid) and its Copolymers

The synthesis of poly((S)-α-malic acid) (α-PMA) (Scheme 2) was realized by anionic ROP of the synthetic dimalide dibenzyl ester **2**, followed by hydrogenolysis of poly((S)-α-benzyl malolactonate) **3** (Ouchi and Fujino, 1989). Stannous carboxylates were utilized as initiators, because of their recognized ability to activate the anionic ROP process of six-membered ring cyclic diesters.

The preparation of the key intermediate **2** was accomplished by zinc oxide-promoted dehydration of malic α-benzyl monoester **1**, which in turn was obtained by a two-step chemical transformation of either (S)-aspartic acid or (S)-malic acid (Scheme 2) (Ouchi and Fujino, 1989). The final polymer displayed a stereochemical configuration analogous to that of the starting (natural) building blocks, since both the reactions affecting the chiral center (1→2, 2→3) proceeded according to a SN2 mechanism. Although the overall yield of **1** from (S)-malic acid was less (11%) than that from (S)-

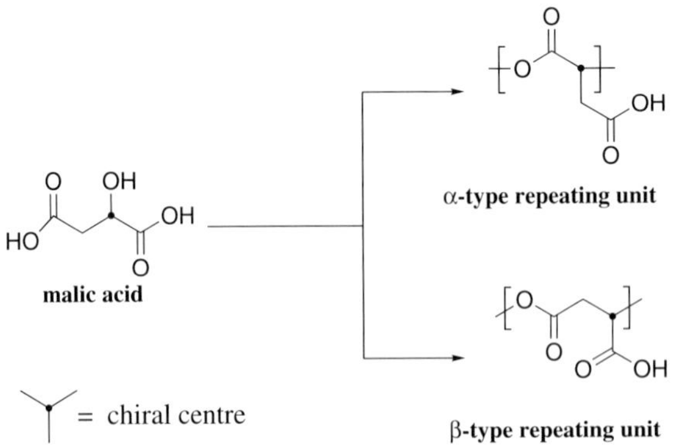

α-type repeating unit

β-type repeating unit

malic acid

Y = chiral centre

Scheme 1   Preparation of carboxyl-containing polyesters of malic acid.

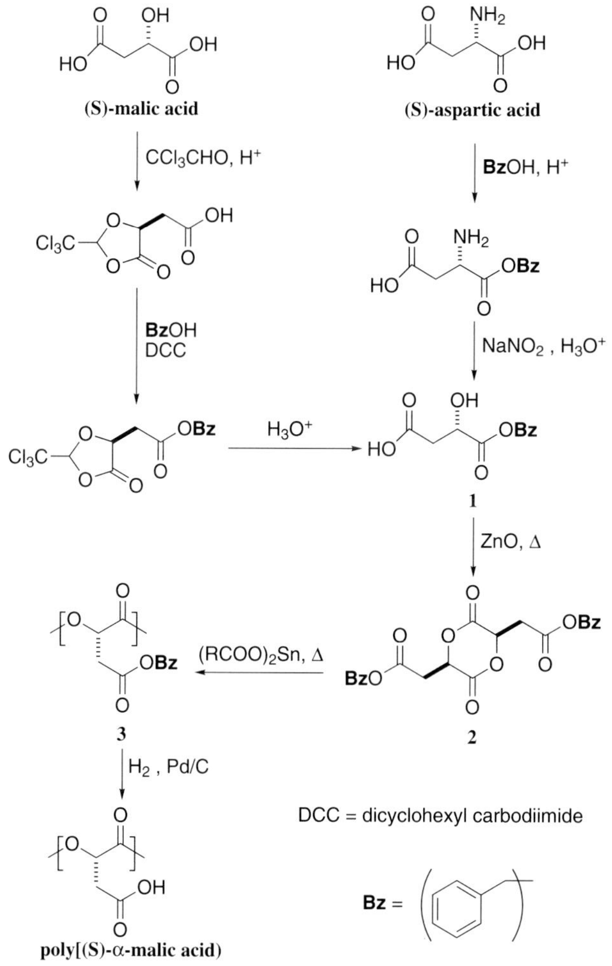
(S)-malic acid
(S)-aspartic acid
CCl3CHO, H+
BzOH, H+
BzOH
DCC
NaNO2 , H3O+
H3O+
1
ZnO, Δ
(RCOO)2Sn, Δ
3
2
H2 , Pd/C
DCC = dicyclohexyl carbodiimide
Bz =
poly[(S)-α-malic acid)

Scheme 2 Preparation of poly[(*S*)-α-malic acid] (Ouchi and Fujino, 1989).

aspartic acid (59%), the first synthetic route to **1** allowed easier purification and greater retention of optical activity (Ouchi and Fujino, 1989).

Synthetic dimalide **2** displayed only moderate chemical reactivity towards homopolymerization due to steric hindrance of the bulky benzyl ester pendant groups; as result, the molecular weights of **3** and poly((*S*)-α-malic acid) were low (2–4 kDa) (Ouchi and Fujino, 1989). In contrast, copolymerization of **2** with the more reactive cyclic diester, L-dilactide, yielded the random copolyester PLA-PMA **4** which displayed appreciably higher molecular weights (14–40 kDa) (Ouchi and Fujino, 1989) (Scheme 3).

The synthesis of the mixed glycolic-malic cyclic diester **5** from (*S*)-aspartic acid (Kimura et al., 1988) allowed the controlled preparation of polymeric materials with complex chain structures. Indeed, the alternating PLA-PMA copolyester **6** was generated from homopolymerization of **5**, whereas its copolymerization with L-dilactide led to

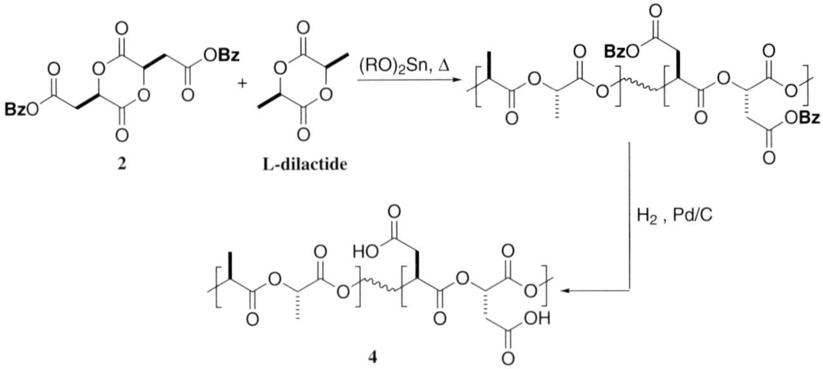

Scheme 3 Preparation of poly[(S)-α-malic acid-*co*-L-lactic acid] acid (Ouchi and Fujino, 1989).

the statistical-alternating PLA-PGA-PMA ter-polymer **7** (Kimura et al., 1988, 1993) (Scheme 4).

The synthesis of copolyesters **4**, **6**, and **7** provided an elegant approach to the modulation of physico-chemical properties of PLA, which represents one of the most important polymeric materials for biomedical applications. Indeed, progressive introduction of glycolic units and carboxyl pendant groups into the PLA chain decreased the semicrystallinity of the chain while simultaneously improving its hydrophilicity and degradation rate in phosphate buffer, as demonstrated by a set of terpolymers **7** with different chain composition (Kimura et al., 1993). In addition, the carboxyl function was shown to provide an effective chemical modification site by the attachment of a *p*-azidophenyl group, a well-known photoaffinity labeling agent (Kimura et al., 1993) (Scheme 5).

In another example of post-functionalization of these copolyesters (carried out subsequent to the polymer preparation), the cell adhesion-promoting tripeptide arginine-glycine-aspartic acid (**RGD**) was linked to the pendant carboxylic groups of **7** by a dicyclohexyl carbodiimide (DCC)-mediated con-

densation reaction (Scheme 6) (Yamaoka et al., 1999). The resulting **RGD**-substituted terpolyesters were shown to have a strong promotional effect on cell adhesion to the polymer surface.

2.2
Poly(β-Malic Acid) and Poly(alkyl malolactonate)s

Poly(β-malic acid), called also poly(malolactonate) (PMLA) is a water-soluble polyelectrolyte (pK1/2 = 4.4) which appears to be nontoxic, completely bioresorbable and non-immunogenic when injected either intramuscularly or intravenously into mammalian organisms (Braud et al., 1982, 1985). The only synthetic method developed so far to produce linear PMLA with significant molecular weight has involved the anionic ROP of benzylmalolactonate **8** to yield poly(benzyl malolactonate) **9**, which in turn was deprotected by hydrogenolysis (Scheme 7) (Vert and Lenz, 1979).

Several ionic salts and amine moieties were investigated as initiators for the ROP process, but triethylammonium benzoate and triethylamine gave the best results in

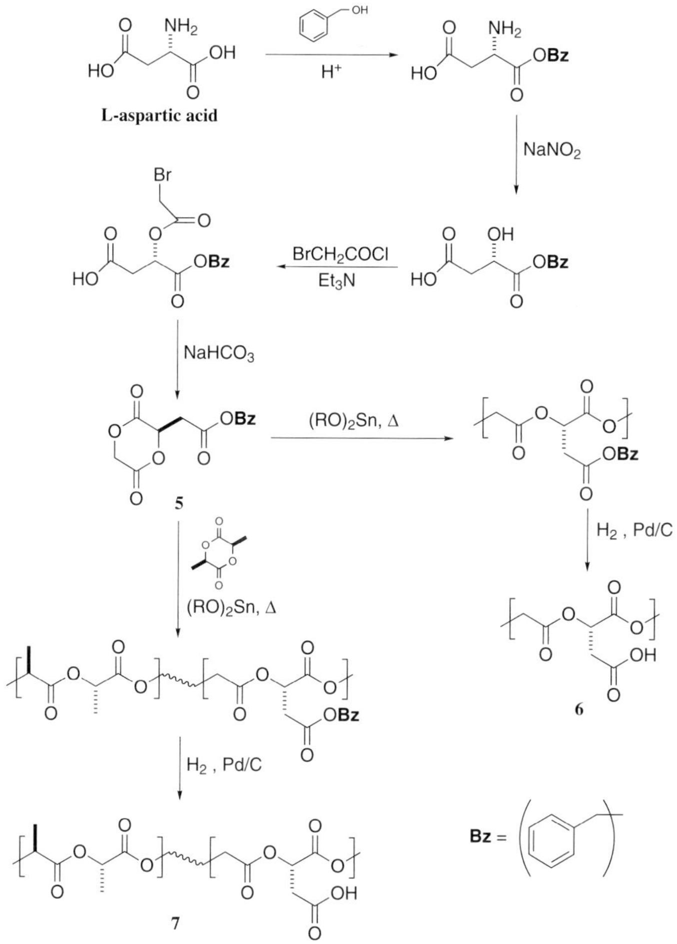

Scheme 4 Preparation of poly[(S)-α-malic acid-co-glycolic acid] and poly[(S)-α-malic acid-co-glycolic acid-co-L-lactic acid] (Kimura et al., 1988, 1993).

terms of polymerization rate and final molecular weight (Lenz et al., 1986).

The preparation of the key monomeric precursor 8, as well as a wide range of alkyl β-malolactonates, was accomplished by using two synthetic strategies which started respectively from either aspartic acid (aspartic acid route) or malic acid (malic acid route) (Vert and Lenz, 1979; Arnold and Lenz, 1986; Cammas et al., 1993, 1994a; Boutault et al., 1995; Moine et al., 1997; Barbaud

et al., 1999; Bizzarri et al., 2001). Both routes permitted the preparation of either racemic or optically active β-lactone structures, depending on the stereochemistry of the initial acid (Guerin et al., 1985; Cammas et al., 1993). The aspartic acid route was also recently used to prepare alkyl malolactonates bearing a methyl group on carbon 2 of the ring (Cammas et al., 1994b; Renard et al., 1996). As an example, the preparation of the generic optically active (R)-alkyl malolacto-

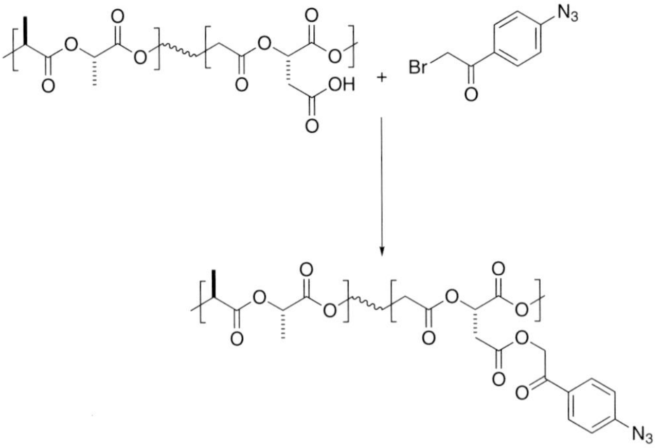

Scheme 5 Preparation of poly[(S)-α-malic acid-co-glycolic acid-co-L-lactic acid] photolabeled with a p-phenylazido group (Kimura et al., 1993).

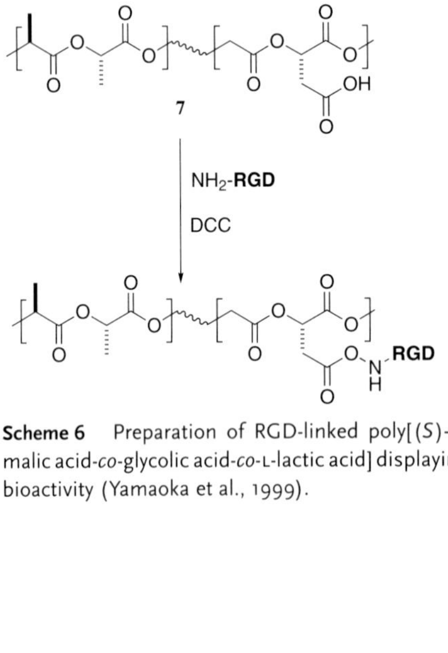

Scheme 6 Preparation of RGD-linked poly[(S)-α-malic acid-co-glycolic acid-co-L-lactic acid] displaying bioactivity (Yamaoka et al., 1999).

nate **14** by either the aspartic acid route (upper part of Scheme 8) or the malic acid route (lower part of Scheme 8) has been reported.

In the aspartic acid route (Vert and Lenz, 1979; Guerin et al., 1985), (S)-aspartic acid is initially converted to (S)-2-bromosuccinic acid **10**, by the classical nitrosation-bromination procedure. The reaction occurs with double inversion at the stereogenic center, and without appreciable racemization (Guerin and Vert, 1987). Dehydration of **10** by means of trifluoroacetic anhydride leads to the formation of (S)-bromosuccinic anhydride **11**, which is in turn reacted with the general alcohol moiety **ROH** to give the (S)-bromosuccinate monoesters **12** and **13**. No purification of this mixture is necessary for

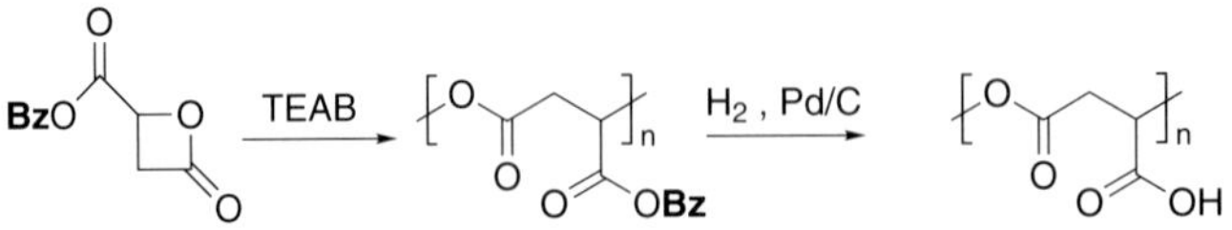

TEAB = tetraethylammonium benzoate

Poly(β-malic acid)

Scheme 7 Preparation of poly(β-malic acid) (Vert and Lenz, 1979).

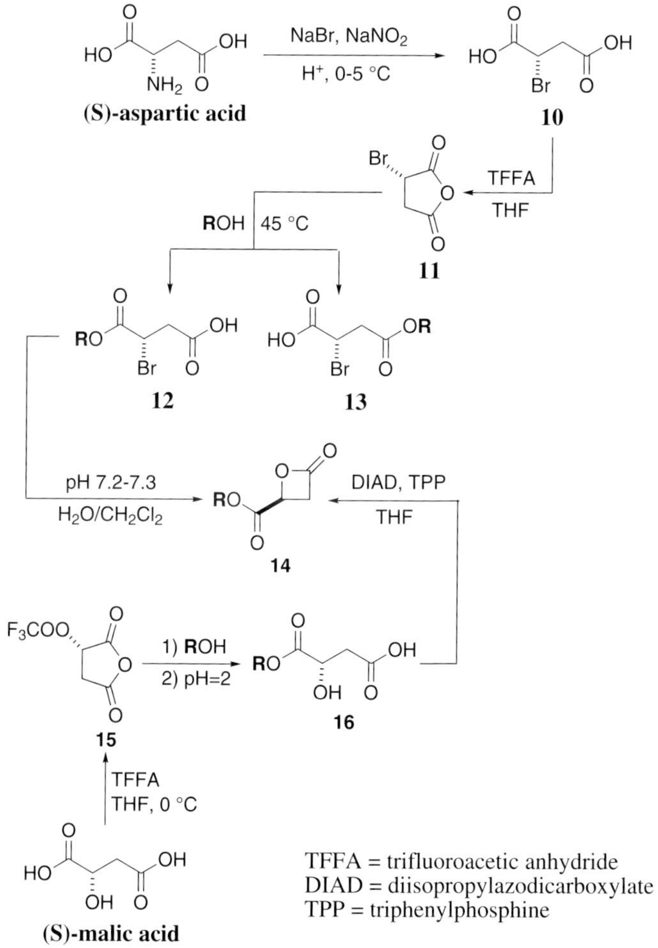

Scheme 8 Preparation of optically active generic alkyl β-malolactonate monomer (Guerin et al., 1985; Cammas et al., 1993).

TFFA = trifluoroacetic anhydride
DIAD = diisopropylazodicarboxylate
TPP = triphenylphosphine

the next synthetic step, although only isomer **12** is able to undergo the internal bromine displacement, and this leads to formation of the β-lactone monomer **14** (Vert and Lenz, 1979; Guerin et al., 1985). This final reaction is accomplished simply by base-titration of a water dispersion of **12** and **13** up to neutral pH, followed by addition of an organic solvent to provide a good partition solvent for the lactone produced (Vert and Lenz, 1979; Guerin et al., 1985). Since an SN2i reaction mechanism is involved in the lactonization process, **14** with (R)-configuration is generated (Guerin et al., 1985).

In the malic acid route, (S)-malic acid is initially converted to (S)-malic trifluoroacetic anhydride **15** by the dehydrating action of trifluoroacetic anhydride. As with the aspartic acid route, the selected alkyl group **R** is incorporated into the molecular structure by reaction of the alcohol **ROH** with the electrophilic cyclic anhydride moiety. However, only one ester isomer is obtained in this case, due to the strong inductive effect of the trifluoroacetic ester group in **15** (Cammas et al., 1993). The acid hydrolysis (pH = 2) allows restoration of the original hydroxyl function, thus yielding the malic monoester

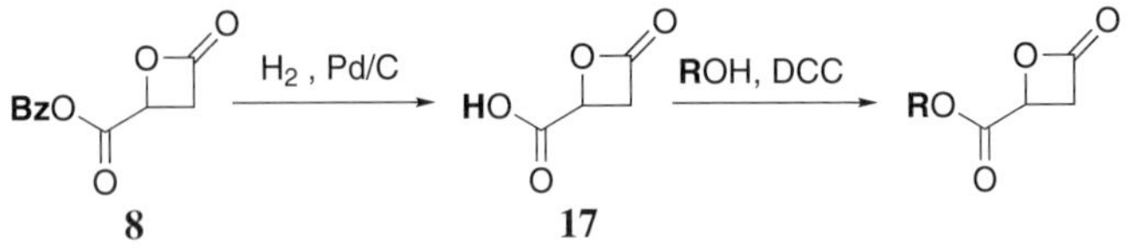

Scheme 9 Preparation of the β-lactone of malic acid (Leboucher-Durand et al., 1996).

DCC = dicyclohexyl carbodiimide

16 (Cammas et al., 1993). Eventually, intramolecular cyclization of **16** is accomplished by a Mitsunobu reaction (Mitsunobu, 1981), the mechanism of which involves inversion of configuration at the asymmetric carbon atom (Mulzer et al., 1979).

Although both synthetic routes were seen to produce the lactone monomer in similar yields (10–30% on **ROH**), some differences were noticed in terms of the final optical purity. For the aspartic acid route the enantiomeric excess of the recovered β-lactone was found to depend critically on the temperature of the lactonization step. In contrast, in the malic acid route the optical purity of the lactone was determined only by the initial value of malic acid (Cammas et al., 1994a). Therefore, the latter synthetic strategy seems better suited for applications where the control of polymer stereoregularity is important (Cammas et al., 1993; 1994a; Boutault et al., 1995).

Attachment of residues with complex structure to the side chain of alkyl malolactonates, by either the aspartic acid or the malic acid route, is generally prevented by the strong dependence of the anhydride ring-opening reaction on steric hindrance. Therefore, a further synthetic method was developed to allow for the incorporation of bulky compounds with potentially high biological activity (Leboucher-Durand et al., 1996; Moine et al., 1997; Bizzarri et al., 2001). According to this strategy (Scheme 9), the benzyl group in **8** was cleaved hydrogenolytically, after which the complex alcoholic moieties were linked to the free carboxyl group of malolactone **17** by using DCC as condensation agent.

As with benzyl malolactonate **8**, the synthesized alkyl malolactonates were homopolymerized or copolymerized (Scheme 10) to yield a large set of polyesters [poly-(alkyl malolactonate)s] whose different phys-

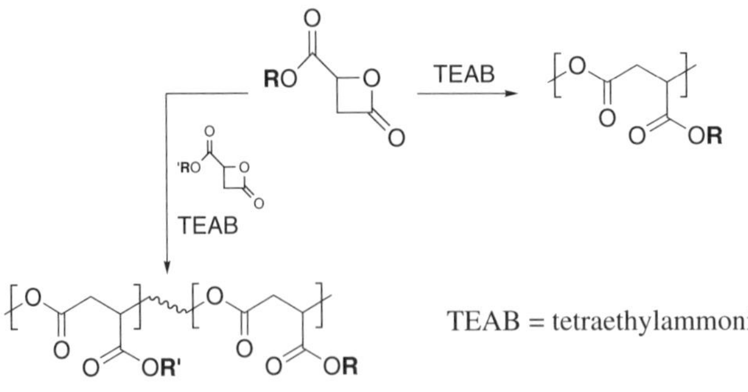

TEAB = tetraethylammonium benzoate

Scheme 10 Preparation of alkyl malolactonate polymers and copolymers (Cammas-Marion and Guerin, 2000; Bizzarri et al., 2001).

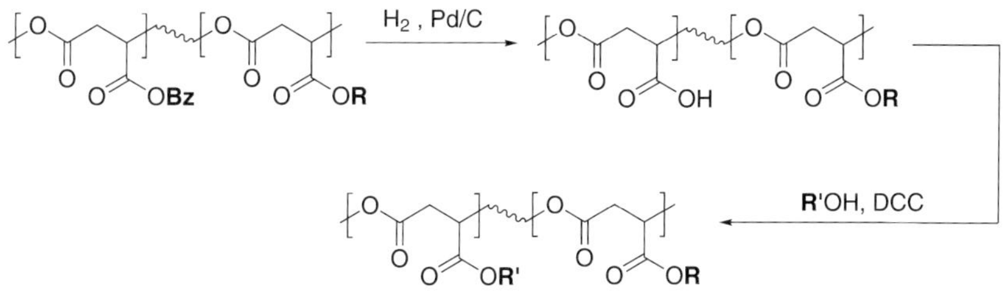

DCC = dicyclohexyl carbodiimide

Scheme 11   Hydrogenolysis of protective benzyl group and selective alkyl attachment to alkyl malolactonate copolymers (Braud et al., 1983).

ico-chemical characteristics depended on the nature as well as the distribution of the alkyl side groups (Vert and Lenz, 1979; Arnold and Lenz, 1986; Cammas et al., 1993, 1994a; Boutault et al., 1995; Moine et al., 1997; Barbaud et al., 1999; Bizzarri et al., 2001).

It is worth noting that pendant benzyl groups were often transformed into free carboxyl functions by hydrogenolysis, as a convenient way either to modulate the material's hydrophilic–hydrophobic balance, or to provide a reactive site for the attachment of other groups (Scheme 11) (Braud et al., 1983; Caron et al., 1990). Benzyl malolactonate was also copolymerized with β-propiolactone and β-butyrolactone, to obtain semicrystalline functional polyesters with high hydrophobicity (Benvenuti and Lenz, 1991; Matsumura et al., 1998).

Some of the side ester chains introduced were chosen for their amenability to further chemical modifications, with a view to tailoring the polymeric structure for specific functions (Barbaud et al., 1999; Cammas et al., 1999; Cammas-Marion and Guerin, 2000). For example, pendant allylic groups allowed the preparation of a heparin-mimicking malic terpolyester **18** for medical applications such as wound healing (Scheme 12).

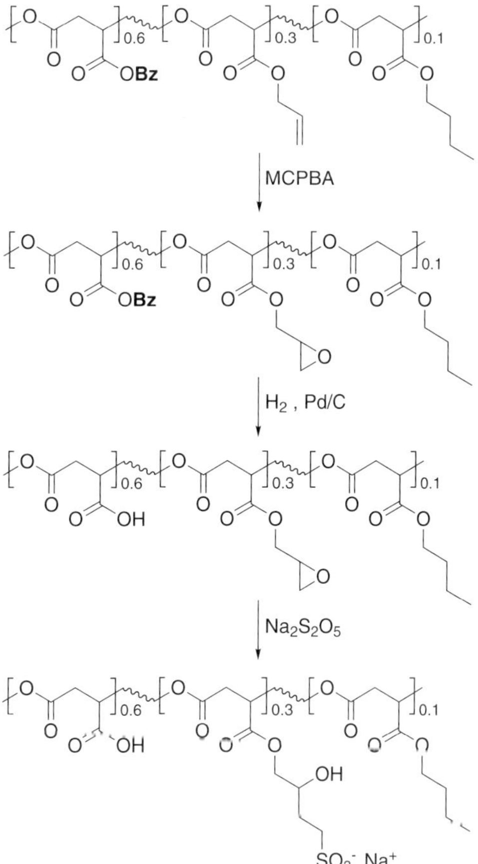

MCPBA = methachloroperbenzoic acid

Scheme 12   Preparation of heparin mimicking malic-based terpolyesters (Cammas et al., 1999).

3

Polyesters with Pendant Amine Groups

3.1

Poly(L-Serine Ester)s

L-Serine is a trifunctional amino acid which has attracted attention as a possible precursor of polyesters with pendant amine groups. The first investigated methodology to obtain such a polymer involved the preparation of N-protected β-lactone monomers from serine. The lactones were successively converted into the respective N-protected polyester by a ROP process analogous to that reported for the polymerization of malic acid (see Section 2.2) (Scheme 13)

N-protected L-serine

poly(L-serine ester)

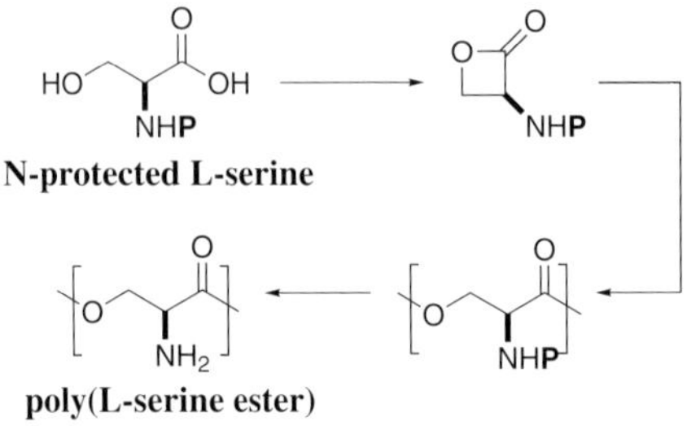

Scheme 13 Preparation of poly(L-serine ester) (Zhou and Kohn, 1990).

(Jarm and Fles, 1977; Fietier et al., 1990; Zhou and Khon, 1990).

In some cases, the nature of the protecting group allowed for its cleavage without affecting the ester bonds in the main chain, thereby yielding poly(L-serine ester) (Zhou and Khon, 1990). In fact, only the protonated form of poly(L-serine) ester could be isolated, because of the lability of the backbone ester linkages in presence of basic amino group.

Until 1985, the synthesis of N-protected serine β-lactones was extremely laborious, and also led to poor ultimate yields (Arnold et al., 1985). In addition, the available methods were often restricted to using particular protecting groups that were not easily cleared after polymerization. For example, (para-substituted benzenesulfonamido) serine β-lactones were obtained from the corresponding protected derivatives of asparagine through a Hoffmann rearrangement, and then submitted to homopolymerization (Jarm and Fles, 1977) (Scheme 14).

However, in 1985, Vederas and co-workers reported a high yield (70–80%) in the preparation of optically pure N-benzyloxycarbonyl (**Z-**) and N-*tert*-butoxycarbonyl (**BOC**) L-serine β-lactones by means of a

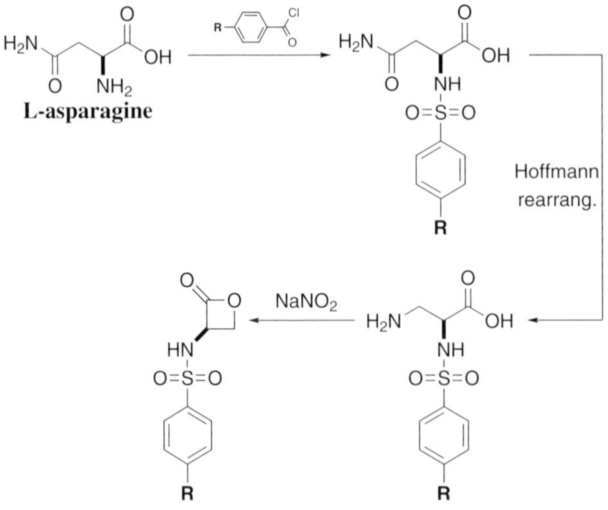

L-asparagine

Hoffmann rearrang.

Scheme 14 Preparation of poly(L-serine ester) amino-protected with *p*-phenylsulfonamido groups (Jarm and Fles, 1977).

modified Mitsunobu reaction. This involved addition of the preformed adduct of trimethyl phosphine and dimethyl azodicarboxylate (DMAD) to the protected L-serine at –78 °C (Arnold et al., 1985) (Scheme 15). This was a major advance, as previous attempts to cyclize N-protected L-serine using a conventional Mitsunobu reaction had produced the

PPh₃ ·DMAD
-78 °C

HO

O

OH

NH
R

O

O

O

NH
R

R =

benzoyloxycarbonyl
(**Z-**)

tert-butoxycarbonyl
(**BOC-**)

TPP = triphenylphosphine
DMAD = dimethyl azodicarboxylate

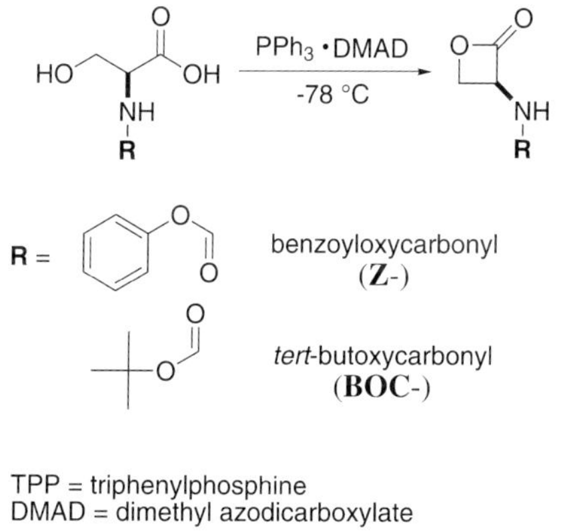

Scheme 15   Preparation of N-protected L-serine β-lactone (Arnold et al., 1985).

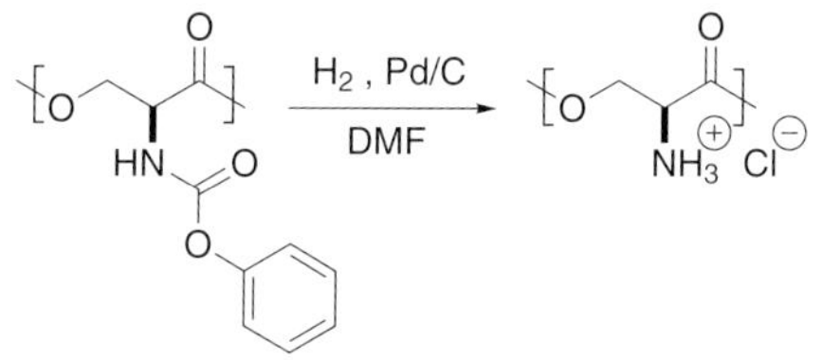

H₂ , Pd/C
DMF

O

O

HN

O

O

O

NH₃⁺ Cl⁻

Scheme 16   Deprotection of the *N*-benzyloxycarbonyl poly(L-serine ester) by hydrogenolysis (Zhou and Kohn, 1990).

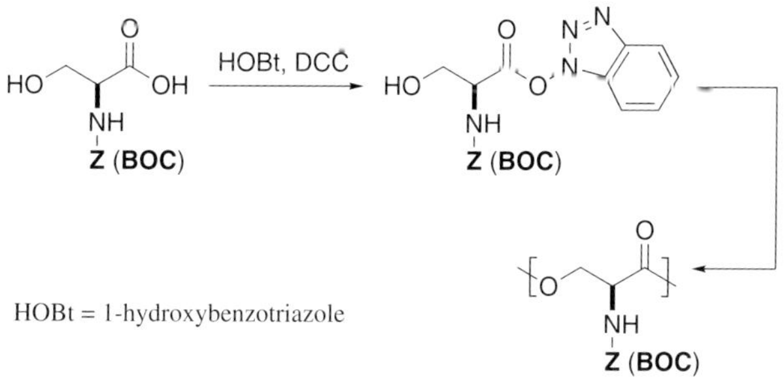

corresponding β-lactones in only 1.4% yield (Arnold et al., 1985).

The recovered β-lactones were homopolymerized either in bulk or in solution, and by using a wide variety of ROP initiators (Zhou and Kohn, 1990). Although the presence of chain transfer mechanisms was clearly demonstrated, protected polyesters and co-polyesters of both lactones with molecular weights up to 50 kDa were easily obtained. Furthermore, no detectable racemization was found upon complete hydrolysis of the polyesters, as determined from the optical rotatory power of the recovered L-serine (Zhou and Khon, 1990). Interestingly, the BOC-protected poly(L-serine ester) showed very poor solubility, even in highly polar solvents such as *N*-methylpyrrolidinone (NMP). By contrast, the high solubility of the Z-protected poly(L-serine ester) in dimethylformamide (DMF) was used to carry out hydrogenolytic cleavage of the Z protecting group to produce the functional poly(L-serine ester) hydrochloride (Scheme 16) (Zhou and Khon, 1990).

The preparation of N-protected poly(L-serine ester)s was also accomplished using synthetic strategies that did not involve β-lactone intermediates. For example, N-Z (BOC) L-serine was converted into the respective homopolymer by preliminary activation of the carboxyl function as 1-hydroxybenzotriazole ester, followed by thermal bulk polymerization (Gelbin and Kohn, 1993) (Scheme 17), the final molecular

HOBt, DCC

HO

O

OH

NH
Z (BOC)

HO

O

O

N-N
N

NH
Z (BOC)

HOBt = 1-hydroxybenzotriazole

O

O

NH
Z (BOC)

Scheme 17   Preparation of poly(L-serine ester) by activation of the carboxyl function of L-serine as 1-hydroxybenzotriazole ester (Gelbin and Kohn, 1993).

weights of the polyesters obtained being about 20 kDa. In fact, this procedure was the first practical method of producing multi-gram quantities of serine-derived polyesters with such acceptable characteristics (Gelbin and Kohn, 1993).

Another methodology for the large-scale production of poly(L-serine ester) was published recently (Rossignol et al., 1999). This involved activation of the N-protected L-serine carboxyl function by means of methanesulfonyl chloride, followed by polymer formation in acetone solution under basic conditions for sodium carbonate. The protecting groups were then removed by the action of a bromidric-acetic acid mixture, and the resulting amino-functionalized polyester was used to form ionic polycomplexes with poly(β-malic acid) (Scheme 18) (Rossignol et al., 1999).

Interestingly, these last two synthetic procedures did not allow for polymer formation when N-protected L-serine was substituted with N-protected L-threonine. This finding was attributed to the steric hindrance of the extra methyl group in the threonine structure, which apparently pre-

vented direct polycondensation (Gelbin and Kohn, 1993; Rossignol et al., 1999).

3.2
Poly(lactic acid-*co*-lysine)

The amino-terminated side chain of the amino acid L-lysine was successfully incorporated into a polyester structure after a series of chemical transformations (Barrera et al., 1993, 1995). This procedure was developed to provide PLA with lateral amino groups, which could act as sites for further chemical functionalization.

The reported synthetic strategy was in fact similar to that adopted for the formation of PLA-PMA copolymers (see Section 2.1) (Scheme 19). Indeed, to produce a copolymer of lactic acid and lysine, it proved necessary to obtain a monomer that was capable of polymerization by the ROP mechanism, as was used in the preparation of PLA. The synthesis of such a compound was accomplished by converting D-alanine into the acyl chloride intermediate **19**, and this in turn was reacted with N-Z-protected L-lysine to afford, after intramolecular cycliza-

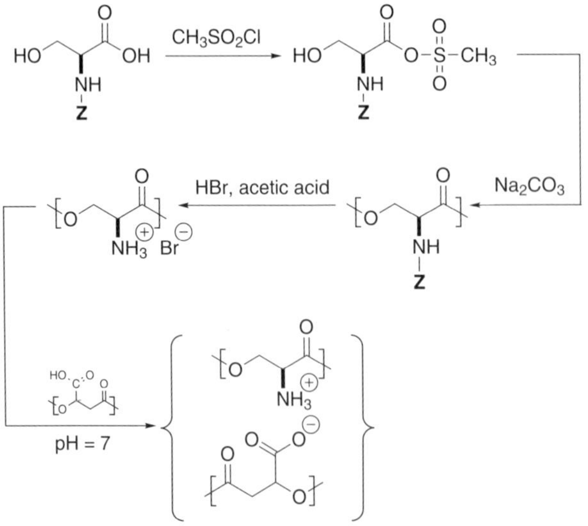

Scheme 18 Preparation of poly(L-serine ester) by activation of the carboxyl function of L-serine as methanesulfonic ester (Rossignol et al., 1999).

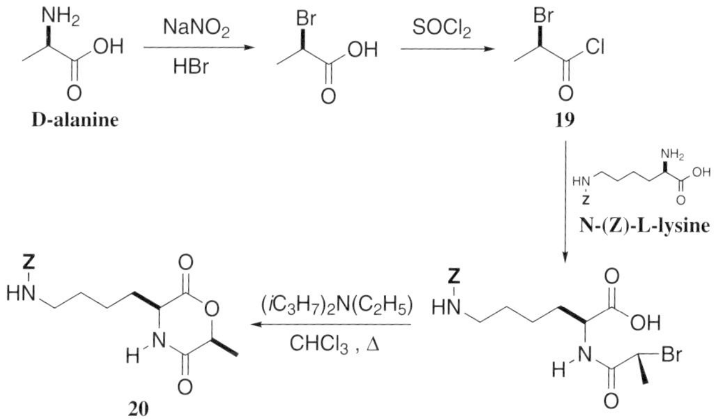

Scheme 19 Preparation of *N*-benzyloxycarbonyl-protected L-lysine-L-lactic acid cyclic lactone dimer (Barrera et al., 1993).

tion, the N-protected diester **20** in 23% yield (Barrera et al., 1993).

A minor amount of racemization at the asymmetric carbon atom of D-alanine occurred during the final ring-closure step, but the final diastereoisomeric excess was as high as 95% (Barrera et al., 1993, 1995). Interestingly, the use of L-alanine instead of D-alanine led to a diastereoisomeric excess of only 75%, perhaps due to the increased steric hindrance of the methyl group linked to the chiral center during the cyclization process (Barrera et al., 1995). However, only the monomer **20** derived from D-alanine was utilized in the next polymerization step.

Copolymers of **20** with L-dilactide were obtained in different stoichiometric ratios by using stannous octoate as anionic ROP initiator (Scheme 20) (Barrera et al., 1993, 1995). Polymerization temperatures around 100 °C gave better results in terms of molecular weights of the final copolyesters (60–90 kDa), although a substantial increase in polymerization time was observed (Barrera et al., 1995). As previously reported for PLA-PMA copolymers (Section 2.1), the introduction of lysine residues into the chain resulted in a reduced semicrystallinity of the resulting polymeric material, whilst at the same time increasing its degradation rate under physiological conditions (Barrera et al., 1995).

Lysine amino group deprotection was carried out using a palladium chloride catalyst system, such that poly(L-lactic acid-*co*-L-lysine) was obtained with a decarbamoylation degree of 75% (Scheme 20). The amenability towards further chemical transformation of the pendant primary amino groups in the deprotected copolyester was exploited for the attachment of the cell adhesion-promoting peptide GRGDY, by using 1,1′-carbonylimidazole as a condensation agent (Barrera et al., 1993). Although some crosslinking also occurred, the copolyester–peptide conjugate proved very active in promoting cell adhesion and cell spreading (Cook et al., 1997), and might be considered a valuable alternative to the previously reported **RGD**-poly(L-malic-*co*-glycolic-*co*-L-lactic acid) terpolyesters (Section 2.1).

4
Functionalization of Poly(ε-caprolactone)

4.1
Synthesis of Functional Monomers by the Bayer–Villiger Oxidation of Cyclohexanones

Poly(ε-caprolactone) (PCL) (Scheme 21) is a polyester which shows a remarkable set of physical properties, such as an extremely low

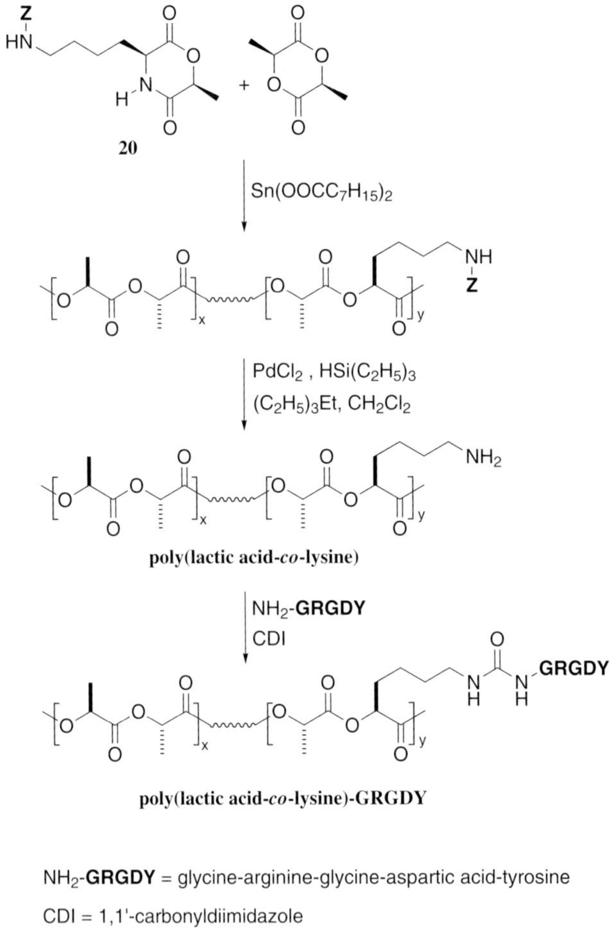

poly(lactic acid-*co*-lysine)

poly(lactic acid-*co*-lysine)-GRGDY

NH$_2$-**GRGDY** = glycine-arginine-glycine-aspartic acid-tyrosine

CDI = 1,1'-carbonyldiimidazole

Scheme 20 Preparation of L-lactic acid and L-lysine copolymers and selective attachment to GRGDY peptide (Barrera et al., 1995).

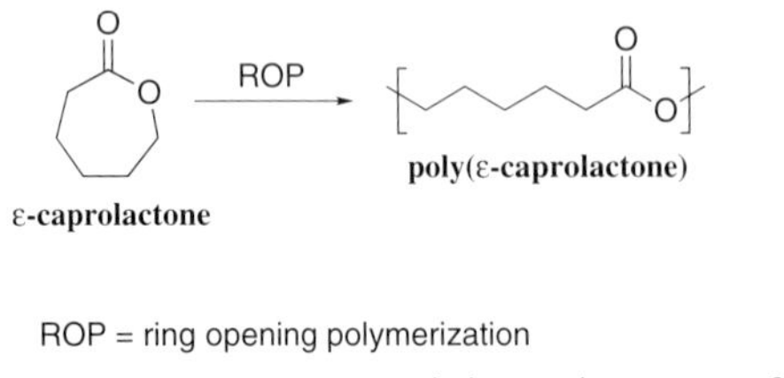

poly(ε-caprolactone)

ε-caprolactone

ROP = ring opening polymerization

Scheme 21 Preparation and chemical structure of poly(ε-caprolactone).

noteworthy thermal stability (up to 350 °C), and an ability to form monophasic blends with a wide range of other polymers (Pachence and Kohn, 1997).

PCL is currently regarded as a non-toxic, tissue-compatible material (Pitt, 1990), the biomedical utilization of which became clear following its commercial use as a (potentially) degradable packaging material. However, the problem with PCL is that its relatively long aliphatic backbone of PCL (five methylene groups) provides neither a way to modulate the hydrophilic–hydrophobic balance (and in turn the hydrolytic

glass transition temperature (T_g) ($\approx$−60 °C), a moderate melting temperature (T_m) ($\approx$60 °C), high solubility in organic solvent,

degradation of the material), nor any active sites for further chemical modification. Consequently, during the past few years much effort has been made to synthesize functional monomeric precursors of PCL (Pitt et al., 1987; Tian et al., 1997a; Detrembleur et al., 2000a,b; Lecomte et al., 2000; Mecerreyes et al., 1999a, 2000a,b; Stassin et al., 2000).

PCL is generally obtained by anionic ROP of ε-caprolactone (ε-CL), initiated by a variety of alkoxide and carboxylate compounds. Bayer–Villiger oxidation offers a convenient means if introducing functional groups into the ε-CL ring, starting from substituted cyclohexanones. The general synthetic strategy is shown in Scheme 21 (Pitt et al., 1987; Tian et al., 1997a; Detrembleur et al., 2000a,b; Lecomte et al., 2000; Mecerreyes et al., 1999a, 2000a,b; Stassin et al., 2000).

Usually, one of the hydroxyl groups of commercially available 1,4- cyclohexanediol is converted into a suitable group, which is not affected by the successive oxidation reaction, to yield the substituted cyclohexanone **21**. Further oxidation, using *m*-chloroperbenzoic acid (Bayer–Villiger reaction), affords the 3-substituted ε-CL monomer **22**, usually with good final yields (50–90%) (Pitt et al., 1987; Tian et al., 1997a; Detrembleur et al., 2000a,b; Lecomte et al., 2000; Mecerreyes et al., 1999a, 2000a,b; Stassin et al., 2000). Some functional groups, such as carbonyl and hydroxyl, were incorporated as protected forms (Scheme 22) in order to avoid side reactions occurring during the monomer preparation and the subsequent polymerization process (Pitt et al., 1987; Tian et al., 1997a; Stassin et al., 2000).

A further chemical transformation was carried out on the bromine-substituted ε-CL (Detrembleur et al., 2000b; Lecomte et al., 2000). Indeed, the non-nucleophilic organic base 1,8-diazabicyclo[5,4,0] undec-7-ene (DBU) was utilized to introduce an alkene bond into the lactone ring via the elimination of bromidric acid. This elimination reaction occurred without control of regioselectivity however, leading to a mixture of unsaturated lactones **22g** (Scheme 23).

Although selective chemical modification of 1,4-dicyclohexanol appears to be the most convenient synthetic method to introduce functional groups into the ε-CL structure, this procedure leads only to 3-substituted lactones. The synthesis of 6-allyl ε-CL **22h** represented an interesting example of ε-CL substituted in a different position of the ring (Scheme 24) (Mecerreyes et al., 2000b). In this case, the starting compound was the commercially available 6-allyl cyclohexanone, which under oxidation conditions provided in addition to the desired allyl derivative a small amount of the relevant epoxide (**23**).

Interestingly, minor formation of the epoxide functional ε-CL **23** was observed after the Bayer–Villiger step, thus indicating the partial instability of the side allyl chain in oxidative experimental conditions (Scheme 24) (Mecerreyes et al., 2000b).

4.2
Post-modification of Functional PCL

Anionic ROP of the described functional ε-CL monomers gave rise to a variety of polyesters of different nature. In many cases, chemical modifications of the original pendant groups were accomplished, thus yielding versatile polyfunctional materials and often complex macromolecular architectures. Furthermore, copolymerization of the functional monomers with ε-CL provided a convenient way of introducing a small amount of reacting sites into the PCL and thereby a means of modulating its ultimate physico-chemical characteristics, similar to that already described for copolymers of PLA (see Sections 2.1 and 3.2).

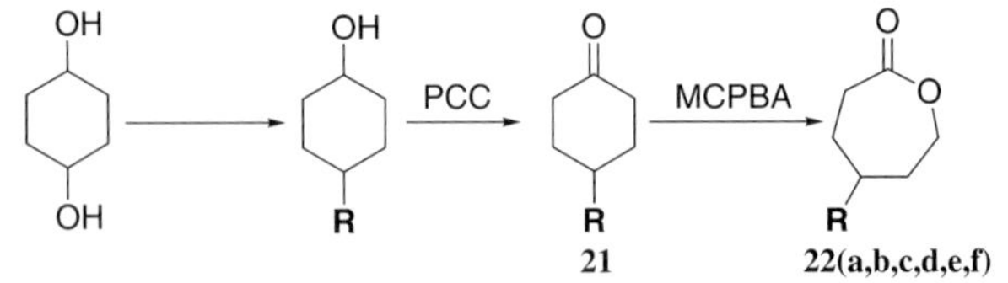

PCC = Pyridinium chlorochromate
MCPBA = *m*-Chloroperbenzoic acid

R	R-εCL	Ref.
Br	22a	[Lecomte et al., 2000; Detrembleur et al., 2000a]
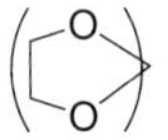	22b	[Tian, 1997a et al.; Lecomte et al., 2000]
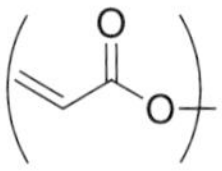	22c	[Mecerreyes et al., 2000a]
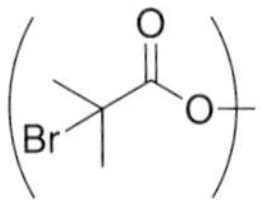	22d	[Mecerreyes et al., 1999a]
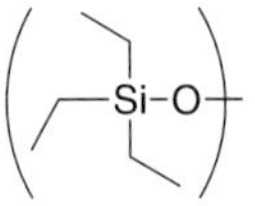	22e	[Lecomte et al., 2000; Stassin et al., 2000]
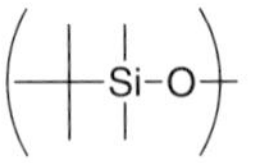	22f	[Pitt et al., 1987]

Scheme 22 Preparation of functional derivatives of ε-caprolactone by the Bayer–Villiger method (Pitt et al., 1987; Tian et al., 1997a; Detrembleur et al., 2000a; Mecerreyes et al., 1999a, 2000a; Lecomte et al., 2000; Stassin et al., 2000).

1,4,8-Trioxaspiro[4,6]-9-undecanone **22b** (TOSUO), was used to prepare homopolymers (Tian et al., 1997a), and also random or block copolymers with ε-CL (Tian et al., 1997b,c). In fact, the living characteristic of the ROP process allowed for the formation of polyesters with narrow molecular weight distribution (Tian et al., 1997a–c, 1998; Lecomte et al., 2000). The ethylene ketal side groups in TOSUO polymers were successfully deacetalized into the corresponding ketones, which were further reduced into hydroxyl groups (Tian et al., 1997a–c, 1998; Lecomte et al., 2000) (Scheme 25). It is worth noting that the diacetalization step was carried out in 100% yield by using the strong electrophile triphenylcarbenium tetrafluoroborate (Barton et al., 1972).

Anionic ROP of **22e** and **22f** provided another means of obtaining hydroxy-functionalized PCL after the removal of the silyl ether protecting groups (Pitt et al., 1987; Tian et al., 1998; Stassin et al., 2000; Lecomte et al., 2000). Actually, the triethylsilyl

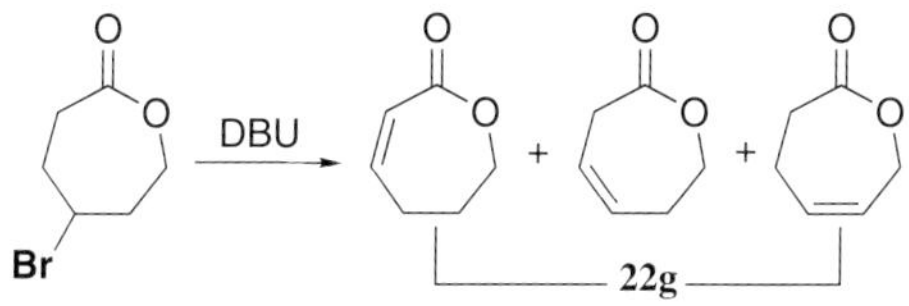

DBU = 1,8 diazabicyclo[5.4.0] undec-7-ene

Scheme 23 Dehydrobromoration of the γ-bromo-substituted ε-caprolactone (Detrembleur et al., 2000b).

ether of **22e** was preferred for its greater reactivity towards hydrolytic cleavage, which for instance allowed selective deprotection of the hydroxyl function in a terpolyester of **22e**, ε-CL, and TOSUO (Scheme 26) (Stassin et al., 2000).

The anionic ROP of the bromine-carrying lactone **22a** initiated by aluminum alkoxides was found to proceed according to a living mechanism (Detrembleur et al., 2000a; Lecomte et al., 2000). Accordingly, some narrowly dispersed random or block copolymers of **22a** with ε-CL were prepared, besides **22a** homopolymer (Detrembleur et al., 2000a). The reactivity of the γ-bromo substituent in the copolymers was then exploited to perform a series of chemical modifications. For example, a polycationic PCL was made available by reaction with nucleophilic pyridine (Scheme 27) (Detrembleur et al., 2000b; Lecomte et al., 2000). After 48 h, about 90% of the bromine groups were

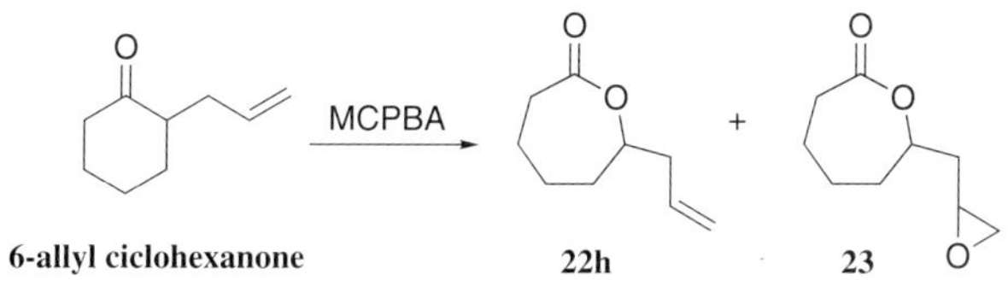

Scheme 24 Preparation of 3-allyl ε-caprolactone (Mecerreyes et al., 2000b).

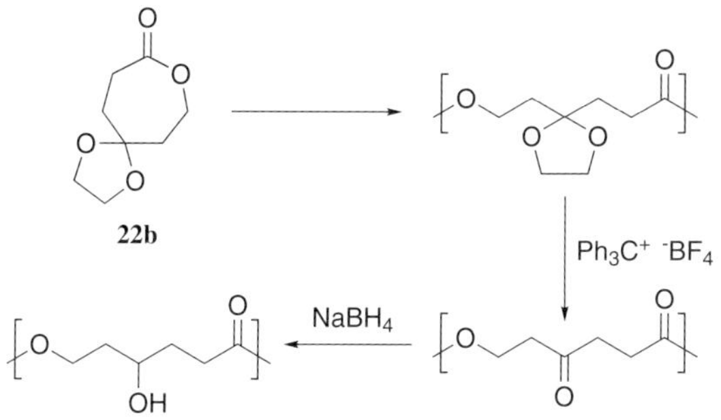

Scheme 25 Preparation and deacetalization of TESUO-based polymers and copolymers (Tian et al., 1997a–c, 1998; Lecomte et al., 2000).

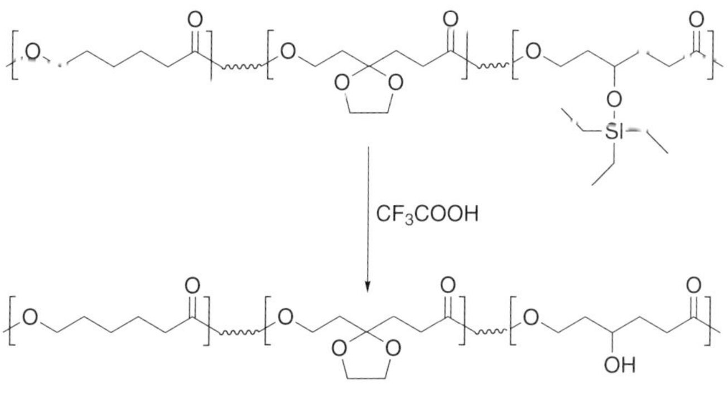

Scheme 26 Preparation and selective deprotection of poly(ε-caprolactone) bearing differently protected hydroxyl groups (Stassin et al., 2000).

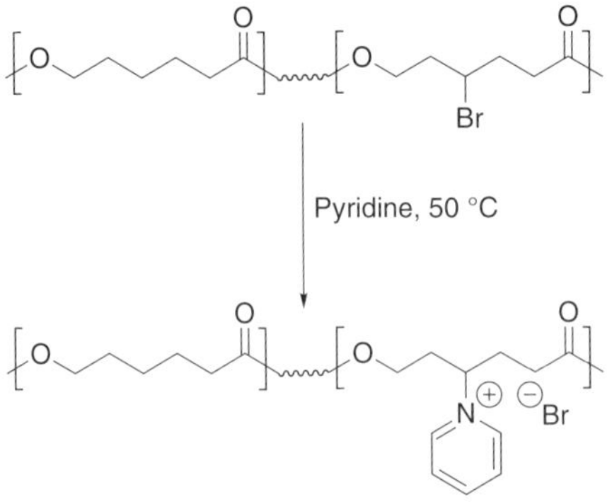

Scheme 27 Preparation of a polycationic form of poly(ε-caprolactone) (Detrembleur et al., 2000b; Lecomte et al., 2000).

displaced, and no unsaturated groups in the main chain were detected, indicating that the elimination reaction did not occur (Detrembleur et al., 2000b; Lecomte et al., 2000). Furthermore, no degradation of the polymer chains was detected. This functionalization may open the way to the synthesis of water-soluble PCL and amphiphilic block copolymers.

The elimination of HBr by means of 1,8-diazabicyclo[5,4,0] undec-7-ene (DBU) allowed for the formation of unsaturated bonds in the main chain of copolymers of **22a** with ε-CL (Detrembleur et al., 2000b; Lecomte et al., 2000) (Scheme 28). As in the case of dehydrobromination of **22a** (Section 4.1), the regioselectivity of the elimination reaction was limited, and a mixture of nonconjugated and conjugated olefinic chain units was generated. Nonetheless, no chain degradation occurred during the process. The alkene linkages were successively oxidized by *m*-chloroperbenzoic acid to form epoxide rings, thus providing highly reactive sites for further functionalization of the polyester backbone (Detrembleur et al., 2000b; Lecomte et al., 2000).

However, only the unsaturated units not conjugated to the carbonyl groups could be transformed into the respective oxiranes. Interestingly, a polyester with double bonds in the main chain was prepared also by anionic ROP of the unsaturated lactone

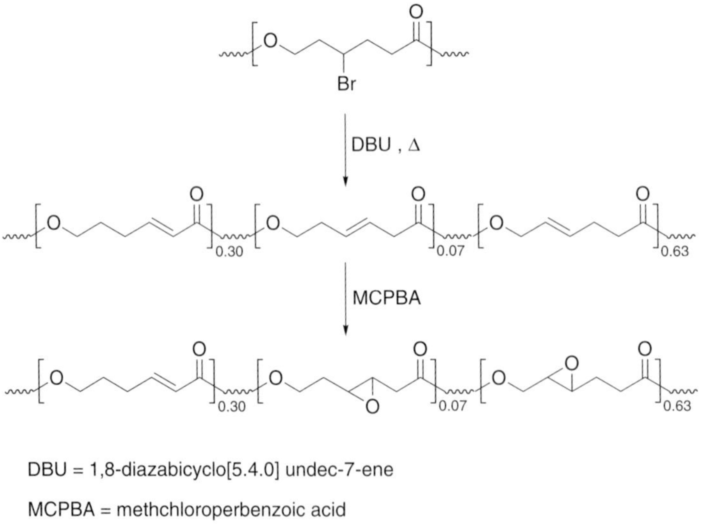

DBU = 1,8-diazabicyclo[5.4.0] undec-7-ene

MCPBA = methchloroperbenzoic acid

Scheme 28 Preparation of poly(ε-caprolactone) with alkene linkages and epoxy groups in the main chain by dehydrobromuration of bromine-bearing poly(ε-caprolactone) followed by peroxidation (Detrembleur et al., 2000b; Lecomte et al., 2000).

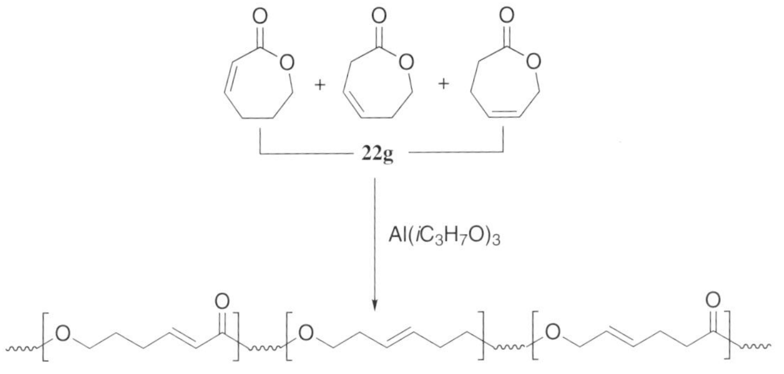

Scheme 29 Preparation of poly(ε-caprolactone) with alkene linkages in the main chain by polymerization of unsaturated ε-caprolactone monomers (Detrembleur et al., 2000b).

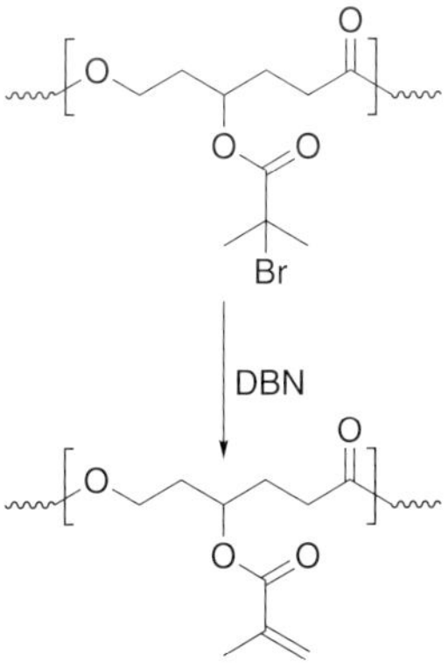

DBN = 1,5-diazabicyclo[4.3.0]non-5-ene

Scheme 30 Preparation of poly(ε-caprolactone) bearing methacrylic pendant groups (Detrembleur et al., 2000b).

mixture **22g** (Scheme 29) (Detrembleur et al., 2000b).

Copolyesters of the bromine-substituted lactone **22d** with ε-CL were subjected to the same chemical transformations as for **22a**-based polymers (Detrembleur et al., 2000b). The dehydrohalogenation of the tertiary alkyl bromide groups was easily carried out in the presence of the non-nucleophilic basic moiety 1,5-diazabicyclo[4.3.0]non-5-ene (DBN), and this led unequivocally to the formation of methacrylic unsaturations (Scheme 30) (Detrembleur et al., 2000b).

It is worth pointing out that the lateral methacrylic groups of these polymers can be extremely useful for a variety of chemical transformations, including free radical cross-linking. A certain amount of unsaturated methacrylic groups also formed upon nucleophilic substitution of the pendant alkyl bromide by pyridine (Scheme 31) (Detrembleur et al., 2000b). Indeed, lower degrees of final quaternization (50–60%) were achieved compared with **22a**-based copolyesters (90%), this being due to the increased reactivity toward the base-promoted elimination reaction (Detrembleur et al., 2000b).

6-Allyl-ε-caprolactone **22h** was successfully converted by anionic ROP in the corresponding homopolymer and in copolymers with either ε-CL or PLA (Mecerreyes et al., 2000b). Three different chemical modifications of the allyl pendant group in the **22h**-(ε-CL) copolyester were subsequently investigated (Mecerreyes et al., 2000b) (Scheme 32). In the first case (Scheme 32/I), the double bonds were transformed into oxirane groups by the oxidative action of metachloroperbenzoic acid. Another possibility was represented by bromination of the alkene linkage, which was found to proceed in quantitative yields and without affecting the length of the backbone chain (Mecerreyes et al., 2000b) (Scheme 32/II). The final

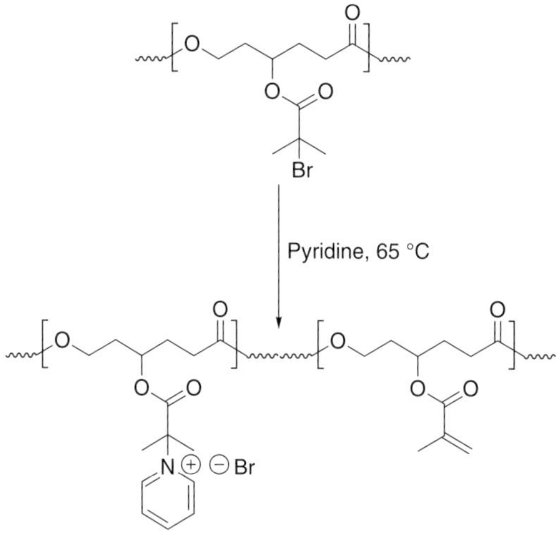

Scheme 31 Preparation of a cationic poly(ε-caprolactone) by reaction of pyridine with the pendant 2-bromo-2-methyl-propyl group (Detrembleur et al., 2000b).

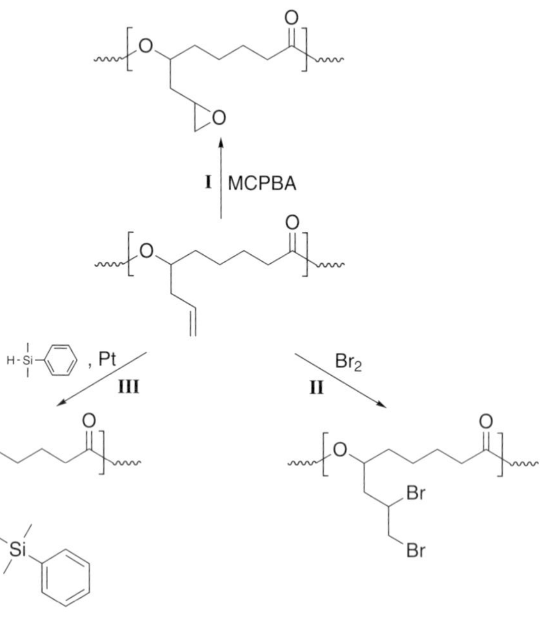

Scheme 32 Conversion of allyl-bearing poly(ε-caprolactone) into different functional polymers (Mecerreyes et al., 2000b).

modification performed on the **22h**-based polyester involved the hydrosilylation of the allyl lateral group (Scheme 32/III) (Mecerreyes et al., 2000b). In fact, hydrosilylation may represent a simple and quantitative way to introduce functionalities into the polymer, bearing in mind the variety of functional hydrosilanes available.

5
Functional Poly(ester amide)s

Polydepsipeptides are copolymers of α-hydroxy acids and α-amino acids which belong to the class of poly(ester amide)s. A convenient procedure for the preparation of polydepsipeptides is represented by the ROP of

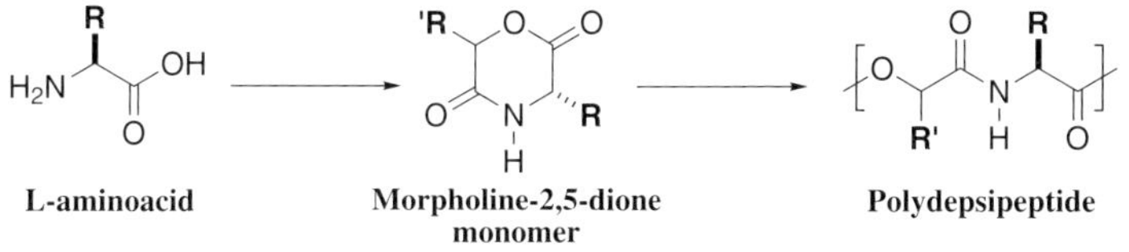

Scheme 33 Preparation of poly(depsipeptide)s by ring-opening polymerization of morpholine-2,5-dione derivatives (in't Veld et al., 1990).

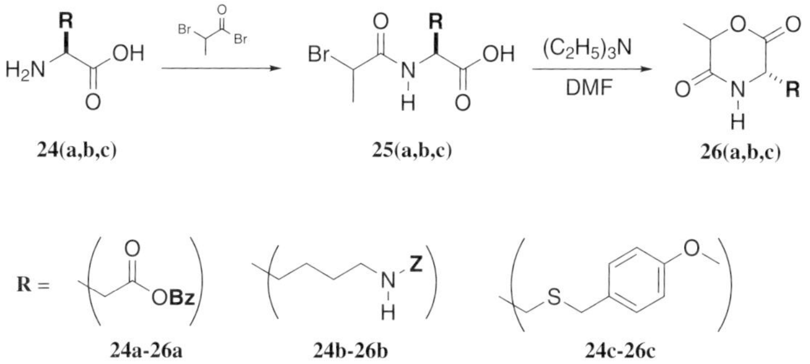

Scheme 34 Preparation of functional morpholine-2,5-dione monomers (in't Veld et al., 1992b).

morpholine-2,5-dione derivatives, which are in turn synthesized from readily available amino acids (in't Veld et al., 1990) (Scheme 33). In addition, copolymerization of morpholine-2,5-dione derivatives with L-dilactide or ε-CL provided a series of biodegradable poly(esteramide)s with a controllable content of ester linkages in the chain and a wide range of physico-chemical properties (in't Veld et al., 1992a). Therefore, the synthesis of functionalized morpholine-2,5-dione monomers and their successive ROP, either alone or with anionically polymerizable cyclic esters or diesters, was investigated as a promising route for the preparation of functional polyesters (in't Veld et al., 1992b).

The synthesis of morpholine-2,5-dione derivatives with functional substituents is outlined in Scheme 34. The trifunctional amino acids L-aspartic acid, L-lysine, and L-cysteine were considered as starting building blocks to introduce respectively carboxyl, amino, and thiol functionalities into the

polyester structure (in't Veld et al., 1992b). However, these functional groups were preliminarily protected, in order to avoid side reactions during the monomer synthesis and polymerization. As for the syntheses of poly(malic acid ester)s and poly(serine ester)s, two groups that are easily clearable by catalytic hydrogenolysis (benzyl and Z) were chosen to protect the γ-carboxyl function of L-aspartic acid polyesters and the ε-amino function of lysine polyesters (in't Veld et al., 1992b). The selection of the p-methoxybenzyl group, as the most suitable and protective group for the thiol function of L-cysteine, proved to be less straightforward (in't Veld et al., 1992b). Indeed, the vast majority of thiol protective groups are unstable in basic or acid conditions; moreover, the cleavage procedure for their removal can affect the ester bonds in the polymer chain.

The first step in the monomer synthesis (Scheme 34) for functional polyesters, involved the reaction of the protected amino acids **24(a,b,c)** with (*R,S*)-bromopropionyl bromide under Schotten–Bauman conditions, to give the N-[(*R,S*)-bromopropionyl]-amino acids **25(a,b,c)** in 54–60% yields.

The morpholine-2,5-dione derivatives **26(a,b,c)** were then obtained by treatment of **25(a,b,c)** with triethylamine in DMF. The fairly high stability of the protective groups of **25(a,b)** allowed the last reaction to be carried out at 100 °C in 3 h, whereas cyclization of **25c** had to be performed at room temperature in order to avoid the formation of side products. Final yields ranged from 30 to 55%. The cyclization step afforded the morpholine-2,5-dione derivatives **26(a,b,c)** as a mixture of (3*S*,6*R*) and (3*S*,6*S*) diastereoisomers, which could not be separated by common isolation procedures for polyesters (in't Veld et al., 1992b). Therefore, the pure (3*S*,6*S*) diastereoisomer of **26a** was prepared by a further synthetic strategy (Scheme 35) which involved ring-closure concomitant with the amide bond formation polyesters (in't Veld et al., 1992b).

In this scheme, β-benzyl N-**BOC**-L-aspartate **27** was coupled with L-lactic acid pentamethylbenzyl ester, by using DCC as condensation agent, to give the intermediate **28**. Then, both the **BOC** and the pentamethylbenzyl (**PMB**) groups were removed by treatment of **28** with HBr and anisole. Cyclization of the derivative **29** was then accomplished by using 1-hydroxybenzotriazole (HOBt) and DCC in the presence of triethylamine. The optically pure (3*S*,6*S*)-**26a** was obtained in 7% overall yield (in't Veld et al., 1992b).

Anionic ROP of the morpholine-2,5-dione derivatives **26(a,b,c)** initiated by stannous octoate failed to give measurable quantities of homopolymers, because of the very low reactivity of the monomers (in't Veld et al., 1992b). In contrast, random copolymerization of the above functional morpholine derivatives with ε-CL and (D,L)-dilactide

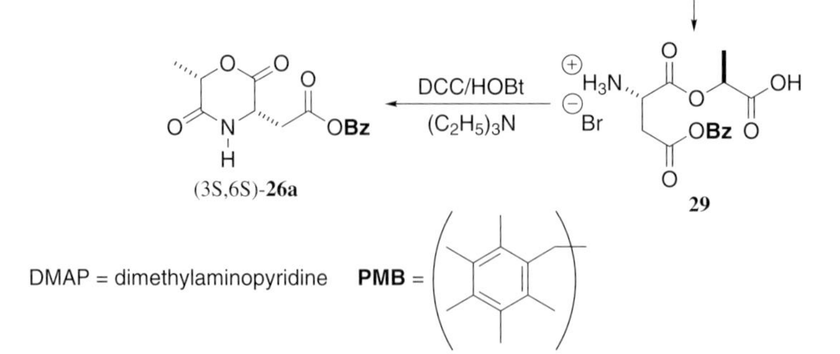

Scheme 35 Preparation of the pure (3*S*,6*S*)-diastereoisomer of 6-methylcarboxybenzyl-morpholine-2,5-dione (in't Veld et al., 1992b).

afforded functional poly(ester-amide)s in good yields (Scheme 36) (in't Veld et al., 1992b).

The minor feed contents of morpholine-2,5-dione were always set (5–20% in molar terms) in order to maintain an adequate reactivity of the monomeric mixture towards polymerization. The benzyl and **Z** protective groups present in the copolymers of **26(a,b)** were completely removed by hydrogenolysis, without affecting the ester and amide bonds present in the polymer backbone. Instead, removal of the *p*-methoxybenzyl thiol-protecting group by the use of trifluoromethanesulfonic acid occurred with severe degradation at the polymer backbone, according to the onset of an acidolysis cleavage of the ester bond (in't Veld et al., 1992b).

6
Functional Polyesters by Miscellaneous Polymerization Methods

6.1
Synthesis of Functional Polyesters by Michael-type Reaction

Michael-type polyaddition reactions between α,ω-bisacrylates and either a secondary amine (Galli et al., 1983; Laus et al., 1988; Angeloni et al., 1985; Chiellini and Solaro, 1992, 1993) or a bisthiol (Chiellini et al., 1987, 1990a; Galli et al., 1989, 1991; Angeloni et al., 1990; Laus et al., 1990; Chiellini and Solaro, 1992, 1993) allowed the preparation of a wide series of functional polyesters containing (in the main chain) amino or thio groups in the β-position with respect to the carboxyl group (Scheme 37). The polymerization reactions could be carried out in anhydrous solvents at room temperature. Triethylamine was however required as catalyst in order to obtain complete conver-

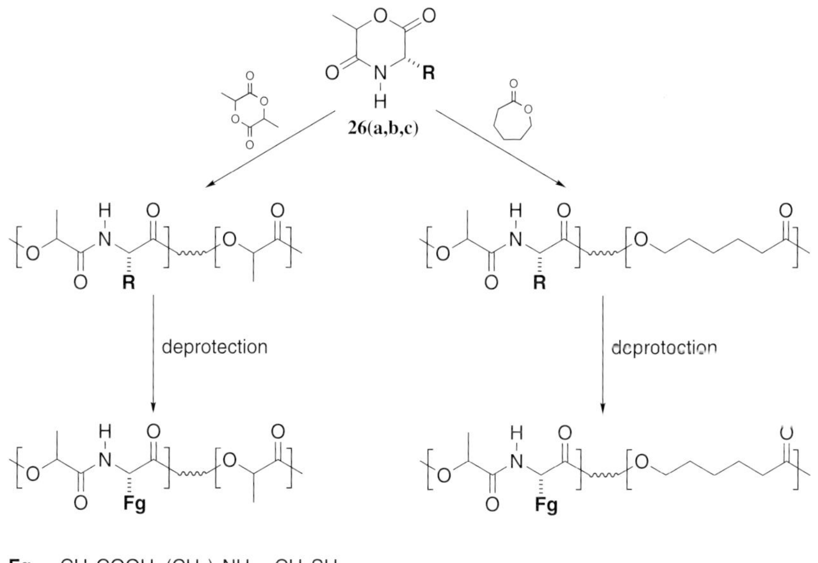

Fg = -CH₂COOH, (CH₂)₄NH₂ , CH₂SH

Scheme 36   Preparation of functional poly(ester amide)s by polymerization of functional morpholine-2,5-dione monomers (in't Veld et al., 1992b).

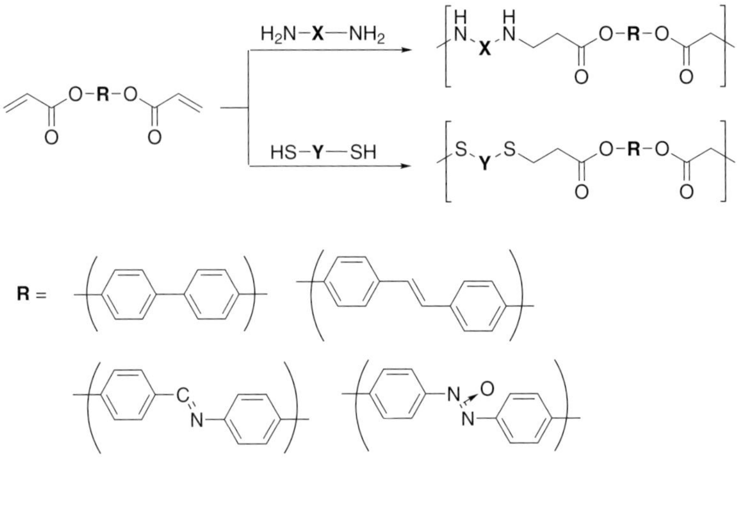

X, Y = spacers of different chemical structure

Scheme 37 Synthesis of functional polyesters by Michael-type polyaddition reactions between α,ω-bisacrylates and a secondary amine or a bisthiol (Chiellini and Solaro, 1992).

sion when bisthiol monomeric precursors were used. It is worth noting that all the prepared poly(ester-β-amine)s were found to be thermally and hydrolytically unstable, whereas the poly(ester-β-sulfide) class displayed fairly good thermal and hydrolytic stability (Chiellini et al., 1987, 1990a; Galli et al., 1989, 1991; Angeloni et al., 1990; Laus et al., 1990; Chiellini and Solaro, 1992, 1993). In addition, some of the prepared poly(ester-β-sulfide) was converted to the corresponding chiral poly(ester-β-sulfoxide)s by asymmetric induction reaction performed in the presence of a chiral oxidant (Angeloni et al., 1990; Chiellini et al., 1990b).

6.2
Synthesis of Functional Polyesters from Protected Polyhydroxylated Monomeric Precursors

A short and straightforward route to produce functional polyesters bearing hydroxyl groups was provided by standard melting

polymerization of D-tartaric acid with alkylene diols containing two, six, eight, 10, and 12 methylene groups (Scheme 38) (Bitritto et al., 1979). As expected, these polyhydroxylated polyesters displayed considerably enhanced thermal properties compared with those of the corresponding wholly aliphatic structures, this being due to the extended hydrogen bonding networks generated by

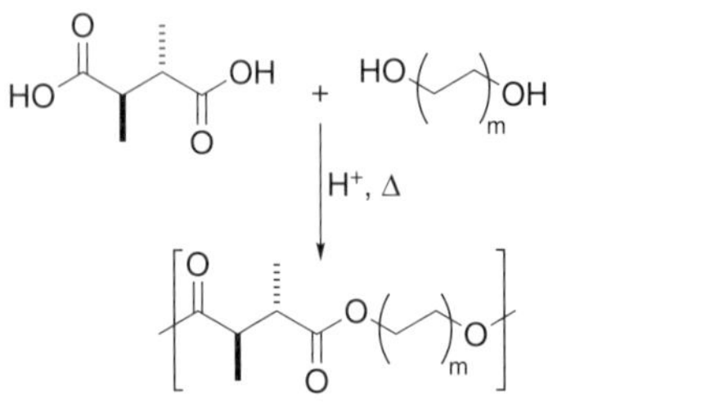

$m = 1, 3, 4, 5, 6$

Scheme 38 Synthesis of functional polyhydroxylated polyesters by melting polymerization of D-tartaric acid with alkylene diols of different length (Bitritto et al., 1979).

the pendant hydroxyl groups. In addition, the mechanical and physico-chemical characteristics of these materials could be further improved through the reaction of their free hydroxyl groups with aliphatic diisocyanate derivatives. All the synthesized poly(alkylene tartrate)s were shown to be fully biodegradable by experiments involving *Aspergillus niger* active principles (Bitritto et al., 1979). Interestingly, poly(alkylene tartrate)s proved to be useful materials for the formulation of microscale (μm to nm) devices for the controlled release of drugs (Di Benedetto and Huang, 1989, 1994).

Stereoregular polyesters containing two or four hydroxyl groups per repeating unit were prepared by polymerization of a recognized powerful mesogenic molecule bearing two reactive acid chloride groups with protected polyol derivatives obtained from readily available natural compounds (Scheme 39) (Chiellini et al., 1990c; Chiellini and Solaro, 1993, 1995). The polymerization reactions were carried out in 1,2-dichloroethane solution in the presence of pyridine as hydro-

chloric acid acceptor. After polymerization, the controlled acidic removal of the *iso*-propylidene protective groups allowed for the preparation of functional polyhydroxy-lated mesophasic polyesters with tunable physico-chemical characteristics.

Rational sequences of protection–deprotection synthetic steps afforded a large class of racemic optically active selectively monoprotected glyceric acid monomeric precursors (Scheme 40) which were subsequently used in the preparation of hydroxyl-protected functional polyesters (Chiellini et al., 1990d; Chiellini and Solaro, 1992).

Polymerization of the monoprotected glyceric acids was performed according to a Fisher-type self-condensation procedure in the presence of *p*-toluenesulfonic acid as catalyst, whereas the monoprotected methyl glycerate derivatives were polymerized by trans-esterification reaction catalyzed by titanium tetrabutoxide (Chiellini and Solaro, 1992, 1993) (Scheme 41). Removal of the hydroxyl-protective groups yielded water-soluble poly(α-glycerate)s and poly(β-glycerate)s.

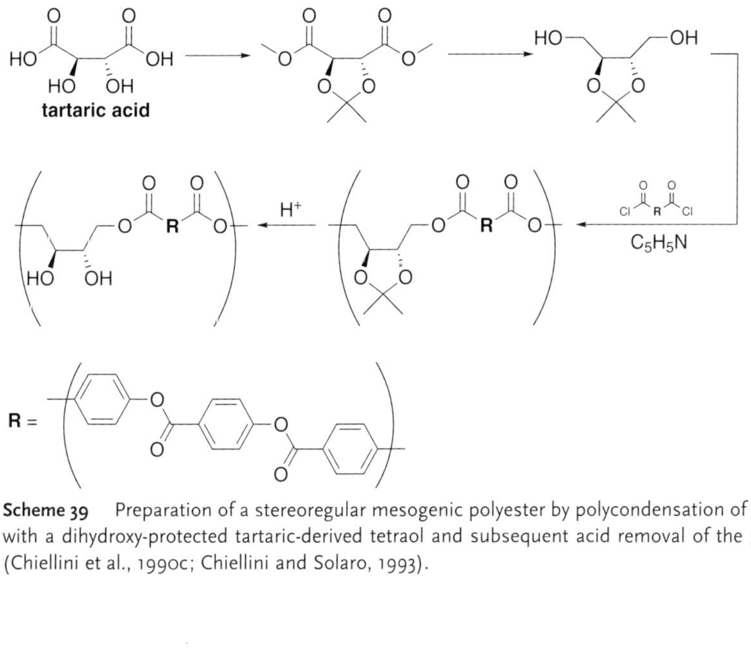

Scheme 39 Preparation of a stereoregular mesogenic polyester by polycondensation of an acid dichloride with a dihydroxy-protected tartaric-derived tetraol and subsequent acid removal of the protecting groups (Chiellini et al., 1990c; Chiellini and Solaro, 1993).

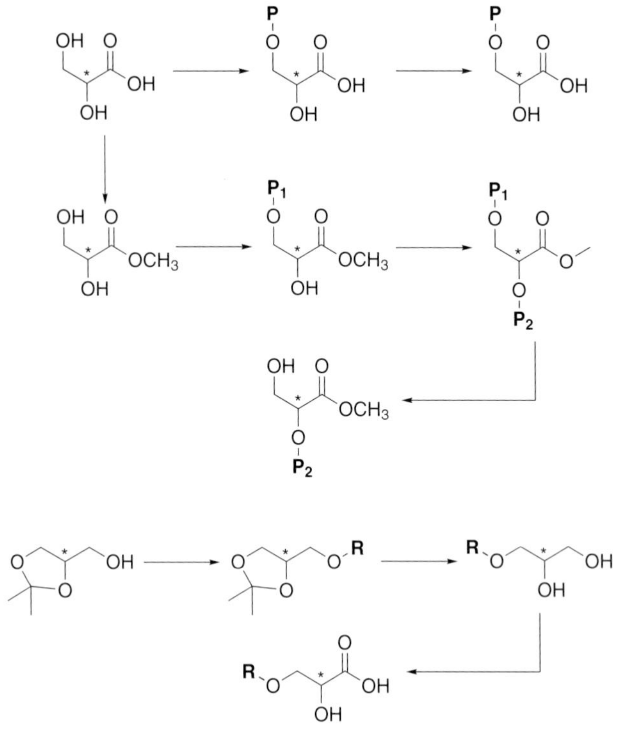

Scheme 40 Preparation of racemic or optically active selectively monoprotected glyceric acid monomeric precursors by sequences of protection–deprotection synthetic steps (Chiellini and Solaro, 1992, 1993).

P   = C$_2$H$_5$-, C$_6$H$_5$CH$_2$-, (C$_6$H$_5$)$_3$C-, THP, CH$_3$C$_6$H$_4$SO$_2$-

P$_1$ = (C$_6$H$_5$)$_3$C-   P$_2$ = C$_6$H$_5$CH$_2$-   R = CH$_3$-, C$_6$H$_5$CH$_2$-

Interestingly, these functional polyesters displayed a high rate of hydrolytic degradation, and hence they seem well suited to provide functional building blocks for the preparation of biodegradable materials tailored towards specific biomedical applications (Chiellini and Solaro, 1993).

Alternating ROP of epoxides and cyclic anhydrides provided a convenient synthetic route to functional polyesters (Solaro et al., 1991). Indeed, functional polyesters containing two or more side hydroxyl groups were obtained from the copolymerization of succinic anhydride with glycidyl ethers of protected polyols (Chiellini and Solaro, 1992, 1993) (Scheme 42). The glycidyl ether

derivatives resulted from the *iso*-propylidenic protection of alditols having an odd number of hydroxyl groups such as glycerol, xylitol, and arabitol, followed by a monoglycidylation reaction with epichloridrine under basic conditions.

Almost quantitative conversions to polymeric compounds were attained when the polymerization reaction was initiated by organoaluminum derivatives such as triisobutyl aluminum and tetra-*iso*-butyltetraalumoxane (Chiellini and Solaro, 1993). The protecting *iso*-propylidene groups could be removed either partially or completely under mild conditions, without appreciably affecting the degree of chain polymerization. Also

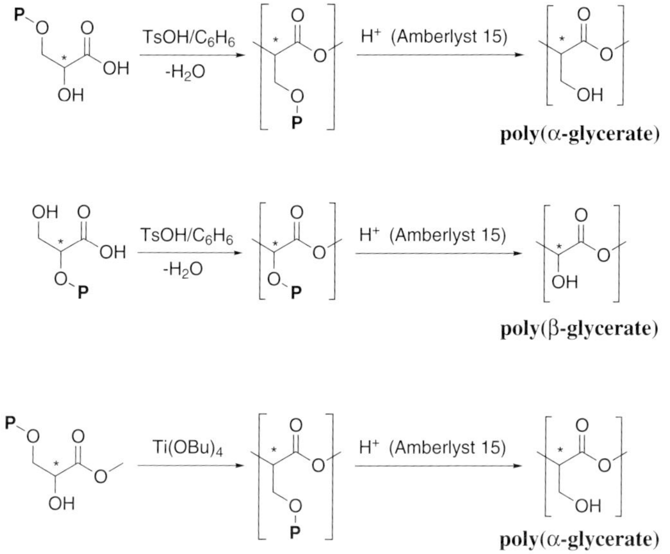

P = C$_2$H$_5$-, (C$_6$H$_5$)$_3$C-, THP, CH$_3$C$_6$H$_4$SO$_2$-

Scheme 41 Polymerization of monoprotected glyceric acid precursors by Fischer reaction of trans-esterification, and subsequent acidic removal of the protective groups (Chiellini and Solaro, 1992).

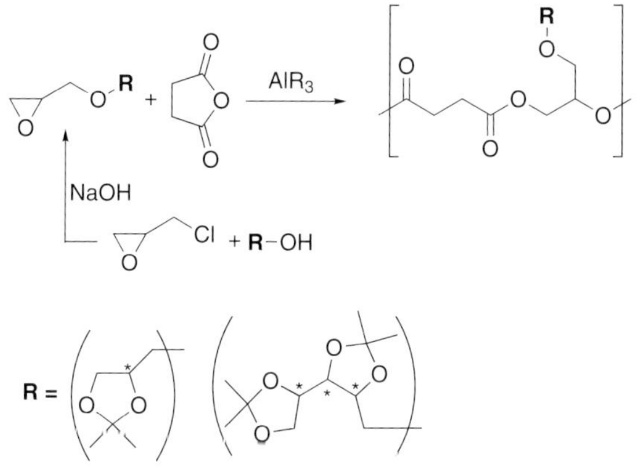

Scheme 42 Preparation of functional polyesters containing two or more side hydroxyl group by alternating ring-opening polymerization of alditol-based glycydyl ethers and succinic anhydride (Chiellini and Solaro, 1992, 1993).

in this case, selective modulation of the deprotection extent allowed the successful preparation of a large variety of functional polyester materials with rheological properties that could be modulated on chemical grounds (Chiellini and Solaro, 1992).

7
Telechelic Polyesters

7.1
Role of Polymerization Mechanism

So far, only the synthesis of polyesters bearing functional groups along the macromolecular chain has been discussed extensively. The incorporation of functional groups at the terminal positions of the polymeric chain (telechelic polyesters) constitutes an alternate option which has proved very attractive. Functionally end-capped telechelic polyesters can participate in a variety of chain extension, branching, and cross-linking reactions (Quirk and Kim, 1993). Therefore, telechelic polyesters were used for the production of complex macromolecular architectures, and are among the preferred starting materials for the engineering of potentially bioerodible and biodegradable macromolecular materials (Goethals, 1993).

The choice of polymerization conditions represents the essential parameter for the synthesis of telechelic polyesters. Indeed, a strict control on chain formation and growth must be achieved, in order to provide the final macromolecular structure with the selected functional end-groups. Many commonly used methods, such as polycondensation or transesterification, do not allow for selective introduction into the chain terminus of suitable functional groups. In fact, only living or "controlled" polymerization mechanisms are compatible with the production of telechelic polymers, and the ROP of lactones and lactides proved to be the best-suited procedure for the preparation of telechelic polyesters (Goethals, 1993).

There are two principal methods to introduce functional end-groups onto polyesters, namely functional initiation, and end-capping. Functional initiation requires an initiator containing a functional group that does not interfere with the polymerization mechanism. If such a functional group is capable of undergoing a different type of polymerization, the final telechelics are better referred to as "macromonomers". End-capping can, in principle, be used only if the polymerization is of the "living" type (Goethals, 1993). The introduction of the

functional group relies on the ability of the growing chain to react with an externally added compound, with concomitant irreversible termination of the polymerization process.

The following sections aim to discuss the most convenient strategies for the synthesis of telechelic polyesters, and are classified according to the polymerization mechanism.

7.2
Anionic ROP

The anionic polymerization of lactones is usually complicated by chain transfer reactions, such as inter- and intramolecular transesterification, which do not allow for strict control of the growing chain structure (Lofgren et al., 1995). For example, polymerization of ε-CL initiated by most alkoxide salts gives rise to a mixture of linear and cyclic polymeric materials and broad molecular weight distributions. Only slightly better results are obtained for the anionic ROP of β-lactones by using weak nucleophilic initiators.

ROP of β-propiolactone and β-butyrolactone by using potassium hydride or naphthalenide complexes with crown ethers as initiators proved to be an interesting example of chain end functionalization with simultaneous accurate control of the macromolecular properties (Jedlinski et al., 1992; Kurcok et al., 1995) (Scheme 43).

In this case, the real initiator of polymerization was shown to be the conjugated carboxylate complex **30**, which was originated upon the abstraction of one α-proton from the monomer and subsequent ring-opening of the β-lactone structure. Polyester macromonomers with narrow polydispersity and possessing a readily polymerizable acrylic end-group, were prepared by this procedure (Kurcok et al., 1995).

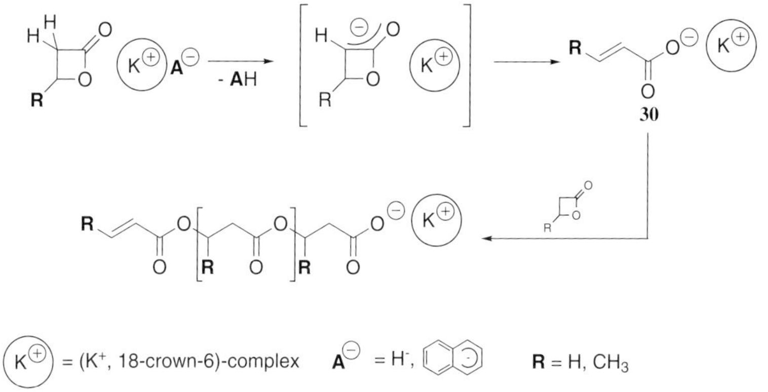

$K^{\oplus}$ = (K⁺, 18-crown-6)-complex $A^{\ominus}$ = H⁻, naphthalene   R = H, CH₃

Scheme 43 Preparation of acryloxy-terminated poly(β-lactone) macromonomers (Jedlinski et al., 1992; Kurcok et al., 1995).

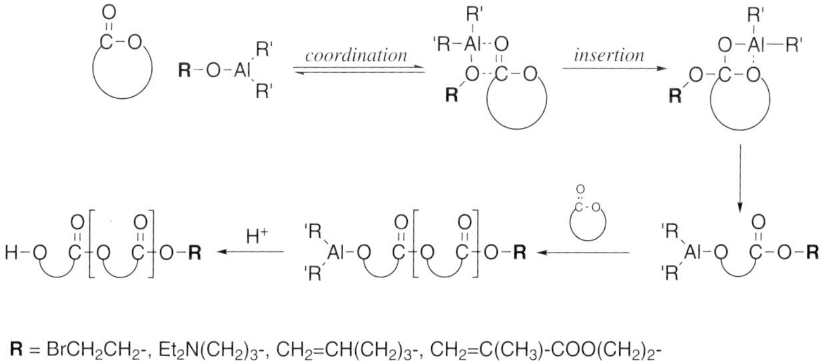

R = BrCH₂CH₂-, Et₂N(CH₂)₃-, CH₂=CH(CH₂)₃-, CH₂=C(CH₃)-COO(CH₂)₂-

Scheme 44 Preparation of semitelechelic polylactones by coordinative ring-opening polymerization with functional aluminum alkoxides as polymerization initiators (Dubois et al., 1989; Barakat et al., 1993).

7.3
Coordinative ROP

In 1962, Chedron and coworkers were the first to show that some Lewis acids are effective initiators of lactone polymerization (Chedron et al., 1962). Since then, many other initiators have been discovered, and the nature of the polymerization mechanism, indicated as coordinative ROP, was thoroughly investigated (Lofgren et al., 1995). Unlike anionic ROP of lactones, which is apparently not easy to control, coordinative ROP was reported to proceed according to a living mechanism and with only minor reversible side reactions – par-

ticularly when aluminum alkoxides are used as initiators. The kinetic control of the polymerization process is commonly attributed to the much lower reactivity of the initiator species compared with ordinary anionic ROP systems.

7.3.1
Coordinative ROP Initiated by Aluminum Alkoxides

When aluminum alkoxides are used as initiators, the polymerization of lactones and lactides proceeds through a general "coordination–insertion" mechanism, which involves insertion of the monomer into the Al–O bond of the initiator (Scheme 44)

(Lofgren et al., 1995; Mecerreyes and Jerome, 1999).

The acyl–oxygen bond of the cyclic monomer is cleaved in such a way that the growing chain remains attached to the aluminum atom by an alkoxide bond. After monomer consumption, the final polyester is obtained by acid hydrolysis of the active aluminum alkoxide bond (Scheme 43). This process leads to the formation of one hydroxyl endgroup, and to introduction of the alkyl radical of the initiator in the ester group at the other chain end. The coordination–insertion mechanism was used to produce several semi-telechelic polyesters by using aluminum alkoxide initiators of general molecular structure $(C_2H_5)_{(3-p)}$ $Al(OCH_2CH_2X)_p$, where X represents an appropriate functional group, such as bromine, tertiary amine, allyl, or methacryloyloxy (Scheme 44) (Dubois et al., 1989, 1991a,b; Barakat et al., 1993; Eguiburu et al., 1995; Mecerreyes et al., 1999b). The substituted initiators were prepared by reaction of triethylaluminum with an equimolar amount of an alcohol derivative (Dubois et al., 1989, 1991a,b; Barakat et al., 1993);

this reaction is clearly favored by the elimination of ethane (Scheme 45, upper part) (Dubois et al., 1991b).

When a dialcohol was utilized for this reaction, a difunctional aluminum alkoxide was obtained; this initiator provided a convenient route to dihydroxy end-capped polyesters (Scheme 45: lower part) (Sosnowski et al., 1991; Dubois et al., 1993). The incorporated end-groups were in some cases submitted for further chemical modification. For example, the two terminal hydroxyl groups of α,ω-dihydroxy-poly(ε-CL) were either exploited for chain extension by using a difunctional aromatic diisocyanate (Dubois et al., 1991b), esterified with chromophores for physico-chemical studies on the polymeric structure (Sosnowski et al., 1991), or converted into carboxyl groups by reaction with succinic anhydride (Dubois et al., 1993). The bromine end-group of α-hydroxy-ω-bromododecanoyl-poly(ε-CL) was converted in two steps in a primary amine function (Degee et al., 1992a), which proved effective as initiator for the ROP of N-carboxyanhydrides (NCA), with formation of poly(ε-CL-b-peptide) copolymers (Scheme

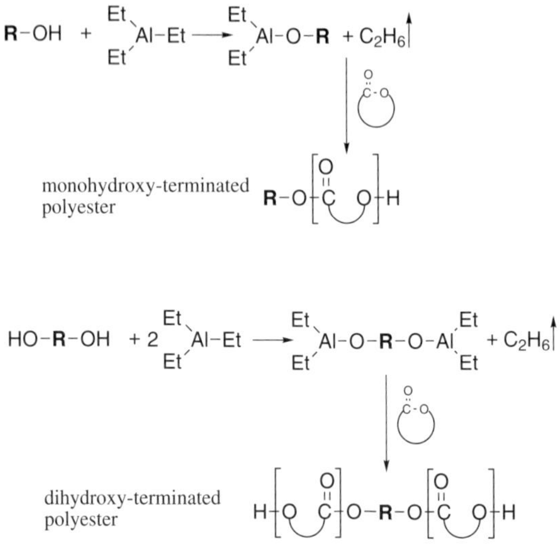

Scheme 45 Preparation of functional aluminum alkoxide initiators in monomer and dimer forms (Dubois et al., 1991b).

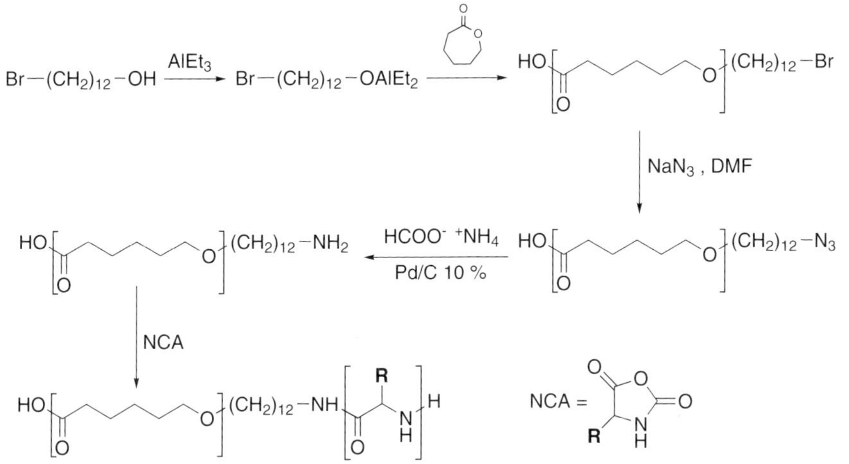

$$Br{-}(CH_2)_{12}{-}OH \xrightarrow{AlEt_3} Br{-}(CH_2)_{12}{-}OAlEt_2 \longrightarrow HO{-}[\ldots]{-}O{-}(CH_2)_{12}{-}Br$$

NaN$_3$, DMF

$$HO{-}[\ldots]{-}O{-}(CH_2)_{12}{-}NH_2 \xleftarrow[\text{Pd/C 10 \%}]{\text{HCOO}^- {}^+NH_4} HO{-}[\ldots]{-}O{-}(CH_2)_{12}{-}N_3$$

NCA

$$HO{-}[\ldots]{-}O{-}(CH_2)_{12}{-}NH{-}[\ldots]{-}H$$

$$NCA = $$

Scheme 46 Preparation of poly(ε-caprolactone) chain terminated with a primary amino group and its use as initiator for the ring-opening polymerization of N-carboxyanhydride derivatives.

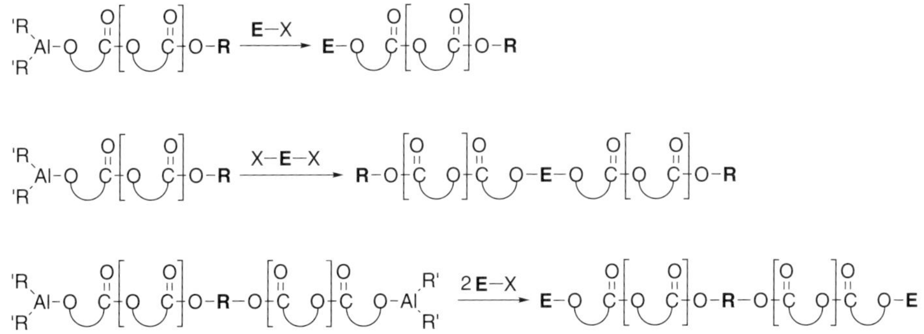

$${}'R{-}Al{-}O{-}[\ldots]{-}O{-}R \xrightarrow{E-X} E{-}O{-}[\ldots]{-}O{-}R$$

$${}'R{-}Al{-}O{-}[\ldots]{-}O{-}R \xrightarrow{X-E-X} R{-}O{-}[\ldots]{-}O{-}E{-}O{-}[\ldots]{-}O{-}R$$

$${}'R{-}Al{-}O{-}[\ldots]{-}O{-}R{-}O{-}[\ldots]{-}O{-}Al{-}R' \xrightarrow{2E-X} E{-}O{-}[\ldots]{-}O{-}R{-}O{-}[\ldots]{-}O{-}E$$

R = functional alkyl group, **E** = electrophilic group

Scheme 47 End-capping of the alkoxide-propagating species with electrophilic derivatives (Dubois et al., 1997).

46) (Degée et al., 1992b). When diethylaluminum benzoate was used as coordinative ROP initiator of ε-CL, a benzyl ester function was introduced into one chain extremity, after which selective cleavage by hydrogenolysis occurred to produce a carboxyl-terminated PCL (Trollsas et al., 1998).

Aluminum alkoxide initiators may allow also for selective chain end-capping with functional compounds. Indeed, the nucleophilic aluminum alkoxide propagating species can be readily reacted with electrophilic derivatives, such as acid chlorides, anhydrides and isocyanates (Scheme 47) (Dubois et al., 1997).

In fact, the simultaneous control of the initiation and termination steps was recognized at an early stage as being an ideal strategy to prepare symmetric telechelic polyesters (Dubois et al., 1994). For instance, the combination of a functional initiator with an effective coupling agent of two propagating chains, such as terephthaloyl chloride, was shown to yield quantitatively α,ω-functional telechelic polyesters with a concomitant two-fold increase in molecular weight (Dubois et al., 1991a, 1994). Another important example of end-capping of the propagating chain was provided by the preparation of α,ω-dimethacryloyl-PCL, a polymeric structure which is used for thermal or photochemical crosslinking processes (Dubois et al., 1991b; Barakat et al., 1999).

As anticipated (see Section 7.1), functional telechelic polyesters can be potentially employed for the realization of many complex multibranched macromolecular architectures. Coordinative ROP initiated by aluminum alkoxide derivatives has proved to be a powerful tool for modulating the polyester structure characteristics. Some examples are worth noting, since the nature of the incorporated functional group into the polyester plays a decisive role, regardless of its actual position along the chain backbone. The terminal methacrylate groups of PLA and PCL chains have been used in the production of polyester-poly(methacrylate) branched block copolymers and crosslinked gels (Scheme 48) (Barakat et al., 1994, 1996, 1999).

α-Norbornenyl-PCL macromonomers were converted to hyperbranched structures by ring-opening methathesis polymerization of the norbornene end-group (Mecerreyes et al., 1999b). The pendant hydroxyl groups of TOSUO homopolymers and copolymers (see Section 4.2), as well as those derived by deprotection of silyl ethers, were reacted with triethylaluminum to yield aluminum alkoxide macroinitiators, which proved to be active towards the coordinative ROP of lactones, lactides and glycolide (Scheme 49) (Lecomte et al., 2000; Stassin et al., 2000). The pendant 2-bromo-methyl propionate chains of **22d**-based homopolymers and copolymers (Section 4.1) were used as macroinitiators for the atom-transfer radical polymerization (ATRP) of methyl methacrylate with formation of PCL-poly(methyl methacrylate) graft copolymers (Mecerreyes et al., 1999a).

7.3.2
Coordinative ROP Initiated by other Metal Compounds

Alkoxides of transition metals such as Sn, Zn, Ti, Zr, or Mg, were shown to initiate ROP according to a coordination–insertion mechanism analogous to that described with aluminum alkoxide. However, zinc gave better results in terms of control of final polyester characteristics (Lofgren et al., 1995) and selective chain end functionalization (Barakat et al., 1991).

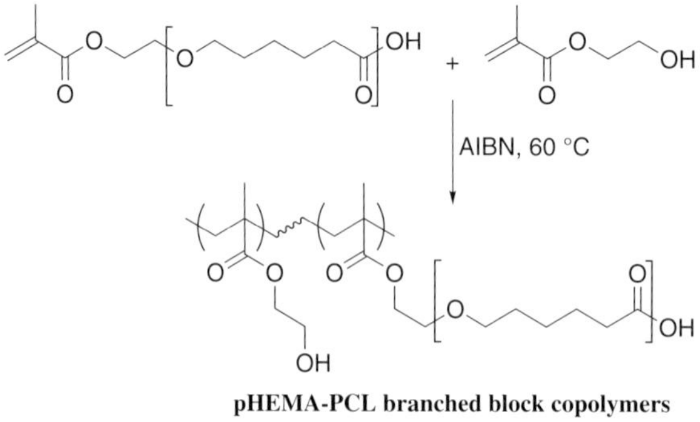

AIBN, 60 °C

pHEMA-PCL branched block copolymers

Scheme 48 Preparation of poly(2-hydroxyethyl methacrylate)-poly(ε-caprolactone) branched-block copolymers (Barakat et al., 1993, 1996, 1999).

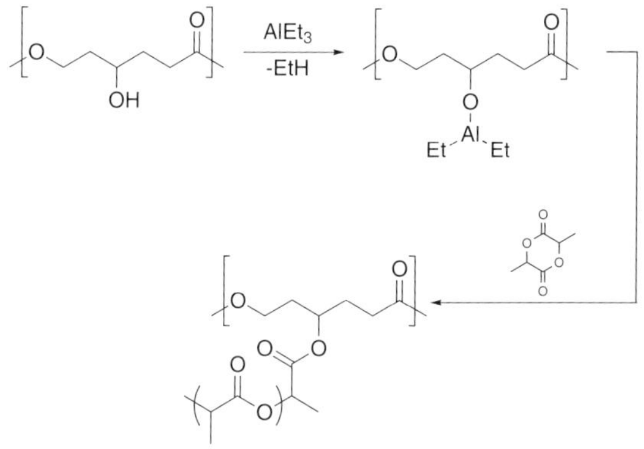

PCL-PLA branched block copolymers

Scheme 49 Preparation of poly(L-lactic acid)-poly(ε-caprolactone) branched-block copolymers (Barakat et al., 1993, 1996, 1999).

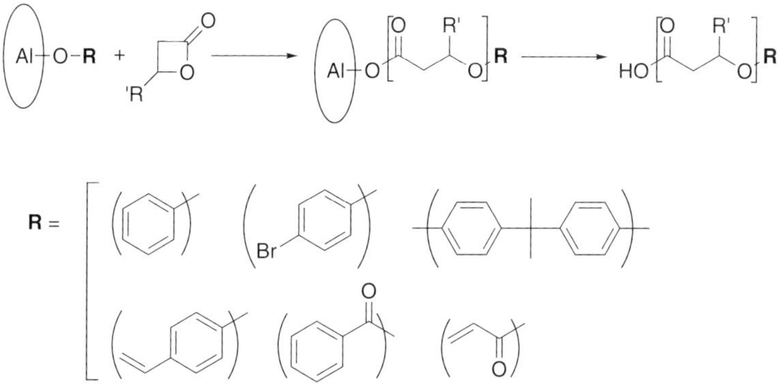

Scheme 50 Preparation of functional poly(β-lactone)s by coordinative ring-opening polymerization initiated by aluminum porphyrin derivatives (Yasuda et al., 1984).

Some aluminum porphyrin derivatives were reported as extremely efficient and versatile initiators for the ROP of lactones and lactides, according to a coordination–insertion mechanism (Lofgren et al., 1995). The nature of nucleophilic group linked to the complexed aluminum atom in the initiator was recognized as a determinant for the effective polymerization of the cyclic ester monomer (Inoue and Aida, 1993). For example, α,β,γ,δ-tetraphenylporphinatoaluminum carboxylates and phenoxides were found to be active as initiators of the polymerization of β-lactones (Yasuda et al., 1983). Modification of the structure of the

carboxylate and phenoxide radical chains in the initiator allowed for the selective introduction of reactive end-chain functionalities in β-propiolactone- and β-butyrolactone-based polyesters (Scheme 50) (Yasuda et al., 1984).

Transition metal carboxylates are also used as initiators of the ROP of cyclic esters. Carboxylates are less nucleophilic than alkoxides, and are generally considered to act as catalysts rather than initiators. In fact, most of these initiators are usually added to the monomer mixture together with active hydrogen compounds, such as alcohols (Lofgren et al., 1995). The alkyl radical of such

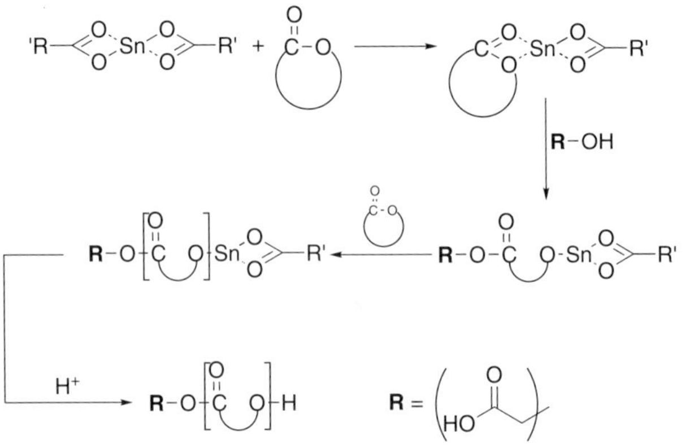

Scheme 51 Preparation of car-
boxyluc-terminated poly(L-lactic
acid) (Storey and Sherman,
1996).

coinitiators is eventually found at one end of the macromolecular chain, because of the participation of the protic moiety in the first step of polymerization process, whereas the other chain terminus is capped with an hydroxyl group (Scheme 51).

Accordingly, the use of a substituted protic coinitiator provided a synthetic route to functional telechelic polyesters, although transition metal carboxylates have an inherent capability to cause transesterification reactions, and therefore poorly defined structures of the final polymers often resulted (Lofgren et al., 1995). For example, PCL with one carboxy-terminal group was synthesized by stannous octoate-catalyzed ROP of ε-CL in the presence of glycolic acid as initiator (Scheme 50) (Storey and Sherman, 1996). However, the terminal hydroxyl groups generated by this polymerization mechanism were also utilized for the functionalization of the obtained polyester by means of suitable chemical transformations, such as esterification with methacryloyl chloride (Storey et al., 1993).

8
Functional Microbial Polyesters

In keeping with the new wave of interest in environmentally degradable polymeric materials and plastics which has developed during the past decade (Barenberg et al., 1990), much attention has been devoted to biodegradable microbial polymers, with particular emphasis on the class of polyesters under the general heading of polyhydroxyalkanoates (PHAs) (Dawes, 1990; Doi and Abe, 1990; Lenz, 1993; Lee, 1996).

The best-known polyester materials in the PHA family are the homopolymer of 3-hydroxybutyric acid, poly(hydroxybutyric acid) (PHB) and its copolymers with 3-hydroxyvaleric acid (PHBV). Once polymer scientists became acquainted with the production technology, isolation and characterization of PHB and PHBV polymeric materials, interest began to focus on the possibility of selecting microbial strains capable (depending upon the nature of the feedstock and growing conditions) of creating PHAs with either functional or easily manipulated moieties in their side chains.

Indeed, a thorough investigation was undertaken and a series of copolyesters of 3-hydroxybutyric acid with cronomomeric 3-hydroxyalkanoates containing a variety of saturated alkyl groups (also including substituents) was prepared (Portier et al., 1995; Steinbüchel and Valentin, 1995). Scheme 52 represents the general structure of copolyesters based on 3-hydroxybutyric acid and 3-hydroxyalkanoates.

By feeding *Pseudomonas oleovorans* with different carbon sources containing functional moieties, copolyesters containing phenyl (Fritzsche, 1990a; Curley et al., 1996), ester (Scholz et al., 1994), bromide (Lenz et al., 1990; Kim et al., 1992), chloride (Doi, 1990), fluoride (Abe et al., 1990; Kim et al., 1996), nitrile (Kim et al., 1991), and terminal alkenyl groups (Doi, 1990; Fritzsche et al., 1990b; Kim et al., 1995; Bear et al., 1999) were obtained.

The presence of unsaturations in the side-chain of microbial copolyesters can allow for a wide variety of post-functionalizations, including controlled oxidation to epoxy or exhaustive oxidation to carboxy, as well as all the 1,2 electrophilic addition reactions. The double bond itself, as well as some of the related functional groups, can be utilized in postcuring reactions for the attainment of inherently biodegradable thermosets or elastomers for which applications in environmental and biomedical fields have been claimed.

No further detailed comments on this intriguing class of polyesters will be reported in this chapter, as they are dealt with extensively in other chapters of this handbook.

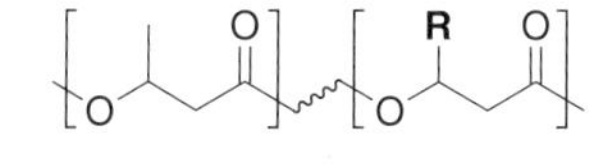

R = linear and branched saturated and unsaturated alkyl groups including also functional groups

Scheme 52 Representation of the general structure of copolyesters based on 3-hydroxybutyric acid and 3-hydroxyalkanoates.

9
Concluding Remarks

Although the present chapter is aimed at highlighting the synthetic strategies currently adopted to produce functional biodegradable polymers, it is clearly not a comprehensive treatise of all activities performed within this area. The most common methodologies to prepare functional polyesters are considered, and the relevant studies analyzed in terms of the practical potential of this approach. Such studies form two basic groups:

1) Chemical transformation of reformed polyesters free of any other functional group beyond the ester present in the repeating units in homopolyesters or copolyesters.
2) Chemical transformation of functional groups present in the repeating units as such, or in a temporally protected form which is susceptible to simple quantitative and selective removal under experimental conditions that are nonstructurally traumatic for the polyester functionality in the main-chain backbone. In the ideal case, the average degree of polymerization of the original polyesters should not be affected by the reaction conditions.

A prerequisite fundamental to the practical use of either procedure is represented by the viability of the functional groups present in the monomeric precursors in free or protected form to not interfere with the polymerization mechanism, as mediated on either chemical or biochemical grounds.

Although the potential and versatility of the two options has been stressed, the second option is clearly the more versatile and effective. However, it is worth noting that whichever route is used in their production, functional polymers in general –

and polyesters in particular – constitute a focal point of demand in keeping with the ongoing modern trends in polymer science and technology. This is increasingly being oriented towards satisfying demands for polymeric materials that are tailored for specific functional roles, and have no negative environmental impact when ultimately they become waste.

Particular attention is devoted to those polymeric materials intended for environmental and biomedical applications, where both functionality and biodegradation remain fundamental attributes for satisfactory use and environmental compliance.

Currently, the introduction of functional groups as pendant moieties that are bound covalently to polyester backbones is under close scrutiny as resorbable scaffolds in tissue engineering (Bizzarri et al., 2001; Chiellini, E. et al., 2001a,b; Chiellini, F. et al., 2001a,b) and the targeted release of both conventional and protein drugs (Chiellini, E. et al., 2001a,b; Chiellini, F. et al., 2001a,b). Consequently, future perspectives in this field appear highly promising, especially when polymeric materials and plastics that are environmentally friendly can be produced from renewable resources as feedstock.

10
References

Abe, C., Taima, Y., Nakamura, Y., Doi, Y. (1990) New bacterial copolyester of 3-hydroxyalkanoates and 3-hydroxy-ω-fluoroalkanoates produced by *Pseudomonas oleovorans, Polym. Commun.* **31**, 404–406.

Angeloni, A. S., Laus, M., Castellari, C., Galli, G., Ferruti, P., Chiellini, E. (1985) Liquid crystalline poly(β-aminoester)s containing different mesogenic groups, *Makromol. Chem.* **186**, 977–997.

Angeloni, A. S., Laus, M., Caretti, D., Chiellini, E., Galli, G. (1990) Chiral liquid-crystalline poly-(ester/β-sulfoxide)s by asymmetric oxidation of prochiral nematic poly(ester/β – sulfoxide)s, *Makromol. Chem.* **191**, 2787–2793.

Arnold, L. D., Kalantar, T. H., Vederas, J. C. (1985) Conversion of serine to stereochemically pure β-substituted α-amino acids via β-lactones, *J. Am. Chem. Soc.* **107**, 7105–7109.

Arnold, S. C., Lenz, R. W. (1986) Synthesis of stereoregular poly(alkyl malolactonates), *Macromol. Symp.* **6**, 285–303.

Barakat, I., Dubois, P., Jerome, R., Teyssie, P. (1991) Living polymerization and selective end functionalization of ε-caprolactone using zinc alkoxides as initiators, *Macromolecules* **24**, 6542–6545.

Barakat, I., Dubois, P., Jerome, R., Teyssie, P. (1993) Macromolecular engineering of polylactones and polylactides. X. Selective end-functionalization of poly(D,L)-lactide, *J. Polym. Sci. Polym. Chem. Ed.* **31**, 505–514.

Barakat, I., Dubois, P., Jérôme, R., Teyssie, P., Goethals, E. (1994) Macromolecular engineering of polylactones and polylactides. XV. Poly(D,L)-lactide macromonomers as precursors of biocompatible graft copolymers and bioerodible gels, *J. Polym. Sci. Polym. Chem. Ed.* **32**, 2099–2110.

Barakat, I., Dubois, P., Grandfils, C., Jerome, R. (1996) Macromolecular engineering of polylac-tones and polylactides. XXI. Controlled synthesis of low molecular weight polylactide macromonomers, *J. Polym. Sci. Polym. Chem. Ed.* **34**, 497–502.

Barakat, I., Dubois, P., Grandfils, C., Jerome, R. (1999) Macromolecular engineering of polylactones and polylactides. XXV. Synthesis and characterization of bioerodible amphiphilic network and their use as controlled drug delivery system, *J. Polym. Sci. Polym. Chem. Ed.* **37**, 2401–2411.

Barbaud, C., Cammas-Marion, S., Guerin, P. (1999) Poly(-malic acid) derivatives with non-charged hydrophilic lateral groups: synthesis and characterization, *Polym. Bull.* **43**, 297–304.

Barenberg, S. A., Brash, J. L., Narayan, R., Redpath, A. E. (1990) *Degradable Materials – Perspectives, Issues, and Opportunities.* Boca Raton, FL, USA: CRC Press.

Barrera, D. A., Zylstra, E., Jr., Lansbury, P. T., Langer, R. (1993) Synthesis and RGD peptide modification of a new biodegradable copolymer: poly(lactic acid-*co*-lysine), *J. Am. Chem. Soc.* **115**, 11010–11011.

Barrera, D. A., Zylstra, E., Lansbury, P. T., Langer, R. (1995) Copolymerization and degradation of poly(lactic acid-*co*-lysine), *Macromolecules* **28**, 425–432.

Barton, D. H. R., Magnus, P. D., Smith, G., Streckert, G., Zurr, D. (1972) Experiments on the synthesis of tetracycline. Part XI. Oxidation of ketone acetals and ethers by hydride transfer, *J. Chem. Soc., Perkin Trans.* **1**, 542–552.

Bear, M. M., Mallarde, D., Langlois, V., Randriamahefa, S., Bouvet, O., Guerin, P. (1999) Natural and artificial functionalized biopolyesters. II. Medium-chain length polyhydroxyoctanoates from *Pseudomonas* strains, *J. Environ. Polym. Degrad.* **7**, 179–184.

Benvenuti, M., Lenz, R. W. (1991) Polymerization and copolymerization of β-butyrolactone and benzyl-β-malolactonate by aluminoxane catalysts, *J. Polym. Sci. Polym. Chem. Ed.* **29**, 793–805.

Bitritto, M. M., Bell, J. P., Brenckle, G. M., Huang, S. J., Knox, J. R. (1979) Synthesis and biodegradation of polymers derived from α-hydroxy acids, *J. Appl. Polym. Sci.-Appl. Polym. Symp.* **35**, 405–414.

Bizzarri, R., Chiellini, F., Solaro R., Chiellini, E., Cammas-Marion, S., Guerin, P. (2001) Synthesis and Characterization of New Malolactonate Polymers and Copolymers for Biomedical Applications (submitted).

Boutault, K., Cammas, S., Huet, F., Guerin, P. (1995) Polystereoisomers with two stereogenic centers of malic acid 2-methylbutyl ester configurational structure/properties relationship, *Macromolecules* **28**, 3516–3520.

Braud, C., Bunel, C., Vert, M., Bouffard, P., Clabaut, M., Delpech, B. (1982) 28th IUPAC International Symposium on Macromolecules, Amherst, MA, USA.

Braud, C., Bunel, C., Garreau, H., Vert, M. (1983) Evidence for the amphiphilic structure of partially hydrogenolyzed poly(β-malic acid benzyl ester), *Polym. Bull.* **9**, 198–203.

Braud, C., Bunel, C., Vert, M. (1985) Poly(β-malic acid): a new polymeric drug-carrier. evidence for degradation *in vitro*, *Polym. Bull.* **13**, 293–297.

Brunelle, D. J. (1993) *Ring-Opening Polymerization.* Munich: Hanser.

Cammas, S., Renard, I., Boutault, K., Guerin, P. (1993) A novel synthesis of optically active 4-benzyloxy- and 4-alkyloxycarbonyl-2-oxetanones, *Tetrahedron: Asymmetry* **4**, 1925–1930.

Cammas, S., Renard, I., Girault, J. P., Guerin, P. (1994a) Poly(β-3-methylmalic acid): a new degradable functional polyester with two stereogenic centers in the main chain, *Polym. Bull.* **33**, 148–158.

Cammas, S., Boutault, K., Huet, F., Guerin, P. (1994b) 4-Alkyloxycarbonyl-2-oxetanones with two stereogenic centers as precursors of malic acid alkyl esters polystereoisomers, *Tetrahedron: Asymmetry* **5**, 1589–1597.

Cammas, S., Bear, M.-M., Moine, L., Escalup, R., Ponchel, G., Kataoka, K., Guerin, P. (1999) Polymers of malic acid and 3-alkylmalic acid as synthetic PHAs in the design of biocompatible hydrolyzable devices, *Int. J. Biol. Macromol.* **25**, 273–282.

Cammas-Marion, S., Guerin, P. (2000) Design of malolactonic acid esters with a large spectrum of

specified pendant groups in the engineering of biofunctional and hydrolyzable polyesters, *Macromol. Symp.* **153**, 167–186.

Caron, A., Braud, C., Bunel, C., Vert, M. (1990) Blocky structure of copolymers obtained by Pd/C-catalysed hydrogenolysis of benzyl protecting groups as shown by sequence-selective hydrolytic degradation in poly(β-malic acid) sequences, *Polymer* **31**, 1797–1802.

Chedron, H., Ohse, H., Korte, F. (1962) Die Polymerisation von Lactonen, *Makromol. Chem.* **56**, 179–194.

Chiellini, E., Solaro, R. (1992) New synthetic functional polyester, *Macromol. Symp.* **54/55**, 483–494.

Chiellini, E., Solaro, R. (1993) New bioerodible-biodegradable hydrophilic polymers, *ChemTech* **7**, 29–36.

Chiellini, E., Solaro, R. (1995) Macromol. Symp., Multifunctional Bioerodible/Biodegradable Polymeric Materials **98**, 803–824.

Chiellini, E., Solaro, R. (1996a) Biodegradable polymeric materials, *Advanced Materials* **3**, 305–313.

Chiellini, E., Solaro, R. (1996b) Functional synthetic and semisynthetic polymers in biomedical applications, in: *Advanced Biomaterials in Biomedical Engineering and Drug Delivery Systems* (Ogata, N., Kim, S. W., Feijen, J., Okano, T., Eds.), Tokyo: Springer-Verlag, 355–368.

Chiellini, E., Galli, G., Angeloni, A. S., Laus, M., Pellegrini, R. (1987) Chiral liquid crystal polymers VIII. Thermotropic poly(ester-β-sulphide)s based on variously spaced twin p-oxybenzoate diads, *Liquid Crystals* **2**, 529–537.

Chiellini, E., Galli, G., Angeloni, A. S., Laus, M. (1990a) Chiral liquid-crystalline polymers. XI. Thermotropic behavior and chiroptical properties in solution of a new series of optically active poly(ester β-sulfide)s, *J. Polym. Sci. Polym. Chem. Ed.* **28**, 3541–3550.

Chiellini, E., Galli, G., Angeloni, A. S., Laus, M. (1990b) Synthesis and chemical modification of chiral liquid-crystalline poly(ester β-sulphide)s, *ACS Symp. Ser.* **6**, 79–92.

Chiellini, E., Galli, G., Po, R. (1990c) Chiral liquid crystalline polymers, *Polym. Bull.* **23**, 397–402.

Chiellini, E., Faggioni, S., Solaro, R. (1990d) Polyesters based on glyceric acid derivatives as potential biodegradable materials, *J. Bioact. Compat. Polym.* **5**, 16–29.

Chiellini, E. E., Giannasi, D., Solaro, R., Chiellini, E., Fernandes, E. G. (1998) Hybrid polymeric hydrogels for the controlled release of protein drugs: DSC investigation of water structure, in: *Frontiers in Biomedical Polymer Application* (Ottenbrite, R. M., Chiellini, E., Cohn, D., Migliaresi, C., Sunamoto, J., Eds.), Lancaster, PA, USA: Technomic Publishing Company, Inc., 13–26.

Chiellini, E., Gil, H., Braunegg, G., Buchert, J., Gatenholm, P., Van der Zee, M. (2001a) *Biorelated Polymers – Sustainable Polymer Science and Technology.* New York, USA: Kluwer Academic/Plenum Publishers.

Chiellini, E., Sunamoto, J., Migliaresi, C., Ottenbrite, R. M., Cohn, D. (2001b) *Biomedical Polymers and Polymer Therapeutics.* New York, USA: Kluwer Academic/Plenum Publishers.

Chiellini, F., Petrucci, F., Ranucci, E., Solaro, R. (2001a) Polymeric hydrogels in drug release, in: *Biomedical Polymers and Polymer Therapeutics* (Chiellini, E., Sunamoto, J., Migliaresi, C., Ottenbrite, R. M., Cohn, D., Eds.), New York, USA: Kluwer Academic/Plenum Publishers, 63–74.

Chiellini, F., Schmaljohann, D., Chiellini, E., Ober, C. K., Solaro, R (2001b) (submitted).

Cook, A. D., Hrkach, J. S., Gao, N. N., Johnson, I. M., Pajvani, U. B., Cannizzaro, S. M., Langer, R. (1997) Characterization and development of RGD-peptide-modified poly(lactic acid-co-lysine) as an interactive, resorbable biomaterial, *J. Biomed. Mater. Res.* **35**, 513–523.

Curley, J. M., Hazer, B., Lenz, R., Fuller, C. (1996) Production of poly(3-hydroxyalkanoate)s containing aromatic substituents by *Pseudomonas oleovorans, Macromolecules* **29**, 1762–1766.

Davis, G. W., Hill, E. S. (1982) Polyester fiber, in: *Encyclopedia of Chemical Technology* (Kirk, R. E., Othmer, D. F., Eds.), New York: John Wiley & Sons, 531–549.

Dawes, E. A. (1990) *Novel Biodegradable Microbial Polymers.* Dordrecht: Kluwer Academic Publishers.

Degee, P., Dubois, P., Jerome, R., Teyssie, P. (1992a) Macromolecular engineering of polylactones and polylactides. 9. Synthesis, characterization, and application of ω-primary amine poly(ε-caprolactone), *Macromolecules* **25**, 4242–4248.

Degee, P., Dubois, P., Jerome, R., Teyssie, P. (1992b) Synthesis and characterization of biocompatible and biodegradable poly(ε-caprolactone-β-γ-benzylglutamate) diblock copolymer, *J. Polym. Sci. Polym. Chem. Ed.* **31**, 275–278.

Detrembleur, C., Mazza, M., Halleux, O., Lecomte, P., Mecerreyes, D., Hedrick, J. L., Jerome, R. (2000a) Ring-opening polymerization of γ-bromo-ε-caprolactone: a novel route to functionalized aliphatic polyesters, *Macromolecules* **33**, 14–18.

Detrembleur, C., Mazza, M., Lou, X., Halleux, O., Lecomte, P., Mecerreyes, D., Hedrick, J. L., Jerome, R. (2000b) New functional aliphatic polyesters by chemical modification of copolymers of ε-caprolactone with γ-(2-bromo-2-methylpropionate)-ε-caprolactone, γ-bromo-ε-caprolactone, and a mixture of β- and γ-ene-ε-caprolactone, *Macromolecules* **33**, 7751–7760.

DiBenedetto, L. J., Huang, S. J. (1989) Biodegradable hydroxylated polymers as controlled release agents, *ACS Polym. Prep.* **30**, 403.

DiBenedetto, L. J., Huang, S. J. (1994) Poly(alkylene tartrate)s as controlled release agents, *Polym. Degrad. Stab.* **45**, 249–255.

Doi, Y. (1990) *Microbial Polyesters.* New York, USA: VCH.

Doi, Y., Abe, C. (1990) Biosynthesis and characterization of a new bacterial copolyester of 3-hydroxyalkanoates and 3-hydroxy-ω-chloroalkanes, *Macromolecules* **23**, 3705–3707.

Dubois, P., Jerome, R., Teyssie, P. (1989) Macromolecular engineering of polylactones and polylactides. I. End-functionalization of poly(ε-caprolactone), *Polym. Bull.* **22**, 475–482.

Dubois, P., Jerome, R., Teyssie, P. (1991a) Aluminium alkoxides: a family of versatile initiators for the ring-opening polymerization of lactones and lactides, *Macromol. Symp.* **42/43**, 103–116.

Dubois, P., Jerome, R., Teyssie, P. (1991b) Macromolecular engineering of polylactones and polylactides. 3. Synthesis, characterization, and applications of poly(ε-caprolactone) macromonomers, *Macromolecules* **24**, 977–981.

Dubois, P., Degee, P., Jerome, R., Teyssie, P. (1993) Macromolecular engineering of polylactones and polylactides. 11. Synthesis and use of alkylaluminium dialkoxides and dithiolates as promoters of hydroxy telechelic poly(ε-caprolactone) and α,ω-dihydroxy triblock copolymers containing outer polyester blocks, *Macromolecules* **26**, 2730–2735.

Dubois, P., Zhang, J. X., Jerome, R., Teyssie, P. (1994) Macromolecular engineering of polylactones and polylactides. 13. Synthesis of telechelic polyesters by coupling reactions, *Polymer* **35**, 4998–5004.

Dubois, P., Degee, P., Ropson, N., Jerome, R. (1997) *Macromolecular Design of Polymeric Materials*, New York: Marcel Dekker, 247.

Eguiburu, J. L., Fernandez-Berridi, M. J., Roman, J. S. (1995) Functionalization of poly(L-lactide) macromonomers by ring-opening polymerization of L-lactide initiated with hydroxyethyl methacrylate-aluminium alkoxides, *Polymer* **36**, 173–179.

Fietier, I., Borgne, A. L., Spassky, N. (1990) Synthesis of functional polyesters derived from serine, *Polym. Bull.* **24**, 349–353.

Fritzsche, X. K., Lenz, R. W., Fuller, R. C. (1990a) An unusual bacterial polyester with a phenyl pendant group, *Makromol. Chem.* **191**, 1957–1965.

Fritzsche, K., Lenz, R. W. and Fuller, R. C. (1990b) Production of Unsaturated Polyesters by *Pseudomonas oleovorans*, *Int. J. Biol. Macromol.* **12**, 85–91.

Galli, G., Laus, M., Angeloni, A. S., Ferruti, P., Chiellini, E. (1983) Thermotropic poly(β-aminoester)s containing azoxy groups, *Macromol. Rapid Commun.* **4**, 681–686.

Galli, G., Chiellini, E., Laus, M., Angeloni, A. S. (1989) Thermotropic liquid crystalline poly(ester β-sulfide)s based on twin hexamethylene-spaced p-oxybenzyl diads, *Macromolecules* **22**, 1120–1124.

Galli, G., Chiellini, E., Laus, M., Angeloni, A. S. (1991) Structural effects on the mesomorphic behaviour of thermotropic liquid crystal poly(ester β-sulphide)s, *Eur. Polym. J.* **27**, 237–241.

Gelbin, M. E., Kohn, J. (1993) Synthesis and polymerization of N-Z-L-serine-β-lactone and serine hydroxybenzotriazole active esters, *J. Am. Chem. Soc.* **1993**, 3962–3965.

Goethals, E. J. (1993) Telechelic polymers by ring-opening polymerization, in: *Ring-Opening Polymerization* (Brunelle, D. J., Ed.), Munich: Hanser, 295–307.

Guerin, P., Vert, M. (1987) Enantiomeric purity of R(+) and S(-) benzyl malolactonate monomers as determined by 250 MHz 1H-nuclear magnetic resonance, *Polym. Commun.* **28**, 11–13.

Guerin, P., Vert, M., Braud, C., Lenz, R. W. (1985) Optically active poly(β-malic acid), *Polym. Bull.* **14**, 187–192.

Inoue, S., Aida, T. (1993) Living polymerization of acrylic monomers with aluminioporphyrin, *Macromol. Symp.* **32**, 255–265.

in't Veld, P. J. A., Dijkstra, P. J., Lochem, J. H. V., Feijen, J. (1990) Synthesis of alternating polydepsipeptides by ring-opening polymerization of morpholine-2,5-dione derivatives, *Makromol. Chem.* **191**, 1813–1825.

in't Veld, P. J. A., Wei-Ping, Y., Klap, R., Dijkstra, P. J., Feijen, J. (1992a) Copolymerization of ε-caprolactone and morpholine-2,5-dione derivatives, *Makromol. Chem.* **193**, 1927–1942.

in't Veld, P. J. A., Dijkstra, P. J., Feijen, J. (1992b) Synthesis of biodegradable polyesteramides with pendant functional groups, *Makromol. Chem.* **193**, 2713–2730.

Jarm, V., Fles, D. (1977) Polymerization and properties of optically active α-(p-substituted benzenesulfonamido)-β-lactones, *J. Polym. Sci. Polym. Chem. Ed.* **15**, 1061–1071.

Jedlinski, Z., Kowalczuk, M., Kurcok, P. (1992) Polymerization of β-lactones initiated by alkali metal naphthalenides. A convenient route to telechelic polymers, *J.M.S.-Pure Appl. Chem.* **A29**, 1223–1230.

Kim, O., Gross, R., Hammar, W. J., Newmark, R. A. (1996) Microbial synthesis of poly(β-hydroxyalkanoates) containing fluorinated side-chain substituents, *Macromolecules* **29**, 4572–4581.

Kim, Y. B., Lenz, R. W., Fuller, R. C. (1991) Preparation and characterization of poly(β-hydroxyalkanoate)s obtained from *Pseudomonas oleovorans* grown with mixtures of 5-phenylvaleric acid and n-alkanoic acids, *Macromolecules* **24**, 5256–5260.

Kim, Y. B., Lenz, R. W., Fuller, R. C. (1992) Poly(β-hydroxyalkanoate) copolymers containing brominated repeating units produced by *Pseudomonas oleovorans*, *Macromolecules* **25**, 1852–1857.

Kim, Y. B., Lenz, R. W., Fuller, R. C. (1995) Poly(3-hydroxyalkanoate)s containing unsaturated repeating units produced by *Pseudomonas oleovorans*, *J. Polym. Sci. Polym. Chem. Ed.* **33**, 1367–1374.

Kimura, Y., Shirotani, K., Yamane, H., Kitao, T. (1988) Ring-opening polymerization of 3(S)-[(benzyloxycarbonyl)methyl]-1,4-dioxane-2,5-dione: a new route to a poly(α-hydroxy acid) with pendant carboxyl groups, *Macromolecules* **21**, 3338–3340.

Kimura, Y., Shirotani, K., Yamane, H., Kitao, T. (1993) Copolymerization of 3(S)-[(benzyloxycarbonyl)methyl]-1,4-dioxane-2,5-dione and L-lactide: a facile synthetic method for functionalized bioabsorbable polymer, *Polymer* **34**, 1741–1748.

Kirk, R. E., Othmer, D. F. (1991) *Encyclopedia of Chemical Technology, Volume 1* (Kroschwitz, J. I., Howe-Grant, M., Eds.), New York: John Wiley & Sons.

Kohn, J., Langer, R. (1996) Bioresorbable and bioerodible materials, in: *Biomaterials Science* (Ratner, B. D., Hoffmann, A. S., Schoen, F. S., Lemons, J. E., Eds.), San Diego, CA: Academic Press, 64–73.

Kurcok, P., Matuszowicz, A., Jedlinski, Z. (1995) Anionic polymerization of β-lactones initiated with potassium hydride. A convenient route to polyester macromonomer, *Macromol. Rapid Commun.* **16**, 201–206.

Laus, M., Angeloni, A. S., Galli, G., Chiellini, E. (1988) Liquid-crystalline poly(β-aminoester)s. Thermotropic mesomorphism and degradability in solution, *Makromol. Chem.* **189**, 743–754.

Laus, M., Angeloni, A. S., Galli, G., Chiellini, E. (1990) The molecular weight effect on the mesomorphic behaviours of liquid-crystalline poly(ester-β-sulfide)s, *Makromol. Chem.* **191**, 147–154.

Leboucher-Durand, M.-A., Langlois, V., Guerin, P. (1996) 4-Carboxy-2-oxetanone as a new chiral precursor in the preparation of functionalized racemic or optically active poly(malic acid) derivatives, *Polym. Bull.* **36**, 35–41.

Lecomte, P., Detrembleur, C., Lou, X., Mazza, M., Halleux, O., Jerome, R. (2000) Novel functionalization routes of poly(ε-caprolactone), *Macromol. Symp.* **157**, 47–60.

Lee, S. Y. (1996) Bacterial polyhydroxyalkanoates, *Biotechnol. Bioeng.* **49**, 1–14.

Lenz, R. W. (1993) Biodegradable polymers, in: *Biopolymers I* (Peppas, N. A., Langer, R. S., Eds.), Berlin, Germany: Springer-Verlag, 1–40.

Lenz, R. W., Johns, D. B., Vert, M., Camps, M., Boileau, S. (1986) Recent investigation on the polymerization reactions of β-lactones and the synthesis of poly(β-malic acid), *ACS Polym. Prep.* **27**, 175–176.

Lenz, R. W., Kim, B. W., Ulmer, H. W., Fritsche, K., Knee, E., Fuller, R. C. (1990) Functionalized poly(β-hydroxyalkanoate)s produced by bacteria, in: *Novel Biodegradable Microbial Polymers* (Dawes, E. A., Ed.), London: Kluwer Academic Publishers, 23–37.

Lofgren, A., Albertsson, A.-C., Dubois, P., Jerome, R. (1995) Recent advances in ring-opening polymerization of lactones and related compounds, *J.M.S.-Pure Appl. Chem.* **C35**, 379–418.

Matsumura, S., Beppu, H., Toshima, K. (1998) Enzymatic preparation of malate-based polycarboxylates having higher molecular weights by copolymerization with lactone, *Chem. Lett.* 249–250.

Mecerreyes, D., Jerome, R. (1999) From living to controlled aluminium alkoxide mediated ring-opening polymerization of (di)lactones, a powerful tool for the macromolecular engineering of aliphatic polyesters, *Macromol. Chem. Phys.* **200**, 2581–2590.

Mecerreyes, D., Atthoff, B., Boduch, K. A., Trollsas, M., Hedrick, J. L. (1999a) Unimolecular combination of an atom transfer radical polymerization initiator and a lactone monomer as a route to new graft copolymers, *Macromolecules* **32**, 5175–5182.

Mecerreyes, D., Dahan, D., Lecomte, P., Dubois, P., Demonceau, A., Noels, A. F., Jerome, R. (1999b) Ring-opening methathesis polymerization of new α-norborneyl poly(ε-caprolactone) macromonomers, *J. Polym. Sci. Polym. Chem. Ed.* **37**, 2447–2455.

Mecerreyes, D., Humes, J., Miller, R. D., Hedrick, J. L., Detrembleur, C., Lecomte, P., Jerome, R., Roman, J. S. (2000a) First example of an unsymmetrical difunctional monomer polymerizable by two living/controlled methods, *Macromol. Rapid Commun.* **21**, 779–784.

Mecerreyes, D., Miller, R. D., Hedrick, J. L., Detrembleur, C., Jerome, R. (2000b) Ring-opening polymerization of 6-hydroxynon-8-enoic acid lactone: novel biodegradable copolymers containing allyl pendent groups, *J. Polym. Sci. Polym. Chem. Ed.* **38**, 870–875.

Mitsunobu, O. (1981) The use of diethyl azodicarboxylate and triphenylphosphine in synthesis and transformation of natural products, *Synthesis* 1–28.

Moine, L., Cammas, S., Amiel, C., Guerin, P., Sebille, B. (1997) Polymers of malic acid conjugated with the 1-adamantyl moiety as lipophilic pendant group, *Polymer* **38**, 3121–3127.

Mulzer, J., Bruntrup, G., Chucholowski, A. (1979) Competition in the three-component synthesis triphenylphosphine/azo ester/3-hydroxy-carboxylic acid: OH-versus COOH-activation, *Angew. Chem. Int. Edn. Engl.* **18**, 622–623.

Ouchi, T., Fujino, A. (1989) Synthesis of poly(α-malic acid) and its hydrolysis behaviour *in vitro*, *Makrom. Chem.* **190**, 1523–1530.

Pachence, J. M., Kohn, J. (1997) Biodegradable polymers for tissue engineering, in: *Principles of Tissue Engineering* (Lanza, R. P., Langer, R., Chick, W. L., Eds.), Georgetown, TX: Landes Co., Academic Press, Inc., 273–293.

Pitt, C. G. (1990) Poly-ε-caprolactone and its copolymers, in: *Biodegradable Polymers as Drug Delivery Systems* (Chasin, M., Langer, R., Eds.), New York: Marcel Dekker, 71–120.

Pitt, C. G., Gu, Z.-W., Ingram, P., Hendren, R. W. (1987) The synthesis of biodegradable polymers with functional side chains, *J. Polym. Sci. Polym. Chem. Ed.* **25**, 955–966.

Platè, N.A. (1973) Some problems in the reactivity of macromolecules, in *Reactions on Polymers* (Moore, J. A., Ed.), Dordrecht: D. Reidel Publishing Co., 169–187.

Portier, Y., Nawrath, C., Somerville, C. (1995) Production of polyhydroxyalkanoates, a family of biodegradable plastics and elastomers, in bacteria and plants, *Biotechnology* **13**, 142–150.

Quirk, R. P., Kim, J. (1993) Macromonomers and macroinitiators, in: *Ring-Opening Polymerization* (Brunelle, D. J., Ed.), Munich: Hanser, 264–293.

Renard, I., Cammas, S., Langlois, V., Bourbouze, R., Guerin, P. (1996) Poly(β-3-methyl malic acid), an hydrolyzable polyester: access to tailor-made derivatives with defined configurational structures and study of *in vitro* compatibility, *ACS Polym. Prep.* **37**, 137–138.

Rossignol, H., Boustta, M., Vert, M. (1999) Synthetic poly(β-hydroxyalkanoates) with carboxylic acid or primary amine pendent groups and their complexes, *Int. J. Biol. Macromol.* **25**, 255–264.

SaintPierre, T., Chiellini, E. (1987) Biodegradability of synthetic polymers used for medical and pharmaceutical applications. Part 2. Backbone hydrolysis, *Bioactive Compatible Polymers* **2**, 4–8.

Scholz, C., Fuller, R. C., Lenz, R. W. (1994) Production of poly(β-hydroxyalkanoate)s with β-substituents containing terminal ester groups by *Pseudomonas oleovorans*, *Macromol. Chem. Phys.* **195**, 1405–1421.

Solaro, R., Bemporad, L., Chiellini, E. (1991) Novel hydroxyl containing polyesters and related polymers viable to the formulation of drug release systems, in: *Synthetic Polymers as Drug Carriers. Interactions with Blood* (Baszkin, A., Ferruti, P., Marchisio, M. A., Tanzi, M. C., Eds.), Brescia, Italy: ALFA, 71–77.

Sosnowski, S., Slomkowsky, S., Penczec, S. (1991) Telechelic poly(ε-caprolactone) terminated at both ends with OH groups and its derivatization, *Makromol. Chem.* **192**, 2730–2738.

Stassin, F., Halleux, O., Dubois, P., Detrembleur, C., Lecomte, P., Jerome, R. (2000) (Triethylsilyloxy)-ε-caprolactone and γ-ethylene ketal-ε-caprolactone: a route to hetero-graft copolyesters, *Macromol. Symp.* **153**, 27–39.

Steinbuchel, A., Valentin, M. E. (1995) Diversity of bacterial polyhydroxyalkanoic acids, *FEMS Microbiol. Lett.* **128**, 219–228.

Storey, R. F., Sherman, J. W. (1996) Novel synthesis of (carboxylic acid)-telechelic poly(ε-caprolactone), *ACS Polym. Prep.* **37**, 624–625.

Storey, R. F., Warren, S. C., Allison, C. J., Wiggins, J. S. (1993) Synthesis of bioresorbable networks from methacrylate-endcapped polyesters, *Polymer* **34**, 4365–4372.

Tian, D., Dubois, P., Grandfils, C., Jerome, R. (1997a) Ring-opening polymerization of 1,4,8-trioxaspiro[4,6]-9-undecanone: a new route to aliphatic polyesters bearing functional pendent groups, *Macromolecules* **30**, 406–409.

Tian, D., Dubois, P., Jerome, R. (1997b) Macromolecular engineering of polylactones and polylactides. 23. Synthesis and characterization of biodegradable and biocompatible homopolymers and block copolymers based on 1,4,8-trioxa[4,6]spiro-9-undecanone, *Macromolecules* **30**, 1947–1954.

Tian, D., Dubois, P., Jerome, R. (1997c) Macromolecular engineering of polylactones and polylactides. 22. Copolymerization of ε-caprolactone and 1,4,8-trioxaspiro[4,6]-9-undecanone initiated by aluminium isopropoxide, *Macromolecules* **30**, 2575–2581.

Tian, D., Dubois, P., Jerome, R. (1998) Ring-opening polymerization of 1,4,8-trioxaspiro-[4,6]-9-undecanone: a route to novel molecular architectures for biodegradable aliphatic polyesters, *Macromol. Symp.* **130**, 217–227.

Trollsas, M., Hedrick, J. L., Dubois, P., Jerome, R. (1998) Synthesis of acid-functional asymmetric aliphatic polyester, *J. Polym. Sci. Polym. Chem. Ed.* **36**, 1345–1348.

Vert, M., Lenz, R. W. (1979) Preparation and properties of poly-β-malic acid: a functional polyester of potential biomedical importance, *ACS Polym. Prep.* **20**, 608–611.

Yamaoka, T., Hotta, Y., Kobayashi, K., Kimura, Y. (1999) Synthesis and properties of malic acid-containing functional polymers, *Int. J. Biol. Macromol.* **25**, 265–271.

Yasuda, T., Aida, T., Inoue, S. (1983) Living polymerization of β-lactone catalyzed by (tetraphenylporphinato)aluminium chloride. structure of the living end, *Macromolecules* **16**, 1792–1796.

Yasuda, T., Aida, T., Inoue, S. (1984) Synthesis of end-reactive polymers with controlled molecular weight by metalloporphyrin catalyst, *J. Macromol. Sci.-Chem.* **A21**, 1035–1047.

Zhou, Q.-X., Kohn, J. (1990) Preparation of poly(L-serine ester): a structural analogue of conventional poly(L-serine), *Macromolecules* **23**, 3399–3406.

14

Production of Chiral and other Valuable Compounds from Microbial Polyesters

Prof. Dr. Sang Yup Lee[1], Dr. Sang Hyun Park[2], M. Eng. Young Lee[3],
M. Eng. Seung Hwan Lee[4]

[1] Metabolic and Biomolecular Engineering National Research Laboratory,
Department of Chemical Engineering and BioProcess Engineering Research
Center, Korea Advanced Institute of Science and Technology, 373-1 Kusong-dong,
Yusong-gu, Taejon 305-701, Korea; Tel.: +82-42-869-3930; Fax: +82-42-869-3910;
E-mail: leesy@mail.kaist.ac.kr

[2] Metabolic and Biomolecular Engineering National Research Laboratory,
Department of Chemical Engineering and BioProcess Engineering Research
Center, Korea Advanced Institute of Science and Technology, 373-1 Kusong-dong,
Yusong-gu, Taejon 305-701, Korea; Tel.: +82-42-869-5970; Fax: +82-42-969-8800;
and ChiroBio Inc., #2324, Undergraduate Building 2, Korea Advanced Institute of
Science and Technology, 373-1 Kusong-dong, Yusong-gu, Taejon 305-701, Korea;
E-mail: sanghpark@bcline.com

[3] Metabolic and Biomolecular Engineering National Research Laboratory,
Department of Chemical Engineering and BioProcess Engineering Research
Center, Korea Advanced Institute of Science and Technology, 373-1 Kusong-dong,
Yusong-gu, Taejon 305-701, Korea; Tel.: +82-42-869-5970; Fax: +82-42-969-8800;
and ChiroBio Inc., #2324, Undergraduate Building 2, Korea Advanced Institute of
Science and Technology, 373-1 Kusong-dong, Yusong-gu, Taejon 305-701, Korea;
E-mail: chirobio@mail.kaist.ac.kr

[4] Metabolic and Biomolecular Engineering National Research Laboratory,
Department of Chemical Engineering and BioProcess Engineering Research
Center, Korea Advanced Institute of Science and Technology, 373-1 Kusong-dong,
Yusong-gu, Taejon 305-701, Korea; Tel.: +82-42-869-5970; Fax: +82-42-969-8800;
E-mail: leesh@mail.kaist.ac.kr

4-AA   (1′R,3R,4R)-3-[1′-hydroxyethyl]-4-acetoxy-2-azetidinone
MCL   medium-chain-length
PHA   polyhydroxyalkanoate
poly(3HB) poly[(R)-3-hydroxybutyrate]
R3HB   (R)-3-hydroxybutyric acid
R3HV   (R)-3-hydroxyvaleric acid

1

Introduction

The production of enantiomerically pure chemicals has long been considered important in various industries including foods, pharmaceuticals, cosmetics, aromatics, and other fine chemicals. Chirality is extremely important in enzyme-catalyzed reactions as, in general, enzymes have specificities for the reaction of only one type of the two enantiomers. Therefore, only one of the two enantiomers is the active compound of interest. In the case of drugs, the presence of an undesirable enantiomer clearly leads to an increase in the dose required, and often also to adverse side effects. Hence, in certain countries the development and registration of a new drug must be based on its pure enantiomer (Crosby, 1992).

Polyhydroxyalkanoates (PHAs) are a class of microbial polyesters which are synthesized by, and accumulate in, numerous microorganisms as a carbon and/or energy storage material under unfavorable conditions (Anderson and Dawes, 1990; Lee, 1996). Indeed, PHAs are regarded as a good candidate to replace petroleum-derived non-degradable (or barely degradable) polymers due to their many advantages, including their complete biodegradability, their production from renewable resources, and the diversity of their possible monomers (Steinbüchel and Valentin, 1995; Lee, 1996) and physical properties (Inoue and Yoshie, 1992). PHAs can also be used as truly biodegradable plastics and elastomers (Lee, 1996). More than 140 types of 3-, 4-, 5- or 6-hydroxylated carboxylic acids have been found to be incorporated into the polymer (Lee, 1996;

Steinbüchel and Valentin, 1995; Curley et al., 1996; Kim et al., 1996a,b; Andújar et al., 1997), and an increasing number of new monomers are being discovered. All of the monomeric units of PHAs are enantiomerically pure, and in (*R*)-configuration if they possess a chiral center (Jendrossek et al., 1996; Lee et al., 1999). Therefore, it was reasoned that various enantiomerically pure (*R*)-hydroxycarboxylic acids might be conveniently prepared by depolymerizing biosynthesized PHAs. For example, poly[(*R*)-3-hydroxybutyrate] [poly(3HB)] was first suggested to be a source for a chiral pool about 10 years ago (Crosby, 1992).

In this chapter, the potential of PHAs as a source of chiral pool and other valuable intermediates is reviewed.

2
Historical Outline

A 'brief history' of PHAs and their uses includes:

Year	Depolymerization of PHAs	Applications
1964	Intracellular enzyme system first proposed	
1980	Unsuccessful pyrolysis of poly(3HB)	
1982	Acidic alcoholysis of poly(3HB)	
1984	Acidic alcoholysis of poly(3HB-*co*-3HV)	
1985		4-AA
1986~1987		Macrolides Dioxanones (*S*)-β-Lactone
1994		Cyclic oligomers
1996		Dendrimer
1999	*In vivo* depolymerization	
2000	Acidic alcoholysis of poly(3HB) with HCl	
2001	Intracellular PHA depolymerase	β-Amino acids

3
(R)-Hydroxycarboxylic Acids

(*R*)-Hydroxycarboxylic acids, the monomeric units of PHAs, contain hydroxyl (-OH) and carboxyl (-COOH) groups within their molecules. These two functional groups can be easily transformed to other functional groups, and are also convenient for carrying out reactions with other monomers. As mentioned briefly above, various chiral hydroxycarboxylic acids can be prepared from PHAs without racemization. Therefore, hydroxycarboxylic acids can be widely used as starting materials for the chemical synthesis of valuable products.

The monomeric units of PHAs can be presented as the general structural formula shown in Figure 1a, with a few exceptions including 3-hydroxy-2-butenoic acid (Figure 1b), 6-hydroxy-3-dodecanoic acid (Figure 1c), and 2-methyl-substituted 3-hydroxycarboxylic acids (Figure 1d). Most of these compounds are 3-hydroxylated (n=1 in Figure 1a), while 4-, 5-, or 6-hydroxycarboxylic acids are also detected (Steinbüchel and Valentin, 1995 and references cited therein). The R-pendant group of the 3-hydroxycarboxylic acids can contain one or more double bonds, a methyl branch, or various functional groups including alkyl or benzyl ester, acetoxy, phenoxy, *para*-cyanophenoxy, *para*-nitrophenoxy, phenyl, cyclohexyl, second hydroxyl, epoxy or cyano groups (Kim et al., 1996a,b; Song and Yoon, 1996; Andújar et al., 1997; Garía et al., 1999; Steinbüchel and Valentin, 1995, and references cited therein). Some examples of ω-halogenated 3-hydroxycarboxylic acids with fluorine,

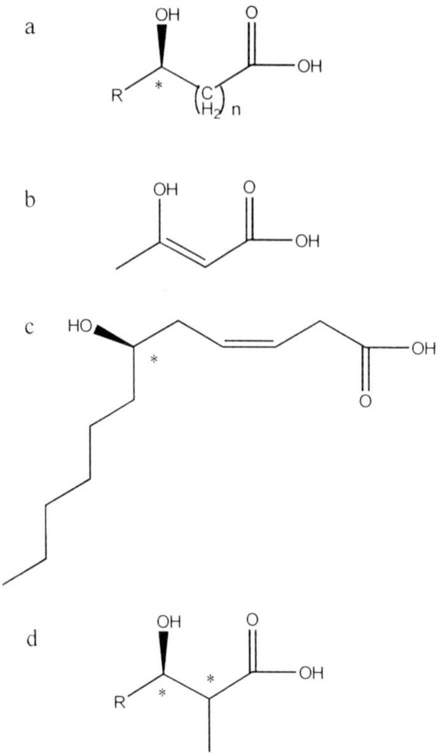

a

b

c HO

d

Fig. 1 Structures of monomeric units of polyhy-droxyalkanoates. (a) General structure and structures of (b) 3-hydroxy-2-butenoic acid, (c) 6-hydroxy-3-dodecanoic acid and (d) 2-methyl-substituted 3-hydroxycarboxylic acids. $1 \leq n \leq 4$, * = chiral center, R = hydrogen atom or alkyl or alkenyl groups containing various functional groups. [See Steinbüchel and Valentin (1995) for various R groups.]

chlorine or bromine were also reported (Steinbüchel and Valentin, 1995 and references cited therein). All of the hydroxycarboxylic acids in PHAs are enantiomerically pure, in (R)-configuration, and have the chiral center at the hydroxylated carbon position (Jendrossek et al., 1996; Lee et al., 1999).

4

Production of (R)-3-Hydroxycarboxylic Acid

Bacterial PHAs can serve as a useful chiral pool for the preparation of (R)-3-hydroxy-

carboxylic acids. Hence, several attempts have been made to produce (R)-3-hydroxy-carboxylic acids either chemically or biologically, as outlined below.

4.1

Chemical Depolymerization

The pyrolysis of bacterial PHAs was first investigated by Morikawa and Marchessault (1980), who observed that alkenoic acids such as crotonic acid and 2-pentenoic acid were produced via intramolecular cis-elimination.

The successful production of (R)-3-hy-droxybutyric acid (R3HB) and (R)-3-hydroxy-valeric acid (R3HV) in alkyl ester form by chemical degradation of the corresponding PHAs was first reported by Seebach and colleagues (Seebach and Züger, 1982, 1984; Seebach et al., 1992), who pioneered this area of research. These authors used an alcoholysis, transesterification reaction with alcohol in the presence of acidic catalysts such as sulfuric acid and para-toluenesul-fonic acid. Although a good final yield of 70% was accomplished in three days of reaction for the production of methyl ester of R3HB from 50 g of purified poly(3HB) (Seebach et al., 1992), the method required large amounts of organic solvents. Another problem was the need for PHA purification before the transesterification reaction, and this resulted in a lower overall yield due to the loss of PHA during purification. These researchers also demonstrated the production of a free acidic form of R3HB by saponification of its alkyl ester produced by alcoholysis of poly(3HB). The saponification was combined with distillation to remove the organic solvent added initially, as well as the alcohol that remained after alcoholysis and also that was produced during saponification. Hydrochloric acid was found to give higher initial reactivity than sulfuric acid as

Tab. 1 Production of methyl (R)-$(-)$-3-hydroxybutyrate (mg) by acid-catalyzed methanolysis of poly(3HB) (initially 10 mg) purified by three digestion methods[a] (Reproduced from Lee et al., 2000.)

Digestion solution		SDS [50 g L^{-1}]		KOH [0.2 M]		NaOH [0.2 M]	
Acid catalyst		Conc.H$_2$SO$_4$[b]	HCl[c]	Conc.H$_2$SO$_4$	HCl	Conc.H$_2$SO$_4$	HCl
Reaction time	1 h	1.29	1.65	1.05	2.20	0.99	2.96
	2 h	1.34	3.77	1.06	3.75	1.18	9.75

SDS, sodium dodecylsulfate; KOH, potassium hydroxide; NaOH, sodium hydroxide; Conc., concentrated.[a]Choi and Lee (1999).[b]Commercially available sulfuric acid contained 93–98% H$_2$SO$_4$, and the remainder is water.[c]Commercially available hydrochloric acid contained 62% water.

an acidic catalyst (Lee et al., 2000). When hydrochloric acid was used as an acidic catalyst, the initial reactivity of poly(3HB) depolymerization was over eight-fold higher than that obtained using sulfuric acid (Table 1). Several inexpensive methods for the purification of poly(3HB) from cells have also been developed for the economic production of R3HB. As shown in Table 1, poly(3HB) purified using a simple digestion method with 0.2 M NaOH (Choi and Lee, 1999) was efficiently alcoholyzed by using hydrochloric acid as an acidic catalyst.

4.2
Biological Depolymerization

The biological depolymerization of PHAs can be divided into two types, namely *in-vitro* depolymerization and *in-vivo* depolymerization of PHAs, these being based on whether the polymer was separated from cells before depolymerization, or not.

4.2.1
In vitro Depolymerization

The depolymerization of poly(3HB) by an intracellular enzyme system was first suggested by Merrick and Doudoroff (1964). Poly(3HB) can be depolymerized to R3HB *in vitro* by a complex enzyme fraction of polymer-depleted cells of *Rhodospirillum rubrum*. These authors proposed the depolymeriza-

tion mechanism mediated by three components: activator; intracellular poly(3HB) depolymerase; and esterase (Figure 2). The proposed depolymerization mechanism requires native amorphous poly(3HB) granules as a starting material (Merrick and Doudoroff, 1961, 1964).

Although the enzymatic depolymerization of poly(3HB) to R3HB was first proposed during the early 1960s (see above), research into the production of R3HB by intracellular poly(3HB) depolymerase was not continued until the mid-1990s, mainly due to difficulties in analysis and pretreatment.

Subsequently, much more extensive research has been carried out into the degradation of PHAs by extracellular PHA depolymerases. Many hundreds of different bacteria possessing extracellular PHA depolymerases were isolated and classified into 11 groups, according to their polymer-degrading ability (Schirmer et al., 1995; Jendrossek et al., 1996). Jendrossek et al. (1996) also noted that the production of (R)-3-hydroxycarboxylic acids from PHAs might be possible by degrading PHAs with extracellular PHA depolymerases.

4.2.2
In vivo Depolymerization

Recently, the efficient production of (R)-3-hydroxycarboxylic acid by the *in-vivo* depolymerization of microbial PHAs has been

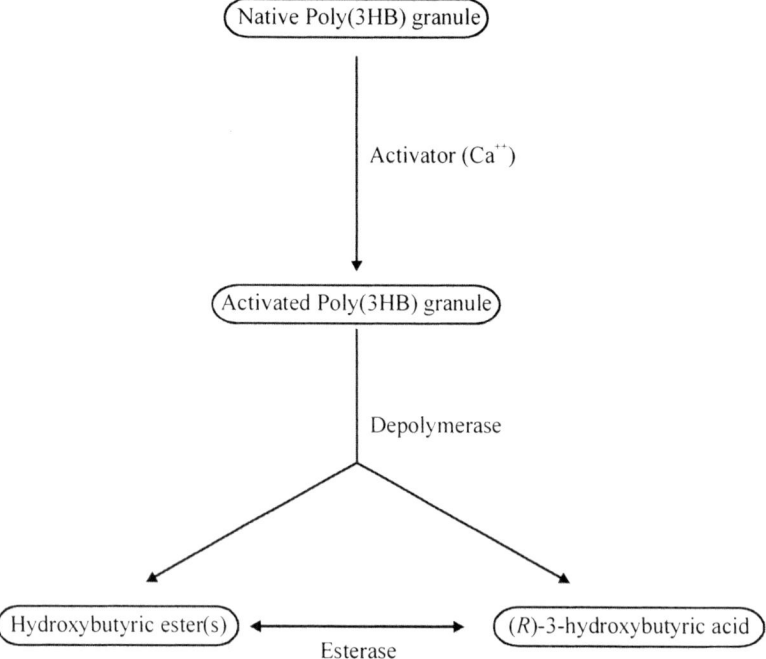

Fig. 2 *In vitro* depolymerization of poly(3HB) by intracellular enzyme system.

reported (Lee et al., 1999). The method employs the intracellular PHA metabolism of natural PHA-producing microorganisms.

The metabolic pathway involved in the biosynthesis and degradation of poly(3HB) is shown in Figure 3. The intracellular depolymerase activity in *Alcaligenes latus* was found to be significantly enhanced at low pH (see Figure 4). Moreover, the activity of (R)-3-hydroxybutyate dehydrogenase was close to zero at low pH. Therefore, by resuspending only the poly(3HB)-containing *A. latus* cells in buffer at pH 4, and incubating for only 30 min at 37 °C, 96% of the accumulated poly(3HB) was depolymerized to R3HB and its dimer, without racemization. By applying this method directly to the cells harvested after the fed-batch culture, concentrations up to 117.8 g L^{-1} of R3HB could be produced. This *in vivo* depolymerization method could be applied to other microorganisms producing PHAs

naturally. The production of other (R)-hydroxycarboxylic acids by the *in-vivo* depolymerization of PHAs was also demonstrated. Several bacteria, including *Ralstonia eutropha*, *Pseudomonas oleovorans* and *Pseudomonas aeruginosa*, each accumulating copolymers such as poly[(R)-3-hydroxybutyrate-co-(R)-3-hydroxyvalerate], PHAs containing medium-chain-length (MCL) hydroxycarboxylic acid monomers of C$_6$ to C$_{14}$ or PHAs containing an unusual monomer, (R)-3-hydroxy-5-phenylvaleric acid, were all successfully employed for the production of corresponding (R)-monomers.

5

Applications of (R)-Hydroxycarboxylic Acids

It is clear that, due to their structures, the enantiomerically pure (R)-hydroxycarboxylic acids have great potential for many

Acetyl-CoA + Acetyl-CoA

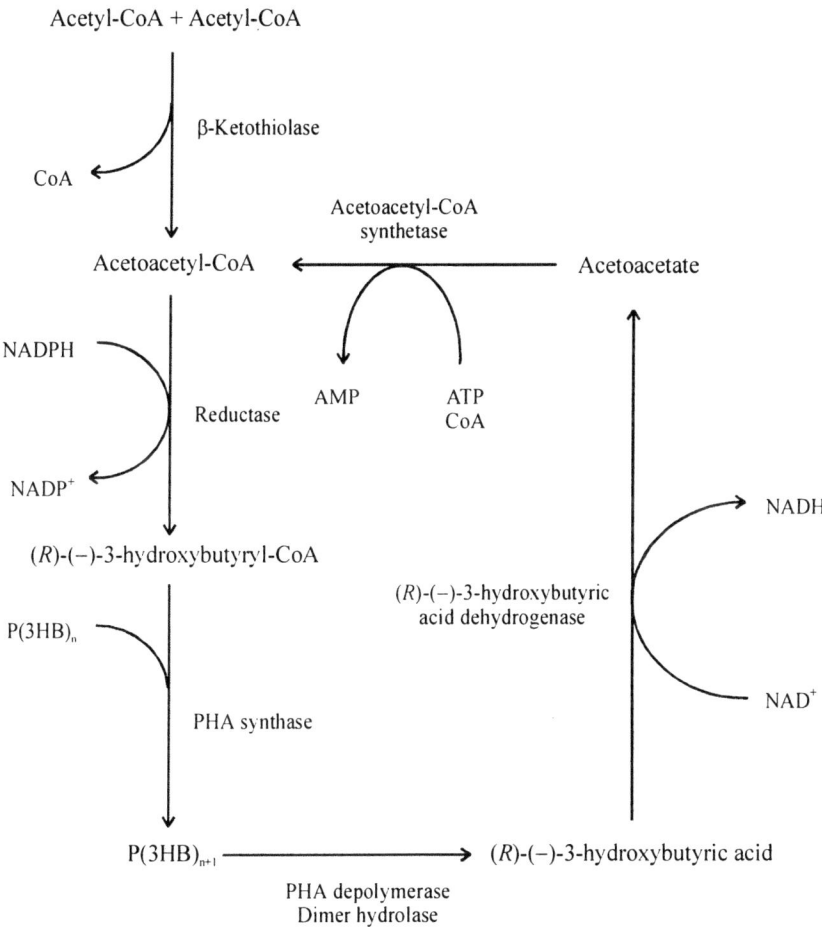

Fig. 3 Metabolic pathway related to the synthesis and depolymerization of poly(3HB).

applications related to the synthesis of useful compounds. Since the economic production method was developed only a few years ago, the proposed applications have as yet been limited to the synthesis of high-cost materials such as pharmaceuticals.

5.1
Chiral Starting Material for the Synthesis of Valuable Compounds

Although (R)-hydroxycarboxylic acids can serve as a starting material for the synthesis of valuable compounds, few uses of enantiomerically pure (R)-3-hydroxycarboxylic acids have been reported due to their limited availability. A well-known application of R3HB is its use as a chiral building block for the synthesis of carbapenem antibiotics (Chiba and Nakai, 1985). The essential precursor for the production of these antibiotics, (1′R, 3R, 4R)-3-[1′-hydroxyethyl]-4-acetoxy-2-azetidinone (4-AA), can be synthesized from methyl R3HB, which provides the correct (R)-configuration of the 3-(hydroxyethyl) side chain of the 2-azetidinone.

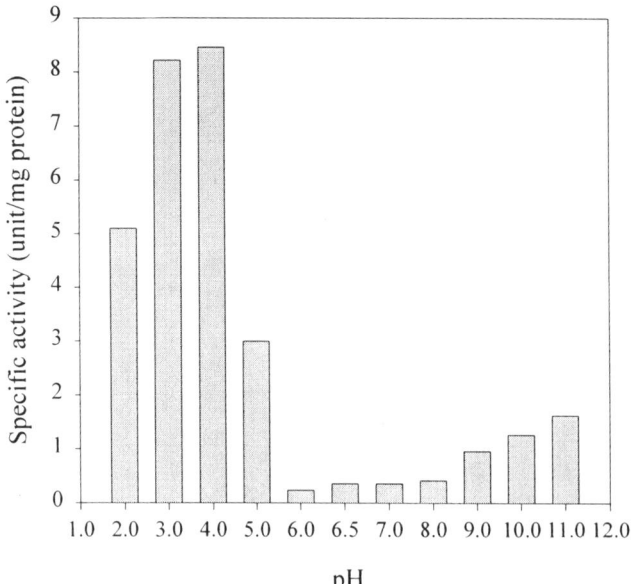

Fig. 4 The specific activities of poly(3HB) depolymerase depending on the pH of depolymerization solution. The unit is defined as µmoles of (R)-(−)-3-hydroxybutyric acid released per minute. [Reproduced from Lee et al. (1999).]

Seebach et al. (1986, 1987b) reported that 3-hydroxy acids can be used as chiral building blocks for the total synthesis of macrolides such as pyrenophorin, colletodiol, grahamimycin A1, and elaiophylidene. Figure 5 shows the structural formulae of these target molecules, and highlights the location of the R3HB units within the molecules.

The hydroxyl group of (R)-3-hydroxycarboxylic acids can be transformed to nitrate or azide by stereoselective nucleophilic substitutions to give β-amino acids (Seebach et al., 1987b, and references cited therein). Recently, an efficient method that allows reduction of the azido group to a primary amine has been developed (S. H. Park and S. Y. Lee, unpublished data). In this way, (R)-3-hydroxycarboxylic acids can be converted stereoselectively to the corresponding (S)-β-amino acids that can be used in the synthesis of biomimetic molecules

such as β-peptides. Some reports have demonstrated interesting biological activities of β-peptides as peptide drugs (Werder et al., 1999; Gademann et al., 2000). β-Amino acids can also be used in the synthesis of β-lactam molecules that are the essential structural units of a group of β-lactam antibiotics.

5.2
Dioxanones

Seebach et al. (1987b and references cited therein) have reported the synthesis of various dioxanones from β-hydroxycarboxylic acids, which can be used as intermediates for a number of enantioselective reactions, including overall enantioselective nucleophilic addition to aldehydes (Seebach et al., 1987a). It was also shown that stereospecific 3-hydroxycarboxylic acid can be easily α-

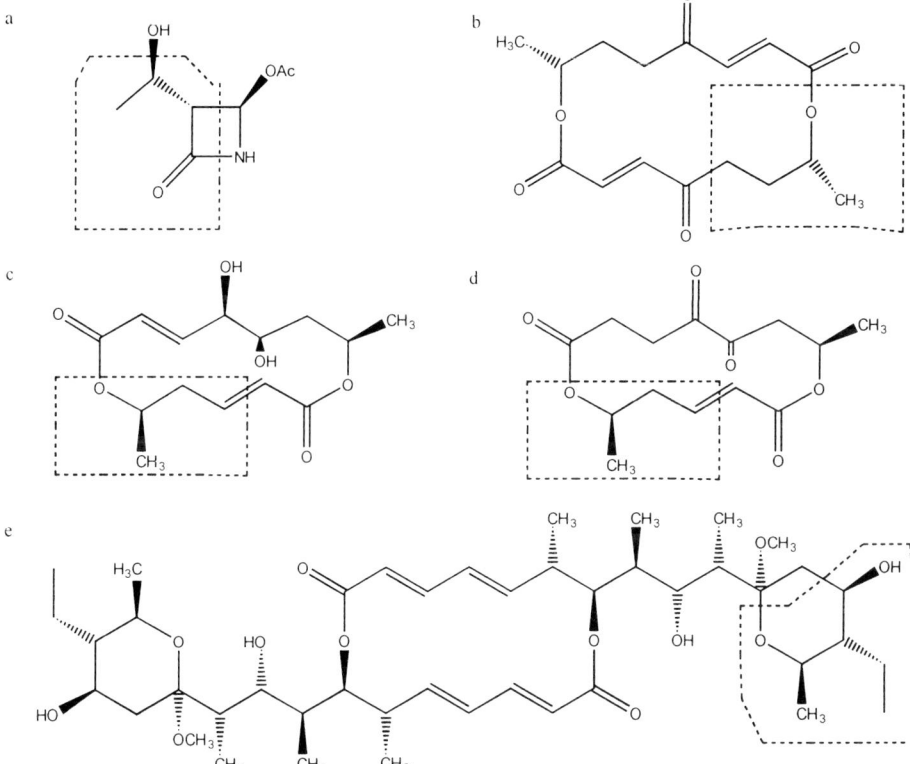

Fig. 5 Structure of (1'R, 3R, 4R)-3-[1'-hydroxyethyl]-4-acetoxy-2-azetidinone (a) and macrolides: pyreno-phorin (b), colletodiol (c), grahamimycin A₁ (d), and elaiophylidene (e). The (R)-3-hydroxybutyric acid units in (a), (b), (c) and (d) are indicated as dashed boxes.

alkylated through dioxanone enolates. Three acetylization methods were employed for the preparation of dioxanones from β-hydroxy-butyric acid or other β-hydroxycarboxylic acids and aldehydes or acetals: (1) the direct acid-catalyzed reaction with azeotropic removal of water (only applicable to aliphatic alde-hydes); (2) the Noyori method of the disilyl derivatization of β-hydroxy acids; and (3) the transacetylization of methyl or ethyl acetals or orthoesters with hydroxybutyric acid, without added catalysts, and with azeotropic removal of the alcohol formed (Seebach et al., 1987b and references cited therein).

The transformation of 3-hydroxycarboxylic acid to its dioxanone enolates allows the subsequent synthesis of various 3-hydroxy acid derivatives. For example, the dioxanone enolates of R3HB can be used as intermedi-ates for the synthesis of β-lactones and 2-alkylated 3-hydroxy acids. Also, the R-pen-dant group of 3-hydroxycarboxylic acid can be easily changed by self reproduction of the stereogenic center.

5.3

Lactones, Cyclic Oligomers and Dendrimer

When the cyclic orthocarbonate of (*R*)-configuration was pyrolyzed at 150–160 °C/60 torr, (*S*)-β-lactone was produced with 90% conversion (Seebach et al., 1987b). This material can be used for the synthesis of high-molecular weight polymers. Cyclic oligomers of R3HB can be prepared in the same manner (Seebach et al., 1994; Melchiors et al., 1996). By using catalytic degradation in solution, poly(3HB) can be depolymerized to yield cyclic oligomers (Melchiors et al., 1996). Since lactones and cyclic oligomers do not produce any small molecules as byproducts during polymerization, high-molecular weight polyester can be synthesized by using these as reactants.

The synthesis of dendrimer (Seebach et al., 1996) by using R3HB has also been reported. This compound, due to its advantages of biodegradability, monodispersity and large numbers of surface functional moieties, is expected to be used in the future as a drug carrier *in vivo*.

As shown above, (*R*)-hydroxycarboxylic acids can be used in a wide range of applications including antibiotics, biomimetic materials, and new polymers. Until recently, (*R*)-hydroxycarboxylic acids were not easily accessible for chemical synthesis, but with the recent development of efficient methods to produce (*R*)-hydroxycarboxylic acids (see Section 5.2), the synthesis of many useful chemicals, using these enantiomerically pure materials as chiral building blocks, is expected.

6
Outlook and Perspectives

The enantiomerically pure hydroxycarboxylic acids have great potential as chiral building blocks in the synthesis of various valuable compounds, for example antibiotics. As a number of structurally diverse, enantiomerically pure (*R*)-hydroxycarboxylic acids can be incorporated into bacterial PHAs as monomeric units, a new group of chiral pool can be generated by the depolymerization of PHAs.

Methods for the depolymerization of bacterial PHAs to produce alkyl esters of enantiomerically pure (*R*)-hydroxycarboxylic acids have been developed by alcoholysis in the presence of acidic catalysts such as sulfuric acid and hydrochloric acid. Moreover, newly developed *in-vivo* depolymerization methods have provided even more efficient and economical production of (*R*)-hydroxycarboxylic acids. With the recent cloning of an intracellular PHA depolymerase gene (Saegusa et al., 2001), metabolic and genetic engineering of various microorganisms for more efficient production of enantiomerically pure (*R*)-hydroxycarboxylic acids is also expected.

As an efficient process for the production of enantiomerically pure hydroxycarboxylic acids has been developed only within the past few years, only a small number of examples are available of the synthesis of valuable compounds using these chiral pools. However, it is expected that many more useful compounds will be synthesized using (*R*)-hydroxycarboxylic acids as building blocks, which are now available in good quantities. This represents an exciting phase in the development of these materials, as we are now able to produce fine chemicals from sugars, via PHAs: "Sugars to plastics, and then to fine chemicals" (Lee et al., 1999) indeed!

7
Patents

Publication number(date)	Assignee	Inventors
	Title of Patent	
US4,365,088 (1982/12/21)	Solvay & Cie	Vanlautem, N., Gilain, J.
	Process for the manufacture of β-hydroxybutyric acid and its oligocondensates	
US5,107,016 (1992/04/21)	Solvay & Cie	Pennetreau, P.
	Process for the preparation of β-hydroxybutyric acid esters	
US5,229,528 (1993/07/20)	E. I. Du Pont de Nemours and Company	Brake, L. D., Subramanian, N.S.
	Rapid depolymerization of polyhydroxy acids	
US5,264,626 (1993/11/23)	E. I. Du Pont de Nemours and Company	Brake, L.D., Subramanian, N.S.
	Rapid depolymerization of polyhydroxy acids	
JP6,086,681A2 (1994/04/29)	Chikyu Kankyo Sangyo Gijutsu Kenkyu Kiko, Kanagawa Univ., Denki Kagaku Kogyo KK	Saito, M., Saegusa, H.
	PHA intracellular degrading enzyme	
JP11,113,574A2 (1999/04/29)	Research Institute of Innovative Technology for The Earth, Kanagawa University	Saito, M., Saegusa, H.
	New intracellular PHA decomposition enzyme	
WO9929889A1 (1999/06/17)	ChiroBio Inc.	Lee, S.Y., Wang, F., Lee, Y.
	A method for producing hydroxycarboxylic acids by auto-degradation of polyhydroxyalkanoates	

Acknowledgments

The studies described in this chapter were supported by the Korean Ministry of Commerce, Industry and Energy.

8
References

Anderson, A. J., Dawes, E. A. (1990) Occurrence, metabolism, metabolic role, and industrial uses of bacterial polyhydroxyalkanoates. *Microbiol. Rev.* **54**, 450–472.

Andújar, M., Aponte, M. A., Díaz, E., Schröder, E. (1997) Polyesters produced by *Pseudomonas oleovorans* containing cyclohexyl groups, *Macromolecules* **30**, 1611–1615.

Chiba, T., Nakai, T. (1985) A synthetic approach to (+)-thienamycin from methylene (*R*)-3-hydroxybutanoate. A new entry to (3*R*,4*R*)-3-[(*R*)-1-hydroxyethyle]-4-acetoxy-2-azetidinone, *Chem. Lett.*, 651–654.

Choi, J., Lee, S. Y. (1999) Efficient and economical recovery of poly(3-hydroxybutyate) from recombinant *Escherichia coli* by simple digestion with chemicals, *Biotechnol. Bioeng.* **62**, 546–553.

Crosby, J. (1992) Chirality in Industry – an overview, in: *Chirality in Industry* (Collins, A.N., Sheldrake, G.N., Crosby, J., Eds), Chichester, UK: John Wiley & Sons, 1–66.

Curley, J. M., Hazer, B., Lenz, R. W. (1996) Production of poly(3-hydroxyalkanoates) containing aromatic substituents by *Pseudomonas oleovorans*, *Macromolecules* **29**, 1762–1766.

Gademann, K., Ernst, M., Seebach, D., Hoyer, D. (2000) The cyclo-β-tetrapeptide (β-HPhe-β-HThr-β-HLys-β-HTrp): synthesis, NMR structure in methanol solution, and affinity for human somatostatin receptors, *Helv. Chim. Acta* **83**, 16–33.

Garía, B., Olivera, E. R., Miñambres, B., Fernández-Valverde, M., Cañedo, L. M., Prieto, M. A., Garía, J. L., Martínez, M., Luengo, J. M. (1999) *J. Biol. Chem.* **274**, 29228–29241.

Inoue, Y., Yoshie, N. (1992) Structure and physical properties of bacterially synthesized polyesters, *Prog. Polym. Sci.* **17**, 571–610.

Jendrossek, D., Schirmer, A., Schlegel, H. G. (1996) Biodegradation of polyhydroxyalkanoic acids, *Appl. Microbiol. Biotechnol.* **46**, 451–463.

Kim, O., Gross, R. A., Hammar, W. J., Newmark, R. A. (1996a) Microbial synthesis of poly(β-hydroxyalkanoates) containing fluorinated side-chain substituents, *Macromolecules* **29**, 4572–4581.

Kim, Y. B., Rhee, Y. H., Han, S.-H., Heo, G. S., Kim, J. S. (1996b) Poly-3-hydroxyalkanoates produced from *Pseudomonas oleovorans* grown with β-phenoxyalkanoates, *Macromolecules* **29**, 2432–3435.

Lee, S. Y. (1996) Bacterial polyhydroxyalkanoates, *Biotechnol. Bioeng.* **49**, 1–14.

Lee, S. Y., Chang, H. N. (1995) Production of poly(hydroxyalkanoic acid), *Adv. Biochem. Eng.* **52**, 27–58.

Lee, S. Y., Lee, Y., Wang, F. (1999) Chiral compounds from bacterial polyesters: sugars to plastics to fine chemicals, *Biotechnol. Bioeng.* **65**, 363–368.

Lee, Y., Park, S. H., Lim, I. T., Han, K., Lee, S. Y. (2000) Preparation of alkyl (*R*)-(–)-3-hydroxybutyrate by acidic alcoholysis of poly-(*R*)-(–)-3-hydroxybutyrate, *Enzyme Microb. Technol.* **27**, 33–36.

Melchiors, M., Keul, H., Höcker, H. (1996) Depolymerization of poly[(*R*)-3-hydroxybutyrate]to cyclic oligomers and polymerization of the cyclic trimer: an example of thermodynamic recycling, *Macromolecules* **29**, 6442–6451.

Merrick, J. M., Doudoroff, M. (1961) Enzymatic synthesis of poly-β-hydroxybutyric acid in bacteria, *Nature* **189**, 890–892.

Merrick, J. M., Doudoroff, M. (1964) Depolymerization of poly-β-hydroxybutyrate by an intracellular enzyme system, *J. Bacteriol.* **88**, 60–71.

Morikawa, H., Marchessault, R. H. (1980) Pyrolysis of bacterial polyalkanoates, *Can. J. Chem.* **59**, 2306–2313.

Saegusa, H., Shiraki, M., Kanai, C., Saito, T. (2001) Cloning of an intracellular poly[D(-)-3-hydroxy-butyrate] depolymerase gene from *Ralstonia eutropha* H16 and characterization of the gene product, *J. Bacteriol.* **183**, 94–100.

Schirmer, A., Matz, C., Jendrossek, D. (1995) Substrate specificities of PHA-degrading bacteria and active site studies on the extracellular poly(3-hydroxyoctanoic acid) [P(3HO)] depolymerase of *Pseudomonas fluorescens* GK13, *Can. J. Microbiol.* **41**, 170–179.

Seebach, D., Züger, M. F. (1982) Uber die depoly-merisierung von poly-(*R*)-3-hydroxy-buttersäure-ester (PHB), *Helv. Chim. Acta* **65**, 495–503.

Seebach, D., Züger, M. F. (1984) On the preparation of methyl and ethyl (R)-(-)-3-hydroxy-valerate by depolymerization of mixed PHB/PHV biopoly-mer, *Tetrahedron Lett.* **25**, 2747–2750.

Seebach, D., Chow, H.-F., Jackson, R. F. W., Sutter, M. A., Thaisrivongs, S., Zimmermann, J. (1986) (+)-11,11'-Di-*O*-methylelaiophylidene – Prepara-tion from elaiophylin and total synthesis from (*R*)-3-hydroxybutyrate and (S)-malate, *Liebigs Ann. Chem.* 1281–1308.

Seebach, D., Imwinkelried, R., Stucky, G. (1987a) Optisch aktive alkohole aus 1,3-dioxan-4-onen: eine praktikable variante der enantioselektiven synthese unter nucleophiler substitution an acetal-zentren, *Helv. Chim. Acta* **70**, 448–464.

Seebach, D., Roggo, S., Zimmermann, J. (1987b) Biological-chemical preparation of 3-hydroxy-carboxylic acids and their use in EPC-syntheses, in: *Stereochemistry of Organic and Bioorganic Transformation*, Workshop Conferences Hoechst (Bartmann, W. and Sharpless, K.B., Eds), Wein-heim, Germany: VCH Verlagsgesellschaft mbH, 85–126, Vol. 17.

Seebach, D., Beck, A. K., Breitschuh, R., Job, K. (1992) Direct degradation of the biopolymer poly[(*R*)-3-hydroxybutyric acid] to (*R*)-3-hydroxy-butanoic acid and its methyl ester, *Org. Synth.* **71**, 39–47.

Seebach, D., Hoffmann, T., Kühnle, F. N. M., Lengweiler, U. D. (1994) Preparation and struc-ture of oligolides from (*R*)-3-hydroxypentanoic acid and comparison with the hydroxybutanoic acid derivatives: a small change with large consequences, *Helv. Chim. Acta* **77**, 2007–2034.

Seebach, D., Herrman, G. F., Lengweiler, U. D., Bachmann, B. M., Amrein, W. (1996) Synthesis and enzymatic degradation of dendrimers from (*R*)-3-hydroxybutanoic acid and trimesic acid, *Angew. Chem. Int. Ed. Engl.* **35**, 2795–2797.

Song, J. J., Yoon, S. C. (1996) Biosynthesis of novel aromatic copolyesters from insoluble 11-phe-nyoxyundecanoic acid by *Pseudomonas putida* BMo1, *Appl. Environ. Microbiol.* **62**, 536–544.

Steinbüchel, A., Valentin, H. E. (1995) Diversity of bacterial polyhydroxyalkanoic acids. *FEMS Microbiol. Lett.* **128**, 219–228.

Werder, M., Hauser, H., Abele, A., Seebach, D. (1999) β-Peptides as inhibitors of small-intesti-nal cholesterol and fat absorption, *Helv. Chim. Acta* **82**, 1774–1783.

15
Index